Driving 5G Mobile Communications with Artificial Intelligence towards 6G

Driving 5G Mobile Communications with Artificial Intelligence towards 6G presents current work and directions of continuous innovation and development in multimedia communications with a focus on services and users. The fifth generation of mobile wireless networks achieved the first deployment by 2020, completed the first phase of evolution in 2022, and started transition phase of 5G-Advanced toward the sixth generation. Perhaps one of the most important innovations brought by 5G is the platform approach to connectivity, i.e., a single standard that can adapt to the heterogeneous connectivity requirements of vastly different use cases. 5G networks contain a list of different requirements, standardized technical specifications, and a range of implementation options with spectral efficiency, latency, and reliability as primary performance metrics. Toward 6G, machine learning (ML) and artificial intelligence (AI) methods have recently proposed new approaches to modeling, designing, optimizing, and implementing systems. They are now mature technologies that improve many research fields significantly.

The area of wireless multimedia communications has developed immensely, generating a large number of concepts, ideas, technical specifications, mobile standards, patents, and articles. Identifying the basic ideas and their complex interconnections becomes increasingly important.

The book is divided into three major parts, with each part containing four to five chapters:

- Advanced 5G communication
- Machine learning-based communication and network automation
- Artificial intelligence towards 6G

The first part discusses three main scenarios and standard specification of 5G use cases (eMBB, URLLC, mMTC), vehicular systems beyond 5G, and efficient edge architecture on NFV infrastructure. In the second part, different AI/ML-based methodologies and open research challenges are presented in introducing 5G-AIoT artificial intelligence of things, scheduling in 5G/6G communication systems, application of DL techniques to modulation, detection, and channel coding as well as 5G open source tools for experimentation and testing. The third part paves the way to deploy scenarios for different innovative services including technologies and applications of 5G/6G intelligent connectivity, AI-assisted eXtended Reality, integrated 5G-IoT architecture in next-generation Smart Grid, privacy requirements in a hyperconnected world, and evaluation of representative 6G use cases and technology trends.

The book is written by field experts from Europe and Mauritius who introduce a blend of scientific and engineering concepts covering this emerging wireless communication area. It is a very good reference book for telecom professionals, engineers, and practitioners in various 5G vertical domains and, finally, a basis for student courses in 5G/6G wireless systems.

Driving 5G Mobile Communications with Artificial Intelligence towards 6G

Edited by
Dragorad A. Milovanovic, Zoran S. Bojkovic,
and Tulsi Pawan Fowdur

CRC Press is an imprint of the
Taylor & Francis Group, an **informa** business

First edition published 2023
by CRC Press
6000 Broken Sound Parkway NW, Suite 300, Boca Raton, FL 33487-2742

and by CRC Press
4 Park Square, Milton Park, Abingdon, Oxon, OX14 4RN

CRC Press is an imprint of Taylor & Francis Group, LLC

© 2023 selection and editorial matter, Dragorad A. Milovanovic, Zoran S. Bojkovic and Tulsi Pawan Fowdur; individual chapters, the contributors.

Reasonable efforts have been made to publish reliable data and information, but the author and publisher cannot assume responsibility for the validity of all materials or the consequences of their use. The authors and publishers have attempted to trace the copyright holders of all material reproduced in this publication and apologize to copyright holders if permission to publish in this form has not been obtained. If any copyright material has not been acknowledged please write and let us know so we may rectify in any future reprint.

Except as permitted under U.S. Copyright Law, no part of this book may be reprinted, reproduced, transmitted, or utilized in any form by any electronic, mechanical, or other means, now known or hereafter invented, including photocopying, microfilming, and recording, or in any information storage or retrieval system, without written permission from the publishers.

For permission to photocopy or use material electronically from this work, access www.copyright.com or contact the Copyright Clearance Center, Inc. (CCC), 222 Rosewood Drive, Danvers, MA 01923, 978-750-8400. For works that are not available on CCC please contact mpkbookspermissions@tandf.co.uk

Trademark notice: Product or corporate names may be trademarks or registered trademarks and are used only for identification and explanation without intent to infringe.

ISBN: 978-1-032-07124-4 (hbk)
ISBN: 978-1-032-07127-5 (pbk)
ISBN: 978-1-003-20549-4 (ebk)

DOI: 10.1201/9781003205494

Typeset in Times
by codeMantra

Contents

Foreword vii
Preface ix
Editors xi
Contributors xiii

PART 1 Advanced 5G Communication

Chapter 1 5G-Advanced Mobile Communication: New Concepts and Research Challenges 3

Dragorad A. Milovanovic, Zoran S. Bojkovic, and Tulsi Pawan Fowdur

Chapter 2 5G Advanced Mobile Broadband: New Multimedia Delivery Platform 83

Dragorad A. Milovanovic and Zoran S. Bojkovic

Chapter 3 5G Ultrareliable and Low-Latency Communication in Vertical Domain Expansion 137

Dragorad A. Milovanovic and Zoran S. Bojkovic

Chapter 4 Vehicular Systems for 5G and beyond 5G: Channel Modeling for Performance Evaluation 183

Caslav Stefanovic, Ana G. Armada, Marco Pratesi, and Fortunato Santucci

Chapter 5 Distribution of NFV Infrastructure Providing Efficient Edge Computing Architecture for 5G Environments 221

Gjorgji Ilievski and Pero Latkoski

PART 2 Machine Learning-Based Communication and Network Automation

Chapter 6 5G-AIoT Artificial Intelligence of Things: Opportunity and Challenges 253

Dragorad A. Milovanovic and Vladan Pantovic

Chapter 7 Machine Learning-Based Scheduling in 5G/6G
Communication Systems 277

M.I. Sheik Mamode and Tulsi Pawan Fowdur

Chapter 8 Application of Deep Learning Techniques to Modulation and
Detection for 5G and beyond Wireless Systems 309

Mussawir A. Hosany

Chapter 9 AI-Based Channel Coding for 5G/6G 327

*Madhavsingh Indoonundon, Tulsi Pawan Fowdur,
Zoran S. Bojkovic, and Dragorad A. Milovanovic*

PART 3 Artificial Intelligence towards 6G

Chapter 10 Enabling Technologies and Applications of 5G/6G-Powered
Intelligent Connectivity 355

*Tulsi Pawan Fowdur, Lavesh Babooram, Madhavsingh
Indoonundon, Anshu P. Murdan, Zoran S. Bojkovic, and
Dragorad A. Milovanovic*

Chapter 11 AI-Assisted Extended Reality Toward the 6G Era:
Challenges and Prospective Solutions 403

Girish Bekaroo and Viraj Dawarka

Chapter 12 An Integrated 5G-IoT Architecture in Smart Grid
Wide-Area Monitoring, Protection, and Control:
Requirements, Opportunities, and Challenges 425

Vladimir Terzija

Chapter 13 Privacy Requirements in a Hyper-Connected World: Data
Innovation vs. Data Protection 445

Myriah Abela

Chapter 14 Evaluation of Representative 6G Use Cases: Identification of
Functional Requirements and Technology Trends 463

Zoran S. Bojkovic and Dragorad A. Milovanovic

Index 481

Foreword

No previous generation of mobile communication has raised so many hopes and fears as the 5G (fifth generation) wireless systems, conceptualized and standardized within the last decade. Fortunately, the fears were largely unfounded, while the hopes are yet to be materialized. As the rollout of 5G is in full swing around the world, we are anticipating a large-scale demonstration of the transformative power that 5G can have on various industries and the society.

Perhaps one of the most important innovations brought by 5G is the platform approach to connectivity, that is, a single standard that can adapt to heterogeneous connectivity requirements by vastly different use cases. This platform has been developed along three dimensions: faster and better mobile Internet access, provision of low latency with high reliability, and support of connectivity for a massive number of IoT devices. While 5G is being rolled out, there is a significant research momentum in the academia and industry toward the next generation, 6G wireless systems. While the detailed concepts are yet to be developed, some contours of the 6G systems start to appear and they reveal that machine learning (ML) and artificial intelligence (AI) methods will play significant roles in the protocols and the network architecture of 6G.

This book is a timely and well-structured effort toward elaborating upon the key ideas and concepts that constitute 5G, the complex interaction among those concepts, the use of ML/AI toward addressing those interactions, and, finally, how they pave the way toward 6G. Several of the chapters provide deep technical insights into the relevant standards and the use cases. This book can be seen as a reference for telecom professionals, engineers, and practitioners in various vertical domains where 5G is poised to have an impact and, finally, a basis for student courses in 5G/6G wireless systems.

Petar Popovski
Professor in Wireless Communications
Aalborg-Ohrid, July 2022

Preface

Generations of mobile communication systems have evolved from the communication infrastructure to the smart infrastructure of a sustainable society. The fifth generation (5G) enables faster and better Internet access, faster network response, and significantly greater connectivity of a large number of devices. The mobile network includes a list of different requirements, standardized specifications, and a number of implementation options. The main purpose of the book is to describe in a well-structured way how AI/ML/DL are applied on different layers of the 5G network and how the paths to 6G are opened. Our goal is to clearly identify and classify the application of artificial intelligence on each layer of 5G communication systems and provide insight into emerging 6G applications. Now, research interest in wireless communication is transitioning to the next evolution of mobile technology, with a paradigmatic shift.

The book brings together a collection of chapters dealing with technical requirements, opportunities, challenges, and recent results in the development of intelligent mobile networks. In addition to presenting the basic concepts in 5G communications, the book describes different levels from physical layer signal processing to applications, where AI can be applied to 5G. This book is divided into three major parts and each part contains four to five chapters. Part 1, *Advanced 5G Communication,* discusses the development of 5G mobile broadband, broadcast, and ultralow latency communication, vehicular systems channel modeling, and performance evaluation, as well as an efficient NFV infrastructure. In Part 2, *Machine Learning-Based Communication,* exclusive chapters have been dedicated on opportunities and challenges of integration of artificial intelligence in 5G-IoT, scheduling in 5G/6G systems, modulation and detection, and channel coding. Part 3, *Artificial Intelligence toward 6G,* opens up the major deployment scenarios for 5G/6G-powered intelligent connectivity, AI-assisted eXtended reality, integrated 5G-IoT architecture in next-generation smart grid, privacy requirements in hyper-connected world, as well as evaluation of representative 6G use cases.

The book is written by experts in the field from Europe and Mauritius, who bring out the intrinsic challenges of 5G multimedia communication. The authors have introduced blends of scientific and engineering concepts, covering this emerging wireless communication area. The book can be read cover to cover or selectively in the areas of the interest for readers. It is a very good reference book for undergraduate students, young researchers, and practitioners in the field of wireless multimedia communications.

Dragorad A. Milovanovic, Zoran S. Bojkovic, Tulsi Pawan Fowdur
Belgrade-Mauritius, July 2022

Editors

Dragorad A. Milovanovic received the Dipl. Electr. Eng. and Magister degree from the University of Belgrade, Serbia. From 1987 to 1991, he was a research assistant, and from 1991 to 2001, he was a PhD researcher at the Department of Electrical Engineering, where his interest includes simulation and analysis of digital communications systems. He has been working as R&D engineer for DSP software development in digital television industry. Also, he is serving as an ICT lecturer in medicine/sports informatics and consultant in development standard-based solutions. He participated in research/innovation projects and published more than 300 papers in international journals and conference proceedings. He also, co-authored reference books and chapters in multimedia communications published by Prentice Hall, Wiley, CRC Press (2009, 2020), Springer, and IGI Global. Present projects include adaptive coding of 3D video and immersive media, intelligent connectivity integration and interoperability, and 5G/6G multimedia wireless communication.

Zoran S. Bojkovic is full professor of Electrical Engineering at University of Belgrade, Serbia; Life Senior Member of IEEE; full member of Engineering Academy of Serbia; member of Scientific Society of Serbia; and member of Athens Institute for Education and Research ATINER. He was and still is a visiting professor worldwide. He is author/co-author of more than 500 publications: monographies, books (Prentice Hall, Wiley, McGraw Hill, Springer, CRC Press/Taylor & Francis Group, IGI Global, and WSEAS Press), book chapters, peer-reviewed journals, conferences, and symposium papers. Some of the books have been translated into China, India, Canada, and Singapore. His research focuses on computer networks, multimedia communications, green communications, 5G, and beyond. He is a highly regarded expert in the IEEE field, contributing to the growth of communication industry and society reviewing process in many journals as well as organizing special sessions and being General Chair and TPC member at numerous conferences all over the world. He is also serving as editor-in-chief and associate editor of several international journals such as World Scientific and Engineering Academy and Society (WSEAS), North Atlantic University Union (NAUN), and International Association of Research and Science (IARAS).

Tulsi Pawan Fowdur, received his BEng (Hons) degree in Electronics and Communication Engineering with first-class honors from the University of Mauritius in 2004. He was the recipient of a Gold Medal for having produced the best degree project at the Faculty of Engineering in 2004. In 2005 he obtained a full-time PhD scholarship from the Tertiary Education Commission of Mauritius and was awarded his PhD degree in Electrical and Electronics Engineering in 2010 by the University of Mauritius. He is also a registered chartered engineer of the Engineering Council of the UK, member of the Institute of Telecommunications Professionals of the UK, and the IEEE. He joined the University of Mauritius as an academic in June 2009 and is presently an associate professor at the Department of Electrical and Electronics Engineering of the University of Mauritius. His research interests include Mobile and Wireless Communications, Multimedia Communications, Networking and Security, Telecommunications Applications Development, Internet of Things, and AI. He has published several papers in these areas and is actively involved in research supervision, reviewing of papers, and has been the general co-chair of four international conferences.

Contributors

Myriah Abela
Betsson Group
Ta' Xbiex, Malta

Ana García Armada
Universidad Carlos III de Madrid
Getafe, Spain

Lavesh Babooram
Department of Electrical and
 Electronics Engineering
University of Mauritius
Réduit, Republic of Mauritius

Girish Bekaroo
Middlesex University Mauritius
Flic-en-Flac, Republic of Mauritius

Zoran S. Bojkovic
University of Belgrade
Belgrade, Republic of Serbia

Viraj Dawarka
Staffordshire University
London, United Kingdom

Tulsi Pawan Fowdur
Department of Electrical and
 Electronics Engineering
University of Mauritius
Réduit, Republic of Mauritius

Mussawir Ahmad Hosany
University of Mauritius
Reduit, Republic of Mauritius

Gjorgji Ilievski
Makedonski Telekom AD
Skopje, RN Macedonia

Madhavsingh Indoonundon
Department of Electrical and
 Electronics Engineering
University of Mauritius
Reduit, Republic of Mauritius

Pero Latkoski
Faculty of Electrical Engineering and
 Information Technologies
Ss. Cyril and Methodius University
Skopje, RN Macedonia

Maryam I. Sheik Mamode
Department of Electrical and
 Electronics Engineering
University of Mauritius
Reduit, Republic of Mauritius

Dragorad A. Milovanovic
University of Belgrade
Belgrade, Republic of Serbia

Anshu Prakash Murdan
Department of Electrical and
 Electronics Engineering
University of Mauritius
Reduit, Republic of Mauritius

Vladan Pantovic
Faculty of Information Technology and
 Engineering
University Union – Nikola Tesla
Belgrade, Republic of Serbia

Marco Pratesi
University of L'Aquila
L'Aquila, Italy

Fortunato Santucci
University of L'Aquila
L'Aquila, Italy

Caslav Stefanovic
Universidad Carlos III de Madrid
Getafe, Spain

Vladimir Terzija
University of Manchester
Manchester, United Kingdom

Part 1

Advanced 5G Communication

1 5G-Advanced Mobile Communication
New Concepts and Research Challenges

Dragorad A. Milovanovic and Zoran S. Bojkovic
University of Belgrade

Tulsi Pawan Fowdur
University of Mauritius

CONTENTS

1.1 Introduction 4
    1.1.1 IMT-2020 Submission and Evaluation Process 6
    1.1.2 3GPP Standardization Activities 7
1.2 5G-Advanced Transformational Phase 11
    1.2.1 5G System Architecture Options and Standardization Process 12
        1.2.1.1 5G System Reference Architecture 14
        1.2.1.2 5G NR Architecture 15
        1.2.1.3 5G Core Network 16
        1.2.1.4 Separation of Control Plane and User Plane 19
        1.2.1.5 RAN Protocol Architecture 20
        1.2.1.6 IAB Protocol and Physical Layer 22
        1.2.1.7 Physical Layer (PHY) 24
        1.2.1.8 Network Slicing (NS) 31
        1.2.1.9 Quality of Service (QoS) 33
        1.2.1.10 5G Security 34
    1.2.2 5G New Radio and Core Network Enhancement and Vertical Expansion 36
        1.2.2.1 Support for Industrial IoT Applications 36
        1.2.2.2 Ultra-Reliable Low-Latency Communication 37
        1.2.2.3 Support for V2X Connections 37
        1.2.2.4 Non-Public Networks 39
    1.2.3 Advanced Service Requirements and Performance Indicators 40
        1.2.3.1 Phase 1 and Phase 2 42
        1.2.3.2 Foundation for the Next Phase 43
        1.2.3.3 5G-Advanced 44

   1.2.4 Novel System AI&ML Paradigm 49
     1.2.4.1 AI-Enabled RAN Architecture..51
     1.2.4.2 Network Automation and Data Analytics Function.............54
     1.2.4.3 5G Advanced Architecture and Technical Trends...............55
1.3 6G Concept, Research, and Transition Technologies 57
   1.3.1 Extreme System Performance and Network Evolution 59
     1.3.1.1 IMT Vision for 2030 and Beyond.......................................62
   1.3.2 Technology Enablers and Research Programs 63
     1.3.2.1 Review of Global Activities and Research Programs..........63
1.4 Open Issues and Concluding Remarks 67
References..68

1.1 INTRODUCTION

Mobile communications are continuously innovating and developing. Mobile networks cover different requirements, standardized technical specifications, and various implementation options with a focus on services and users. The new fifth generation (5G) supports faster and better mobile Internet access, low-latency network response, and significantly higher connectivity for huge number of Internet of things (IoT) devices. The overall performance and capacity of the network have been significantly expanded compared to previous generations. The primary performance metrics are latency, spectral efficiency, and reliability. 5G system provides operators with optimal resources and capacity as well as quality for different use cases [1–11].

The first phase of the 5G standard is currently being commercially deployed around the world, and the technology continues to evolve toward the second transformational phase of fully backward-compatible 5G-Advanced. Preparation was carried out during 2021, work on the specification started in 2022, and is anticipated to result in completion in late 2023 and early 2024 for various items. The first 5G-Advanced networks are due to be utilized in a commercial setting in 2025 [12–22].

The key 5G use cases are separated into two groups: mobile Internet services enable streaming, broadcasting, conversation, interaction, transmission, and exchange of messages, while IoT services support the industrial application of acquisition and control.

- Mobile broadband focuses on user's needs and gives more attention to the quality of experience. Continuous improvement of multimedia interaction capabilities on mobile devices enables commercial applications of UHD video, 3D video, and immersive reality AR/VR. Users expect immersive audiovideo and personalized experiences, which requires the performance of 5G wireless networks comparable to fiber-optic access networks. At the same time, users also expect a real-time online experience and an imperceptible network delay.
- The Internet of connected objects (IoT) facilitates new demands and differentiated experience in service delivery. Rapid development requires that 5G network connect applications, services, and devices, such as people, objects,

processes, content, knowledge, information, products, and the like. With the coming large number of industrial applications, variety of IoT deployment require 5G network to support a vast array of services with radically distinct performance requirements of various industries. The rapid expansion of connected devices is being driven by the establishment of multiple connections, as well as, the development of a wide range of industrial applications, so it is necessary that the 5G network provides ultra-large capacity and massive number of connections.

The use cases were adopted as guidelines in development of a new 5G system concept based on software virtualization and separation of network functions from underlying hardware, network slicing, and multiplexing independent logical subnets on common physical infrastructure, open source that supports the development of new ecosystems involving all stakeholders as well as moving data centers to edges of the network as close to users as possible.

A flexible system 5GS integrates new radio access (5G NR) network, core network (5G CN), transport network (TN), and application layer. The new reference architecture based on separation of 5G base transceiver station NodeB (gNB) into two physical entities, CU (centralized unit) and DU (distributed unit), enables flexible stack protocol in the radio-access network as well as centralized control and cooperation of radio resources. A flexible NR frame structure in the time domain is also supported to reduce radio-interface delays and improve the user experience. Simultaneously, subcarrier bandwidth increases and minimum scheduling resource time decreases. In addition, the capacity of the 5G network is improved by the use of new antenna beam control, new reference signal, new coding, advanced antenna systems, and larger bandwidth [23–27].

5G TN transport network introduces software-defined networking (SDN) to achieve global programmable planning and maintenance. The new approach is realized through concept of softwarization. The core network CN has been transformed and differs significantly from previous reference models. The system design concept is based on the application of Internet technology, simplicity, and service. A service-based architecture (SBA) separates complex single network element into modular service. Each individual network function (NF) consists of several services that enable the form network architecture on demand. The SBA interface between NFs includes multiple layers of protocol selection in transmission and application, program interface design method (API), serialization method, and interface description language (IDL). Network function virtualization (NFV) technology enables decoupling of software and hardware in 5G network using general-purpose computer servers and software virtualization technique.

The first implementation of 5G network intelligence AI/ML was opened by the introduction of network data analysis functions (NWDAF) in the CN. Next, network slicing (NS) provides end-to-end (E2E) independent subnetworks on unified infrastructure with logical isolation of resources, customization of functions, and ensures quality. 5G is first communications network, which inherently integrate the network virtualization along with cloud computing and control technologies, providing reconfigurable and smart wireless system [28].

The suggested 5G system concept generalizes fundamental features of the use case, harmonizes requirements, and combines technological components into three generic communication services:

- eMBB enhanced mobile broadband enables extremely high data rates and low-latency communication, as well as extreme coverage in the service area. It includes mobile broadband access and mobile video streaming, with bandwidth and availability as the main requirements.
- mMTC massive machine-type communication gives wireless connectivity to millions of devices on the network, scalable connectivity for a growing number of devices, efficient small data-packet transmission, wide area coverage, and deep penetration take priority over data rates. The requirements are concurrently and mostly based on bandwidth, latency, and reliability.
- uMTC ultra-reliable machine-type communication provides ultra-reliable low-latency communication connections (URLLC) for network services with stringent prerequisites in terms of availability, latency, and reliability. Typical applications are vehicle-to-everything (V2X) communications and industrial IoT (IIoT) applications. Reliability and low latency are a priority over data transfer speeds. The requirements are targeted at providing bandwidth for a large number of devices and reliability.

Individual 5G use cases can be observed as linear combination of the basic functions. Each generic service (eMBB, mMTC, uRLLC) emphasizes different subset of given requirements, but all are relevant to some degree. The generic communications services contain service-specific functions, and the main drivers include functions that are common to more than one generic service [29–35].

1.1.1 IMT-2020 Submission and Evaluation Process

5G NR wireless access technology is being developed based on the requirements of selected usage scenarios and it is necessary in order to attain the standards set by the International Telecommunication Union (ITU) for wireless networks. The IMT-2020 tried-and-tested process consists of four basic phases: ITU-R vision and definition; minimal prerequisites and evaluation criteria; call pertaining to proposal, appraisal, and consensus building; technical specification, approval, and implementation. The ITU-R set out the goals that need to be achieved at the beginning of each IMT process. After that, the candidates who support the goals develop functional technology that fulfills the specifications. When standardization organizations have submitted IMT candidate technologies, an evaluation process is initiated through the cooperation of ITU member states, equipment manufacturers, network operators, standards development organizations (SDOs), and academia [36–39].

The 5G network standardization process is launched after the ITU-T Focus group for IMT-2020 completed its preparatory standardization activities in December 2016 and about at the same period, ITU-R published its recommendation for IMT vision beyond 2020 [10–12].

The report ITU-R M.2410 specifies the minimal technical performance standards of the IMT-2020 candidate technology for new radio-interface (NR). Report ITU-R M.2411 describes in detail the service requirements, spectrum, and technical performance of radio-interface technologies, as well as evaluation criteria and submission forms for the development of IMT-2020 recommendations and reports. Report ITU-R M.2412 elaborates the procedure, methodology, and criteria for evaluating candidate technologies. Report ITU-R WP 5D IMT Systems M.2410 also contains the necessary background information on individual requirements and clarification for the selected items and values chosen. Providing such basic information is necessary to fully understand the requirements. The report was based on external research and technology organizations' research and development efforts [13].

ITU-R has now released recommendation M.2150 detailing the radio interface parameters for IMT-2020. Late last year, a variety of IMT-2020 radio technology choices were evaluated, and this suggestion provides a set of terrestrial radio-interface specifications consolidated in a single document. The specification contains three radio interface technologies 3GPP 5G-SRIT, 3GPP 5G-RIT, and Telecommunications Standards Development Society of India 5Gi, which provide the foundation for the worldwide deployment of 5G networks. Following a seven to eight years interval of dedicated development, the examination of IMT-2020 technologies has concluded in 193 ITU member states' acceptance. Radio interface technology (RIT), or a set of radio interface technologies (SRIT), is becoming part of the IMT-2020 radio-interface for which frequency bands are recognized under ITU radio regulations.

Given the global success of 5G networks, ITU has started work on report about the forthcoming technological developments of terrestrial IMT systems through 2030 and beyond (June 2022). The study offers a list of IMT design motivating elements for new technologies and a listing of potential technologies for improving the efficiency of the radio-interface and radio-network. The report is an input to the ITU recommendation on vision of IMT beyond 2030 and very general guidelines for the development of new 6G generation of networks.

1.1.2 3GPP Standardization Activities

The 3GPP (Third Generation Partnership Project) consortium of seven national or regional standards organizations has been developing technical specifications and proposed a 5G standard. Following initial studies, in March 2017, the 3GPP approved a work item for the NR standard as part of R15. At the same working meeting, the proposal to accelerate the 5G work plan and complete non-stand-alone (NSA) standard by December 2017, while stand-alone SA-NR system architecture option by June 2018. NSA development is based on LTE technical specifications for initial access and handling of mobility, while the SA version can be applied independently of LTE. The last step of the Phase 1 was completed in March 2019 by including a number of design alternatives, for example, the ability to connect 5G NodeB (gNB) with evolved packet core (EPC) and operating NR and LTE multiconnectivity mode where NR is the main node and the LTE secondary [40–46].

3GPP technical specifications are organized into Releases. A release includes a collection of features that are internally consistent. Timeframe of each release defines freezing date after which additions and alterations to functions are prohibited. The 3GPP approach of parallel releases offers 5G developers a solid framework for the deployment of features at a particular time and then permits the inclusion of additional capabilities in future releases. The 3GPP standardization efforts are divided into the following technical specifications groups (TSG) [7,12,32,36,37]:

- TSG-SA 1, 2, 3, 4, 5, 6 group for service and system aspects have the responsibility for defining, developing, and maintaining the overall system architecture and service capabilities. They additionally support the coordination of groups.
- TSG RAN 1, 2, 3, 4, 5, 6 group is responsible for radio access networks and defines the functions, requirements, and interface of the access network in the physical layer, Layers 2 and 3 of the protocol stack. In addition to this, it is accountable for the conformity testing of UEs and base stations that implement the solutions that have been established.
- TSG CT 1, 3, 4, 6 group for core network and terminals is responsible for the specification of terminals, interfaces and capabilities, core network development, and interconnection with external networks in end-to-end networking.

5G NR has been in focus of RAN standardization for efficient and effective wireless network access of user equipment (UE) to different types of services/verticals. The 38.xxx document series is relevant to NR radio technology. The work results of TSG technical groups are documented as reports and specifications:

- Technical report (1,110 TRs in version R15, R16, R17, R18) is the result of an initial study (SI) in the initial phase of the topic being considered for the specification procedure. Reports intended to be issued by the organizational partners as their own publications have specification numbers of the form XX.9XX. Reports that are not intended for publication but are just 3GPP internal working documents have specification numbers of the form XX.8XX (feasibility study reports) or 30.xxx of 50.xxx (planning and scheduling).
- An SI becomes a working item (WI) after the preliminary research and consensus. Then, the finally compliant WIs are covered by the technical specification (3,754 TSs in R15, R16, R17, R18, and R19 version). The specifications are published up to four times a year after the quarterly TSG plenary meetings.

In release R15, the specification of the eMBB and uRLLC service scenarios, which support commercial needs in the initial phase of 5G implementation, has been largely completed. The R15 wireless base station focuses on the technical directions of architecture, frequency range, antenna system, new design, and construction of 5G intelligent networks. The new architecture is capable of supporting both distributed and centralized access network design; the new frequency range includes mid- and high frequencies, spectra above 6 GHz, and new antennas with improved system efficiency. By beamforming, antenna system generates concentrated energy beams, larger 3D coverage zones, and more precise targeting. The cellular base station will determine the path that will transmit data to a specific user in the most effective manner while simultaneously reducing the amount of interference experienced by other users in the process. 5G provides multiple structure options for system-frame in radio-interface design, to meet the different needs of vertical industries and diverse and complex application scenarios. The structure of the radio-frame is flexibly configured in accordance with the different requirements of 5G main communication services, mobile Internet, and IoT connections.

5G CN in release R15 has undergone the most significant transformation. A software-based CN enables configuration and planning based on a unified IT infrastructure. The SBA of the 5GS system allows for dynamic customization as contrast to the fixed rigid networks of the previous generation. The global HTTP2.0 Internet protocol has been adopted for the protocol system between network elements. Improved performances are supported by the introduction of a minimum forwarding data plane (DP), and a centralized yet adaptable control plane (CP) improves efficiency. Flexible use of technologies such as NS and mobile edge computing (MEC) support different network requirements and scenarios through a service-oriented architecture.

3GPP has initiated almost 70 standardization research projects in R16 Phase 2, focusing on uRLLC and mMTC scenarios as well as expand vertical industrial applications. In addition, improvements of 5G standard in areas such as intelligent networks, extreme performance, expanded spectrum, and applications were completed in December 2019 (Figure 1.1) [16–19].

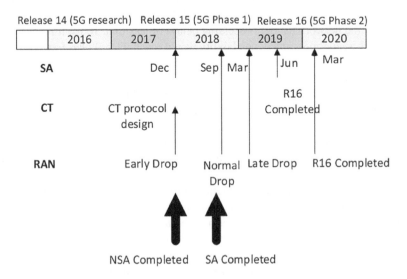

FIGURE 1.1 5G standardization process and timeline of 3GPP R15 Phase 1 and R16 Phase 2.

3GPP R15 was delivered in 2018 and was the first official specification that described a full 5G network. To accelerate the adoption of 5G globally and taking into consideration that many mobile operators aimed to launch 5G in 2018, 3GPP decided to split R15 in two *drops*. An early drop included NSA, allowing 5G radio networks to be connected to 4G core networks, thus allowing the faster deployment of the new generation. A late drop included stand-alone (SA), which also fully specified the 5G CN as well. R16 was frozen in June 2020 and includes various improvements on top of R15. Over a period of three years, starting from mid-2018 to mid-2021, 3GPP has also launched the next phase of 5G-Advanced evolution. While R17 edition was completed and finalized at the end of 2021, work on the R18 edition will be launched in the second quarter of 2022.

The evolution of 5G NR has progressed rapidly since 3GPP standardized the first version of R15. The next step in evolution, R16 contains several significant expansions and improvements. The R17 version work items approved in December 2019 lead to the introduction of new functionalities for the three key use cases of eMBB, uRLLC, and mMTC. 3GPP has published the R17 edition at the end of the first quarter of 2022. The 5G-Advanced roadmap was launched with the R17 edition. Discussion on the scope of the R18 was launched in June 2021 with the aim of approving detailed scope by December 2021. Initial 5G-Advanced planning indicates significant development of 5G systems in AI/ML and augmented XR reality. The final version of the R18 standard is expected by the end of 2023, so the first networks appear after 2025. R19 and subsequent iterations will specify forward-looking concepts and technologies that will support for much higher frequencies (up to THz), more advanced radio interfaces (including full duplex), joint sensing and communication, energy harvesting and passive IoT, and cognitive access across many wireless technologies. Furthermore, future 6G mobile communications networks are predicted to provide peak data rates in excess of 1 Tbps. There will be no discernible end-to-end delays below 0.1 ms. Access to very advanced AI will be made possible via 6G networks at the edge of the network having processing delays of less than 10 ns. It is anticipated that network availability and dependability would reach 99.99999%. Extremely high connection densities of over 10^7 devices/km^2 for IoE will be supported. Spectrum efficiency increases 5× over 5G, while support for severe mobility up to 1,000 km/hour can be anticipated. Many new research works and projects are being launched in the direction of technology development, use cases, applications, and standards [38–46].

Based on the planning transformational technical development of 5G-Advanced, the collaborative efforts of industry and academia focus on evolution in the first five years, while setting 6G specifications. The first test platforms can be expected post-2025. Everyone agrees that the new architecture will be fully software and flexible. 6G communications networks will be the first generation of networks with native artificial intelligence (AI), AI will not only be an application tool, but an integral component of infrastructure, network management, and operations.

AI technology makes a significant contribution to the solutions for issues posed by a high level of complexity in network operation and maintenance, a high level of need for flexibility, and a varied range of network conditions. Increasing the performance and flexibility of mobile applications contributes to a significant growth in complexity of network operation, which causes new challenges in operation and maintenance. Due to its advantages in solving high complexity problems, frequency of changes, and significant uncertainty, the application of AI can improve network efficiency, reduce operating costs, and achieve better user experience closer to the requirements. The network is becoming smarter, more efficient, more convenient, and more secure. Realization of deep integration of AI and 5G sets new requirements of network structure and unified planning on several levels, such as terminal, network element, network management, AI mechanism, and capabilities [28].

Machine learning (ML) allows system to build models based on large amounts of data. In particular, deep learning (DL), reinforcement learning (RL), and federated learning (FL) are key functions. DL has made considerable strides in the last decade, and it is now being used in contexts where large quantities of easily accessible data are readily available for training. FL is a learning approach in which multiple clients with local models collaborate with a data center that integrates and develops a global model and then sends it out to all of the user devices as a broadcast. RL is structured as a recursive learning of an agent who interacts with the environment and thus learns how to take action. In general, ML becomes appropriate in situations where an accurate mathematical model of the system is not available, but a sufficiently large amount of training data is available, the system changes slowly over time, and numerical analysis is acceptable.

The specification of 5G evolution toward 6G and associated technological needs are currently in the early stages of investigation, so different institutions are forming their visions and views without a single definition being adopted. Future vertical sectors are predicted to benefit from 5G evolution's new architecture and post-AI age capabilities by becoming more integrated, generating new apps, and spawning a new 6G ecosystem. The existing centrally controlled and layered cellular mobile network is unlikely to be able to meet the requirements of future omni-directional communication. Flexible network can be optimized, perceived autonomously, managed automatically, and made completely intelligent and secure [47–60].

1.2 5G-ADVANCED TRANSFORMATIONAL PHASE

The fifth generation of mobile communication technology offers significantly increased data transfer rates with almost no lag time, compared to previous generations. Furthermore, 5G can accommodate multiple devices that communicate simultaneously. The standard is in line with the strict requirements of ITU-R IMT-2020. The capabilities of 5G evolution have been significantly improved, supporting multidimensional requirements that far exceed key performance indicators (KPIs) and on-demand network services [61–63].

The community of standardization organizations SDO has anticipated trends and defined a new generation of specifications. As a result, in early 2019, 3GPP published a set of Phase 1 of 5G technical specifications based on completely updated system architecture (Table 1.1). R15 edition specifies an initial, simpler version of the system. Phase 2, as defined in R16 specification, adds remaining functionality and increases performance. Full 5G delivers innovative solutions including virtualized network services and compatibility for a vast array of IoT devices that communicate

TABLE 1.1
3GPP 5G NR Workplan [64–67]

3GPP Standards	Capacity and Operational Efficiency	Vertical Expansion
Phase 1 Release 15	NSA (non-stand-alone) and SA (stand-alone) Carrier aggregation operation Intel-RAT between NR and LTE FR1: 450 MHz–7.125 GHz	eMBB (enhanced mobile broadband) URLLC (ultra-reliable low latency)
Phase 2 Release 16	MIMO enhancements MR-DC (multi-RAT dual connectivity) IAB (integrated access and backhaul) Mobility enhancements CLI/RIM (cross link interface/remote interference management) UE power saving FR: 24.25–52.6 GHz mmWave	IIoT (industrial IoT) URLLC (ultra-reliable low latency) 2-step RACH (random access channel) UE positioning US (unlicensed spectrum) V2X (vehicle-to-everything)
Release 17	MIMO enhancements Sidelink enhancements DSS enhancements IIoT/URLLC enhancements Coverage enhancements IAB enhancements	NR up to 71 GHz NR up to NTN RedCap (NR-light) MBMS (multicast and broadcast services)
5G-Advanced Release 18	**eMBB use cases:** UL enhancement, FR2 mobility enhancement, DL MIMO enhancement, smart repeaters **Non-eMBB use cases:** XR enhancements, sidelink enhancements, positioning enhancements **Cross-functionality use cases:** Evolution of duplex operation, network energy savings, AI/ML RAN enhancements	

simultaneously, ultra-reliable low-latency communications, NS, edge computing, and an optimized SBA model. R16 edition is finally in line with the requirements of IMT-2020 for global interoperable and uniform 5G mobile communication systems. 3GPP has also launched the next phase of evolution toward 5G-Advanced. The R17 edition was finalized at the beginning of 2022, and work on the R18 edition will be launched in the second quarter of 2022 [64–67].

1.2.1 5G System Architecture Options and Standardization Process

5G system (5GS) is continuously developing and improving intelligent connections. The first version of the 5GS includes network core and new radio according to UE customer equipment, and is commercially implemented worldwide in the frequency ranges FR1 below 6 GHz and FR2 mmWaves 24–40 GHz. Based on various requirements and potentials of new technologies, 3GPP has launched the technical specification of the new 5G NR (new radio) as well as the new 5G CN (core network). Improvements have been made to the radio access network (RAN) and network

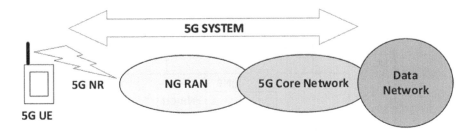

FIGURE 1.2 Overall 5GS system architecture: Radio access network (NR) and core network (CN).

core architecture, including the separation of functions between the two networks (Figure 1.2).

3GPP defines the initial architecture of the 5G system and functionality in the R15 version. Some of the most important technical specifications and procedures describe the interconnection of data and services (TS 23.501) as well as the set of interactions of NFs (TS 23.502). Details on the eMBB data service, subscriber authentication and authorization, application support, edge computing (EC), and interoperability with 4G and other conceivable access systems are provided [68,69].

5G RAN is in charge of all radio-related network tasks, including scheduling, radio resource management, retransmission protocols, coding, and various multiantenna systems. RAN represents a set of 5G gNB nodes connected to 5GC via the NG interface. The gNB nodes are interconnected point-to-point with an Xn interface, which supports node functionality and mobility, dual connectivity (DC). The NG and Xn interfaces consist of separate management and user-level components. The 5G node gNB provides the completion of the NR, user plane (UP), and CP protocol according to the UE, connected via the NG interface to the 5GC. Base stations are composed of gNB elements and house radio equipment such as transmitters, receivers, batteries, and power supplies that let mobile devices to communicate wirelessly with the network. The housing may take the form of a SA structure, a chamber within an already established structure, or a straightforward box that is installed on a wall or tower. Since the 3GPP specifications allow for the split of gNB functions, it is not required to have all of them in the same physical place. Instead, part of the functions can be relocated, for example to a cloud that is physically located in the data center (Figure 1.3) [70].

Transport network (TN) is a set of optical fibers, other cables, and radio connections between NR and the core network. In tandem with rising quality standards, the transportation network is becoming increasingly intelligent based on services and levels of geo-redundancy. A modern TN based on function virtualization enables optimization of transport management.

The 5G core network, also known as the 5G CN, is in charge of performing tasks that are not directly related to radio access but are nonetheless essential to support the complete network. For example, authentication, functionality charging for authorized network features, and end-to-end connection setup. Separate management of these functions, instead of being integrated into the RAN, is useful and allows more

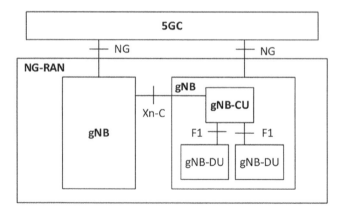

FIGURE 1.3 3GPP RAN Xn interface between two gNB–gNB nodes and NG interface gNB–5GC.

than one radio network to be served by a single core network. 5G CN has all the essential capabilities for establishing, sustaining, and executing data and voice calls, as well as establishing any other communication connections, such as messaging between users and network signaling. 5G standards ensure that these characteristics are applicable in both domestic and foreign situations, therefore interoperability is essential. Compared to prior reference models, the new network core is drastically different. NF is processed by dedicated elements owned by the operator, as separate elements, each of which possesses their own software and hardware. 5G is predicated on NF virtualization. Therefore, each function is comprised of a collection of software instances that are executed on general-purpose hardware that has been virtualized. The 5G core network can now function in operator or third-party cloud on resources provided by data centers. Virtualized networks significantly improve 5G performance compared to previous generations [71–73].

Instead of storing computer resources in a central location, as is done in traditional cloud computing, EC moves those resources to the periphery of a network, also known as the edge cloud (instead in the cloud core). The operator is enabled to reallocate the desired content or part of the computer processing closer to the user within core network. It is becoming increasingly significant for real-time latency-sensitive applications because it achieves low latency for demanding applications, speeds up operational data transfer, and saves transportation costs.

1.2.1.1 5G System Reference Architecture

The 5G network's design is built on NFV and SDN, which allow for seamless data communication and the rollout of new services. The specification addresses roaming and non-roaming scenarios in their entirety, covering topics such as mobility within the 5G system, quality of service, policy management and invoicing, as well as general system operations (SMS, location services, emergency services). The use of service-based interactions between the CP functions of a network is made possible by architectural concepts. The basic principles are the separate functions of UP from the functions of CP, so that independent scalability, evolution, and flexible

5G-Advanced Mobile Communication 15

application are enabled. The properties of the modular design enable NS that is both flexible and efficient. In order to make reuse possible, NFs are typically specified as services. Procedures are a set of interactions between these services. There is a significant reduction in the amount of dependence on the RAN access network and the CN. A single authentication framework is supported. Computer resource and storage resources (stateless NFs operate by storing the current state of the end-user device UE in a remote database) are separated. Capability exposure is supported as well as simultaneous access to local and centralized services [74–80].

5G reference architecture is service-based SBA and the interaction between NFs is presented in two ways:

- Representation based on services, wherein network components (such as AMF) at the CP control level grant access to their services to other network components that have been granted permission to do so. Reference points from point-to-point where necessary are included.
- When there is an interaction between NF services in network functions indicated by a point-to-point reference point (e.g., N11) between any two network functions, reference point representation indicates that there is an interaction between the NF services (e.g., AMF and SMF).

1.2.1.2 5G NR Architecture

5G NR radio access technology (RAT) includes a variety of usage scenarios, from eMBB to URLLC and mMTC. Ultralow power transmission, low latency support, cutting-edge antenna technology, and spectrum adaptability (including high-frequency band operation and low-frequency band interoperability) are all vital technologies.

3GPP explores RAN virtualization and different functional divisions between CU and DU, in a way that CU is centralized and potentially virtualized. Depending on the functional split choice, some gNB functions may be completely allocated to the DU, while others, such as user data transfer, mobility control, RAN sharing, positioning, session management, and the like, may be included in the central unit as a logical node. The central unit controls the operation of the DU via the front-hall (Fs) interface. The central unit is also marked as BBU/REC/RCC/C-RAN/V-RAN. The logical node of DU includes subset of gNB functions, depending on the functional separation option. The DU is also referred as RRH/RRU/RE/RU [81–88].

In parallel, 3GPP also considered the possibility of a horizontal division between the UP and the CP (TR 38.801). The resulting overall architecture has one gNB-CU-CP per gNB node that allows a wide range of applications in the virtualization of all or individual CU components. Scalability, low-cost hardware implementations, coordination of performance characteristics, load control, real-time performance optimization, and the ability to use NFV/SDN in a wide range of use cases are all benefits of this design. The split of CU-DU into higher layers is suitable for application in wide area with significant processing at the base location. The split of CU-DU into lower layers further reduces such processing near transmission points, with increasing backhaul requirements (less delay and much higher bandwidth). The choice of the way in which the NR functions will be split depends on the scenario of the

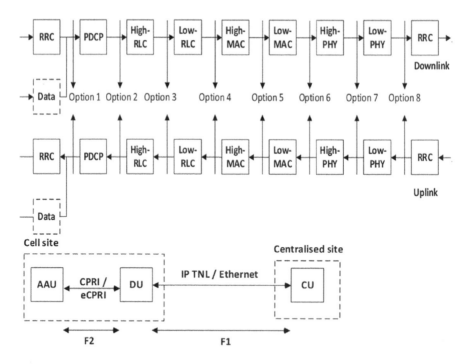

FIGURE 1.4 5G NR functional split options between centralized and distributed units. Option 2 is selected for standardization with RRC-PDCP protocols in CU, and RLC-MAC-PHY-RF functions in DU.

implementation of the radio-network, the limitations and the envisaged supported services, i.e., support for specific QoS or specific user density and load demand in a given geographical area, availability of transport networks with different levels of performance, and real-time applications (Figure 1.4).

Media access control and radio resource management are two examples of the higher-level layers that make up the 5G NR radio interface. Specifications for the higher layers are defined in the 3GPP TS 38.300 series, whereas the physical layer is covered in the 3GPP TS 38.200 series. TS 38.201 specification presents the architecture of the NR radio interface protocol on the physical layer (Layer 1). Media access control (MAC) in Layer 2 and radio resource control (RRC) in Layer 3 are linked by the physical layer (PHY). Service access points (SAPs) are highlighted between different layers/sublayers. The physical layer allows the transport channel to MAC sublayer. The transport channel characterizes the way information is transmitted on radio interface. MAC provides different logical links of radio link control (RLC) sublayer of Layer 2. The format of the information that is sent defines the kind of logical channel that exists (Figure 1.5) [81–88].

1.2.1.3 5G Core Network

Control of UE user devices is handled centrally by the 5G CN. In terms of functionality, it sits atop the RAN and oversees tasks such as authentication and the setup,

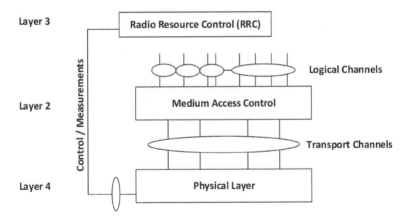

FIGURE 1.5 RAN architecture around the physical layer PHY.

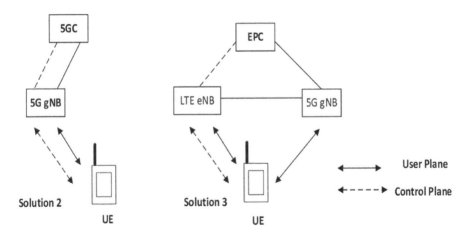

FIGURE 1.6 Stand-alone (SA) Solution 2 and non-stand-alone (NSA) Solution 3.

maintenance, and teardown of communication links that are essential to the network's operation but have nothing to do with radio access. 3GPP enabled gradual transition to completely autonomous 5G network, specifying two primary architectures NSA, whereby there are a great deal of variants depending on the data routing, and SA. Figure 1.6 shows the standardized SA Solution 2 and NSA Solution 3. The NSA architecture uses the EPC LTE core and the eNodeB LTE base station. CP data is connected to eNB, and DP to/from UE is connected to eNodeB or NR gNB base stations or used by two base stations at the same time. In the second case, double connection (DC) is established, which increases the speed of user data and reliability. NSA architecture only supports eMBB service. However, dual connection is also possible within 5G network so that the UE user device communicates with two gNB nodes simultaneously [73].

5G CN connects to RAN access network. One of the main goals is the separation of different functionalities and possibility to separate the network logic with more granularity and enable modular implementation. All NFs in the core network that offer a set of service-based interfaces are denoted by *N* and the name of the corresponding NF (Namf for AMF). Using the interface, any other NF can communicate and retrieve data with the NF. Access and mobility function (AMF) is a 5GC entry function that connects to 5G access network for 3GPP access or N3IVF for non-3GPP. AMF interrupts the UEs NAS session, performs access authentication and mobility management. It is also responsible for route session management messages to the correct session management function (SMF). Function UPF (user-level function) implements packet forwarding and routing for UP data as an anchor point for intra/inter-RAT mobility (when applicable), assigning UE IP address/prefix (if supported) in response to an SMF request also as external PDU point of the interconnection session to the data network. Protocol data unit (PDU) session forms virtual pipe between the UE and the data network identified by DNN (data network name). Policy control function (PCF) supports the unified policy framework for managing network behavior, provides policy rules for the CP function(s) to implement them, and accesses subscription information relevant to unified repository (UDR) policy decisions. Network exposure function (NEF) supports APIs for exposure capabilities, events, and analytics.

The list of functions is extensive, total of more than 20 functions: NEF network exposure function, NRF network repository function, AUSF authentication server function, AMF access and mobility management function, DN data network operator services, Internet access or third-party services, UDSF unstructured data storage function, NSSF network slice selection function, PCF policy control function, SMF session management function, UDM unified data management, UDR unified data repository, UPF user plane function, AF application function, UE user equipment, RAN radio access network, 5G-IER 5G-equipment identity register, NWDAF network data analytics function, CHF charging function (Figure 1.7) [75–77].

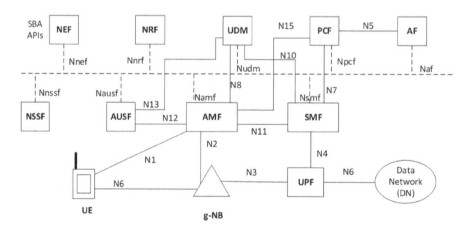

FIGURE 1.7 5G CN service-based architecture (SBA) of network functions that can operate independently from one another and are free from hardware dependencies.

5G-Advanced Mobile Communication

Virtualized network functions (VNFs) are the actual implementation of NFs as software processes. In the 5G SBA, two VNFs can communicate with one another using SBI, which is an API for doing so. A given VNF can utilize an API call over the SBI in order to invoke particular service or service operation.

Additionally, the 5G system includes the following network entities: the service communication proxy (SCP), the security edge protection proxy (SEPP), the non-3GPP interworking function (N3IWF), the trusted non-3GPP gateway function (TNGF), and the wireline access gateway function (W-AGF).

1.2.1.4 Separation of Control Plane and User Plane

In a conventional mobile network, the active network element possesses both user-level and management-level functionalities at the same time. With the development of software-defined network SDN technologies, the design of architectural functions has been re-examined thanks to the 5G architecture. The functional separation of the control plane (CP) from the user plane (UP) is the fundamental idea behind software-defined networking SDN that allows the 5G network to centralize application and management control plane, and optimize re-organization. The UP simplifies functions, deploys flexibly, and forwards efficiently. By separating the functions of the CP and UP, optimization and re-organization have been achieved, which further enables flattening of the network architecture. A nonhierarchical architecture brings for shorter paths, lower delays, and higher network efficiency (Figure 1.8) [77–79].

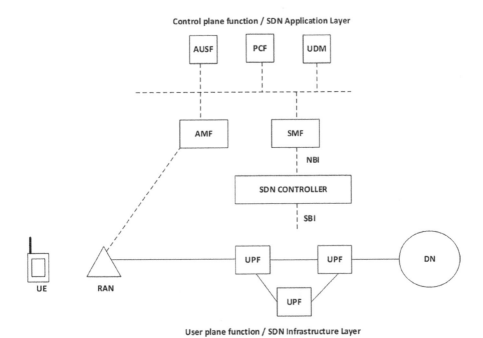

FIGURE 1.8 Separation of control plane (CP) and user plane (UP) functions.

1.2.1.5 RAN Protocol Architecture

A 3GPP base station is the physical representation of a virtual node in a radio access network (RAN). The network node B (NB) and its evolved counterpart, the eNB (evolved node B), are used in both 3G UMTS and 4G LTE. 3GPP marked the gNB node of the next-generation 5G NR. To be clear, gNB is not a real base station but rather a conceptual entity that acts as one. On the basis of the standard protocol, the base station can be implemented in several ways.

One can separate the CP from the UP in the NR radio protocol architecture. While the CP is primarily in charge of connection, mobility, and security, the UP is in charge of delivering data to end users. RAN protocol stack is divided into layers. The first layer, or PHY, is the physical layer. The radio link control (RLC), packet data convergence protocol (PDCP), service data adaptation protocol (SDAP), and media access control (MAC) all fall under Layer 2. Layer 3 includes RRC control of radio resources. It is pointed out that the user and control planes share a lot of protocols in common. Layer-to-layer and sublayer-to-sublayer linkages are labeled in Figure 1.9. The MAC layer supplies the RLC layer with logic channels, which are distinguished by the nature of the data they convey. Transport channels, which physical layer offers to MAC sublayer, are characterized by how and with what characteristic information they are transmitted via RAN. Transport blocks (TB) are used to organize the data that travels via the transport channel. The physical channel is

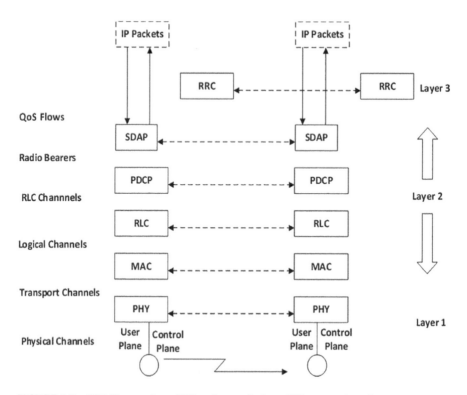

FIGURE 1.9 5G NR user plane (UP) and control plane (CP) protocol stacks.

responsible for transporting data between the gNB node and the user equipment UE. This data comes from higher levels [83–87].

The main functionalities of the layers are briefly described as follows:

- SDAP layer manages service quality QoS flow and radio carrier mapping. Radio carriers are assigned IP packets based on the QoS needs of those packets.
- PDCP layer's primary function is to compress and decompress IP headers, as well as to reorganize and identify duplicates, encrypt and decrypt, and guard against integrity breaches. Sending fewer bytes over RAN is made possible thanks to a header compression method. Encryption provides protection against interception of communications and protection of confidential information. Data units may be sent sequentially, and duplicates can be eliminated, thanks to reordering and duplication detection algorithms.
- RLC layer is responsible for delivering data units to higher levels, segmenting and re-segmenting (compressed IP packet header), and error correction through the automatic repeat request (ARQ) mechanism.
- Scheduling uplink UL and downlink DL, as well as error correction through the hybrid ARQ (HARQ) mechanism are primarily the purview of the MAC layer. Transmission time and frequency resources at the UL and DL levels are divided up by the scheduler. When using carrier aggregation, the MAC layer additionally multiplexes data over the various component carriers.
- PHY layer is responsible for transforming signals into the physical time-frequency resources of the network, including encoding, decoding, modulation, multiantenna processing, and signal mapping.

Scheduling is key function of the NR multi-user mobile system and significantly determines the overall behavior of the system. At the start of each transmission time interval (TTI), the scheduler decides which users will share the available time-frequency resource and at what data rate they will transmit. There are two types of schedulers in a gNB, one for downlink transmission and one for uplink transmission. Allocation of transmission resources, such as resource blocks (RB) in the frequency domain and OFDM (orthogonal frequency-division multiplexing) symbols in the time domain, are managed by the scheduler, which is part of the MAC layer. NR introduces the concept of slot-based scheduling (Type A) and mini-slot scheduling (Type B). Therefore, TTI concept for the data channel in NR becomes much more flexible.

A key NR feature is the ability to configure for low latency, mainly important in the low-latency URLLC service. The user plane UL/DL delay is the amount of time it takes for an IP packet to go from the RAN to the user device. The desired UL/DL latency for eMBB is 4 ms. The minimum acceptable latency for URLLC is 0.5 ms in both UL and DL. With 4G LTE, the lowest equivalent latency is around 5 ms. Therefore, the goal of NR is to reduce latency by a factor of 10 or more. For low delay requests, the NR mostly uses the 0.125 ms transmission time period TTI as the minimum standard slot length to be allowed. Further, by using mini-slots, the interval can be further reduced. For 4G LTE, the transmission time interval is 1 ms.

Reduced minimum term TTI in NR implies packets may be sent and received more quickly, cutting down on processing and transmission buffering times. Utilizing preloaded reference signals and control signaling that conveys scheduling information at the beginning of the slot enables the receiving device to immediately begin processing incoming data without the need for buffering. The detailed structure of the UP and the CP Layer 2 protocols (MAC: media access control; RLC: radio link control; PDCP: packet data convergence protocol; SDAP: service data adaption protocol sublayers) is shown in Figure 1.10 [83–88].

The control plane CP is primarily accountable for the control signaling that governs the establishment of connections, as well as mobility and security. The gNB base node's radio resource control (RRC) layer or the core network provides the control signals. Broadcasting system information, paging messages, security management including key management, handovers, cell selection/re-selection, QoS management, and radio fault detection and recovery are all primary services provided by the RRC layer. The RRC layer utilizes the user plane's PDCP, RLC, MAC, and PHY layers to send and receive messages. For this reason, there is no essential technical distinction from a physical layer point of view in service delivery to the upper layers in control layer and user layer protocol stacks [81–88].

1.2.1.6 IAB Protocol and Physical Layer

Wireless backhaul may now be supported by the 5G NR thanks to the integrated access backhaul (IAB). It offers functionality that enables the use of NR radio access technology (RAT) for wireless feedback in addition to the connection between base stations and devices, also known as the access connection. IAB enables self-backhauling to eliminate wired backhaul dependence. It is based on Layer 2 architecture with PDPC end-to-end layer (from donor IAB node to UE for CP and UP). For the NR, the IAB study was approved in March 2017 and conducted during the R15 and R16 editions [89–92].

The basic structure of the network using the IAB is illustrated in Figure 1.11. A donor node, which is simply a standard base station that employs traditional non-IAB backhaul, provides the connection between the IAB node and the network. Every IAB node may generate its own cells and appear to connected devices as a standard base station. Since IAB is a network-only function, it has no direct effect on the device. The fact that R15 devices may now connect to the Internet using IAB nodes is a major development. Multiple-hop wireless backhauling is made possible by the IAB node's ability to produce cells that may be used to link other nodes to the network.

IAB's general design is based on gNB's CU/DU separation into two functionally distinct parts with standardized interface between. The IAB lists two types of network nodes:

- A fiber-based backhaul, for example, links an IAB donor node, which comprises of CU and DU capabilities, to the remainder of the network. Although the DU donor node typically provides service to UE user devices such as traditional gNB, it will also provide service to IAB nodes that are wirelessly linked to it.

5G-Advanced Mobile Communication

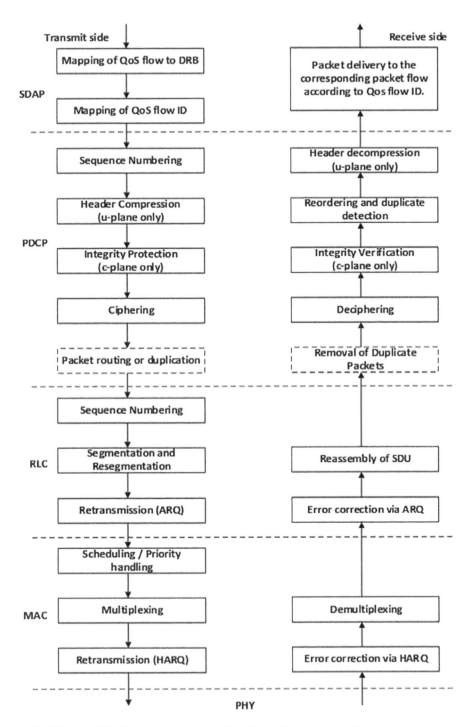

FIGURE 1.10 5G NR Layer 2 structure of user/control plane protocols.

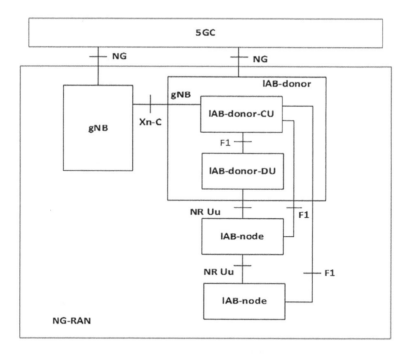

FIGURE 1.11 Integrated access and multi-hop (2 hops) wireless backhauling.

- A node that is considered to be an IAB node is one that is dependent on the IAB for its backhaul. In the case of multi-hop IAB, this comprises the DU functionality supporting the UE and, perhaps, additional IAB nodes. There is a UE function (IAB-MT) at the opposite end of the IAB node that communicates with the DU of the parent node.

The whole spectrum of NR is supported by the IAB. However, mmWave spectrum is most relevant for the IAB. One of the characteristics that greatly influences design of physical layer is half-duplex constraint of the IAB node. In general, simultaneous operation of the remote control (communication with children of the nodes in downstream connection) and mobile terminal (MT; communication with the parent node in the upstream connection) is not possible due to self-interference. As a result, IAB node operation is defined with TDM operation between DU and MT. The parent DU controls subordinate MT resources through planning. The basic issue is the coordination of resources for DU in time domain. A large number of physical layer extensions have been identified to support IAB modes.

1.2.1.7 Physical Layer (PHY)

5G NR relies heavily on the physical layer, as do all wireless technologies. Numerous frequencies between 1 and 100 GHz are supported by the physical layer, as well as several applications (pico cells, micro cells, and macro cells). To successfully meet the challenges, 3GPP has developed a flexible physical layer. Radio wave propagation mechanisms and hardware flaws in networks and devices have been explored to

5G-Advanced Mobile Communication

improve the performance of adaptable parts. By June of 2018, the first R15 edition had been finalized and all future versions of NR are compatible with the first edition [82].

The frequency range in which 5G NR is expected to function spans around 400 MHz to roughly 90 GHz and includes both licensed and unlicensed spectrum as well as shared spectrum. A shared band is one in which 5G shares the spectrum with a service that isn't a mobile operator. Frequency division duplex (FDD) and time division duplex (TDD) modes are supported by NR. For NR, the 3GPP specifies two frequency bands, FR1 and FR2, with respective ranges of 410–7,125 MHz and 24.25–52.6 GHz (mmWave). In terms of radio spectrum, the FR2 band is a new feature of 5G over 4G. It offers large capacity and data speeds by transmitting in bands with substantially more spectrum than is accessible in FR1. TDD operation is used in the C-band/3.5 GHz band (FR1), which spans 3.3–4.2 GHz and has a maximum spectrum allocation of roughly 100 MHz per operator. Since the quantity of spectrum in the C-band is relatively substantial and the loss in propagation is only around 5 dB larger than the 2 GHz band, it plays a pivotal role in 5G networks. High-gain beamforming antennas can compensate for this extra path loss and provide coverage in the 2 GHz bands that is on par with the 1 GHz bands. 5G NR is characterized primarily by its extremely high data rates [85–88].

Support for spectrum above 52.6 GHz opens the V to W band (60–114.25 GHz) for NR will be specified in R17 and R18. The expansion is notable because it not only provides ultra-wide spectrum up to 15 GHz, but also new features for back/fronthaul, relay, industrial IoT, private network, advanced V2Xs communications, tightly connected use of licensed/unlicensed spectrum, and minimizing form factors for future wireless devices. However, there are challenges that need to be overcome and achieve mobile communication in the V–W range: high attenuation due to atmospheric gases, inefficient RF devices such as power amplifier, switcher, and mixer due to interference and attenuation inside the device, and strong shading by reflection and diffraction due to short-wave length. The solutions require improving the radio interface from a completely new perspective (Table 1.2).

Modulation systems, waveforms, channel codes, multiantenna methods, frame structure, duplexing schemes, control and reference signals, and so on are all components of the NR physical layer (Table 1.3).

TABLE 1.2
Maximum 5G NR Data Rates per Layer per Component Carrier [85–88]

Frequency Range	Subcarrier Spacing (kHz)	Bandwidth (MHz)	Downlink Rate (Mb/s)	Uplink Rate (Mb/s)
FR1	15	20	113	121
FR1	15	50	290	309
FR1	30	100	584	625
FR1	60	100	578	618
FR2	60	200	1,080	1,180
FR2	120	400	2,150	2,370

TABLE 1.3
5G NR Physical Layer [85–88]

Concept	5G NR
Frame structure (FS)	Single and highly configurable frame structure for all use cases. Notion of DL, UL, and flexible resources.
Scheduling flexibility	Slot bases (A type scheduling); mini-slot based (B type scheduling)
Waveform	DL: OFDM (orthogonal frequency-division multiplexing) UL: SC-OFDM (for single-layer transmission); OFDM FR1:15, 30, 60 (data) 15, 30 kHz SSB (synchronization signal block) FR2: 60, 120 kHz (data) 120, 240 kHz SSB (synchronization signal block)
Forward compatibility	Reserved resources
Always ON signal	No CRS (cell reference signal)
Transmission modes	Single-transmission mode for data channels transparent TxDiv scheme
Multibeam operation	FR1: up to 8 SSB beams; FR2: up to 64 SSB beams
Channel coding	Data: LDPC codes; Control: Polar codes
Bandwidths and bandwidth part concept	BWs FR1: 5, 10, 15, 20, 25, 30, 40, 50, 60, 80, 90, 100 MHz; FR2: 50, 100, 200, 400 MHz BWP (contiguous resource blocks configured inside a channel) Up to four configured BWPs. Single active BWP. Additional BWs could be added in future releases of NR

Uplink and downlink transmissions in NR are compatible with QPSK and 16/64/256 types of quadrature amplitude modulation (QAM). In addition, mMTC services benefit greatly from the uplink's support of $\pi/2$ BPSK since it allows for a lower peak-to-average power ratio and higher power amplifier efficiency at lower data rates.

The UL and DL links in NR operate at frequencies of at least 52.6 GHz using the cyclic prefix CP-OFDM. LTE uses CP-OFDM for DL transmission, and DFTS-OFDM is used for UL transmission. The overall 5G architecture is simplified by the use of a waveform that is same in both directions, especially in terms of wireless transport and communication between devices (D2D). DFTS-OFDM in UL, which uses a single stream transmission method (no spatial multiplexing), is another solution for situations with constrained coverage. In practice, it is possible for the gNB to select the UL link waveform CP/DFTS-OFDM and the UE is able to support both OFDMs. The NR waveform may have any receiver-independent function, such as windowing or filtering, added on top of it to better limit or constrain the operating frequency range. NR mandates a flexible OFDM numerology that makes it possible to provide several services across many frequency bands and use cases. One subcarrier spacing correlates to numerology in the frequency domain. Various numerologies may be established by scaling the reference subcarrier spacing by an integer N.

Multiantenna techniques are already important in wireless communication, and they play a more integral part in the overall system design in NR. The expansion of the spectrum for mobile communication toward mmWave influenced beam-oriented NR design to support the formation of an analog beamforming for achieving adequate coverage. Moreover, multiantenna techniques are key for the 5G performance requirements in traditional cellular frequency bands [91–95].

5G-Advanced Mobile Communication

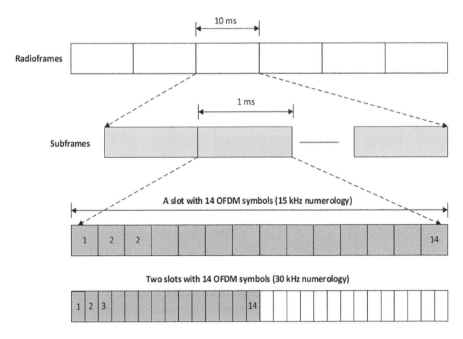

FIGURE 1.12 Frame structure configuration for the physical layer of NR.

5G NR uses advanced error-correction techniques for channel coding, low-density parity check (LDPC) codes to transmit eMBB mobile broadband access data, and Polar linear block code for control signaling. At data speeds of many gigabits per second in rate-compatible structure, LDPC codes are appealing from an implementation standpoint. This permits HARQ operation with incremental redundancy and transmission at multiple code rates.

Radio transmissions are divided into radio-frames, sub-frames, slots, and mini-slots, all of which exist in the time domain. Each radio-frame, seen in Figure 1.12, lasts for 10 ms and is divided into 10 sub-frames, each of which lasts for 1 ms. One or more adjacent slots, each of which contains 14 OFDM symbols, together make up a subframe. Greater flexibility in terms of delays is achieved by using less than 14 transmission symbols, and the resulting smaller effective slots are called mini-slots. Mini-slot structures may have 2, 4, or 7 symbols allocated to them, and they can have a variable beginning position. This enables the transfer to begin as soon as feasible, rather than having to wait for the slot boundary to begin. Since the duration of an OFDM signal is inversely proportional to its subcarrier spacing, the slot and mini-slot durations are scaled with the specified numerology (subcarrier spacing). One subcarrier in the frequency domain and one OFDM symbol in the time domain make up a resource element (RE) in 3GPP nomenclature. RB is defined only for frequency domain. Bandwidth part (BWP) is a set of attached common resource blocks (RBs). The term "physical channel" refers to a collection of resource elements REs that transport information that is generated at a higher layer, whereas "physical signals" refers to a collection of resource elements that are utilized by the physical layer but do not contain information generated at a higher layer.

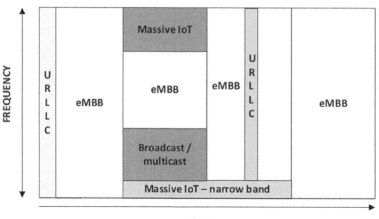

FIGURE 1.13 Flexible frame structure in 5G NR.

Basic principle of flexible frame structure in 5G NR is shown in Figure 1.13. It is possible to set each of the many components of the frame differently depending on the application. This allows for the efficiency to be improved in a variety of ways, including the achievement of QoS QPI, the use of resources, and the consumption of device power. All BWP-capable devices need to do is decode a few lines of common control information and the BWP-specific control information [91–95].

In the temporal and frequency domains, the NR radio resources are divided as shown in the transmission structure (Figure 1.14). Resource grid consists of number of subcarriers in frequency axis and number of OFDM symbols in time axis. The term "resource element" refers to any one of the squares in the grid used to set up things such as antenna ports and subcarrier distances. Frequency-domain resource blocks RBs have a total of 12 consecutive subcarriers. Bandwidth portion, abbreviated as BWP, is a collection of shared RBs that are linked to one another.

Both time division duplex (TDD) and frequency division duplex (FDD) are supported by NR. The duplex scheme usually depends on the allocation of the spectrum. The majority of lower-frequency spectrum is allocated in pairs, suggesting FDD operation. It is common for higher-frequency spectrum allocations to be unpaired, indicating the need for TDD. NR also enables dynamic TDD, in which UL and DL allocations alter dynamically over time. As a major upgrade over LTE, this is especially helpful in situations where traffic patterns are subject to sudden shifts. The gNB planner makes the decisions about the transmission schedule, and the UE conforms to those schedules. The network is able to coordinate scheduling choices across nearby network sites in the event that this becomes essential in order to prevent interference. There is also the possibility of semi-statically configuring TDD by alternating between UL and DL at certain intervals.

In order to promote backward compatibility and simplify interactions between various features, the NR frame structure is built on three core design concepts. The fundamental premise is that there is no dependence between transmissions. The data in a bundle may be decoded separately since each slot contains all of the reference

5G-Advanced Mobile Communication

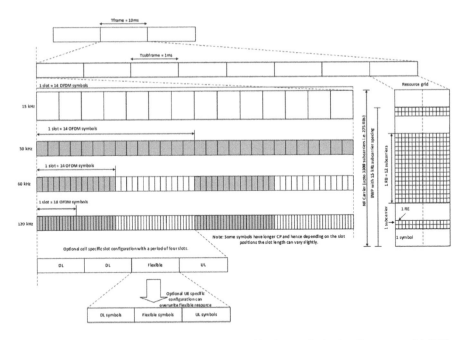

FIGURE 1.14 5G NR transmission structure and basic terminologies: Resource grid (RG), resource element (RE), resource block (RB), bandwidth part (BWP).

signals needed for demodulation. The second point is that transmissions are time and frequency constrained, so that introduction of new types of transmissions is simplified. To prevent static and/or rigid timing correlations between slots and multiple transmission routes is the third principle [81–88].

NR frame structure supports varied length transmissions. For instance, the URLLC has a short duration, but the eMBB has a lengthy duration. With respect to TDD operation, Figure 1.15 gives examples of the structure of the NR framework for different scenarios. In TDD mode, a slot can be scheduled for all DL transmissions, all UL transmissions, or a mixture of both DL and UL, where guard periods for UL/DL switching are inserted [81–88].

The smallest specified physical NR resource element RE consists of one subcarrier during one OFDM symbol. Resource block RB is a collection of 12 subcarriers with the same spacing in the frequency domain, with the block's width proportional to its numerology. Allocations are made on a per-physical-RB basis, with one RB corresponding to a block of 12 consecutive subcarriers with the same spacing in the frequency domain, and one OFDM symbol corresponding to a block of time. On the same carriers, many numerologies are supported, and the precise positions of RBs are defined such that their borders are in perfect alignment. Figure 1.16 depicts a physical resource grid (RG) with the resource elements RE (dark shade) and the resource block RB (bright shading). As a function of numerology, NR outlines the minimum and maximum values of RB for each carrier, which are also the same for DL and UL. A resource network is specified for each numerology and carrier, covering the complete carrier bandwidth in the frequency domain and one subframe in

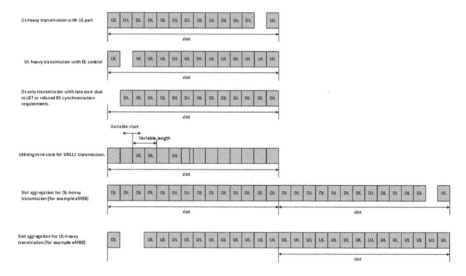

FIGURE 1.15 Examples of NR physical layer frame structure in different scenarios with respect to TDD operation.

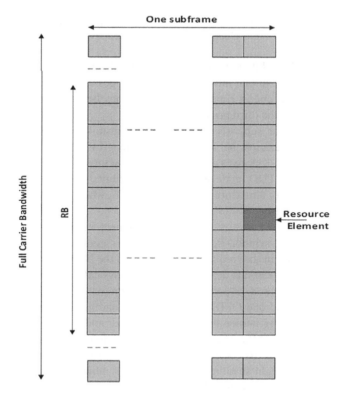

FIGURE 1.16 Resource grid in frequency domain structure: Resource element (RE) and resource block (RB).

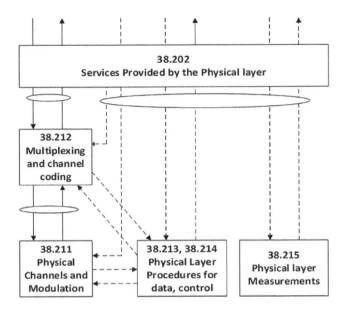

FIGURE 1.17 Relation between 3GPP physical layer specifications.

the time domain. There is just one set of RGs for each transmission direction, and the grids for all subcarrier spacings overlap. For a particular subcarrier distance, the resource grid specifies the signal space as viewed by the UE.

The 3GPP New Radio (NR) standards specify many physical layer processes, including those for cell search, power control, UL synchronization and timing control, random access and beam management, and operations relating to channel status information. Measurements for the physical layer, including those for timing and radio resource management, as well as intra- and interfrequency and inter-RAT handovers, are all included in the NR standard.

Six papers make up the 3GPP physical layer definition in addition to the overarching TS 38.201 (TS 38.202 and 38.211 through 38.215). Figure 1.17 depicts the connection between the technical standards in the context of the upper layers [81–88].

1.2.1.8 Network Slicing (NS)

Network slicing NS enables segmentation of network control plane CP and user plane UP functions, RAN, and CN into several logical planes that potentially share the same physical infrastructure, but have autonomous operation and control. The network operator is now in a position to dedicate part of network to third-party verticals. The control, generation of statistics, and management of these parts of the network are separated from each other, so it is easier to separately negotiate the level of service (SLA) between MNO and vertical. 3GPP defines only signaling framework and allows the UE and several NFs to assign specific policies to a particular NS. The network operator specifies the actual policies and how the NS is used [93–98].

NS is a key 5GS technology that enables implementation of on-demand networks. This system creates numerous virtual E2E networks atop a single physical network to meet the needs of various services. From the 3GPP edition of the R13 DECOR dedicated core network, methods are specified that allow different types of devices to serve different instances of packet core subnetworks. DECOR has been improved by the standards in the R14 edition and fully realized by working on the slicing within the R15 system architecture. The NS has also been significantly updated in the R16 edition.

A network slice in NS is a logical network that makes use of existing nodes in RATs, core networks, clouds, and even at the very periphery of the network, known as the EC nodes, for edge computing [99]. Cross-slice isolation can be performed depending on the layer by physical isolation, virtual machine-based insulation, and language-based isolation. The Network Service architecture is composed of three layers: the network infrastructure layer (which provides physical-based virtual machine and language-based isolation), the network slice instance layer (which provides virtual machine-based isolation), and the service instance layer (IDL-based isolation). Flexibly design is necessary to support QoS users and optimization of overall network expenditure. In addition, elasticity is required for adaptive availability of slice resources, so it is avoided under-utilization and overutilization due to variations in user demands (Figure 1.18) [93–102].

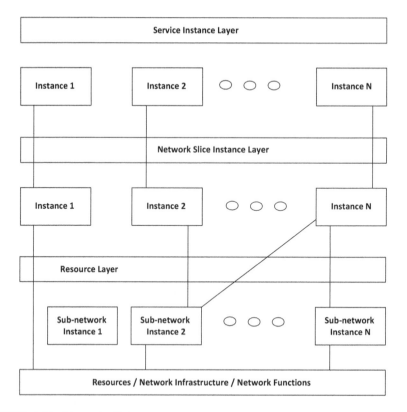

FIGURE 1.18 Network slicing components: Resource layer, NS instance layer, service instance layer.

5G-Advanced Mobile Communication   33

The physical network is able to be partitioned into a number of separate virtual networks, also known as logical segments, thanks to NS. These virtual networks are able to serve a wide variety of services and may be tailored to a particular RAN or user group. More efficient use of infrastructure significantly reduces the expenses associated with constructing separate networks. When it comes to generating NSs, different combinations of NFs are required for various use cases. The development of network software has greatly contributed to the adaptability of end-to-end 5G networks. In CN, NFV and SDN are technological enablers for NS because they allow network parts and functions to be virtualized and controlled, making it easy to tailor and reuse in each slice to match the needs of individual services. Slicing in the RAN is based on radio resources that are either physically present or logically represented in some way. As a result, there will have to be a wide variety of network slice configurations. Agility and adaptability on a very high level are therefore required.

1.2.1.9 Quality of Service (QoS)

5G service quality model uses Quality-of-Service (QoS) flows as its foundation as the lowest granularity of quality of service distinction within a session data protocol (PDU), an abstraction for user-level services that provides connectivity between UE applications and DN such as Internet or private corporate networks. The user plane function of UPF in DL and UE in UL map service data streams SDF to QoS streams. Mapping is based on packet filters provided by policy control function PCF for session management function SMF and signal either when establishing a PDU session or dynamically after application interactions (Figure 1.19) [100–106].

Each QoS stream is identified using QFI. It also has a rule that contains the QoS profiles as a set of parameters and flow template as a set of UL/DL packet filters that represent the flow of service data mapped to the given flow. All PDU session UP communication with the same QFI is sent in the exact same way (scheduling and admission threshold). The SMF is responsible for managing the 5GS system's

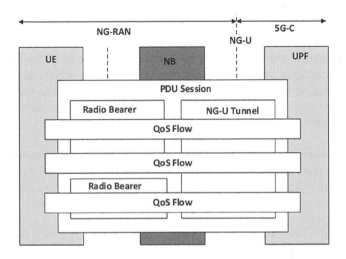

FIGURE 1.19   5G QoS flow model.

QoS stream, which may be set up in advance or created through a PDU session setup operation or a PDU session modification procedure.

3GPP defines the following QoS parameters: QoS identifier (5GQI), allocation and retention priority (ARP), reflective QoS attribute (RQA), notification control, bit rate, aggregate baud rate, default values, and maximum packet loss rate. The 5G quality of experience index (5GQI) is a scalar that may be used to refer to QoS features, access node-specific metrics that regulate flow forwarding treatment (scheduling weights, receive thresholds, queue management thresholds, link layer protocol configuration). Values on the standard 5GQI are converted directly into a set of QoS metrics that have also been standardized. A 5GQI result indicates standard or pre-configured features, but does not signal them on any interface until certain configuration changes are made. Streams that use a 5GQI that is dynamically allocated will have their QoS features notified as part of the profile.

QoS parameters signaled are 5QI, priority allocation and retention ARP, for every QoS flow and guaranteed flow bitrate GFBR, and maximum flow bitrate (MFBR) for GBR and delay-critical GBR QoS flows. From these parameters, the 5QI of 5G system points to combination of QoS parameters: packet delay budget, packet error rate, averaging window, and priority for all types of 5QIs. For delay-critical GBR 5QI is in addition the maximum data burst volume is also defines.

Both QoS flows that need guaranteed flow bit rate GBR and those that do not require guaranteed flow bit rate non-GBR are supported by the model. The first one is the average bit rate over the averaging time frame that the network commits to provide the flow. The highest possible flow bitrate MFBR restricts the data transfer rate to the maximum bit rate that can be supported by the flow while at the same time discarding any surplus traffic via the use of a rate-shaping or rate-policing function at the UE, RAN, or UPF. It is possible to grant relative priority for bit rates that are higher than the GFBR value and up to the MFBR value. The QoS parameters GFBR and MFBR are sent to the RAN in the profile, and they are communicated to the UE as the flow level parameter for each unique QoS flow [100–106].

1.2.1.10 5G Security

The 5G mobile architecture is the first of its kind to be built to concurrently serve use cases with specified needs. The 3GPP SA3 working group develops the protocols for 5G's privacy, security, and network architecture. All aspects of NR and CN network security, including architecture, features, methods, and procedures, are detailed in the specifications. System requirements for secure end-to-end core network connections are follows:

- mechanisms for adding, removing, and modifying message components by intermediary nodes, which exist at the application layer
- integrity and privacy of transmitted data from one network to another, from the point of origin to the point of reception, for predetermined message components
- reduced impact with very minimum changes to already established 3GPP network components
- standard security protocols

5G-Advanced Mobile Communication

- considerations on performance and overhead
- account for operational aspects of key management.

Mutual authentication between UE and network, security context creation and dissemination, UP data confidentiality and integrity protection, CP signaling confidentiality and integrity protection, and user identity secrecy are only some of the 5G system's security characteristics [103–106].

There are several network architectural components and ideas included into the 3GPP 5G security framework:

- Protecting against attacks on the (radio) interface is an important aspect of network access security, which is a feature set that enables the UE to authenticate and securely access services across a network. Access security is also provided by the secondary node (SN) delivering security context to the access node (AN).
- The term "network domain security" refers to a group of attributes that enable network nodes to safely communicate signal and user-level data.
- Access to mobile devices is ensured by the user domain security measures.
- Applications in both the user domain and the provider domain may safely communicate with one another thanks to application domain security.
- SBA domain security is a set of features that support SBA architecture NFs to securely communicate within network service domains. Included are safeguards for service-based interfaces and the registration, discovery, and authorization processes that occur inside a network.
- The capability to monitor the status of security measures and alert the user when they are disabled or disabled is part of the visibility and security configurability feature set.

The normative development on standards for 3GPP's System architecture group SA2 and Security group SA3 is now underway. Figure 1.20 [103–106] shows 5G architecture with security associations: security anchor function (SEAF) for primary authentication, AUSF authentication server function interrupts SEAF requests and continues to interact with authentication credential repository and processing function ARPF, authentication credential repository and ARPF processing function, security context management function (SCMF) takes over SEAF policy key, and security control function (S)PCF provides security policy to network entities (SMF, AMF) and/or UE. The security of VNFs of 3GPP network is based on network domain security NDS.

3GPP's standardization work involves addressing the following security concerns raised by 5G's technological and architectural evolution:

- UP endpoint security
- authentication and authorization (including identity management)
- RAN security
- security within the UE/secure storage and processing, eSIM
- NS security.

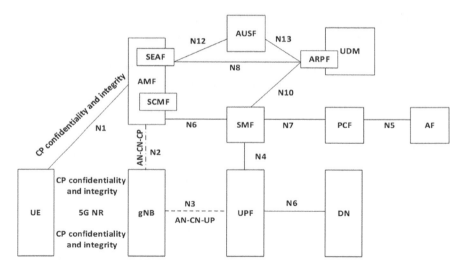

FIGURE 1.20 5G system architecture with security associations.

1.2.2 5G New Radio and Core Network Enhancement and Vertical Expansion

The latest revision of the 3GPP standard, version R16, enhances the 5G network's infrastructure and adds support for several fresh service possibilities [107]:

- 5G core architecture has been improved to achieve greater operational efficiency and enable further application optimization.
- 3GPP develops various tools to support the use of 5G systems for industrial use and new verticals (factory of the future, building automation, eHealth, smart city, intelligent transportation system, electric power distribution, smart agriculture, critical applications).

1.2.2.1 Support for Industrial IoT Applications

Industrial Internet of things (IIoT) is a primary area of concentration for improving the 5G architecture. Availability of connectivity, security, accuracy, automation, and interoperability are all highly prioritized in the IIoT. The 3GPP R15 version provides very low radio-interface latency and high reliability, and further delays and reliability improvements have been introduced in the R16 edition. These enhancements allow for more applications in the IIoT and are a response to the growing need for these types of systems in fields including manufacturing, electricity distribution, and transportation [108].

Improvements are two major benefits to operators. It allows complete radio technology to be provided under a single network core. Then, it provides features that are inherently available in the 5G core. The following main new features have been added to 5GC: support for optimizing energy savings, definition of data delivery

TABLE 1.4
3GPP Progress Toward 5G mMTC Requirements [108]

	Release 14	Release 15	Release 16
	↓ Non-anchor carrier	↓ New PRACH format	↓ Improve multicarrier operation
	↓ Release assistance indicator	↓ Small cell support	
	↓ Re-connection with RLF	↓ TDD support	↓ Inter-RAT cell selection
	↓ Maximum TX power 14 dBm		
NB-IoT	eNB-IoT	FeNB-IoT	NB-IoTenh3
	> Enhance TBS/HARQ	> Enhance cell acquisition	> Coexistence with 5G
	> Positioning	> Wake-up signal	> Group wake-up signal
	> Single cell multicast	> Early data transmission	> Early DL/UL transmission
LTE-M	FeMTC	eFEMTC	eMTC5
	↑ VoLTE improvement	↑ 64 QAM for spectral efficiency	↑ CE for non-BL UEs
	↑ 5 MHz bandwidth		↑ Improve measurement for mobility
		↑ CE mode for velocity	

procedures without IP and reliable data service (RDS), definition of various exposure services, support for congestion control, and overload in the core network and RAN.

The general structure and service capacities of 5G-IoT systems are currently being developed by the 3GPP TSG-SA technical standard group. Both enhanced machine-type communication (eMTC) and narrowband internet of things (NB-IoT) have been around since Release 13 (R13), with following releases improving upon these two separate standards. The 5G NR and 5G CN URLLC specifications debuted in R15 and continue to mature in subsequent versions (R16 and R17) (Table 1.4).

1.2.2.2 Ultra-Reliable Low-Latency Communication

Ultra-reliable low-latency communications (URLLC) include IoT communications for industrial IoT and critical applications. Reliability is defined in this context as the rate of successful and timely data transfer. URLLC provides data exchange even in conditions of heavy network load. 3GPP has solved the requirements for ultra-reliable and low latency by updating the architecture of 5G system and specifying end-to-end solutions. The 3GPP creates solutions based on the concept of redundancy from beginning to end, including parts of the user plane UP and the transport network TN. For URLLC communication between UE user equipment and UPF user plane function or inside network entities such as 5G-NR and UPF, 3GPP is also working on a system to monitor packet delay using the 5QI QoS identifier linked with URLLC service. QoS may be implemented if the operator, network, and endpoint all support it. It is possible for operators or real URLLC services (AF) to make adjustments and further decrease packet latency [109–111].

1.2.2.3 Support for V2X Connections

3GPP connectivity supports use cases for vehicular communications (V2X) device-to-device as well as communication via the infrastructure. Intelligent transportation services (ITS) offers intelligent messaging and other services to 5G system users.

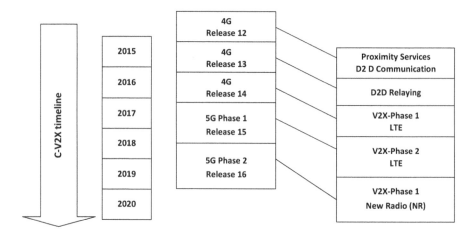

FIGURE 1.21 NR-V2X evolution in 3GPP releases [112].

In order to let devices, cars, and units positioned along roadways gather and share contextual data, the system facilitates communication between them and with an underlying application server. Further intelligent improvements and security of service are being developed. Basic applications of road safety and traffic efficiency for the implementation of ITS were considered at the beginning [112–114].

V2Xs concept includes vehicle communication with the network (V2N), between vehicles (V2V), vehicle to pedestrian (V2P), vehicle to infrastructure (V2I) represented by stationary roadside unit RSU that supports V2Xs applications with other vehicles or UE user devices. Communication vehicles–vehicles and pedestrians directly (not through the network but with or without support of gNB) requires the development of PC5 interface and use of communication LTE side link SL and 5G NR. To determine if a vehicle or UE is in or out of the monitoring and control of RAN nodes gNB, the coverage/out-of-coverage concept is used.

Device setup, direct communication support, and PC5 compatibility are the three pillars of V2X communication. MBMS support has not yet been implemented for 5GS.

The following is a brief overview of the 3GPP versions of the standards that encompass the V2X service (Figure 1.21):

- The first V2Xs communication was researched and standardized in the R14 and R15 editions. Four ways of allocating resources for D2D communication in LTE were considered, where Modes 3 and 4 were reserved for use with V2X applications. All coverage situations (in-coverage, out-of-coverage, and partial-coverage) were supported in LTE-V2Xs, however only the broadcast communication was possible. LTE-V2X was designed primarily to meet the stringent latency and reliability needs of applications related to road safety.
- When it comes to enhancing the dependability, latency, capacity, and adaptability of various forms of communication, NR-V2X R16 is the first NR

SL standard to place a priority on V2X. Advanced use cases, such as platooning, extended sensors, advanced driving, and remote driving, have been targeted as well, expanding beyond their original focus on road safety. For V2X communication, in addition to broadcast, different types of casts are also specified, such as single and multiple (group cast). Moreover, the NR-V2Xs in the R16 edition support hybrid automatic HARQ reception request that improves the reliability of SL communication.

3GPP R16 edition for the first time specified SL communication based on NR with focus on the V2X application. In addition, public safety service is possible if it meet the requirements of these services. The need for further improvements has been identified to support operational scenarios for V2X, business applications, including public security, and the like. In order to enhance coverage, reliability, latency, and energy efficiency for UE user devices with rechargeable batteries, the attendees at the 3GPP RAN 86 plenary conference for R17 edition in December 2019 unanimously approved a list of research and working items in future development.

In R16, the NR SL definition was introduced, and in R17, it has been enhanced. Nonetheless, a number of novel use cases are anticipated, many of which involve machine-to-machine and robot-to-robot communication in the industry. The high-level SL research in R17 marks the beginning. For instance, many businesses consider relative positioning with SL, which provides the distance between cars, to be a fundamentally important feature. Therefore, SL positioning is expected to be one of the main operating items in the new edition of R18 [112–114].

1.2.2.4 Non-Public Networks

Support for NPN using 5G system architecture is a main evolutionary step of mobile ecosystems based on 3GPP standards. The NPN feature set allows nontraditional mobile network operators to deploy private networks for different purposes and uses. NPN networks are effectively based on 5G system that does not necessarily support traditional MNO services for general subscribers. Private networks can operate in any spectrum, licensed or unlicensed, owned by any entity that has access to the spectrum. NPN networks are divided into two basic categories. The SA NPN does not require additional interaction with the PLMN public terrestrial mobile network. NPNs integrated with PLMN are logically a part of the public terrestrial mobile network (Figure 1.22) [114].

5G NPN private networks use dedicated resources that can be managed independently. The concept's advantages lie in its low latency for local end users and the fact that communication occurs securely and privately on-site. NPNs may be placed in a variety of settings, including commercial and industrial buildings, for a variety of purposes [113,114].

NPN offers secluded businesses with access to 5G networks. Benefits of the application include personalized QoS, potentially improved network performance based on simpler load prediction, dedicated security and protection from other users, and protection of the infrastructure through physical isolation. NPN can also simplify maintenance and operation.

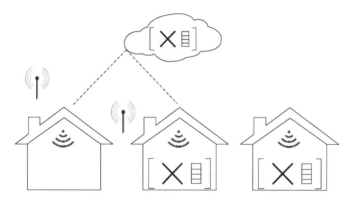

FIGURE 1.22 Examples of NPN deployment options: (a) public network integrated NPN, (b) integrated NPN with provider infrastructure deployed locally, and (c) stand-alone local NPN with dedicated spectrum.

Within a particular location, a private 5G network is a dedicated network that provides enhanced communication capabilities, a unified connection, enhanced services, and individualized security measures. Industry, commerce, utilities, and even the public sector have all found uses for private 5G networks, which take into consideration the benefits that are offered by both public and non-public 5G networks. In the course of developing Industry 4.0, the idea of a private 5G network has lately gained substantial research interest from both industry and academia [115–117].

1.2.3 Advanced Service Requirements and Performance Indicators

Design goals in work on technical specification 5G are based on requirements of service performance for existing and new use cases [118–122]:

- By maximizing network capacity and allowing for an enhanced user experience over more of the network, the requirements for mobile broadband are primarily established to meet the demands of efficiently processing extremely large and rising volumes of data in the network.
- High energy efficiency, for instance, is needed to maximize the battery life of devices, and high connection density is needed to service a large number of devices in a small space, both of which are requirements for IoT applications.
- For industrial IoT applications, the most important requirements are very low latency and very high reliability.

Requirements for 5G services have been formulated by several industries and regulators worldwide since 2015. The requirements are summarized in the ITU-R report M.2410-0/2017 as requirements for IMT-2020 networks and have served as input for relevant technical studies. 3GPP has formulated requirements in the report of TR 38.913 for next-generation access technologies. Some of the most fundamental prerequisites, briefly described for 5G services is shown in Table 1.5.

TABLE 1.5
ITU-R IMT-2020 Minimum Performance Targets [119–122]

System performance targets	Connection density	• Indoor (75 per 1,000 m^2 office) • Dense area (2,500/km^2) • Crowd (30,000 per stadium) • Everywhere (400 km^2 suburban and 100/km^2 rural) • Ultralow-cost broadband (16/km^2) • Airplane (80/plane)
	Traffic density	• Indoor (15 Gbps DL and 2 Gbps UL in 1,000 m^2 office) • Dense area (750 Gbps DL and 125 Gbps UL in km^2) • Crowd (0.75 Gbps DL and 1.5 Gbps UL stadium) • Everywhere (20 Gbps DL and 10 Gbps UL km^2) • Ultralow-cost broadband (16 Mbps/km^2) • Airplane (1.2 Gbps DL and 0.6 Gbps UL per plane)
End-user performance targets	User data rate	• Indoor (1 Gbps DL and 0.5 Gbps UL) • Dense area (300 Mbps DL and 150 Mbps UL) • Crowd (25 Mbps DL and 50 Mbps UL) • Everywhere (50 Mbps DL and 25 Mbps UL) • Ultralow-cost broadband (10 Mbps DL and 10 Mbps UL) • Airplane (15 Mbps DL and 7.5 Mbps UL)
	End-to-end latency	• Low latency case (1 ms) • Low most cases (10 ms) • Ultralow-cost (50 ms)
	Mobile speed	• Airplane (1,000 km/h) • Mobile and ultra-reliable cases (500 km/h) • Typical cases (up to 120 km/h) • Indoor and crowd (3 km/h)
Other targets	Device power efficiency	• Smartphone battery lifetime (3 days) • Low-cost IoT battery lifetime (15 years)
	Network energy efficiency	• 1,000× more traffic than today with 50% lower energy usage
	Resilience and reliability	• Network availability 99.999%
	Ultra low-cost networks	• Low-cost solution for low average revenue per user IoT services

Use cases and methods for meeting the ITU's vision requirements have been the primary focus of 3GPP's efforts. The technical specifications of version R15 define Phase 1 and R16 Phase 2. Version R16 provides sufficient performance that is fully compliant with IMT-2020 requirements, and future releases further improve both performance and feature set. Continuation of the second phase of R17 is finished in 2022. The complete version of 5G provides sophisticated technologies such as network slicing NS, virtualized network functions VNFs, support for multiple IoT devices that connect simultaneously, ultra-reliable low-latency communication URLLCs, edge computing EC, and an improved service-based architectural model.

Version R17 takes movement toward industrial verticals to whole new level. The R18 edition, which is the starting point of the new 5G-Advanced standard, deals with new concepts such as AI&ML, full-duplex operation, energy-savings online. Work on this edition will start in the third quarter of 2022 [119–122].

1.2.3.1 Phase 1 and Phase 2

The evolution of 5G focuses on three main areas: improvements to the features presented in R15 and R16, features needed for operational improvements, and new features to further expand applicability of 5G systems to new markets and use cases. The R15 version is primarily focused on enhanced mobile broadband eMBB, offering ultra-gigabit speeds, and more importantly, extreme capacity. Version R16 focuses on two other aspects, massive IoT machine-type mMTC communications and critical ultra reliable low-latency communication URLLC service. Devices that support R16 release are expected to be available in early 2022 (Table 1.6) [123–124].

3GPP R15 version is a major step in 5G definition. Although it included elements of all three services MBB, massive IoT, and critical services, the main motivation was improved eMBB. A new NR radio interface was introduced and made functional. The primary focus is the work of the new system for mobile broadband access and smartphones.

With the introduction of R16, the emphasis has switched to IoT and industrial sectors, such as time-sensitive communication (TSC), enhanced location services, and support for non-public networks NPNs. Additionally, the R16 version introduces many major new features, including NR in unlicensed bands (NR-U), integrated access and backhaul IAB, and NR V2X. The version also includes improvements to massive MIMO antenna systems, technologies such as service-based architecture SBA, network slicing NS, and wireless/wired convergence. Last but not least, the number of use cases, connection kinds, users, and apps using 5G networks is likely to rise, necessitating the installation of new security measures to deal with the corresponding rise in potential dangers.

TABLE 1.6

NR Specification Timeline in 3GPP RAN Phase 1 (R15) and Phase 2 (R16) [123,124]

2016	2017	2018	2019	2020	2021	2022
Q4	Q1 Q2 Q3 Q4	Q1 Q2 Q3 Q4	Q1 Q2 Q3 Q4	Q1 Q2 Q3 Q4	Q1 Q2 Q3 Q4	Q1 Q2
R15					R17	
		R16				
- NSA and SA		- IAB			- NR up to 71 GHz	
- eMBB		- UE power savings			- NTN	
- URLLC		- IIoT			- NR light (RedCap)	
- Carrier aggregation operation		- UE positioning			- Enhancements	
- Inter-RAT between NR and LTE		- Unlicensed spectrum			- …	
		- V2X				

The R16 version supports a number of enhancements, many of which are exclusively conceived in cellular technology. Many aspects of R16 version are specifically designed for IIoT. One such feature is time-sensitive networking. Precise indoor positioning is another basic feature usable for IIoT. The R16 version also supports massive IoT by using and enhancing LTE-IoT (eMTC) and narrowband NB-IoT. Thanks to the scalable and flexible 5G framework, R16 edition introduces in-band, native support for both of these technologies within the same 5G operator, so that they work seamlessly even when the infrastructure is upgraded and the spectrum allocated for 5G systems. In this way, the investments of the entire ecosystem are protected, including operators, original equipment manufacturers (OEMs), and users.

In addition, the R16 version supports many more configurations for 5G SA deployment mode with direct connectivity to 5GCN and provides even greater efficiency through features such as group wake-up signal, pre-configured UL, multi-block scheduling, early data transmission, and multiplexing between services with different QoS requirements (eMBB, URLLC).

Although the extension of 5G to industrial verticals is the focus of the R16 version, it brings equally significant improvements for eMBB: MIMO improvements, device power savings, mobility enhancements, bandwidth extensions, interference mitigation, single/dual-link switching, and efficient signaling.

Initial 5G implementations have gained considerable attention; the success of 5G ultimately depends on the rapid increase in coverage and expansion of 5G services. Release R16 introduces many advanced features to simplify 5G deployment: integrated access IAB and unlicensed spectrum NR-U. Since R16 version has a heterogeneous combination of features, its commercialization is diverse and achievable in stages. Operators typically implement groups of relevant functions together, based on specific applications and services being introduced or improved. With the completion of initial 5G deployments and operators seeking to expand their networks, the focus is shifting to optimization and network performance improvement [123–125].

1.2.3.2 Foundation for the Next Phase

R17 version lays foundation for the next phase of 5G, starting with R18, launched by 3GPP as 5G-Advanced. The final set of functions for the R17 was selected in December 2019, Stage 3 was released in September 2021, the ASN.1 specification in December 2021, and the specification phase is finalized in the first quarter of 2022.

TABLE 1.7

3GPP General Timeline for R17 and R18 (5G Evolution Roadmap to 5G-Advanced) [126–128]

12020	2021				2022				2023
Q4	Q1	Q2	Q3	Q4	Q1	Q2	Q3	Q4	Q1
			R17 completion						
		Email discussions, workshops, consolidation		R18 Package approval			R18 Work		

The R18 sets the direction for the development of the 6G system, which will begin to materialize later by the end of the decade (Table 1.7).

The R17 edition has approximately 40 specific features, divided into three main groups: new concepts, improvements to the features presented in R16 edition, fine-tuning R15 edition features [126–128].

- new device category called reduced capability (RedCap), support for extended reality (XR), satellite-based non-terrestrial networks (NTN), extending mmWave support beyond 52.6 GHz, introduction of multicast/broadcast (5GMBS)
- enhancements to integrated access backhaul IAB, precise positioning, side-link SL, small data transmission, URLLC in all spectrum types
- performance improvements to MIMO, multi-radio dual connectivity, dynamic shared spectrum (DSS), coverage extension, multi-SIM, RAN slicing, self-organizing networks (SON), and many others.

From 5G architecture perspective, R17 includes but not limited to enhanced IIoT and NPN support, enhanced support for wireless and wireline convergence, multicast and broadcast architecture, proximity services, better multi-access EC support, and improved network automation support. 3GPP SA WG2 working group has approved a set of system projects related to verticals and technology enablers:

- NTN architecture (mobility management with large/moving coverage areas, delay/QoS in satellite access, RAN mobility with satellite access, regulatory services with super-national satellite ground station)
- enhancements to non-public networks
- enhanced support of IIoT
- unmanned aerial vehicle (drone) identification and control
- proximity services in 5G System
- further enhancements in architecture from advanced V2X services
- enhancement of NS
- technology enablers (architectural enhancements for 5G multicast-broadcast services, location services enhancements, enhancements for support for EC in 5GC, enablers for network automation for 5G Phase 2).

The primary RAN issues of interest for discussion in R17 were determined by the 3GPP community in June 2019. NR-Light (RedCap) enables easy communication for sensors used in industry and applications of a similar nature. IIoT, MIMO, SLs for V2Xs and public safety are being improved. Support for satellite NTNs and methods for better coverage are introduced, as well as the start of work on a 5G NR extension to operate at frequencies above 52 GHz, culminate in the specifications in R18.

1.2.3.3 5G-Advanced

The content of R18 edition was mainly decided at the TSG working meeting #94 in December 2021. The arrival of revolutionary ideas such as AI and ML were declared. It is necessary to decide on a few additional details, but it is now known what 5G-Advanced will look like when it is commercially implemented starting in 2025. Standardization organizations are preparing a detailed plan to work on specifications that will be included in the final edition of R18 standard in 2024 [129].

R18 was officially introduced under the name 5G-Advanced in May 2021 by the project coordination group (PCG). The standardization process was launched at the 3GPP online workshop held in June 2021 with more than 500 proposals from 80 companies, and 1,200 participants. The proposals were originally divided almost equally into three groups: eMBB, non-eMBB evolution, and cross-functionalities. After 1-week discussion, the plenary session identified 17 topics of interest, including 13 general topics and three sets of RAN-specific topics WG 1–3, and one set of RAN WG4. Most of the topics are significant feature enhancements introduced in the R16 and R17 releases, such as MIMO, uplink, mobility, precision positioning. Also included is the evolution of network topology, augmented reality XR, satellite networks, broadcast/multicast services, SL, RedCap, and other topics. The most interesting and exciting topics were AI&ML, full-duplex operation, and lower electric power consumption (Table 1.8).

Many of the proposed research and work topics include system improvements that allow network operators to maximize the efficiency of spectral and radio resources in addition to enhancing the overall quality of the experience for an increasingly diverse combination of connected devices. DL/UL links for MIMO radio systems are being improved, IAB nodes are becoming mobile, smart repeaters are becoming aware of the situation, mobility has improved, we are approaching full-duplex transmission, UE power consumption and basic bandwidth processing are being optimized, as well as network design guided by AI&ML techniques (Table 1.9).

3GPP RAN working group discussed the content of the R18 edition in June 2021 with the aim of approving a detailed plan by December 2021. Nominal work was launched in the second quarter of 2022. The approved packages contain wide range of projects:

- continuing to evolve 5G MIMO performance and efficiency
- driving higher 5G uplink performance and efficiency
- further optimization of 5G device mobility management (cost-effective expansion of 5G coverage and capacity using new network topologies, full-duplex wireless system development, R18 lays foundation for future full-duplex)
- 3GPP R18 scope for wireless ML projects (R18 targets to expand ML to E2E system across RAN, device, and air interface)
- driving toward greener 5G networks
- extend 5G to almost all devices and use cases (expanding 5G SL capabilities in R18 for V2Xs, public safety, commercial use cases)
- 5G NR unified, scalable air interface enabling coexistence of a wide range of 5G device classes, continued scaling of 5G NR-Light for reduced-capability devices (RedCap) further enhancing the 5G NR system (other RAN projects).

TABLE 1.8
3GPP Release 18 Priorities in TSG-SA Service and System Aspects [129,130,133,134]

SA2 system architecture and services	• XR (extended reality) & media services • Edge computing Phase 2 • System support for AI/ML-based services • Enablers for network automation for 5G Phase 3 • Enh. support of non-public networks Phase 2 • Network slicing Phase 3 • 5GC location services Phase 3 • 5G multicast-broadcast services Phase 2 • Satellite access Phase 2 • 5G System with satellite backhaul • 5G timing resiliency and TSC&URLLC enhancement • Evolution of IMS multimedia telephony service • Personal IoT networks • Vehicle mounted relays • Access traffic steering, switching, and splitting • Support in the 5G system architecture Phase 3 • Proximity-based services in 5GS Phase 2 • UPF enh. for exposure & SBA • Ranging-based services & sidelink positioning • Generic group management, exposure, & communication enhancement • 5G UE policy Phase 2 • UAS, UAV, & UAM Phase 2 • 5G AM policy Phase 2 • RedCap Phase 2 • Support for 5WWC Phase 2 • System enabler for service function chaining • Extensions to TSC framework to support DetNet • Seamless UE context recovery • MPS when access to EPC/5GC is WLAN
RAN1 radio Layer 1 (physical layer)	• NR-MIMO evolution • AI/ML–air interface • Evolution of duplex operation • NR sidelink evolution • Positioning evolution • RedCap evolution • Network energy savings • Further UL coverage enhancement • Smart repeater • DSS • Low power WUS • CA enhancements
RAN4 radio performance and protocol aspects	• RAN4-led spectrum items • <5 MHz in dedicated spectrum

TABLE 1.9
List of Topics Considered for R18 Edition [129,130,133,134]

Evolution for downlink MIMO	• Further enhancements for CSI (e.g., mobility, overhead, etc.) • Evolved handling of multi-TRP (transmission reception points) and multibeam • CPE (customer premises equipment)-specific considerations
Uplink enhancements	• >4 Tx operation • Enhanced multi-panel/multi-TRP uplink operation • Frequency-selective precoding • Further coverage enhancements
Mobility enhancements	• Layers 1/2-based inter-cell mobility • DAPS (dual active protocol stack)/CHO (conditional handover) improvements • FR2 (frequency range 2)-specific enhancements
Additional topological improvements (IAB and smart repeaters)	• Mobile IAB (integrated access backhaul)/vehicle mounted relay (VMR) • Smart repeater with side control information
Enhancements for XR (eXtended reality)	• KPIs/QoS, application awareness operation, and aspects related to power consumption, coverage, capacity, and mobility (note: only power consumption/coverage/mobility aspects specific to XR)
Sidelink enhancements (excluding positioning)	• SL enhancements (unlicensed, power saving, efficiency enhancements) • SL relay enhancements • Coexistence of LTE V2X & NR V2X
RedCap evolution (excluding positioning)	• New use cases and new UE bandwidths (5 MHz?) • Power saving enhancements
NTN (non-terrestrial networks) evolution	• Including both NR & IoT (internet of things) aspects
Evolution for broadcast and multicast services	• Including both LTE-based 5G broadcast and NR MBS (multicast-broadcast services)
Expanded and improved Positioning	• Sidelink positioning/ranging • Improved accuracy, integrity, and power efficiency • RedCap positioning
Evolution of duplex operation	• Deployment scenarios, including duplex mode (TDD only?) • Interference management
Network energy savings	• KPIs and evaluation methodology, focus areas and potential solutions
Additional RAN1/2/3 candidate topics, Set 1	• UE power savings • Enhancing and extending the support beyond 52.6 GHz • CA (carrier aggregation)/DC (dual-connectivity) enhancements • Flexible spectrum integration • RIS (reconfigurable intelligent surfaces) • Others (RAN1-led)
Additional RAN1/2/3 candidate topics, Set 2	• UAV (unmanned aerial vehicle) • IIoT/URLLC • <5 MHz in dedicated spectrum • Other IoT enhancements/types • HAPS (high-altitude platform system) • Network coding

(Continued)

TABLE 1.9 (*Continued*)
List of Topics Considered for R18 Edition [129,130,133,134]

Additional RAN1/2/3 candidate topics, Set 3	• Inter-gNB coordination network slicing enhancements • MUSIM (multiple universal subscriber identity modules) • UE aggregation • Security enhancements • SON (self-organizing networks)/MDT (minimization of drive test) • Others (RAN2/3-led)
AI (artificial intelligence), ML (machine learning)	• Air interface (use cases to focus, KPIs and evaluation methodology, network and UE involvement, etc.) • NG-RAN

Key enhancements for eMBB use cases. Beamforming/MIMO, enhanced mobility, and reduced energy consumption are among R18 edition's most prominent new features. In order to increase the spectral efficiency of wireless networks, advanced antenna systems (AAS) have been and will continue to be a primary enabler of 1/2 layer mobility, additional enhancements to MIMO uplink, and advancements related to fixed wireless access (FWA) applications. Dynamic spectrum sharing DSS is extremely useful in the transition to 5G. Release R18 also includes research in reducing electricity consumption in network. Proposals have been made to investigate the viability of full duplex, in which gNBs broadcast and receive concurrently on the TDD frequency bands, notwithstanding the difficulties in practice and the ambiguity of the performance potential. The research looks at the relationship between the improvements that can be achieved and the amount of interference that can be eliminated by eliminating cross-links and self-interference.

Key enhancements for non-eMBB use cases. RedCap, XR, and NSPS (national security and public safety) are three of the most important developments for future or current vertical applications. Many potential use cases will benefit greatly from the inclusion of RedCap UE. Based on R17, the new edition of RedCap solution further reduces device costs and power consumption. Solutions that enable energy collection, such as energy-efficient activation of radio devices, will also be explored. In R17, 3GPP RAN working group investigates variations of AR and XR services and estimate their performance in 5G systems. The primary difficulty is providing a very high data rate and low/limited latency at the same time. In R18, RAN group will also work on traffic management for resource-efficient allocation of low-latency radio resources, support for mobility with constant data transfer rates, energy-efficient UE compatible with XR traffic, and latency requirements. National and public safety (NSPS) is the most notable of the new verticals making use of the 5GS. Improvements in RAN for remote control of drones and detection of fake drones in better perception of the situation are considered. The use of UE-to-UE relays and other out-of-coverage methods are two of the ways in which R18 improves 5G's already impressive capabilities [130–132].

Cross-domain functionalities for both MBB and non-MBB use cases. The two cross-domain functionalities are AI/ML for physical layer enhancements (PHY) and RAN. AI and ML are widely believed to greatly boost PHY performance. For this reason, RAN standardization will investigate the potential by establishing a generic framework for AI and ML related PHY enhancements. This framework will include suitable modeling, assessment techniques, and performance criteria and testing. The first area for AI/ML improvement may be beam management or channel estimation/prediction. Part of the research for R17 edition involves identifying applicable use cases and acceptable AI/ML-based solutions for RAN. The selective usage enhancements made in R17 will be made normative in R18, allowing for better traffic management and load balancing. The present architecture's interfaces will be the primary target for enhancement. In order to intensify competitiveness of suppliers, one of the goals is to ensure that models of AI remain specific for implementation [129–134].

1.2.4 Novel System AI&ML Paradigm

The relatively new concept of AI and ML has the potential to become a platform for completely new possibilities of evolution, and even for future 6G systems. Wireless networks are extremely complex, very dynamic, and very heterogeneous. There is no better approach than using AI/ML to solve difficult wireless challenges (Table 1.10) [132,135–140].

5G systems are increasingly complex in a variety of scenarios and applications. The evolution of performance improvements and solving new use cases makes 5G more sophisticated. In addition, 5G systems generate huge amounts of data. Applying rapidly advanced AI techniques, such as ML and data analytics, is key to managing complexity, identifying patterns in data, optimizing network design, improving system performance, and reducing operating costs.

5G RAN radio-interface is complex due to network topology, multiple numerology, network coordination schemes, and different nature of use cases. The main motives for designing mobile networks based on AI are lack of models and deficit of algorithms [141,142].

TABLE 1.10
Industry Initiatives on AI/ML for 5G [132,135–140]

ITU	• Runs multiple programs on AI
	• ITU-T focus group on machine learning for future networks
	• AI/ML in 5G challenge is focused on university advanced R&D projects
	• Developing frameworks for ML data collection, modeling, and evaluation
3GPP	• Evolution of analytics architectures (NWDAF, SON, MDT)
	• Studies on AI/ML for RAN in R17; for example:
	• Study on AI/ML functional frameworks and use cases
	• Study on AI/ML model transfer performance requirements over 5G
	• Study on enhancement for data collection for NR and EN-DC dual connectivity
	• Potential inclusion of AI/ML adaptations to air interface in R18 as part of 5G-advanced
O-RAN	• O-RAN architecture for radio intelligent controller RIC and interfaces
	• Technical report on AI/ML workflow description and requirement
	• O-RAN software community to develop ML-based xApps and rApps network automation tools

- Achieving any performance optimum in such complex application scenarios is probably computationally impossible. However, application of AI enables pragmatic and competitive performance.
- Modern mobile systems are designed under the assumption that the overall system behavior is modeled by means of simple modeling techniques that are amenable to rigorous mathematical investigation. In contrast, new solutions based on AI overcome unknown complex analytical model.
- In mobile networks, there are a number of situations where optimum algorithms have been identified but are too complicated to be used in reality, forcing system designers to depend on heuristics based on basic decision-making principles. AI provides an attractive trade-off between performance and complexity in such scenarios.

The development of ML technology for system-level optimization and network intelligence becoming one of the leading trends in academic, research, and industrial communities. It is possible to classify recently developed applications as:

- ML-based network intelligence and system-level adaptive optimization
- ML-based transmission intelligence
- spectrum intelligence and adaptive radio-resource management
- adaptive baseband signal processing.

Physical (PHY) and media access control (MAC) layers are the basic layers of mobile networks in which many technical innovations have been applied. The application of AI potentially enables improved performance within these layers for channel quality assessment and prediction, reception processing, channel decoding, random access, and dynamic spectrum access [141–145].

Application of AI in 5G network layer reduces additional resources for planning, operations, and troubleshooting. It is possible to implement AI fault detection and self-healing system within self-organized network SON, so that MNO mobile network operators reduce their OPEX operating costs, speed up recovery, and enhance the quality of services provided to its end users.

Although application of AI indicates significant improvements, there are still many obstacles to be overcome:

- ML training issues
- lack of bonding performance
- lack of explainability
- uncertainty in generalization
- lack of interoperability.

Reducing training costs is critical issue for sustainability of AI models based on PHY and MAC layers. Furthermore, it is challenging to get labeled ML training data due to the separation of information in network protocol layers.

Unlike some other areas, it is crucial to anticipate user actions in mobile networks, in the worst case. Traditional model-based approaches provide generally good understanding of system output distribution, in response to particular input distribution. To ensure at least some level of QoS or performance, the designer of the system accounts for the worst-case scenario. In contrast, because to the nonlinear nature of the AI methodology, it is complicated or even impossible, no matter how well they work in real networks – to provide any guarantee of performance in the worst case. For the smooth integration of AI into mobile networks, it is essential to provide bearable and progressive deterioration in the worst case.

There is a tendency to see AI technologies as black boxes because it is difficult to formulate an analytical model to test the correctness or simply explain their behavior. In situations when AI is used for decision-making in real time, the lack of explainability might be a problem. An example of such a real-time system is communication between vehicles. Chronologically, mobile networks and wireless standards are designed based on theoretical analysis, measurement of channel characteristics, and human intuition and understanding. When creating AI models for use in mobile networks, it is preferable for them to have a high degree of explainability.

It is not always obvious whether the data set used to train an AI model to execute a communication job is comprehensive enough to account for the input distribution actually encountered by the model. It is necessary for the learning mechanism to generalize cases that did not appear in the training data set.

Today's mobile networks are becoming more complicated, thus interoperability is more important than ever. Any incompatibility between AI modules made by various manufacturers has the potential to reduce the network's efficiency as a whole. Finally, mobile network based on artificial intelligence contains complex dependencies; it is not easy to determine which vendor equipment/AI module is responsible in case of any degradation of performance indicators KPI.

From network design perspective, maximizing complexity of AI model application requires clear definition of interface, both within and between protocol stack layers. Given initial nature of AI application in wireless systems and high levels of service guarantees required by MNOs, AI is being implemented in phases. System designers initially implement AI and then refine their testing tools and methodology. It is desirable that AI models operate in longer timescales (order of minutes or hours), so that experts can change model recommendations if necessary. Additional robustness can be added by adjusting the performance of AI model based on expert feedback.

1.2.4.1 AI-Enabled RAN Architecture

3GPP explored RAN networks supported by AI in R17 release, including principles, functional framework, use cases, and solutions. 3GPP adopts set of key principles to focus on AI applications with the potential impact of RAN. One basic principle is that detailed AI model is not standardized. Instead, research focuses on identifying the basic information needed or produced by the AI model for each use case. The use of traffic management has been investigated both from the perspective of UE and from the perspective of RAN. From the UE perspective, goal is to find the best radio cell(s) for UE, which is often called mobility optimization. From the RAN perspective, the goal is to balance load between the network nodes. The use of energy efficiency of RAT is closely related to traffic management since it also refers to distribution of UE

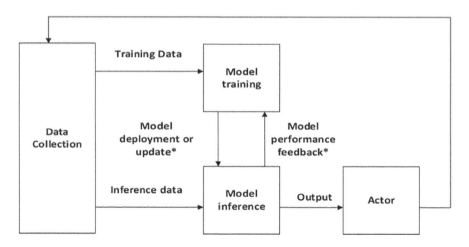

FIGURE 1.23 Functional framework for AI-enabled RAN (* further study in 3GPP).

between network nodes. The goal is to reduce electricity consumption while maintaining satisfactory QoS (Figure 1.23) [146–149].

There are several challenges facing the process of standardizing radio-interface with AI in 3GPP. Existing technical specifications are formed on the principle of specifying a set of minimum performance requirements. Typically, requirements are defined in limited set of conditions and it is assumed that algorithms generalize well in real-world applications. The approach is challenged by algorithms based on AI that can be trained and adapted to a set of requirements and specifications, and therefore not generalized.

Telecommunications industry has shown great interest in the above issues and has conducted initial research within Open-RAN Alliance. O-RAN Alliance was created in 2018 by 3GPP and five mobile network operators with a goal for an open, efficient radio access network RAN with AI applications to automate various network functions NFs and reduce operating costs. O-RAN Alliance consists of over 160 member businesses, publishes standards, and distributes open source software under the aegis of the Linux Foundation. O-RAN reference architecture is the basis for building virtualized RAN on open hardware and cloud computing, with built-in radio control supported by AI. The alliance's standards provide the basis of the architecture, and they are entirely compatible with and supportive of the standards advocated by 3GPP and other organizations in the field. O-RAN reference architecture uses clear, well-established interfaces that enable an open and interoperable ecosystem. It contains the following functional blocks:

- orchestration/network management system layer with non-RT RAN intelligent controller
- RAN intelligent controller near-RT function layer
- multi-RAT control unit CU protocol stack function
- DU and remote radio unit RRU function.

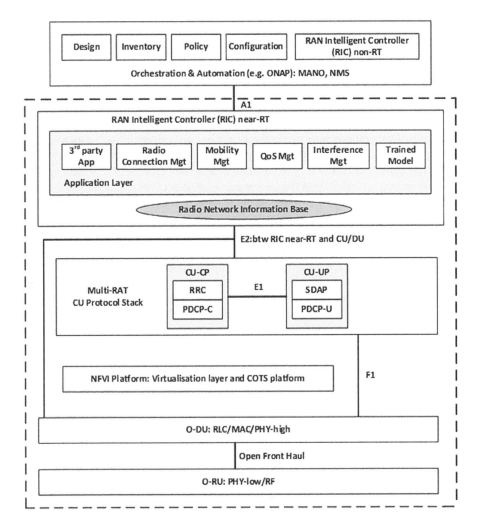

FIGURE 1.24 O-RAN reference architecture.

O-RAN reference architecture shown in Figure 1.24 includes RAN radio intelligent controller (RIC) with non-real-time (non-RT) and near-real-time (near-RT) AI, multi-radio access protocol technology. Non-RT control functions (>1 seconds) and near-RT control functions (<1 seconds) are separated in RAN RIC. Non-RT responsibilities consist of policy and service management, optimization of higher-level procedures, and model training. ML-generated messages and trained real-time models are distributed to near-RT RIC, which accepts and executes the AI model and changes functional behavior of network. The near-RT version of RIC is backward-compatible with existing methods of radio resource management, and it improves upon difficult operational elements such as handover control, QoS management, and AI-powered connection. O-RAN Alliance has established two working groups that standardize A1 interface (between non-RT and near-RT RIC) and E2 interface (between near-RT RIC and DU). ETSI (EU Telecommunications Standards Institute) has also launched ENI (Experienced Networked Intelligence) industry specification group to define cognitive network management architecture that uses AI and context-aware policies to improve operator experience [146–149].

1.2.4.2 Network Automation and Data Analytics Function

Standards organizations have taken the first steps toward providing framework for integrating AI models into planning, operations, and healing of mobile networks. 3GPP has defined a function of network data analytics (NWDAF) for data collection and analytics (including AI) in automated mobile networks. The standard specifies only interfaces for NWDAF block. By leaving development of the AI model to implementation, 3GPP allows adequate flexibility for network providers to implement use cases with AI-enabled. Operational, administrative, maintenance, network function (NF), application function (AF), and data repository inputs are all taken in through the inbound interfaces. The outbound interfaces forward analytics feedback to NF and AF blocks, respectively [150–153].

3GPP promotes network intelligence starting with R16 and continues to advance in standardization of both the Network Infrastructure (SA2) and Network Management (SA5) from a technical standpoint. NWDAF is a standard network element that specified SA2 as AI+BigData engine with standardized capabilities, network data aggregation, improved real-time performance, and closed-loop controllability support. The 3GPP specifies the deployment options for the NWDAF, where it should be located in the network, how it should interact and coordinate with other network services. NWDAF is implemented in certain network functional units through the incorporation of functions and can also be coordinated across functional network units, in order to complete the work in the closed loop of network intelligence. Management data analysis system (MDAS) is described by 3GPP SA5, which, when combined with AI and ML, provides automation and cognition in the management and orchestration of networks and services. The MDAS processes and analyzes network management data, creates reports with findings and recommendations for network administration, and formulates procedures in intelligent automated closed-loop management and orchestration (Figure 1.25) [150–153].

3GPP investigates and specifies collection of data and transmission of analytics feedback to NFs. All possible solutions share the following architectural presumptions:

- Data collection and data analytics are performed in centralized method.
- Instance specific to analytic could be collocated with 5G system NF.
- NF and OAM operating, administering, managing, and maintaining processes determine how to optimize your network with the help of NWDAF's data analytics.
- To interact with the other NFs and OAM procedures in the 5G core network, NWDAF makes use of the already established service-based interfaces.
- Data analytics performed by 5G core network operations may be made available to any user NF that makes use of the SBI service-based interface.

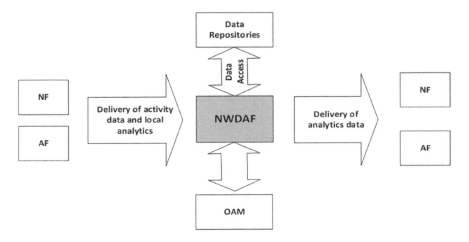

FIGURE 1.25 General framework for 5G network automation of OAM (operations, administration, and maintenance).

- The public land mobile network benefits from NWDAF and NFs working together effectively since it leads to standardized policies, analytics output findings, and, ultimately, decisions.

1.2.4.3 5G Advanced Architecture and Technical Trends

5G advanced networks continue to develop at architectural and technical level based on various requirements to improve network capabilities. At the architectural level, the cloud-native concept, edge network, network-as-a-service, and the continuous enhancement of network capabilities all need to be taken into careful consideration (Figure 1.26) [151–154].

New capabilities and technologies in 5G, including NS and EC, as well as the virtualization of network resources, have introduced new challenges to 5G operations and commercial purposes. Telecommunications networks benefit from the deployment and integration of intelligent technology since it increases network efficiency while decreasing O&M expenses. Intelligent development of a 5G advanced network requires that AI monitoring technologies be used as a point of reference.

Intelligent RAN can be achieved by applying AI/ML in terms of placing on top of existing RAN applications or embedding. The first model allows ML use without major changes in the existing RAN. New algorithms are created by offline learning that can be applied to existing architecture. Other models are created based on data collected, analyzed, and implemented within RAT. The approach requires changes in functional nodes and interfaces and represents a much deeper integration with RAN, which requires more time to implement (Table 1.11) [151–154].

AI/ML contributions continuously improve 5G RAN depending on 3GPP versions of technical specifications. The R16 expands the existing NWDAF framework, minimizes the drive test MDT that provides the operator with tools to optimize network performance in cost-effective way, and enables SON to use these new techniques. R17 edition contains technical studies on how to access AI/ML in the long-term

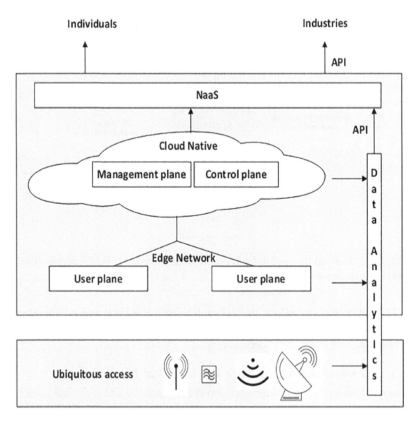

FIGURE 1.26 5G advanced network architecture.

TABLE 1.11
Appling ML to RAN

RAN ML Type	Implementation	Example Benefits
Real-time air interface transmission	L1–L3 radio parameters in the centralized unit (CU), distributed unit (DU), and radio unit (RU) – e.g., to apply new RAN algorithms in seconds/ms time cycles	Better resource utilization and congestion control, power management, handover optimization, etc.
RAN management & optimization	Ongoing changes to cell or cell cluster configurations – e.g., in response to changes in demand on weekly/daily/hourly time cycles	Enhanced network assurance and performance; better security via anomaly detection
RAN planning & deployment	Network design – e.g., to determine placement and configuration of gNBs	Better coverage, greater system capacity, optimized capex, faster deployment

evolution of R18. AI/ML is probably the primary focus and is expected to cover areas of radio interface improvement and complete system framework.

AI/ML is evolving rapidly. It is necessary for mobile industry to use the technology and apply it to the mobile network architecture from 5G CN to 5G NR and end-user device UE itself. The first application of AI/ML is centralized, offline learning for planning and network operations. There are examples of the positive impact of technology, so ML is on the roadmap to further improve 5G network architecture and enable advanced services. In the time frame of the next 6G generation, possibilities for AI are bright. For example, AI-native radio interface opens up the possibility of designing a radically simpler RAN with potentially large gains in efficiency and performance [155–161].

1.3 6G CONCEPT, RESEARCH, AND TRANSITION TECHNOLOGIES

Upcoming 6G communication systems are anticipated to support the greatest number of applications. to date. Extreme system performance and novel use case combinations are the primary drivers. Apart from the establishment of brand-new performance indicators and benchmarks, it is necessary to redefine the types of services in 6G by improving and possibly integrating traditional URLLC, eMBB, and mMTC in new services [162–169]:

- Mobile broadband reliable communication with low-latency (MBRLLC) allows the 6G system to achieve the desired performance goal in large dimensions of speed–reliability–latency. The new service may be seen as an extension of existing ones, such as URLLC and eMBB. In addition, energy efficiency is becoming a major challenge because of not only its impact on reliability and speed of data transfer, but also continuous reduction in the size of 6G devices and increased functionality, which requires highly energy-efficient design.
- Massive URLLC combines URLLC with conventional mMTC service. A compromise in reliability–delay–scalability is necessary, and therefore requires a departure from system design based on average data rate or latency. A principled and scalable framework, on the other hand, takes into consideration factors such as latency, dependability, packet size, network architecture, topology (across edge, access, and core), and decision-making in the face of uncertainty. In addition to this, mURLLC contends with challenging networking conditions.
- Human-centered services (HCS) are a new class services that impose physical experience performance quality (QoPE) objectives closely integrated with users and their body/physiology, rather than technical metrics of speed–reliability–latency. An excellent example is wireless brain–computer interactions (BCI) in which the performance of a service is dependent on the user's thoughts, actions, and even physiological state. It is necessary to define a set of QoPE performance indicators and quantify as a function of traditional (technical) perceived quality of user experience (QoE) and QoS performance metrics.

- MPS multipurpose control, localization, and sensing and energy services (3CLS) are central to connecting robotics and autonomous systems. There is need for common UL–DL design and fulfillment of desirable performance control objectives in terms of stability (computer latency, amount of transmission power), localization accuracy, as well as mapping and sensing accuracy. The MPS service is also suitable for cyber–physical operations on wireless infrastructure.

New types of 6G services require more stringent KPIs and their simultaneous fulfillment in an end-to-end perspective. Indicators include data rate experienced by the user, end-to-end delay, mobility, connection density, traffic density, spectrum efficiency, coverage, resources, and signaling efficiency. Then, the concept of KPIs is considered incomplete and it is necessary to compare them with key value indicators (KVI). Indicators are grouped into three basic categories: growth, sustainability, and efficiency. The first category is primarily connected to economic expansion and the formation of new values, ecosystems, and models. The second and third are mainly related to digital inclusion of individuals and groups, zero-energy devices, resource efficiency, and user privacy. Additionally, different definitions of term quality have been presented. The idea of a high-quality physical experience (QoPE) is an effort to complete and solidify the assessment offered independently by QoS and QoE by including additional physical attributes of users like cognition, bodily traits, and gestures (Table 1.12) [162–169].

Here are some primary findings:

- As a first, intermediate step toward 6G systems, enabling mobile broadband services in mmWave bands is essential for maintaining high-speed communication at high frequencies.
- The next step toward 6G is understanding basics of URLLC, with focus on notion of reliability. It is necessary to explore new statistical tools for a distribution-based, rather than average-based, measurement of wireless system performance.
- In contrast to the standard, short-packet, and slow-speed URLLC services, the requirements of future wireless services include high dependability, low latency, and high data rates. A fresh perspective on the fundamentals of the trade-offs are required to regulate space–speed–reliability–latency.
- Analyzing and optimizing the performance of 6G demands working in three-dimensional space and researching a complicated system that incorporates drones, satellites, and conventional wireless infrastructure.
- AI plays a crucial function in 6G systems, ranging from enabling a self-sustaining network to embedding collective network intelligence. The new paradigm of combined learning–communication co-design is particularly important for wide application of new AI algorithms.

The coming decade offers a plethora of prospects for wireless research that spans several disciplines. Driving 5G mobile communications with AI toward 6G will be an exciting era of convergence of wireless technologies ranging from artificial

TABLE 1.12
6G Service Classes, Their Performance Indicators, and Example Applications

MBRLLC	• Stringent requirements in rate-reliability-latency space • Energy efficiency • Rate-reliability-latency for mobile scenarios • Handover failures	• XR/AR/VR • Autonomous vehicular systems • Autonomous drones • Legacy eMBB and URLLC
mURLLC	• Ultra-high reliability • Massive connectivity • Massive reliability • Scalable URLLC	• Conventional IoT • Device tracking • Distributed blockchain and DLT • Massive sensing • Autonomous robotics
HCS	• QoPE (quality-of-physical-experience) capturing raw wireless metrics as well as human and physical factors	• BCI (brain–computer interactions) • Haptics • Empathic communication • Affective communication
MPS	• Control system stability • Computing delay • Localization precision and accuracy • Accuracy of sensing and mapping functions • Delay and reliability for communications • Energy	• CRAS (connected robotics and autonomous systems) • Telemedicine • Environmental mapping and imaging • Some special cases of XR services

intelligence to computer, control, and cyber–physical systems within the new system (Table 1.13) [170,171].

In AI for wireless communication, there is a need for deploying self-sustainable NFs: optimization, administration, and decentralized control of networks using artificial intelligence. In wireless communication for AI, algorithm performance at the edge may be affected by a number of different wireless characteristics, such as fading, mobility, and interference, which can be better understood when learning algorithms are designed in tandem with wireless protocols [172–186].

1.3.1 Extreme System Performance and Network Evolution

Technological advances have reached significant critical mass in the evolution of mobile networks:

- continued architectural enhancements and cloudification
- 6G system is based on a data-driven architecture that will use huge amount of data to support AI in the cloud, core, RAN, and devices
- hardware acceleration and possibly different types of meta-materials
- higher reliance on open source
- Internet and IP network development and the confluence of the Internet, telecommunications, and media.

TABLE 1.13
Key 6G Research Areas

Research Area	Challenges	Open Problems
3D rate-reliability-latency fundamentals	• Fundamental communication limits • 3D nature of 6G systems	• 3D performance analysis of rate-reliability-latency region • Analysis of achievable rate-reliability-latency performance targets • SEE analysis in 3D space • Quantification of spectrum and energy needs
Leveraging integrated, heterogeneous high-frequency bands	• Challenges of operation in highly mobile systems • Susceptibility to molecular absorption and blockages • Short range • Lack of propagation models • Need for high-fidelity hardware • Presence of frequency bands with different characteristics	• Mobility and handover management for high-frequency THz and mmWave systems • Cross-band physical, link, and network layer optimization • Coverage and range improvement • Design of mmWave and THz tiny cells • Design of new high-fidelity hardware for THz • Propagation characterization for mmWave and THz bands
3D networking	• Presence of users and base stations in 3D • High mobility	• 3D propagation modeling • 3D performance metrics • 3D mobility management and network optimization
Communication with RISs	• Complexity of metasurfaces and RIS • Absence of precise models of performance • Absence of precise models of performance • Heterogeneity of 6G devices and services • RIS capability to offer various functions (reflectors, BSs, etc.)	• Optimal deployment and location of RIS surfaces • RIS reflectors vs. RIS transceiver BSs • Energy transfer using RISs or other means • AI-enabled RIS • RIS across 6G services • Fundamental analysis of the performance of RIS transmitters and reflectors across frequency bands
New QoPE metrics	• Incorporate raw metrics with human perceptions • Accurate modeling of human perceptions and physiology	• Theoretical development of QoPE metrics • Empirical QoPE designs • Practical psychophysics experiments • Need for precise and realistic QoPE performance targets and measures

(Continued)

TABLE 1.13 (*Continued*)
Key 6G Research Areas

Research Area	Challenges	Open Problems
Joint communication and control	• Integration of communication and control performance indicators • Handling dynamics and multiple time scales	• Co-design of communication and control systems • Control-aware wireless communication metrics • Wireless-enabled control metrics • Joint system optimization for CRAS
3CLS	• Integration of multiple functions (communications, computing, control, localization, sensing) • Lack of prior models	• Design of 3CLS metrics • Joint 3CLS optimization • AI-enabled 3CLS • Energy-efficient 3CLS
Design of 6G protocols	• 3D network-enabled 6G protocols that can also handle diverse propagation environments • Need to serve different devices with heterogeneous capabilities and mobility patterns • Need for adaptive and self-learning protocols across the network stack	• Design of signaling, scheduling, and network coordination protocols that do not rely on pre-fixed and rigid frame structures • Development of adaptive multiple access protocols • Design of adaptive and proactive handover schemes that can handle 3D mobility • Novel identification and authentication techniques suitable for new 6G devices • Development of AI-inspired edge protocols for multiple 6G functions
RF and non-RF link integration	• Different physical nature of RF/non-RF interfaces	• Hardware for joint RF/non-RF systems • System-level analysis for systems with joint RF and non-RF capabilities • Use of RF/non-RF systems for various 6G services
Holographic radio	• Lack of existing models • Hardware and physical layer challenges	• Design of holographic MIMO using RISs • Performance analysis of holographic RF • 3CLS over holographic radio • Network optimization with holographic radio
AI for wireless	• Design of low-complexity, edge AI solutions • Small but massively distributed data	• SON using reinforcement learning techniques • Data analytics for both big and small data • AI-guided network management • Edge AI operating on wireless networks

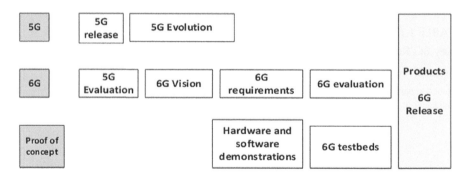

FIGURE 1.27 Roadmap enabling 6G deployment by 2030.

There are obvious synergies between technology groups in enabling various use cases. One example is the introduction of ML in all network components based on open source development environments and cloudification trends.

From the standpoint of mobile network operators, who offer not only connection but also services, performance insight, and interfaces with third-party network computing capabilities, the objective of technical advancement is to accomplish the following characteristics:

- Unparalleled efficiency for all uses, both traditional and otherwise.
- Smarter, more cognitive networks may provide the necessary capacity without considerably increasing the associated costs or levels of complexity.
- As networks evolve, they become more flexible, able to deploy quickly, and open to adding new services in a timely manner.
- 6G for sustainability is enabled by the wireless system, and sustainable 6G within the wireless system.

Prediction alone is not enough, and in order to achieve the goal of implementing 6G by 2030, it is necessary that the research of 6G system follows the roadmap shown in Figure 1.27. Joint efforts focus on 5G assessment and evolution over the first 5 years, with the definition of 6G specifications. The first combination of hardware and software environment for 6G testbeds will appear after 2025 [187–193].

1.3.1.1 IMT Vision for 2030 and Beyond

ITU-T SG 13 study group for standardization established in July 2018 the focus group FG NET-2030 networks. The working group identified 6G mobile system specification guidelines, network architecture, requirements, use cases, and network capabilities. The FG NET-2030 concluded its activity on July 2020 [194–199].

ITU-R SG5 WP5D study group is in charge of everything concerning the IMT-2030 infrastructure. At the start of each IMT procedure, a vision of what needs to be achieved is established. After that, the candidates who support the requirements start development of necessary functional technology. When standardization

5G-Advanced Mobile Communication 63

organizations submit candidate technologies, an evaluation process is initiated through the cooperation of ITU member states, equipment manufacturers, network operators, and academia. The single global framework serves as a venue for debate and consensus about the potential of new radio technologies. Approval and implementation bring the process to a close after the radio interface has been finished and an agreement has been obtained.

In line with described IMT process, ITU-R is launching 6G vision study as first step and then publishing the minimum technical requirements and evaluation criteria for IMT-2030. At a working meeting in February 2020, WP5D working group decides to launch a study on future technological trends and present study at working meeting in June 2022.

The World Radiocommunication Conference (WRC) is also organized by ITU-R, which regulates allocation of frequencies, and is held every 3–4 years. Spectrum allocation for the 5G technology was decided during the 2019 World Radiocommunication Conference. WRC-23 is expected to be scheduled for 2023 and discuss spectrum issues for 6G, so allocating spectrum for 6G communications could be officially decided in 2027 (WRC-27). At the same time, 3GPP is expected to launch 6G research around 2025, followed by a specification phase, to ensure the first commercial 6G implementation by 2030.

1.3.2 Technology Enablers and Research Programs

6G wireless communications networks are anticipated to offer worldwide coverage, improved spectral/energy/cost efficiencies, higher levels of intelligence and security. To accommodate these needs, 6G networks depend on new technologies that enable radio interface and transmission technologies and a new network architecture.

The design of physical layer (PHY) communication systems conventionally relies on wireless channel models and extensive mathematical analysis. The approach has led to well-defined methods for modulation, channel estimation, equalization, code design, and so forth. However, models usually fail to describe all the complexity of the actual communication medium. Moreover, approximations are made in the design of communication algorithms for tractability, which causes suboptimal performance. Over the last decade, ML has made breakthroughs in communications and significant gains in bandwidth and reliability. The goal is to apply DL to develop improved physical layer (PHY) transceiver algorithms for new modulations and waveforms, as well as for channel estimation and detection.

In higher-layer applications, through the big data learning functions generated by the wireless network infrastructure and sensor devices, it is possible to optimize network configurations and achieve better network performance. Predicting and managing network traffic and network performance has been researched using AI. ML techniques enable optimization and management of mobile networks and reduction of operating costs.

1.3.2.1 Review of Global Activities and Research Programs

The most important participants in mobile communications market have already officially started their national 6G research or have announced their ambitions and

TABLE 1.14
Core Research Areas of 6GFP Program

Wireless Connectivity Solutions and RAN Technologies	Distributed Intelligent Wireless Computing	Devices and Circuit Technologies and Implementation	Vertical Applications and Services
Optimization of wireless networks	Mobile edge, cloud, and fog computing	RF transceiver implementations and integrated circuit	Vertical applications
Network capacity			An experimental wireless network
Physical layer (PHY) technologies			
Network densification and use of dynamic infrastructure	Augmented reality (AR)/virtual reality (VR) over wireless	IC design	
Ultra-reliable low-latency communications (URLLC)		Materials and components	Techno-economic and business considerations
Communications concept with embedded positioning	Intelligent distributed computing and data analytics	Embedded systems and software	
Software-defined networking			
Wireless networks security		Ubiquitous sensing	

plans. EU programs are the most transparent and open with valuable information available to the public. Europe has made progress with 6G initiatives involving governments and academia at regional and national levels. Under framework program for research and technological development FP8 Horizon 2020, EC has launched further 5G research through ICT-20-2019 5G long-term evolution. A total of 66 high-quality ideas were submitted by consortia of suppliers, mobile carriers, academics, research organizations, small firms, and verticals, and eight of these pioneering initiatives will begin in early 2020. A second round of 6G research projects, funded by the ICT-52-2020 Smart Connectivity program, got underway in early 2021. HorizonEurope's next PPP Smart Network & Services public-private partnership, which is part of the FP9 framework program, will carry on the continent's 6G R&D efforts after the success of the Horizon 2020 5G-PPP initiative. In early 2018, Academy of Finland selected the University of Oulu to lead the national research *6Genesis Enabled wireless Smart Society and ecosystem*, a 6GFP leading flagship program. After the inaugural 6G Wireless Summit in September of 2019, a series of white papers were released addressing 12 different topics, including ML, edge intelligence, localization, sensing, and security. The 6GFP explores in detail four interrelated key research areas (Table 1.14) [200–208].

The first generation of communication networks that include AI from the ground up will be 6G networks. AI is not just an application; it is an inherent part of infrastructure, network management and operations (Table 1.15).

The study is undertaken along the lines of information and communication theory, implementation and verification of the idea of 6G technologies, and test networks, among others: the following is a summary of these research lines' specifics [209–213]:

TABLE 1.15
Taxonomy of ML Applications across Different Layers of Future Communication Network

Network Layer	Supervised	Unsupervised and Partially Supervised	Reinforcement
Physical layer	Channel equalization, decoding, and prediction, path loss and shadowing, prediction, localization, sparse coding, filtering, adaptive signal processing, beamforming	Modulation, interference cancellation, mobility prediction, spectrum sensing, radio resources optimization, localization, security, transmission optimization, nodes clustering, duplexing configuration, multiple access, beam switching	Link preservation, channel tracking, on-demand beamforming, secure transmission, energy harvesting, transmit power selection, nodes selection, channel access management, modulation mode selection, coverage optimization, anti-jamming radio identification
Network and other layers	Caching, traffic classification, network anomalies identification, throughput optimization and adaptation, latency minimization, optimization of other KPIs	Multi-objective routing, traffic control, network state prediction, source encoding and decoding, network parameters prediction, intrusion detection, fault detection, anomaly detection, etc.	Multi-objective routing, packet scheduling, access control, adaptive rate control, network security, capacity and latency demand prediction, traffic prediction and classification, network slicing
Application layer	Smart health care, smart home, smart city, smart grid, query, processing, data mining, crime detection, etc.	Data processing, data ranking, data analysis (spatial, temporal, etc.), data flow prediction, dimension reduction, malware detection and classification, network anomaly prediction, demographics features, extraction and prediction, fraud detection	Proactive caching, data offloading, error prediction, traffic rate determination and allocation, data rate selection for segments

- **Information & communications theory & optimization for mmWave and (sub-)THz bands.** Physically inspired models of materials, circuits, and wave propagation are used to provide fundamental information and communication theory that may be used in the development of fundamental design tools and recommendations for the optimization of RAN architectures. It will be realized as a system optimization framework that enables different design goals.
- **Superfast wireless broadband connectivity.** System optimization framework is used for high data rate and capacity technology solutions in the Tbps range, with a focus on solutions that save energy. The research also considers RF&BB requirements of transceivers and their architecture based on C-RAN and basic technological options, as well as RAN methods and algorithms.
- **Ultralow latency communications.** A framework for optimizing low latency highly reliable connections at the expense of spectral efficiency is being developed, in order to define RAN protocols and transceivers. The architecture of the system and the limitations of C-RAN and other implementations will be explored in order to specify cost and energy-efficient implementations.
- **Security for 6G.** For the next generation of RAN technology, from the physical layer on up, a comprehensive security analysis and solution architecture is now in development. The RAN and SDN frameworks will include software security concerns at the network level.
- **THz technology enablers toward 6G.** Expertise in mmWave device design is combined with sub-THz imaging circuits and the relevant channel and propagation models, systems theory, and communications to build sub-THz wireless transceivers. In order to provide sufficient system performance, researchers are developing new low-loss materials and packaging techniques.
- **High-risk technology enablers toward 6G.** Active research on promising and relevant developments for the period 2030 is being conducted in order to innovate and take risks for research. For example, revolutionary developments in quantum theory and quantum communication are monitored, and feasibility studies are initiated when there is justification (Table 1.16).

6G communications technologies are still at a very early stage of development [245–256]. Solutions for the full exploitation of communication capacities, which are theoretically predicted at frequencies up to 100 GHz and higher, also require the realization of evidence of proof of concept (PoC) before trials of 6G network are justified. With the advent of very fast radio links and enhanced sensing possibilities on THz frequencies, the groundwork has been laid for the creation of critical new capabilities on the path to 6G, and a new idea has emerged in which physical restrictions are considered with more attention than ever before. Radically new developments necessary for system-level demonstration include feasibility studies of very large and efficient antenna arrays (with low-losses), as well as processing large amounts of available bandwidth in the sub-THz and THz regions.

TABLE 1.16
Summary of Key Open Challenges and Potential Solutions [145,158,170,214–244]

Challenges	Potential 6G Solution	Open Research Question
Stable service quality in coverage area	User-centric cell-free massive MIMO	Scalable synchronization, control, and resource allocation
Coverage improvements	Integration of space-borne layer, ultra-massive MIMO from tall towers, IRS	Joint control of space and ground-based APs, real-time control IRS
Extremely wide bandwidths	Sub-THz, VLC	Hardware development and mitigation of impairments
Reduced latency	Faster FEC, wider bandwidths	Efficient encoding and decoding algorithms
Efficient spectrum utilization	Ultra-massive MIMO, waveform adaptation, interference cancellation	Holographic radio, use case-based waveforms, full duplex, rate-splitting
Efficient backhaul infrastructure	Integrated access and backhauling	Dynamic resource allocation framework using space and frequency domains
Smart radio environment	Intelligent reflectance surfaces IRS	Channel estimation, hardware development, remote control
Energy efficiency	Cell-free massive MIMO, suitable modulation techniques	Novel modulation methods with limited hardware complexity
Modeling or algorithmic deficiencies in complex and dynamic scenarios	ML/AI-based model-free, data-driven learning and optimization techniques	End-to-end learning/joint optimization, unsupervised learning for radio resource management

1.4 OPEN ISSUES AND CONCLUDING REMARKS

The introduction of 5G technology into the telecoms market is a very exciting development. The International Telecommunication Union defines 5G as having greater speeds, reduced latency, and huge connectivity. There are more possibilities for how applications may be used, such as improved mobile broadband, high dependability with low latency, and low power with high connectivity. With minimum technical criteria and assessment process provided by ITU-R, 3GPP began collaborative process in 2015 and completed its first release of technical specifications on 5G system in June 2018, the second in June 2020, and the third in March 2022 and lastly, transitional 5G-Advanced started in the second quarter of 2022.

Although efforts to develop a worldwide standard for 5G are still in their infancy, the focus of wireless communication researchers is rapidly turning to the next-generation mobile technology. It is a radical departure from previous mobile technology paradigms. 6G is the convergence of physical space, cyberspace, and intelligent connection. In addition, interaction is fundamental to providing consumers with a fully immersive experience. Here are some key observations: first short-term step is to make possible broadband wireless services in the mmWave bands required to maintain high-speed communication at high frequencies. The next stage is to prioritize space–speed–dependability–latency design space trade-offs that emphasize high reliability, low latency, and high data rates. In order to analyze and optimize performance, one must work in three-dimensional space and investigate intricate systems. AI is crucial to 6G systems, ranging from enabling a self-sustaining network to embedding collective network intelligence in joint learning–communication co-design.

The sixth generation of mobile communications is anticipated to offer worldwide coverage, improved spectral/energy/cost efficiencies, higher levels of intelligence and security. 6G offers considerably faster data speeds, reduced latency, and much enhanced terminal device densities, while natively harnessing artificial intelligence AI inspired by vast quantities of unexplored data and intrinsic complexity of innovative use cases. Numerous scholars have predicted that the technological environment would undergo significant changes by 2030. The world is now undergoing a transition toward a more data-driven, highly digitalized, and intelligent environment. Challenges and opportunities in technology will arise with the next revolution.

Technologies for 6G communications are just in their infancy. The open issues in achieving an ideal mobile network infrastructure are interference, user localization, traffic management, network control, channel optimization, power consumption, and security. As for multimedia communications, issues that need more study are multimedia operating over mobile environment, network slicing optimization, mobile edge computing, mobile cloud computing, as well as immersive media delivery, mobile video multicast/broadcast, QoE/QoS management.

The dominant tendencies in the academic, scientific, and business communities are development of ML technology for system-level optimization and network intelligence. New applications include ML-based network intelligence and adaptive optimization at the system level, ML-based transmission intelligence, spectrum intelligence, and adaptive radio-resource management, adaptive baseband signal processing. However, significant challenges remain to be overcome: training challenges, lack of scalability, lack of explainability, ambiguity in generalization, and interoperability concerns.

The moment has come for researchers in both academia and business to investigate the potential of next-generation mobile communication systems. In the coming decade, there will be a plethora of possibilities for wireless research that spans several fields. The chapter elaborates system concept based on network virtualization, architecture options, technology enhancements, advanced service requirements, and new paradigms. The main aim is to provide well-structured description on how AI is being applied at the different layers of the 5G system and how it can pave the way for 6G networks. Driving 5G mobile communications with AI toward 6G will be an exciting era of technology convergence.

REFERENCES

1. ITU Broadband Commission for Sustainable Development, *The state of broadband 2021*, Sept. 2021.
2. EC Directorate-General for Communications Networks, Content and Technology, *5G observatory*, Quarterly Report 13, Oct. 2021.

3. GSM Association Report, *The mobile economy 2021*, Jun. 2021.
4. GSM Association Report, *Global mobile trends 2022*, Dec. 2021.
5. M.A. Marsan, N.B. Melazzi, S. Buzzi, *5G Italy: From research to market*, White eBook, Consorzio Nazionale Interuniversitario per le Telecomunicazioni, pp. 1–476, Dec. 2018.
6. J.M.C. Brito, L.L. Mendes, J.G.S. Gontijo, Brazil 6G project – An approach to build a national-wise framework for 6G networks, in *Proc. 2nd 6G Wireless Summit*, pp. 1–5, 2020.
7. 5G Americas White Paper, *3GPP releases 16 & 17 and beyond*, pp. 1–81, Jan. 2021.
8. 5G Americas White Paper, *Mobile communications beyond 2020*, pp. 1–60, Dec. 2020.
9. 5G Americas White Paper, *Mobile communications towards 2030*, pp. 1–56, Nov. 2021.
10. NTT DOCOMO Technical Journal, *Real issues and future vision 5G*, vol. 2, no. 2, 2020.
11. NTT DOCOMO White Paper, *5G evolution and 6G*, vols. 1/3, Jan. 2020/Feb. 2021.
12. A. Benjebbour et al., 3GPP defined 5G requirements and evaluation conditions, *NTT DOCOMO Technical Journal*, vol. 10, no. 3, pp. 13–23, 2018.
13. W. Yamada, Recent standardization activities in ITU-R SG 3, *NTT Technical Review*, vol. 20, no. 3, pp. 46–51, 2022.
14. Huawei, *5G-advanced technology evolution from a network perspective 2.0*, 2022.
15. Huawei, *Communication networks 2030: Building a fully connected, intelligent world*, Industry Report, pp. 1–43, Sept. 2021.
16. Huawei, *Intelligent world 2030: Building a fully connected, intelligent world*, Industry Report, pp. 1–317, Sept. 2021.
17. Ericsson White Paper, *Ever-present intelligent communication: A research outlook towards 6G*, 2020.
18. IEEE Future Networks WG, *IEEE 5G and beyond technology roadmap*, White Paper, 2020.
19. Nokia Bell Labs, *Communications in the 6G era*, White Paper, 2020.
20. 6G Flagship, *White paper on 6G networking*, 2020.
21. NGMN, *6G use cases and analysis*, Feb. 2022.
22. China Mobile, *6G network architecture*, White Paper, 2022.
23. Z. Li, X. Wang, T. Zhang, *5G+ how 5G change the society*, Springer, 2021.
24. W. Tong, P. Zhu (Eds.), *6G the next horizon: From connected people and things to connected intelligence*, Cambridge University Press, 2021.
25. A. Ebrahimzadeh, M. Maier, *Toward 6G: A new era of convergence*, IEEE Press & Wiley, 2021.
26. Y. Wu, S. Singh, T. Taleb, A. Roy, H.S. Dhillon, M.R. Kanagarathinam, A. De (Eds.), *6G mobile wireless networks*, Springer, 2021.
27. H. Kim, *Design and optimization for 5G wireless communications*, Wiley, 2020.
28. H. Kim, *Artificial intelligence for 6G*, Springer, 2022.
29. A. Osseiran, J.F. Monserrat, P. Marsch, *5G mobile and wireless communications technology*, Cambridge University Press, 2016.
30. Y. Yang, J. Xu, G. Shi, C. Wang, *5G wireless systems: Simulation and evaluation techniques*, Springer, 2018.
31. D.H. Morais, *5G and beyond wireless transport technologies*, Springer, 2021.
32. J.T.J. Penttinen, *5G Second phase explained: The 3GPP Release 16 enhancements*, Wiley, 2021.
33. W. Lei et al., *5G system design: An end to end perspective*, Springer, 2020.
34. E. Dahlman, S. Parkvall, J. Skold, *5G NR: The next generation wireless access technology 2E*, Academic Press imprint of Elsevier, 2021.
35. S. Rommer, P. Hedman, M. Olsson, L. Frid, S. Sultana, C. Mulligan, *5G core networks: Powering digitalization*, Academic Press imprint of Elsevier, 2020.

36. H. Holma, A. Toskala, T. Nakamura (Eds.), *5G technology: 3GPP new radio*, Wiley, 2020.
37. A. Zaidi, F. Athley, J. Medbo, U. Gustavsson, G. Durisi, X. Chen, *5G physical layer: Principles, models and technology components*, Elsevier, 2018.
38. D.H. Morais, *Key 5G physical layer technologies enabling mobile and fixed wireless access*, Springer, 2020.
39. M. Mandloi, D. Gurjar, P. Pattanayak, H. Nguyen (Eds.), *5G and beyond wireless systems: PHY layer perspective*, Springer, 2021.
40. R. Dangi et al., Study and investigation on 5G technology: A systematic review, *MDPI Sensors*, vol. 22, no. 26, pp. 1–32, 2022.
41. J.G. Andrews et al., What will 5G be? *IEEE Journal on Selected Areas in Communications*, vol. 32, no. 6, pp. 1065–1082, 2014.
42. J. Hansryd, 5G wireless communication beyond 2020, in *Proc. 45th European Solid State Device research Conference (ESSDERC)*, pp. 1–3, 2015.
43. M. Agiwal, A. Roy, N. Saxena, Next generation 5G wireless networks: A comprehensive survey, *IEEE Communications Surveys & Tutorials*, vol. 18, no. 3, pp. 1617–1655, 2016.
44. K. David, H. Berndt, 6G vision and requirements: Is there any need for beyond 5G? *IEEE Vehicular Technology Magazine*, vol. 13, no. 3, pp. 72–80, 2018.
45. K. Samdanis, T. Taleb, The road beyond 5G: A vision and insight of the key technologies, *IEEE Network*, vol. 34, no. 2, pp. 135–141, 2020.
46. M. Chen et al., Cloud-based wireless network: Virtualized, reconfigurable, smart wireless network to enable 5G technologies, Springer *Mobile Networks and Applications*, vol. 20, pp. 704–712, 2015.
47. S. Dang, O. Amin, B. Shihada, M.S. Alouini, What should 6G be? *Nature Electronics*, vol. 3, no. 1, pp. 20–29, 2020.
48. B. Bertenyi, 5G evolution: What's next? *IEEE Wireless Communications*, vol. 28, no. 1, pp. 4–8, Feb. 2021.
49. F. Tariq, M.R.A. Khandaker, K.-K. Wong, M.A. Imran, M. Bennis, M. Debbah, A speculative study on 6G, *IEEE Wireless Communications*, vol. 27, no. 4, pp. 118–125, 2020.
50. R. Bassoli et al., Why do we need 6G? ITU *Journal on Future and Evolving Technologies*, vol. 2, no. 6, pp. 1–31, 2021.
51. W. Jiang et al., The road towards 6G: A comprehensive survey, *IEEE Open Journal Communication Society*, vol. 2, pp. 334–366, Feb. 2021.
52. Q. Bi, Ten trends in the cellular industry and an outlook on 6G, *IEEE Communications Magazine*, vol. 57, no. 12, pp. 31–36, 2019.
53. W. Jiang, H.D. Schotten, The KICK-OFF of 6G research worldwide: An overview, in *Proc. International Conference on Computer and Communications (ICCC)*, pp. 1–6, 2021.
54. P. Popovski et al., A perspective on time towards wireless 6G, *Proceedings of the IEEE*, pp. 1–31, 2022. (Early access).
55. C. Yeh, G.D. Jo, Y.-J. Ko, H.K. Chung, Perspectives on 6G wireless communications, *ICT Express*, pp. 1–10, Jan. 2022.
56. T.R. Raddo et al., Transition technologies towards 6G networks, *EURASIP Journal on Wireless Communications and Networking*, article no. 100, pp. 1–22, 2021.
57. ICT-317669 METIS Project, *Final report on the METIS 5G system concept and technology roadmap*, Deliverable D6.6, 2015. Available Online: https://www.metis2020.com/documents/deliverables/.
58. UN ITU Specialized Agency for Information and Communication Technologies, *Beyond 5G: What's next for IMT?* Available Online: https://www.itu.int/hub/2021/02/beyond-5g-whats-next-for-imt/.

59. J. Gozalvez, Tentative 3GPP timeline for 5G, *IEEE Vehicular Technology Magazine*, vol. 10, no. 3, pp. 12–18, 2015.
60. J.M.C. Brito, L.L. Mendes, J.G.S. Gontijo, Standardization of 5G mobile networks: A systematic literature review and current developments, *International Journal of Standardization Research*, vol. 15, no. 2, pp. 1–5, 2017.

5G-Advanced Evolution

61. A. Ghosh, A. Maeder, M. Baker, D. Chandramouli, 5G evolution: A view on 5G cellular technology beyond 3GPP Release 15, *IEEE Access*, vol. 7, pp. 127639–127651, 2019.
62. Y. Kim et al., New radio (NR) and its evolution toward 5G-advanced, *IEEE Wireless Communications*, vol. 26, no. 3, pp. 2–7, Jun. 2019.
63. E. Dahlman et al., 5G evolution and beyond, *IEICE Transactions on Communications*, pp. 1–10, Sept. 2021.
64. W. Chen, P. Gaal, J. Montojo, H. Zisimopoulos, *Fundamentals of 5G communications: Connectivity for enhanced mobile broadband and beyond*, McGraw Hill, 2021.
65. X. Lin, N. Lee (Eds.), *5G and beyond: Fundamentals and standards*, Springer, 2021.
66. W. Chen, J. Montojo, J. Lee, M. Shafi, Y. Kim, The standardization of 5G-advanced in 3GPP, *IEEE Communications Magazine*, pp. 1–7, Jun. 2022.
67. X. Lin et al., An overview of 5G advanced evolution in 3GPP Release 18, *arXiv:2201.01358*, 2022.

System Architecture Options and Standardization Process

68. C.-X. Wang et al., Cellular architecture and key technologies for 5G wireless communication networks, *IEEE Communications Magazine*, vol. 52, no. 2, pp. 122–130, 2014.
69. Q.C. Li, H. Niu, A.T. Papathanassiou, G. Wu, 5G network capacity: Key elements and technologies, *IEEE Vehicular Technology Magazine*, vol. 9, no. 1, pp. 71–78, 2014.
70. M. Jaber, M.A. Imran, R. Tafazolli, A. Tukmanov, 5G backhaul challenges and emerging research directions: A survey, *IEEE Access*, vol. 4, pp. 1743–1766, 2016.
71. E. Dahlman, S. Parkvall, A. Furuskar, M. Frenne, NR – The new 5G radio-access technology, *IEEE Communications Standards Magazine*, vol. 1, no. 4, pp. 24–30, 2017.
72. X. Lin et al., 5G new radio: Unveiling the essentials of the next generation wireless access technology, *IEEE Communications Standards Magazine*, vol. 3, no. 3, pp. 30–37, Sept. 2019.
73. G. Liu et al., 5G deployment: Standalone vs. non-standalone from the operator perspective, *IEEE Communications Magazine*, vol. 58, no. 11, pp. 83–89, 2020.
74. F. Rinaldi, A. Raschella, S. Pizzi, 5G NR system design: A concise survey of key features and capabilities, Springer *Wireless Networks*, vol. 27, pp. 5173–5188, 2021.
75. 3GPP NG-RAN TS 38.401, *Architecture description*, Release 16 V16.0.0, Dec. 2019.
76. 3GPP NG-RAN TR 38.801, *Radio access architecture and interfaces*, V14.0.0, Mar. 2019.
77. 3GPP Services and System Aspects TS 23.501, *System architecture for the 5G system (5GS)*, Release 16 V16.0.0, Mar. 2019.
78. 3GPP Services and System Aspects TS 23.502, *Procedures for the 5G system (5GS)*, Release 16 V16.0.0, Mar. 2019.
79. 3GPP Services and System Aspects TS 23.503, *Policy and charging control framework for the 5G system (5GS)*, Release 16 V16.0.0, Mar. 2019.
80. E. Dahlman, S. Parkvall, J. Skold, *5G NR: The next generation wireless access technology*, Academic Press, 2020.

81. 3GPP Radio Access Network TR 38.801, *Study on new radio access technology: Radio access architecture and interfaces*, 14.0.0, Mar. 2017.
82. 3GPP Radio Access Network TR 38.201, *NR; Physical layer; General description*, Release 15 15.0.0, Jan. 2018.
83. 3GPP Radio Access Network TR 38.202, *Study on new radio access technology, physical layer aspects*, Release 16 16.0.0, Nov. 2020.
84. 3GPP Radio Access Network TR 38.912, *Study on new radio (NR) access technology*, Release 16 16.0.0, Jul. 2020.
85. 3GPP Radio Access Network TS 38.211, *NR physical channels and modulation*, Release 16 V16.0.0, Dec. 2019.
86. 3GPP Radio Access Network TS 38.213, *NR physical layer procedures for control*, Release 16 V16.0.0, Dec. 2019.
87. 3GPP Radio Access Network TS 38.214, *NR physical layer procedures for data*, Release 16 V16.0.0, Dec. 2019.
88. 3GPP Radio Access Network TS 38.215, *NR physical layer measurements*, Release 16 V16.0. 16.0.0, Dec. 2019.
89. C. Saha, H.S. Dhillon, Millimeter wave integrated access and backhaul in 5G: Performance analysis and design insights, *IEEE Journal on Selected Areas in Communications*, vol. 37, no. 12, pp. 2669–2684, 2019.
90. L. Chen et al., QoS assurance in IAB network, in *Proc. International Wireless Communications and Mobile Computing (IWCMC)*, pp. 1860–1865, 2020.
91. M. Polese et al., End-to-end simulation of integrated access and backhaul at mmWaves, in *Proc. IEEE 23rd International Workshop on COMPUTER AIDED MODELING and Design of Communication Links and Networks (CAMAD)*, pp. 1–7, 2018.
92. C. Madapatha et al., On integrated access and backhaul networks: Current status and potentials, *IEEE Open Journal of the Communications Society*, vol. 1, pp. 1374–1389, 2020.
93. X. Foukas, G. Patounas, A. Elmokashfi, M.K. Marina, Network slicing in 5G: Survey and challenges, *IEEE Communications Magazine*, vol. 55, no. 5, pp. 94–100, 2017.
94. P. Popovski, K.F. Trillingsgaard, O. Simeone, G. Durisi, 5G wireless network slicing for eMBB, uRLLC, and mMTC: A communication-theoretic view, *IEEE Access*, vol. 6, pp. 5765–5779, 2018.
95. I. Afolabi, T. Taleb, K. Samdanis, A. Ksentini, H. Flinck, Network slicing and softwarization: A survey on principles, enabling technologies, and solutions, *IEEE Communications Surveys & Tutorials*, vol. 20, no. 3, pp. 2429–2453, 2018.
96. R. Su, D. Zhang, R. Venkatesan, Z. Gong, C. Li, F. Ding, F. Jiang, Z. Zhu, Resource allocation for network slicing in 5G telecommunication networks: A survey of principles and models, *IEEE Network*, vol. 33, no. 6, pp. 172–179, 2019.
97. S. Zhang, An overview of network slicing for 5G, *IEEE Wireless Communications*, vol. 26, no. 3, pp. 111–117, 2019.
98. F. Debbab et al., Algorithmics and modeling aspects of network slicing in 5G and beyond network: Survey, *IEEE Access*, vol. 8, pp. 162748–162762, 2020.
99. Y. Mao, C. You, J. Zhang, K. Huang, K.B. Letaief, A survey on mobile edge computing: The communication perspective, *IEEE Communications Surveys & Tutorials*, vol. 19, no. 4, pp. 2322–2358, 2017.
100. L. Pierucci, The quality of experience perspective toward 5G technology, *IEEE Wireless Communications*, vol. 22, no. 4, pp. 10–16, 2015.
101. Q. Ye, J. Li, K. Qu, W. Zhuang, X.S. Shen, X. Li, End-to-end quality of service in 5G networks: Examining the effectiveness of a network slicing framework, *IEEE Vehicular Technology Magazine*, vol. 13, no. 2, pp. 65–74, 2018.
102. G. Zhu, J. Zan, Y. Yang, X. Qi, A supervised learning based QoS assurance architecture for 5G networks, *IEEE Access*, vol. 7, pp. 43598–43606, 2019.

103. 3GPP Services and System Aspects TS 33.501, *Security architecture and procedures for 5G system*, Release 16 V16.0.0, Sept. 2019.
104. D.P.M. Osorio et al., *Physical layer security for 5G and beyond*, Wiley, 2021.
105. D.P.M. Osorio et al., Safeguarding MTC at the physical layer: Potentials and challenges, *IEEE Access*, vol. 8, pp. 101437–101447, 2020.
106. A.K. Yerrapragada et al., Physical layer security for beyond 5G: Ultra secure low latency communications, *IEEE Open Journal of the Communications Society*, vol. 2, pp. 2232–2242, 2021.

5G New Radio & Core Network Enhancement and Vertical Expansion

107. K. Flynn, Progress on 5G Releases 16/17 in 3GPP, *IEEE Communications Standards Magazine*, vol. 4, no. 2, pp. 4–5, Jun. 2020.
108. A. Mahmood et al., Industrial IoT in 5G-and-beyond networks: Vision, architecture, and design trends, *IEEE Transactions on Industrial Informatics*, 2021. (Early access).
109. M. Bennis, M. Debbah, H.V. Poor, Ultra-reliable and low-latency wireless communication: Tail, risk, and scale, *Proceedings of the IEEE*, vol. 106, no. 10, pp. 1834–1853, 2018.
110. X. Jiang, H. Hokri-Ghadikolaei, G. Fodor, E. Modiano, Z. Pang, M. Zorzi, C. Fischione, Low-latency networking: Where latency lurks and how to tame it? *Proceedings of the IEEE*, vol. 107, no. 2, pp. 280–306, 2019.
111. K. Chen, T. Zhang, R.D. Gitlin, G. Fettweis, Ultra-low latency mobile networking, *IEEE Network*, vol. 33, no. 2, pp. 181–187, 2019.
112. M. Harounabadi et al., V2X in 3GPP standardization: NR sidelink in Release-16 and beyond, *IEEE Communications Standards Magazine*, vol. 5, no. 1, pp. 12–21, Mar. 2021.
113. M. Wen et al., Private 5G networks: Concepts, architectures, and research landscape, *IEEE Journal of Selected Topics in Signal Processing*, vol. 16, no. 1, pp. 7–25, 2021.
114. J. Prados-Garzon et al., 5G non-public networks: Standardization, architectures and challenges, *IEEE Access*, vol. 9, pp. 893–908, Nov. 2021.
115. D. Xu, A. Zhou, X. Zhang, G. Wang, X. Liu, C. An, Y. Shi, L. Liu, H. Ma, Understanding operational 5G: A first measurement study on its coverage, performance and energy consumption, in *Proc. Annual Conference of the ACM SIG on Data Communication on the Applications, Technologies, Architectures, and Protocols for Computer Communication*, pp. 479–494, 2020.
116. A. Narayanan, E. Ramadan, J. Carpenter, Q. Liu, Y. Liu, F. Qian, Z.-L. Zhang, A first look at commercial 5G performance on smartphones, in *Proc. The Web Conference*, pp. 894–905, 2020.
117. J. Rischke, P. Sossalla, S. Itting, F.H.P. Fitzek, M. Reisslein, 5G campus networks: A first measurement study, *IEEE Access*, vol. 9, pp. 121786–121803, 2021.

Advanced Service Requirements and Performance Indicators

118. ITU-R Recommendation M.2083, *IMT vision: Framework and overall objectives of the future development of IMT for 2020 and beyond*, Sept. 2015.
119. ITU-R Report M.2410-0, *Minimum requirements related to technical performance for IMT-2020 radio interface(s)*, Nov. 2017.
120. ITU-R Report M.2411, *Requirements, evaluation criteria and submission templates for the development of IMT-2020*, Nov. 2017.

121. ITU-R Report M.2412, *Guidelines for evaluation of radio interface technologies for IMT-2020*, Oct. 2017.
122. S. Henry, A. Alsohaily, and E.S. Sousa, 5G is real: Evaluating the compliance of the 3GPP 5G new radio system with the ITU IMT-2020 requirements, *IEEE Access*, vol. 8, pp. 42828–42840, 2020.
123. 3GPP Radio Access Network TR 38.913, *Study on scenarios and requirements for next generation access technologies*, Release 15 V15.0.0, Jun. 2018.
124. 3GPP Radio Access Network TR 38.913, *Study on scenarios and requirements for next generation access technologies*, Release 16 V16.0.0, Jul. 2020.
125. J. Navarro-Ortiz, P. Romero-Diaz, S. Sendra, P. Ameigeiras, J.J. Ramos-Munoz, J.M. Lopez-Soler, A survey on 5G usage scenarios and traffic models, *IEEE Communications Surveys & Tutorials*, vol. 22, no. 2, pp. 905–929, 2020.
126. 3GPP, *Advanced plans for 5G*, Jul. 2021. Available Online: https://www.3gpp.org/news-events/2210-advanced_5g.
127. 3GPP SA1 – Services TR 33.852, *Study on traffic characteristics and performance requirements for AI/ML model transfer in 5G systems (5GS)*, Release 18, DRAFT.
128. 3GPP SA1 – Services TR 22.874, *5G system (5GS); Study on traffic characteristics and performance requirements for AI/ML model transfer*, Release 18, DRAFT.
129. I. Rahman et al., 5G evolution toward 5G advanced: An overview of 3GPP Releases 17 and 18, *Ericsson Technology Review*, vol. 14, pp. 1–12, Oct. 2021.
130. 3GPP Services and System Aspects TR 22.867, *Study on 5G smart energy and infrastructure*, Release 18 V18.0.0, Jul. 2021.
131. J. Lorincz, A. Capone, J. Wu, Greener, energy-efficient and sustainable networks: State-of-the-art and new trends, *Sensors*, vol. 19, no. 22, p. 4864, 2019.
132. A. Adamatzky, Ö. Bulakci, D. Chatterjee, M. Shafik, S.K.S. Tyagi (Eds.), Artificially intelligent green communication networks for 5G and beyond, Elsevier *Computer Communications*, vol. 169, no. 1, Mar. 2021.
133. 3GPP Services and System Aspects TS 22.261, *Service requirements for the 5G system*, Release 18 V18.0.0, Sept. 2020.
134. 3GPP Services and System Aspects TR 22.847, *Study on supporting tactile and multimodality communication services*, Release 18 V18.0.0, Sept. 2021.

Novel System AI/ML Paradigm

135. ITU Journal Future and Evolving Technologies, Special issue, AI and machine learning solutions in 5G and future networks, vol. 2, no. 4, 2021.
136. R. He, Z. Ding (Eds.), *Applications of machine learning in wireless communications*, IET, 2019.
137. F.-L. Luo (Ed.), *Machine learning for future wireless communications*, Wiley – IEEE Press, 2019.
138. Y. Sun et al., Application of machine learning in wireless networks: Key techniques and open issues, *IEEE Communications Surveys & Tutorials.*, vol. 21, no. 4, pp. 3072–3108, 2019.
139. R. Li, Z. Zhao, X. Zhou, G. Ding, Y. Chen, Z. Wang, H. Zhang, Intelligent 5G: When cellular networks meet artificial intelligence, *IEEE Wireless Communications*, vol. 24, no. 5, pp. 175–183, Mar. 2017.
140. C.X. Wang, M. Di Renzo, S. Stanczak, S. Wang, E.G. Larsson, Artificial intelligence enabled wireless networking for 5G and beyond: Recent advances and future challenges, *IEEE Wireless Communications*, vol. 27, no. 1, pp. 16–23, Mar. 2020.
141. S.M. Aldossari, K.-C. Chen, Machine learning for wireless communication channel modeling: An overview, *Wireless Personal Communication*, vol. 106, no. 1, pp. 41–70, Mar. 2019.

142. T. O'Shea, J. Hoydis, An introduction to deep learning for the physical layer, *IEEE Transactions on Cognitive Communications and Networking*, vol. 3, no. 4, pp. 563–575, Dec. 2017.
143. C. Zhang et al., Deep learning in mobile and wireless networking: A survey, *IEEE Communications Surveys & Tutorials*, vol. 21, no. 3, pp. 2224–2287, 2019.
144. H. Fourati, R. Maaloul, L. Chaari, A survey of 5G network systems: Challenges and machine learning approaches, Springer *International Journal of Machine Learning and Cybernetics*, vol. 12, pp. 385–431, 2021.
145. Z. Zhu et al., Research and analysis of URLLC technology based on artificial intelligence, *IEEE Communications Standards Magazine*, vol. 5, no. 2, pp. 37–43, 2021.
146. A. Arnaz et al., Toward integrating intelligence and programmability in open radio access networks: A comprehensive survey, *IEEE Access*, vol. 10, pp. 67747–67770, 2022.
147. Y. Cao et al., User access control in open radio access networks: A federated deep reinforcement learning approach, *IEEE Transactions on Wireless Communications*, vol. 21, no. 6, pp. 3721–3736, 2022.
148. B. Brik et al., Deep learning for B5G open radio access network: Evolution, survey, case studies, and challenges, *IEEE Open Journal of the Communications Society*, vol. 3, pp. 228–250, 2022.
149. A. Giannopoulos et al., Supporting intelligence in disaggregated open radio access networks: Architectural principles, AI/ML workflow, and use cases, *IEEE Access*, vol. 10, pp. 39580–39595, 2022.
150. S. Sevgican et al., Intelligent network data analytics function in 5G cellular networks using machine learning, *KICS Journal of Communications and Networks*, vol. 22, no. 3, pp. 269–280, 2020.
151. 3GPP Services and System Aspects TR 23.791, *Study of enablers for network automation for 5G*, Release 16 V16.0.0, Dec. 2018.
152. 3GPP Services and System Aspects TR 23.700–91, *Study of enablers for network automation for 5G – Phase 2*, Release 17, Dec. 2020.
153. ETSI Group Report Experiential Networked Intelligence, *ENI use cases*, Apr. 2018.
154. A. Pouttu et al., *Validation and trials for verticals towards 2030's*; 6G Research Visions, Jun. 2020.
155. A. Mourad et al., A baseline roadmap for advanced wireless research beyond 5G, *MDPI Electronics*, vol. 91, no. 351, pp. 1–14, 2020.
156. J. Tanveer, A. Haider, R. Ali, A. Kim, Machine learning for physical layer in 5G and beyond wireless networks: A survey, *MDPI Electronics*, vol. 11, no. 121, pp. 1–44, 2022.
157. M.S. Murshed, C. Murphy, D. Hou, N. Khan, G. Ananthanarayanan, F. Hussain, Machine learning at the network edge: A survey, *ACM Computing Surveys*, vol. 54, no. 8, pp. 1–37, 2021.
158. S. Han et al., Artificial-intelligence-enabled air interface for 6G: Solutions, challenges, and standardization impacts, *IEEE Communications Magazine*, vol. 58, no. 10, pp. 73–79, Oct. 2020.
159. M.N. Mahdi et al., From 5G to 6G technology: Meets energy, internet-of-things and machine learning: A survey, *MDPI Applied Sciences*, vol. 11, 8117, pp. 1–58, 2021.
160. X. Lin et al., Fueling the next quantum leap in cellular networks: Embracing AI in 5G evolution towards 6G, *IEEE Wireless Communications*, vol. 27, no. 2, pp. 212–217, Apr. 2020.
161. A.T.H. The et al., Artificial intelligence for the Metaverse: A survey, *arXiv:2202.10336*, 2022.

6G Concept, Research and Transition Technologies

162. A. Yazar et al., 6G vision: An ultra-flexible perspective, *ITU Journal on Future and Evolving Technologies*, vol. 1, no. 1, pp. 1–20, 2020.
163. S. Sambhwani, Z. Boos, S. Dalmia, A. Fazeli, B. Gunzelmann, A. Ioffe, M. Narasimha, F. Negro, L. Pillutla, J. Zhou, Transitioning to 6G: Part 1 – Radio technologies, *IEEE Wireless Communications*, vol. 29, no. 1, pp. 6–8, Feb. 2022.
164. S. Sambhwani, A. Bharadwaj, C. Drewes, K. Hamidouche, A. Naguib, D. Nickisch, S. Roessel, M. Sauer, Y. Schoinas, T. Tabet, S. Vallath, Y.-T. Yu, Transitioning to 6G Part 2 – Systems and network technology areas, *IEEE Wireless Communications*, vol. 29, no. 2, pp. 6–8, Apr. 2022.
165. A.A. Hakeema, H.H. Hussein, H.W. Kim, Vision and research directions of 6G technologies and applications, Elsevier *Computer and Information Sciences*, vol. 34, no. 6, pp. 2419–2442, Part A, Jun. 2022.
166. M.W. Akhtar et al., The shift to 6G communications: Vision and requirements, Springer *Human-Centric Computing and Information Sciences*, vol. 10, no. 53, pp. 1–28, 2020.
167. A. Dogra, R.K. Jha, S. Jain, A survey on beyond 5G network with the advent of 6G: Architecture and emerging technologies, *IEEE Access*, vol. 9, pp. 67512–67547, 2020.
168. K. David et al., Defining 6G: Challenges and opportunities, *IEEE Vehicular Technology Magazine*, vol. 14, no. 3, pp. 14–16, 2019.
169. B. Zong et al., 6G technologies: Key drivers, core requirements, systems architectures and enabling technologies, *IEEE Vehicular Technology Magazine*, vol. 14, no. 3, pp. 18–27, 2019.
170. H. Yang, A. Alphones, Z. Xiong, D. Niyato, J. Zhao, K. Wu, Artificial-intelligence-enabled intelligent 6G networks, *IEEE Network*, vol. 34, no. 6, pp. 272–280, 2020.
171. L. Loven et al., EdgeAI: A vision for distributed, edge native artificial intelligence in future 6G networks, in *Proc. 1st 6G Wireless Summit*, pp. 1–2, 2019.
172. E. Basar, M. Di Renzo, J. De Rosny, M. Debbah, M.-S. Alouini, R. Zhang, Wireless communications through reconfigurable intelligent surfaces, *IEEE Access*, vol. 7, pp. 753–773, 2019.
173. M. DiRenzo et al., Smart radio environments empowered by reconfigurable AI meta-surfaces: An idea whose time has come, *EURASIP Journal on Wireless Communications and Networking*, no. 129, pp. 1–20, 2019.
174. R. Liu, Q. Wu, M. DiRenzo, Y. Yuan, A path to smart radio environments: An industrial viewpoint on reconfigurable intelligent surfaces, *IEEE Wireless Communications*, vol. 29, no. 1, pp. 202–208, 2022.
175. A. Bansal et al., Rate-splitting multiple access for intelligent reflecting surface aided multi-user communications, *IEEE Transactions on Vehicular Technology*, vol. 70, no. 9, pp. 9217–9229, Sept. 2021.
176. Y. Pei, X. Yue, Y. Yao, X. Li, H. Wang, D.-T. Do, Secrecy communications of intelligent reflecting surfaces aided NOMA networks, Elsevier *Physical Communication*, vol. 52, pp. 1–9, 2022.
177. Vision, requirements and network architecture of 6G mobile network beyond 2030, *China Communications*, vol. 17, no. 9, pp. 92–104, 2020.
178. K. David et al., Laying the milestones for 6G networks, *IEEE Vehicular Technology Magazine*, vol. 15, no. 4, pp. 18–21, 2020.
179. H. Viswanathan, P.E. Mogensen, Communications in the 6G era, *IEEE Access*, vol. 8, pp. 57063–57074, 2020.
180. H. Jiang, M. Mukherjee, J. Zhou, J. Lloret, Channel modeling and characteristics for 6G wireless communications, *IEEE Network*, vol. 35, no. 1, pp. 296–303, 2021.
181. M. Matthaiou, O. Yurduseven, H.Q. Ngo, D. Morales-Jimenez, S.L. Cotton, V.F. Fusco, The road to 6G: Ten physical layer challenges for communications engineers, *IEEE Communications Magazine*, vol. 59, no. 1, pp. 64–69, 2021.

182. K. David et al., 6G networks: Is this an evolution or a revolution? *IEEE Vehicular Technology Magazine*, vol. 15, no. 4, pp. 14–15, 2021.
183. N. Adem et al., How crucial is it for 6G networks to be autonomous? *arXiv:2106.06949*, 2021.
184. L. DeNardis, M.-G. DiBenedetto, Mo3: A modular mobility model for future generation mobile wireless networks, *IEEE Access*, vol. 10, pp. 34085–34115, Mar. 2022.
185. T. Huang, W. Yang, J. Wu, J. Ma, X. Zhang, D. Zhang, A survey on green 6G network: Architecture and technologies, *IEEE Access*, vol. 7, pp. 175758–175768, 2019.
186. A. Affan et al., Performance analysis of orbital angular momentum (OAM): A 6G waveform design, *IEEE Communications Letters*, vol. 25, no. 12, pp. 3985–3989, 2021.

EXTREME SYSTEM PERFORMANCE AND NETWORK EVOLUTION

187. N. Rajatheva et al., White paper on broadband connectivity in 6G, *arXiv:2004.14247*, pp. 1–46, 2020. (Published at NGNA 2020).
188. B. Han et al., An abstracted survey on 6G: Drivers, requirements, efforts and enablers, *arXiv:2101.01062v1*, 2021.
189. Z. Zhang et al., 6G wireless networks: Vision, requirements, architecture, and key technologies, *IEEE Vehicular Technology Magazine*, vol. 14, no. 3, pp. 28–41, 2019.
190. P. Yang, Y. Xiao, M. Xiao, S. Li, 6G wireless communications: Vision and potential techniques, *IEEE Network*, vol. 33, no. 4, pp. 70–75, 2019.
191. V. Ziegler, S. Yrjola, 6G indicators of value and performance, in *Proc. 2nd 6G Wireless Summit*, pp. 1–5, 2020.
192. S. Chen, Y. Liang, S. Sun, S. Kang, W. Cheng, M. Peng, Vision, requirements, and technology trend of 6G: How to tackle the challenges of system coverage, capacity, user data-rate and movement speed, *IEEE Wireless Communications*, vol. 27, no. 2, pp. 218–228, 2020.
193. M. Giordani, M. Polese, M. Mezzavilla, S. Rangan, M. Zorzi, Toward 6G networks: Use cases and technologies, *IEEE Communications Magazine*, vol. 58, no. 3, pp. 55–61, 2020.
194. ITU-T Technical Specification FG-NET2030, *Terms and definitions for network 2030*, Focus Group on Technologies for Network 2030, Jun. 2020.
195. ITU-T, *Representative use cases and key network requirements for network 2030*, 2020.
196. ITU-R Recommendation M.2150, *Detailed specifications of the radio interfaces of IMT-2020*, Feb. 2021.
197. ITU-R WP, *5D draft recommendation, IMT vision for 2030 and beyond*, Mar. 2021.
198. ITU-R WP, *5D draft report, IMT future technology trends*, Mar. 2021.
199. ITU-R IMT-2030 (6G) Promotion group, *6G vision and candidate technologies*, Jun. 2021.

TECHNOLOGY ENABLERS AND RESEARCH PROGRAMS

200. W. Saad, M. Bennis, M. Chen, A vision of 6G wireless systems: Applications, trends, technologies, and open research problems, *IEEE Network*, vol. 34, no. 3, pp. 134–142, 2020.
201. M.Z. Chowdhury et al., 6G wireless communication systems: Applications, requirements, technologies, challenges, and research directions, *IEEE Open Journal of the Communications Society*, vol. 1, pp. 957–975, 2020.
202. X. You et al., Towards 6G wireless communication networks: Vision, enabling technologies, and new paradigm shifts, Springer *Science China: Information Sciences*, vol. 64, p. 110301, 2021.

203. H.H.H. Mahmoud, A.A. Amer, T. Ismail, 6G: A comprehensive survey on technologies, applications, challenges, and research problems, Wiley *Transactions on Emerging Telecommunications Technologies*, vol. 32, no. 4, Apr. 2021.
204. H. Tataria, M. Shafi, A.F. Molisch, M. Dohler, H. Sjöland, F. Tufvesson, 6G wireless systems: Vision, requirements, challenges, insights, and opportunities, *Proceeding of the IEEE*, vol. 109, no. 7, pp. 1166–1199, 2021.
205. M.Z. Asghar et al., From 5G to 6G: Key drivers, applications and research directions, in *Proc. IEEE 18th International Conference on Smart Communities: Improving Quality of Life Using ICT, IoT and AI (HONET)*, 2021.
206. S. Alraih et al., Revolution or evolution? Technical requirements and considerations towards 6G mobile communications, *MDPI Sensors*, vol. 22, no. 762, pp. 1–29, 2022.
207. M. Katz, M. Matinmikko-Blue; M. Latva-Aho, 6Genesis flagship program: Building the bridges towards 6G-enabled wireless smart society and ecosystem, in *Proc. IEEE 10th Latin-American Conference on Communications (LATINCOM)*, pp. 1–6, 2018.
208. 6Genesis Flagship Program, White Paper, Key drivers and research challenges for 6G ubiquitous wireless intelligence, in *Proc. 1st 6G Summit*, pp. 1–36, 2019.
209. G. Wikström, J. Peisa, P. Rugeland, N. Johansson, S. Parkvall, M. Girnyk et al., Challenges and technologies for 6G, in *Proc. 2nd 6G Summit*, pp. 1–5, 2020.
210. D. Soldani, 6G fundamentals: Vision & enabling technologies – Towards trustworthy solutions & resilient systems, in *6GW02, 6GWorld*, pp. 1–18, 2021.
211. R. Prasad, A.R. Prasad (Eds.), 6G enabling technologies – Innovation 6G Part 1, *Journal of ICT Standardization*, vol. 9, no. 3, 2021.
212. R. Prasad, A.R. Prasad (Eds.), 6G enabling technologies – Innovation 6G Part 2, *Journal of ICT Standardization*, vol. 10, no. 1, 2022.
213. Nidhi et al., Trends in standardization towards 6G, *Journal of ICT Standardization*, vol. 9, no. 3, pp. 327–348, 2021.
214. A. Clemm et al., Toward truly immersive holographic-type communication: Challenges and solutions, *IEEE Communications Magazine*, vol. 58, no. 1, pp. 93–99, 2020.
215. J. Hooft et al., From capturing to rendering: Volumetric media delivery with six degrees of freedom, *IEEE Communications Magazine*, vol. 58, no. 1, pp. 49–55, 2020.
216. Y. Zhou, L. Liu, L. Wang, N. Hui, X. Cui, J. Wu, Y. Peng, Y. Qi, C. Xing, Service aware 6G: An intelligent and open network based on convergence of communication, computing and caching, Elsevier *Digital Communications and Networks*, vol. 6, no. 3 pp. 253–260, 2020.
217. M. Giordani, M. Zorzi, Non-terrestrial networks in the 6G era: Challenges and opportunities, *IEEE Network*, vol. 35, no. 2, pp. 12–19, 2020.
218. 5G from space: An overview of 3GPP non-terrestrial networks, *IEEE Communications Standards Magazine*, pp. 1–8, 2022.
219. X. Huang, J.A. Zhang, R.P. Liu, Y.J. Guo, L. Hanzo, Airplane-aided integrated networking for 6G wireless: Will it work? *IEEE Vehicular Technology Magazine*, vol. 14, no. 3, pp. 84–91, 2019.
220. M. Polese, J.M. Jornet, T. Melodia, M. Zorzi, Toward end-to-end, full-stack 6G terahertz networks, *IEEE Communications Magazine*, vol. 58, no. 11, pp. 48–54, 2020.
221. K. Rikkinen, P. Kyosti, M.E. Leinonen, M. Berg, A. Parssinen, THz Radio communication: Link budget analysis toward 6G, *IEEE Communications Magazine*, vol. 58, no. 11, pp. 22–27, 2020.
222. R. Minerva, G.M. Lee, N. Crespi, Digital twin in the IoT context: A survey on technical features, scenarios, and architectural models, *Proceedings of the IEEE*, vol. 108, no. 10, pp. 1785–1824, 2020.
223. A. ElSaddik, Digital twins: The convergence of multimedia technologies, *IEEE MultiMedia*, vol. 25, no. 2, pp. 87–92, 2018.

224. R. Saracco, Digital twins: Bridging physical space and cyberspace, *IEEE Computer*, vol. 52, no. 12, pp. 58–64, 2019.
225. A. Rasheed, O. San, T. Kvamsdal, Digital twin: Values, challenges and enablers from a modeling perspective, *IEEE Access*, vol. 8, pp. 21980–22012, 2020.
226. F. Tao, H. Zhang, A. Liu, A.Y.C. Nee, Digital twin in industry: State-of-the-art, *IEEE Transactions on Industrial Informatics*, vol. 15, no. 4, pp. 2405–2415, 2019.
227. S.H. Khajavi, N.H. Motlagh, A. Jaribion, L.C. Werner, J. Holmström, Digital twin: Vision, benefits, boundaries, and creation for buildings, *IEEE Access*, vol. 7, pp. 147406–147419, 2019.
228. B.R. Barricelli, E. Casiraghi, D. Fogli, A survey on digital twin: Definitions, characteristics, applications, and design implications, *IEEE Access*, vol. 7, pp. 167653–167671, 2019.
229. H. Laaki, Y. Miche, K. Tammi, Prototyping a digital twin for real time remote control over mobile networks: Application of remote surgery, *IEEE Access*, vol. 7, pp. 20325–20336, 2019.
230. T.R. Wanasinghe, L. Wroblewski, B.K. Petersen, R.G. Gosine, L.A. James, O. DeSilva, G.K.I. Mann, P.J. Warrian, Digital twin for the oil and gas industry: Overview, research trends, opportunities, and challenges, *IEEE Access*, vol. 8, pp. 104175–104197, 2020.
231. F. Laamarti, H.F. Badawi, Y. Ding, F. Arafsha, B. Hafidh, A.E. Saddik, An ISO/IEEE 11073 standardized digital twin framework for health and well-being in smart cities, *IEEE Access*, vol. 8, pp. 105950–105961, 2020.
232. J. Du, C. Jiang, J. Wang, Y. Ren, M. Debbah, Machine learning for 6G wireless networks: Carrying forward enhanced bandwidth, massive access, and ultra-reliable/low-latency service, *IEEE Vehicular Technology Magazine*, vol. 15, no. 4, pp. 122–134, 2020.
233. C. She, R. Dong, Z. Gu, Z. Hou, Y. Li, W. Hardjawana, C. Yang, L. Song, B. Vucetic, Deep learning for ultra-reliable and low-latency communications in 6G networks, *IEEE Network*, vol. 34, no. 5, pp. 219–225, 2020.
234. Y. Dong et al., Joint source-channel coding for 6G communications, *China Communications*, vol. 19, no. 3, pp. 101–115, 2022.
235. A. Salh et al., A survey on deep learning for ultra-reliable and low-latency communications challenges on 6G wireless systems, *IEEE Access*, vol. 9, pp. 55098–55131, 2021.
236. R. Ali et al., URLLC for 5G and beyond: Requirements, enabling incumbent technologies and network intelligence, *IEEE Access*, vol. 9, pp. 67064–67095, 2021.
237. C. Jiang et al., Machine learning paradigms for next-generation wireless networks, *IEEE Wireless Communications*, vol. 24, no. 2, pp. 98–105, Apr. 2017.
238. K.B. Letaief, W. Chen, Y. Shi, J. Zhang, Y.A. Zhang, The roadmap to 6G: AI empowered wireless networks, *IEEE Communications Magazine*, vol. 57, no. 8, pp. 84–90, 2019.
239. E.C. Strinati et al., 6G the next frontier: From holographic messaging to artificial intelligence using subterahertz and visible light communication, *IEEE Vehicular Technology Magazine*, vol. 14, no. 3, pp. 42–50, 2019.
240. R. Shafin et al., Artificial intelligence-enabled cellular networks: A critical path to beyond-5G and 6G, *arXiv*, 2019.
241. K. Sheth et al., A taxonomy of AI techniques for 6G communication networks, Elsevier *Computer Communications*, vol. 161, pp. 279–303, 2020.
242. N. Kato, B. Mao, F. Tang, Y. Kawamoto, J. Liu, Ten challenges in advancing machine learning technologies toward 6G, *IEEE Wireless Communications*, vol. 27, no. 3, pp. 96–103, 2020.
243. Y. Xiao, G. Shi, Y. Li, W. Saad, H.V. Poor, Toward self-learning edge intelligence in 6G, *IEEE Communications Magazine*, vol. 58, no. 12, pp. 34–40, 2020.

244. Y. Liu et al., Federated learning for 6G communications: Challenges, methods, and future directions, *China Communications*, vol. 17, no. 9, pp. 105–118, 2020.
245. G. Gui et al., 6G: Opening new horizons for integration of comfort, security and intelligence, *IEEE Wireless Communications*, vol. 27, no. 5, pp. 126–132, 2020.
246. X. Shen et al., Holistic network virtualization and pervasive network intelligence for 6G, *IEEE Communications Surveys & Tutorials*, 2022. (Early access).
247. E. Calvanese Strinati, S. Barbarossa, 6G in the sky: On-demand intelligence at the edge of 3D networks, *ETRI Wiley Computer Networks Journal*, pp. 1–15, 2022.
248. X. Luo, H.-H. Chen, Q. Guo, Semantic communications: Overview, open issues, and future research directions, *IEEE Wireless Communications*, vol. 29, no. 1, pp. 210–219, 2022.
249. S. Ali et al., Machine learning in wireless communication networks, 6G Research Visions, White Paper, no. 7, Jun. 2020.
250. I.F. Akyildiz, M. Pierobon, S. Balasubramaniam, Y. Koucheryavy, The internet of bio-nano things, *IEEE Communications Magazine*, vol. 53, no. 3, pp. 32–40, 2015.
251. I.F. Akyildiz, A. Kak, S. Nie, 6G and beyond: The future of wireless communications systems, *IEEE Access* vol. 8, pp. 133995–134030, 2020.
252. R. Bassoli, H. Boche, C. Deppe, R. Ferrara, F.H.P. Fitzek, G. Janßen, S. Saeedinaeen, *Quantum communication networks*, 1st ed., Springer, 2021.
253. P.S.R. Henrique, R. Prasad, 6G networks orientation by quantum mechanics, *Journal of ICT Standardization*, vol. 10, no. 1, pp. 39–62, 2022.
254. C. Wang, A. Rahman, Quantum-enabled 6G wireless networks: Opportunities and challenges, *IEEE Wireless Communications*, vol. 29, no. 1, pp. 58–69, Feb. 2022.
255. O.B. Akan, H. Ramezani, T. Khan, N.A. Abbasi, M. Kuscu, Fundamentals of molecular information and communication science, *Proceedings of the IEEE*, vol. 105, no. 2, pp. 306–318, 2017.
256. J.R. Vacca (Ed.), *Nanoscale networking and communications handbook*, CRC Press, 2019.

ADDITIONAL READING

IEEE Communications Society, Best readings in machine learning in communications: Signal detection; Channel encoding and decoding; Channel estimation, prediction, and compression; End-to-end communications and semantic communications; Resource allocation; Distributed and federated learning and communications; Standardization, policy, and regulation; Selected topics. (Available Online: https://www.comsoc.org/publications/best-readings/machine-learning-communications)

IEEE Communications Society, *Best Readings in Machine Learning in Communications: Special Issues*.

Machine learning for cognition in radio communications and radar, *IEEE Journal of Selected Topics in Signal Processing*, vol. 12, no. 1, pp. 3–247, Feb. 2018.

Robust subspace learning and tracking: Theory, algorithms, and applications, *IEEE Journal of Selected Topics in Signal Processing*, vol. 12, no. 6, Dec. 2018.

Machine learning and data analytics for optical communications and networking, *IEEE/OSA Journal of Optical Communications and Networking*, vol. 10, no. 10, Oct. 2018.

Artificial intelligence and machine learning for networking and communications, *IEEE Journal of Selected Areas in Communications*, vol. 37, no. 6, Jun. 2019.

Machine learning in wireless communication Part I, *IEEE Journal of Selected Areas in Communications*, vol. 37, no. 10, Oct. 2019; Part II, *IEEE Journal of Selected Areas in Communications*, vol. 37, no. 11, Nov. 2019.

Leverage machine learning in SDN/NFV-based networks, *IEEE Journal of Selected Areas in Communications*, vol. 38, no. 2, Feb. 2020.

Advances in artificial intelligence and machine learning for networking, *IEEE Journal of Selected Areas in Communications*, vol. 38, no. 10, Oct. 2020.

Artificial intelligence for cognitive wireless communications, *IEEE Wireless Communications*, vol. 26, no. 3, Jun. 2019.

Intelligent radio: When artificial intelligence meets the radio networks, *IEEE Wireless Communications*, vol. 27, no. 1, Feb. 2020.

Artificial-intelligence-driven fog radio access networks: Recent advances and future trends, *IEEE Wireless Communications*, vol. 27, no. 2, Apr. 2020.

Edge intelligence for beyond 5G networks, *IEEE Wireless Communications*, vol. 28, no. 2, Apr. 2021.

Machine learning in communications and networks, *IEEE Journal of Selected Areas in Communications Series*, vol. 39, Jan./Jul./Aug. 2021.

ITU Journal Future and Evolving Technologies, Special Issue, *Wireless Communication Systems in Beyond 5G Era*, vol. 2, no. 6, 2021.

ITU Journal Future and Evolving Technologies, Special Issue, *Terahertz Communications*, vol. 2, no. 7, 2021.

China Communications, 6G towards 2030: From key technology to network architecture, AI-empowered future communication networks, theories & security. *Journal Emerging Technologies & Applications*, vol. 19, no. 3, Mar. 2022.

6G Flagship has facilitated the work of 12 expert groups on selected 6G themes with the goal of writing 6G White Papers in a collaborative manner. From the participants who had voluntarily signed up for the joint 6G vision building through an open call, 250 experts from 100 organizations in over 30 countries significantly contributed to the 6G white papers. (Available Online: https://www.6gflagship.com/white-papers/)

Key Drivers and Research Challenges for 6G Ubiquitous Wireless Intelligence, 2019.

White Paper on 6G Drivers and the UN SDGs, 2020.

White Paper on Business of 6G, 2020.

6G White Paper on Validation and Trials for Verticals towards 2030's, 2020.

6G White Paper on Connectivity for Remote Areas, 2020.

White Paper on 6G Networking, 2020.

White Paper on Machine Learning in 6G Wireless Communication Networks, 2020.

6G White Paper on Edge Intelligence, 2020.

6G White Paper: Research Challenges for Trust, Security and Privacy, 2020.

White Paper on Broadband Connectivity in 6G, 2020.

White Paper on Critical and Massive Machine Type Communication towards 6G, 2020.

6G White Paper on Localization and Sensing, 2020.

White Paper on RF Enabling 6G – Opportunities and Challenges from Technology to Spectrum, 2020.

2 5G Advanced Mobile Broadband
New Multimedia Delivery Platform

Dragorad A. Milovanovic and Zoran S. Bojkovic
University of Belgrade

CONTENTS

2.1	Introduction	84
2.2	Extensions of 5G Architecture for Common Media Delivery Platform	86
	2.2.1 Enhanced Mobile Broadband eMBB and Media Streaming 5G MS	88
	2.2.1.1 Downlink Data Transfer and Control Operation	90
	2.2.1.2 Uplink Data Transfer and Control Operation	92
	2.2.1.3 Media Streaming Principles and Architecture	93
	2.2.2 Evolution of Multimedia Multicast and Broadcast Services	97
	2.2.2.1 5G Broadcast Over Dedicated Terrestrial Network	99
	2.2.2.2 5G MBS Multicast and Broadcast Services	102
	2.2.3 Multimedia Production and Distribution Ecosystem	106
2.3	Immersive Communication	108
	2.3.1 Immersive Media Over 5G	111
	2.3.1.1 Architectural Enhancements for Immersive Media Support	113
	2.3.1.2 Levels of Immersion and Technological Complexity	115
	2.3.2 Extremely Interactive Communication and Low Latency	118
	2.3.2.1 Immersive XR Communication	118
	2.3.2.2 Holographic Communication	121
	2.3.3 AI-Based Multimedia Tools	123
	2.3.3.1 DNNVC Video Codec	126
2.4	Concluding Remarks	129
Bibliography		129

DOI: 10.1201/9781003205494-3

2.1 INTRODUCTION

The development of 5G network technology enables ultrahigh speeds and low-latency communication such as no previous generation of mobile networks. On the other hand, the delivery of new media formats to various devices including smartphones, tablets, smart TVs, or connected cars is increasingly important. Digital media is expected to represent 70%–80% of mobile data traffic by 2025. The ever-growing amount of video content available and the demand for higher resolutions pushing the limits of current networks. Immersive video communication also places unprecedented demands on network bandwidth. 3D volumetric formats Multiview/360-video/Point clouds and XR (extended reality) require much higher throughput than standard video streams as well as low latency and extreme interactive communication. 5G wireless technology solutions support new vertical industry services as well as business opportunities in the global media ecosystem along the entire value chain, from media production, distribution, and media consumption. 5G enables the delivery of media with enhanced broadband mobile access (eMBB), 5G media streaming, 5G broadcast, and 5G MBS multicast/broadcast services.

The basic requirements of 5G mobile networks are to support massive capacity and connectivity, diverse set of services, applications, and users with extremely different requirements, as well as flexible and efficient use of the available spectrum supporting very different network deployment scenarios. 5G presents an opportunity to achieve improved throughput and reduced latency through the use of newly specified radio and core technologies, improved flexibility and mobility, and increased reliability thanks to security and integrity features.

3GPP is the global organization for standardization of mobile technologies, developing new media distribution solutions based on the high potential of improved mobile broadband (eMBB) connectivity, increased data speed and reduced latency, as well as new service-based network architecture. The goals are to support the latest media formats that enable improved quality of service (QoS) for traditional services, as well as new immersive formats for augmented reality XR. There are three ways to deliver multimedia content.

- Unicast transports media from the content server to the end user's device on a dedicated network connection. There are as many two-way connections as there are end-user devices. Unicast effectively covers all services that require a two-way connection. Unicast is also effective in serving users spread across multiple radio cells, consuming different content at different times, such as streaming video on demand.
- Broadcasting transports media from content server to an end-user device via single unidirectional link shared by several users in single radio cell. Broadcasting more efficiently covers all services for which multiple users located in one area consume the same content at the same time. Broadcasting is efficient for quickly forwarding the same content to multiple devices at the same time without user interaction.

5G Advanced Mobile Broadband

- Multicast further enables activation of areas where certain number of end-user devices have previously joined the service or charging users based on actually received services.

3GPP standardizes the specifications of various technical solutions for media distribution. The most significant are terrestrial 5G Broadcast, multicast, and broadcast 5G MBS services as tool for network optimization, and 5G MS architecture for streaming media that enables different cooperation scenarios between different stakeholders. 5G technologies support the distribution of media services as a combination of linear (public broadcasting services) and nonlinear (on demand, podcast) components with a high degree of control and guaranteed end-to-end QoS.

- 5G media streaming (5G MS) architecture is a state-of-the-art solution that enables cooperation of different CDN providers, broadcasters, and mobile network operators (MNOs). The 5G MS exposes network and device functions to third-party providers, enabling the best use of 5G capabilities to provide increased QoS levels for connected users. 3GPP is developing new architecture from 5G standard R16. It introduces the concept of reliable media functions, which are implemented on both the network and the user device, and also defines an API interface with external media servers and functions. Functions that are normally implemented outside the network domain can be integrated into the 5G MS. It is now possible to allocate ABR encoders, streaming manifest generators, segment packagers, CDN servers and caches, DRM servers, and content servers for advertisement replacement, manifest modification servers, or even metrics servers within a 5G network to improve service delivery.
- 5G broadcast (LTE-based 5G terrestrial broadcast) technical specifications meet the requirements for DTV and digital radio broadcasting. The system gives service providers control over linear content delivery, allows radio carriers to be configured with nearly 100% capacity for broadcast services, and supports large-area SFN networks with topologies outside of cellular networks. To receive broadcast content, UL link and online registration are not required, which eliminates the need for SIM card and effectively enables free-to-air reception. 5G broadcast can be fully integrated into any 3GPP devices, with the same chipset architecture, and even be complemented by mobile broadband data.
- 5G multicast and broadcast services (MBS) support the on-demand distribution mechanism of multimedia content, thus ensuring sustainable quality of experience for large audience. 3GPP introduced multiple capabilities for 5G system architecture in R17, initially targeting architectures that meet requirements related to IoT, public safety, V2X, or IPTV, among others.

3GPP initiated the transition of media verticals to common distribution platform. The mobile industry, with its rapid cycles of innovation and replacement, opens up new opportunities for new services. The fundamentals of 5G media are outlined in R16, June 2020. A fundamental improvement has been developed in the 5GMS architecture, protocols, and codecs for streaming media. There are several improvements under consideration for R17, the release scope was approved in December 2019, and the system design was completed by March 2022. Development plans for the next 3GPP standard specifications R18 and R19 have been agreed and include innovations in XR and media services, computer vision, ML/AI-based media, and network automation. The specification and standardization of new technologies enable global application in order to achieve economies of scale and offer the technologies to markets at competitive prices. In this context, 3GPP and MPEG standards are relevant to enable interoperable multivendor XR services, allowing different market participants to develop applications and services. The standardization of the next generation (6G) is not limited to the communication part, but also to the deep integration of communications, intelligence, and computing. 3GPP may start general study on 6G systems in late 2025 (release 20), while research on technical specifications will begin in late 2027.

The main sections of this chapter are an introduction to 5G media platform and immersive media over 5G. In the first part, key characteristics of eMBB, media streaming 5G MS and multimedia broadcast/multicast services (5G MBS), as well as multimedia production and distribution ecosystem are presented. In the second part, immersive media over 5G, extremely interactive and low-latency 6G services as well as AI video are outlined. We conclude with remarks on common 5G media platform.

2.2 EXTENSIONS OF 5G ARCHITECTURE FOR COMMON MEDIA DELIVERY PLATFORM

Mobile communication and networks have advanced rapidly over the past four decades, adopted innovative technologies, and developed efficient system architectures in terms of system capacity. However, it is not an optimal solution in terms of other system parameters (latency, energy efficiency, connection density). 5G system supports enhanced mobile broadband communication (eMBB), ultrareliable low-latency communication (URLLC), and mass machine-type communication (mMTC), whose main performance indicators are system capacity, delay, reliability, and connection density. The 5GS system also introduces the concept of network slicing (NS), enabling the creation of dedicated and isolated network infrastructures that are adapted to the needs of services. A NS is a logical network infrastructure identified by network selection assistance information (S-NSSAI), consisting of a slice/service type (SST) and slice differentiator (SD). The 5G system defines a set of standardized SSTs, one for each service vertical. Media distribution and streaming services can use the slice eMBB.

eMBB services, such as the radio access network (RAN), core network (CN), and user equipment (UE), have been the focus of 3GPP R15 5GS technical specification. The categorization of RAN operating bands allows separate set of requirements for each frequency band: FR1 from 450 MHz to 6 GHz, and FR2 from 24.25 to 52.60 GHz (Figure 2.1). The base station gNB channel bandwidth supports a single RF-carrier at

5G Advanced Mobile Broadband

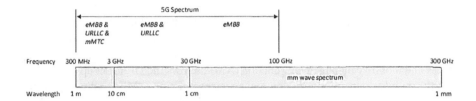

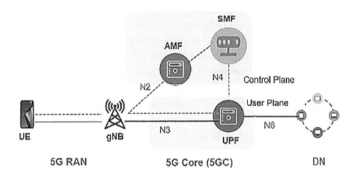

FIGURE 2.1 5G spectrum (FR1, FR2) and 5GS system architecture (SBA).

the base station. The channel raster defines the spacing between the allowed center frequencies of a channel. Transmissions sharing the same antenna port experience the same propagation channel. The waveform OFDM is the baseband signal, which is mixed with RF before being radiated across the air interface. Beamforming requires an antenna array with multiple antenna elements, which improves the uplink (UL) and downlink (DL) budgets by increasing the antenna gain. 5G radio-frame has a fixed duration of 10 ms. The 5G subframe has a fixed duration of 1 ms, so there are always 10 subframes within each radio-frame.

The 5G architecture is based on modular design and separation of control plane (CP) and user plane (UP) to enable scalability and flexible implementation, which is accelerated by the application of virtualization of network functions (VNFs) and software-defined networking (SDN). In addition, the interaction of network functions in the 5G architecture is based on the SBA service model, and the interaction of 5G NF network functions and external third parties is enabled through the concept of capability exposure. The 5G system architecture is guided by the following design principles:

- separation of user plane (responsible for transferring application data between the end user and the application server) from control plane (responsible for transferring signaling messages) functionality;
- service-based architecture (SBA), where network functions (NFs; UPF, AMF, SMF, ...) offer services to other network functions and consumers;
- support for stateless network functions, optimized for speed and large load;
- scalability through virtualization and distribution, allowing for multiple instances of each network function to be created.

The data plane UP includes the user equipment UE, the radio access network RAN, the functions of the user plane UPF; essentially a network switch or router) and the data network (DN; Internet). The control plane CP consists of several network functions necessary to support the UE mobile connection. The most relevant are the access and mobility management function (AMF; allows the UE to set up and terminate the mobile connection), session management function (SMF; configures traffic routing for a specific UE session and chooses the UPF that the session will use), the network exposure function (NEF; exposes some network capabilities to external parties), and the application function (AF; affects traffic routing and policy management for specific application). The session data protocol unit (PDU) supports the end-to-end user plane connection between the UE and a specific DN via the UPF. A PDU session supports one or more QoS streams.

The 5GC core network applies quality of service rules to QoS flows. As part of a PDU session, the stream is identified by a unique QFI in the 5G system. All user plane traffic within a PDU session with the same QFI receives the same treatment, i.e., traffic forwarding, scheduling, and admission control. All flows are controlled by the session management function (SMF). The QoS flow can be preconfigured, established during the PDU session establishment procedure or by the PDU session modification procedure. A flow is assigned a QoS profile with a 5QI flow identifier that specifies a set of static characteristics, including GBR or non-GBR label, priority level, delay budget, packet error rate and averaging window, and maximum data burst size.

Significant progress has been made in both standardization and commercial application since 2016 with the start of 3GPP work on the new 5G NR radio. The first version of the 5G standard in R15 2018 built the foundation for NR by considering different service requirements, a wide range of spectrum (from hundreds of MHz to tens of GHz) and different application scenarios (indoor/outdoor, macro/small cells). 5G NR is standardized to extend service in the vertical domain in R16 2020 and R17 2022, adopting or enhancing new use cases. The evolution of 5G NR is now entering its second phase with 5G-Advanced 2024. There is tremendous interest in continuing to improve the 5G NR platform in R18 and beyond based on the increasing need for eMBB evolution and vertical domain expansion.

2.2.1 ENHANCED MOBILE BROADBAND eMBB AND MEDIA STREAMING 5G MS

The eMBB use case is aimed at the user accessing multimedia content, services, and data. The usage scenario is expanding into new application areas in addition to existing ones for improved performance and an increasingly seamless user experience. Use cases can be classified as broadband access in dense areas (urban centers, stadiums, and malls), uninterrupted access in suburban and rural areas, and high-speed mobility (high-speed trains and airplanes). Each use case imposes different requirements. For example, broadband access in dense areas requires higher connection density and traffic density. The basic requirement is high throughput. Design approaches of eMBB are:

5G Advanced Mobile Broadband   89

- The bandwidth represents the capacity of a medium for data delivery. However, it is a theoretical rate for data transmission over a medium. The throughput is of more practical rate because it considers noise, error, interference, and so on.
- The network throughput (or area throughput) can be expressed as a network throughput (bit/s/km^2), cell density (cell/km^2), and spectral efficiency (bit/s/Hz/cell).
- A high network throughput can be obtained by allocating more available bandwidth, increasing cell density, or improving spectral efficiency.
- 5G techniques for improving throughput are summarized as small cell, heterogeneous networks, mmWave, MIMO, flexible TDD, full duplex, multi-carrier techniques, LDPC error correcting channel code, D2D link, carrier aggregation, and so on.

eMBB support has been continuously improved since R15. Major parts of the eMBB functions introduced in R15, R16, and R17 are for operation in the download (DL) direction. On the contrary, improvements in the upload (UL) direction are now emphasized in R18. Of course, DL operations will also be improved to meet the demand for the evolution of NR applications.

MIMO evolution. Enhancements for R18 UL MIMO have been identified and specified especially for non-smartphone devices such as vehicles, and industrial devices. The use of 8 antennas is supported for transmission and support of 4 or more streams per UE and simultaneous transmission from multiple panels (relevant for FR2 band).

UL coverage enhancements. UL direction coverage is recognized as one of the essential elements for the advancement of 5G NR, since UL performance can be a bottleneck in real applications. In addition, demand for UL intensive applications (video uploading, cloud storage) is constantly increasing. While R17 introduced coverage improvements for UL data and control channels, R18 further improves the coverage of the physical random access channel (PRACH), especially in the FR2 band, including the use of the same beam for the 4-step RACH procedure. Finally, the use of available UL transmission power can be improved, e.g., by increasing the UE transmission power limit for carrier aggregation and dual connectivity as long as it complies with the relevant regulations.

Dynamic spectrum sharing enhancements. Dynamic spectrum sharing (DSS) between LTE and NR in the same frequency range plays a significant role in accelerating the application of NR in existing LTE networks. R18 introduces the use of rate matching pattern to avoid interference caused by the transmission of a common reference signal (CRS) of neighboring LTE cells. In addition, the support of physical DL control channel (PDCCH) in symbols with LTE CRS can be introduced to increase the capacity of NR PDCCH.

Carrier aggregation enhancements. Working with multiple mobile operators through aggregation (operator aggregation) takes significant place in increasing data transfer speed and improving overall system performance. Different frequency bands are expected to become available for the application of 5G-Advanced systems by re-farming spectrum from previous generation systems. It is necessary to further improve carrier aggregation for efficient use of available spectrum blocks. In R18, the reduction of control signaling costs is achieved by a single control grant that schedules multiple data transmissions on different carriers. Moreover, for UEs with dual-antenna simultaneous transmission capability, support for dynamically selecting two transmission bands among 3 or 4 configured bands is introduced to improve UL capacity and spectrum utilization via faster adaptation by considering, e.g., data traffic, bandwidths, and channel conditions of each band.

Mobility enhancements. R18 seeks to reduce latency, signaling overhead, and downtime due to higher layer procedures in the current mobility mechanism. For a serving cell change, L1 and L2 protocol-based intercell mobility is specified, which includes configuration and maintenance for multiple candidate cells, dynamic switching between serving candidate cells based on L1/L2 signaling, and L1-based BM intercell beam management. To change the secondary cell group in NR dual connectivity, selective activation of cell groups is introduced to avoid unnecessary reconfiguration of the higher layer.

Topology enhancements. Integrated access and backhaul IAB and radio-frequency (RF) repeaters are introduced in the R16 and R17, respectively, to extend coverage in a cost-effective manner. As an improvement over conventional RF repeaters, work will be done on network-controlled repeaters with the ability to receive and process lateral control information from the network. With lateral control information, the network-controlled repeater performs an amplify-and-forward operation with better spatial directivity through beamforming, thereby reducing unnecessary noise amplification. Examples of possible side control information include beamforming information, UL-DL time division duplex (TDD) configuration, and on–off control information.

The 5G NR system framework facilitates its evolution including R16 and R17 where additional services and use cases are introduced or improved. R16 extended the 5G standard to new verticals such as 5G broadcast (as an evolution of enhanced DTV based on LTE EnTV), NR sidelink (focusing vehicles on all V2X communication and public safety), non-public networks (NPN), and industrial Internet of things (IIoT) applications. Further evolution in R17 includes multicast and broadcast services (MBS), support for satellite communications (non-terrestrial networks NTN) for smartphones and IoT devices, introduction of UE with reduced capabilities (RedCap) for new types of devices (wearables, surveillance cameras, industrial sensors), and extension FR2-2 frequency range. R18 study provides further vertical domain improvements in NTN, IoT, and RedCap evolution, as well as XR communication.

2.2.1.1 Downlink Data Transfer and Control Operation

In the case of application data for DL connection, the shared channel DL-SCH is selected as the transport channel. The protocol data unit of the MAC layer PDU then becomes a transport block that will be processed by the physical layer before transmission over-the-air interface. DL-SCH channel also carries other pieces of information, such as the different types of system information blocks (SIB) downlink shared channel chain, which includes LDPC coding, the physical downlink shared channel chain, layer mapping, how resource elements are allocated for physical DL shared channel (PDSCH) transmission, and the different types of PDSCH mapping to physical resources. A special PDSCH mapping type is used for mini-slots or partially allocated slots, a feature that allows for reduced latency in 5G NR transmission.

The 5G NR physical-layer model captures those characteristics of the 5G-NR physical layer that are relevant from the point of view of higher layers. The physical-layer model for DL shared channel SCH transmission is described based on the corresponding PDSCH physical-layer-processing (Figure 2.2). Processing steps that are relevant for the physical-layer model, e.g., in the sense that they are configurable by higher layers, are highlighted: higher layer data passed to/from the physical layer; CRC and transport-block-error indication; FEC and rate matching; data modulation; mapping to physical resource; multi-antenna processing; support of L1 control, and HARQ-related signaling.

Downlink control information (DCI) contains the scheduling information for the UL or DL data channels and other control information for one UE or a group of UEs. DCI is carried by PDCCH. Control information is based on different types of messages, including downlink assignment and uplink grant, and how they are encoded and modulated and then mapped to the 5G NR slot via the PDCCH. In the process, the concepts of resource element groups and control channel elements, the basic units to map control to the OFDM grid, are important. The procedure for generating a PDCCH is illustrated in Figure 2.3.

User equipment needs to decode DCI before it can decode or transmit data. DCI are carried in PDCCH and used to schedule user data (PDSCH or PUSCH) indicating modulation and coding scheme, HARQ-related aspects, antenna ports and number of layers, and channel state information (Figure 2.4).

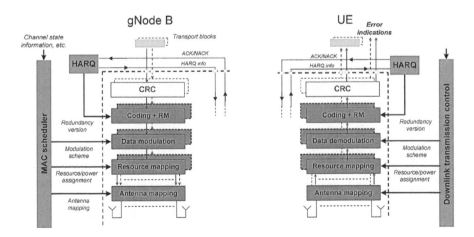

FIGURE 2.2 Downlink DL data transmission in 5G NR: Physical-layer model for DL shared channel (SCH) transmission based on the corresponding PDSCH physical-layer-processing chain.

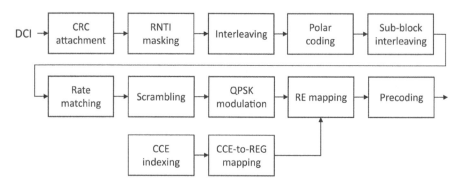

FIGURE 2.3 Procedure for generating physical downlink channel (PDCCH) from control information DCI.

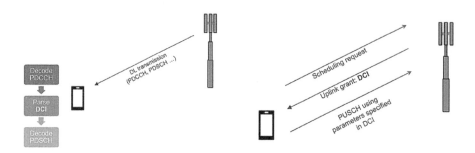

FIGURE 2.4 Downlink control information (DCI): (a) PDSCH channel scheduling and (b) PUSCH channel scheduling.

The physical-layer model for broadcast channel (BCH) transmission is characterized by a fixed predefined transport format. The BCH is used in the downlink only for transmitting the broadcast control channel (BCCH) system information and specifically the master information block (MIB). In order that the data can be utilized, it has a specific format. There is one transport block for the BCH every 80 ms.

2.2.1.2 Uplink Data Transfer and Control Operation

NR data channel coding is common between uplink and downlink. Similar to DL control design, efficient UL control design is critical for wireless multimedia communication as well. Generally, UL control is motivated by either DL transmission, or UL transmission, or a combination of both. Uplink control should provide sufficient information for DL and UL resource management. At the same time, the performances of UL control, including both link-level performance and system-level performance, should be satisfactorily enhanced by mobile broadband. The physical-layer model for uplink shared channel transmission is described based on the corresponding physical-layer-processing chain PUSCH (Figure 2.5).

5G Advanced Mobile Broadband

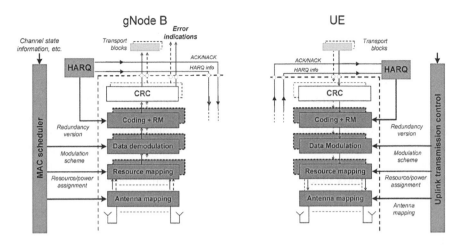

FIGURE 2.5 Uplink (UL) data transmission in 5G NR: Physical-layer model for UL shared channel (SCH) transmission based on the corresponding PUSCH physical-layer-processing chain.

FIGURE 2.6 Uplink UL control information (UCI): (a) UE request for UL data transmission, (b) channel quality measurements request, and (c) data transmission ACK/NACK acknowledgment.

5G NR uplink control information (UCI) is based on different types of messages, including hybrid automatic repeat request (HARQ), channel quality information (CQI), and scheduling request (SR) (Figure 2.6).

2.2.1.3 Media Streaming Principles and Architecture

3GPP defines streaming in its specifications as the delivery of time-continuous media as the dominant media. Streaming refers to the fact that media is mostly sent in one direction only and consumed immediately after reception. Media content can be streamed as it is produced, which is called live streaming. The 3GPP architecture specification refers to two main scenarios: DL streaming (the network is the media source and the UE acts as a consumption device) and UL streaming (the UE is the media source and the network acts as the consumption entity).

5G MS media streaming supported services include mobile network operator MNO and third-party downlink media streaming services, and MNO and third-party uplink media streaming services. The 5G MS architecture is functionally divided into independent components enabling different deployment and collaboration scenarios with various degrees of integration between 5G MNOs and content providers.

5G MS architecture has replaced packet switched streaming (PSS) architecture, specified in TS 26.233 and TS 26.234 for 3G/4G and was tailored for MNO-managed streaming services. 5G MS offers simpler and more modular design enabling services with different degrees of cooperation between third-party content and service providers, broadcasters, and MNOs. The focus of 5G MS is leveraging the concept of network exposure via APIs, in order to provide external service providers an easy way to interact with the 5G network and device functionalities and use the capabilities offered by 5G to deliver superior media services. The initial version of new architecture supports unicast downlink media distribution and uplink streaming, broadcast and multicast are currently being integrated.

The approach taken for 5G MS is specified in general description and architecture document TS 26.501 where framework is aligned with modern over-the-top media distribution practices (Figure 2.7). The application provider uses 5G MS for streaming services. It provides a 5G MS-aware application on the UE to make use of client and network functions using interfaces and APIs. The main 5G MS functions specified within the 5GS are application function (AF), application server (AS) dedicated to media streaming, and UE internal function client dedicated to media streaming.

The reference architecture for 5G media streaming defines the following functions to support the abovementioned features:

- 5G MS AF deployed in the 5G core or in an external DN that manages a 5G MS system. This logical function embodies the CP aspects of the system, such as provisioning, configuration and reporting.

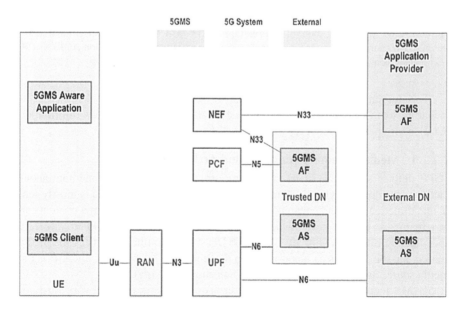

FIGURE 2.7 5G MS overall high-level architecture within the 5GS system.

- 5G MS AS deployed in the 5G core or in an external DN that provides media streaming services to clients. This logical function embodies the data plane aspects of the system that deal with media content.
- 5G MS client deployed in the UE that consumes media streaming services. The 3GPP specifications are silent on whether this logical function is realized as shared UE middleware components or provided piecemeal by individual applications.

The 5G media streaming services offered by the 5G MS system are provided by the application provider for use by the application running on the user equipment UE. The reference architecture and basic functional procedures are defined in 3GPP TS 26.501, and detailed protocols are specified in 3GPP TS 26.512. The basic video codecs and packaging standards that compliant UEs must support as a minimum are specified in 3GPP TS 26.511.

3GPP document TS26.512 specifies requirements related to UE, media application servers, application AF functions, and content provider capabilities related to protocols such as content input and distribution interface, uplink streaming, use of NS slicing, QoS setup, network media processing, quality metrics collection and reporting, network support, consumption reporting, and so forth.

3GPP document TS 26.511 specifies requirements related to UE, AS(s), and content provider capabilities related to encoding, encapsulation, and packetization of media content. The codec and format recommendations defined for each profile apply to the 5G MS client components in the UE as well as the media AS. Profiles are defined to address specific service scenarios. 5G protocols and media streaming formats are based on the common media application format (CMAF) standard media application format.

ISO/IEC 23000-19 packing standard specifies the use of segment formats that are based on the CMAF. By using this format, 5G media streaming is compatible with a broad set of segment-based streaming protocols including dynamic streaming over HTTP DASH and HTTP live streaming (HLS). For example, ISO/IEC 23009-1 defines a detailed DASH profile for delivering CMAF content within a DASH media presentation using a converged format for segmented media content.

5GMS profiles are associated with a set of codec capability requirements in specific service scenarios. Service operation points identify long-lived profiles that will be used by streaming sessions. A default profile is defined for minimum media requirements to be supported in case no other profile is claimed to be supported. The television TV Profile in TS 26.116 covers live and on-demand streaming of audiovisual TV services. The virtual reality (VR) Profile in TS 26.118 covers the live and on-demand streaming of omnidirectional media including spherical video and 3D audio. The codec and format recommendations defined for each profile apply to the client components in the UE as well as the media AS are specified in TS 26.511. Document TR 26.955 presents relevant interoperability requirements, performance characteristics, and implementation constraints of video codecs in 5G services, and to characterize existing 3GPP video codecs, in particular H.264/AVC, H.265/HEVC, and H.266/VVC in order to have a benchmark for the addition of potential future video codecs.

3GPP has identified a set of core research topics for potential 5G MS extensions, building on existing architecture and design principles. The goal is to simplify and optimize the deployment of media streaming services and applications by offering providers access to 5G system functions. Potential extensions can be classified into the following categories:

- extensions that leverage existing and new 5G system features, e.g., edge computing, for media streaming,
- extensions to 5G media streaming protocols and procedures,
- extensions to improve accuracy and for analytics,
- extensions to integrate LTE-based 5G broadcast to 5G MS.

3GPP R16 specification range is limited to single-sided streaming media only. A content hosting capability similar to CDNs for content delivery is defined. The following high-level features are listed for streaming media in R16. Each feature is optional and only available for the 5G MS application if explicitly provisioned by the application provider:

- Content hosting. This may be deployed inside the 5G CN in the form of an operator CDN. Alternatively, an external third-party CDN may be integrated into the 5G media streaming system.
- Media consumption reporting. A random subset of clients can be configured to periodically report media session usage information to the system.
- Quality of experience (QoE) metrics reporting. A random subset of clients can be configured to periodically report QoE metrics to the system. These may be relayed to the application provider.
- Dynamic network QoS policies. Specific network QoS policies are provisioned in advance, expressed as policy templates. During streaming sessions, these templates can then be instantiated on demand by individual clients. The AF negotiates with the policy and charging function (PCF) in the 5G core to apply the requested QoS policy to the relevant packet flow. Policy templates represent long-term agreements made between the application provider and the MNO.
- Network assistance. Two forms of assistance are currently defined. Neither requires any special configuration at the provisioning stage.

3GPP R17 Technical Report TR 26.804 contains the following potential improvements and extensions: 5G multicast and broadcast support, cloud and edge media processing, content preparation, traffic identification, additional/new transport protocols, uplink media streaming, background traffic, content-aware streaming, network event usage, per-application-authorization, support for encrypted and high-value content, and scalable distribution of unicast live services.

2.2.2 Evolution of Multimedia Multicast and Broadcast Services

The 5G system is designed to support unicast communication in which the network sends/receives data to/from individual UEs. However, there are cases where identical content is delivered to multiple devices, and in those cases, improving system efficiency is necessary. Broadcast and multicast communications enable resource-efficient transmission to multiple end users requesting to receive the same content. Multimedia service is a set of audiovisual media assets offered to the consumer by the service provider and delivered to the consumer by a delivery chain comprising one (or more) network operators.

- Broadcast communication service. A communication service in which the same service and the same specific content data are provided simultaneously to all UEs in a geographical area (all UEs in the broadcast coverage area are authorized to receive the data).
- Multicast communication service. A communication service in which the same service and the same specific content data are provided simultaneously to a dedicated set of UEs (not all UEs in the multicast coverage are authorized to receive the data).

3GPP specified an initial set of standardized functions to support multiple eMBMS transmissions in R9–R12 LTE versions. Enhanced multimedia broadcast and multicast services are based on SFN network, and it utilizes synchronized multicell transmissions from many eNBs together providing for over-the-air combining of multicast broadcast SFN (MBSFN) signals, enhancing the reliability and coverage areas for services. MBSFN transmissions are time-interleaved with unicast transmission with pre-assigned and dedicated subframes over the radio frame and utilize full system bandwidth.

To overcome eMBMS limitations, R13 single-cell point-to-multipoint SC-PTM is specified. A more flexible approach of dynamic use of time and frequency resources (and within subframes) is adopted for broadcasting services in a small dense geographic region (hotspot) within the coverage of a single cell. Dynamic resource utilization effectively supports the integration of broadcast service delivery with unicast physical channels.

3GPP Release R14 and R15 specify broadcast services with a dedicated carrier for MBMS transmission that supports up to 100% resource utilization and longer duration OFDM symbol that supports wider coverage in order of tens of kilometers. R14–R16 specifies further enhanced multicast FeMBMS multimedia service as a new broadcast/multicast enhancement for dedicated and mixed modes. R16-introduced 5G terrestrial broadcast for enhanced television (enTV) services in large and static transmission areas with dedicated broadcast infrastructure (high-power HPHT deployments from high towers).

5G terrestrial broadcast supports two distinct modes:

- 5G stand-alone broadcast. A dedicated broadcast-only network (independent of mobile networks) that can respond to the emerging needs of broadcasters and content providers, giving them access to broader audience through the efficient delivery of content to both fixed and mobile devices. R16 enTV meets all the basic requirements of 5G broadcasting defined in the study on scenarios and requirements for next-generation access technologies TR 38.913. It can be deployed in existing UHF spectrum (470–698 MHz) that broadcasters already own or have access to, and the enTV design allows the reuse of existing cellular modem building components.
- 5G mixed mode multicast. A new 5G capability in R17 mobile operator networks that can support dynamic switching between unicast and broadcast modes, which can improve system capacity and efficiency in local deployment (Figure 2.8 and Table 2.1).

There has been significant debate on the broadcast evolution path for 5G NR. It has been decided that a two-track approach will be taken. Dedicated broadcast enhanced enTV evolution of LTE specified in R16 has already been designed to meet the 5G broadcast requirements. Therefore, it was seen that there is no urgent need to add NR-based version of dedicated broadcast. NR-terrestrial broadcast will not happen before R18. Consequently, based on the study on architectural enhancements for 5G MBS TR 23.757 R17, 3GPP started to build functional support of multicast and broadcast services MBS over an existing 5G standards framework. The standardization has been conducted for overall 5G system architecture from both NG-RAN and 5G CN perspectives. It was seen that there is an urgent need to add mixed mode unicast/multicast operation support in R17. The mixed mode will be tightly integrated with the NR eMBB system and there will be maximum commonality between the two. The mixed mode unicast/multicast operation can address use cases such as C2VX, public safety, IIoT, and IP multicast uses. The mixed mode unicast/multicast system will not support SFN transmission, other than the SFN implemented in transparent manner.

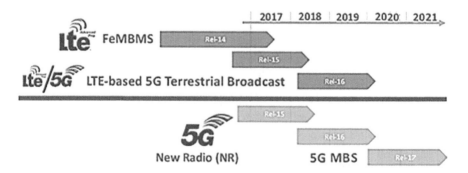

FIGURE 2.8 Two-track evolution approach in R16 and R17 for 5G broadcast and 5G MBS services.

5G Advanced Mobile Broadband

TABLE 2.1
Evolution of 5G Terrestrial Broadcast and Multicast Services

Enhanced MBMS Release 8–12	Focus on mobile network operators: EMBMS basic framework, multiple frequency operation, system enablers for group communications.
SC-PTM single cell point-to-multipoint Release 13	Dynamic time/frequency allocation, group scheduling through unicast channels.
Enhanced TV Release 14	Expansion to terrestrial broadcast by meeting 7/10 requirements: receive-only mode, shared broadcast, use of a larger cyclic prefix for longer intersite distances, improved spectral efficiency for rooftop/car-mounted antennas.
Further enhanced MBMS Release 15	Dedicated carrier for MBMS, large OFDM symbol.
5G terrestrial Broadcast Release 16	enTV enables terrestrial broadcast by meeting all requirements: dedicated broadcast infrastructure (HPHT), static transmission area.
5G MBS multicast and broadcast services Release 17	Requirements: new service scenario and requirements, unique 5G NR/5GC characteristics, wider service area (including legacy network).
	Network architecture: new and enhanced network functions, shared delivery and individual delivery, MBS over nonsupporting node.
	RAN protocol: multicast mode vs. broadcast mode, MRB types including split bearer, ARQ, PTM/PTP RLC.
	Physical layer: MBS CFR for BWP operation, HARQ feedback and retransmissions, group common SPS.
	Service continuity: packet-level PDCP SN synchronization, lossless handover, MBS interest indication.

2.2.2.1 5G Broadcast Over Dedicated Terrestrial Network

With the completion of the 3GPP R16 specifications and the ETSI TS 103720 LTE-based 5G broadcast specification, digital TV delivery with 5G terrestrial broadcast is ready for deployment. The 3GPP specification for 5G broadcasting meets all the basic requirements for DTV delivery. The spectrum is available in the UHF band (470–698 MHz) and can be used for digital TV broadcasting in the EU and other regions. The system is designed for low deployment cost, high efficiency, quick time to market, reuse of broadcast infrastructure (high-power high-tower), and existing 3GPP receiver functions. 5G broadcasting is designed for broadcasters and supports free-to-air services, receive-only mode reception, downlink-only distribution, and delivery in a dedicated broadcast spectrum in order to replicate functionality of existing digital TV services. The IP-based service layer enables the deployment of applications and service layers such as DVB-I, DASH/HLS, and CMAF, on top of 5G broadcast and seamlessly integrates with unicast. And finally, 5G broadcast technology promises continued development in future releases, bringing new services and functionalities, better performance and efficiency, as well as rapid replacement cycles. 5G broadcast is ready for prime-time TV to mobile and stationary rooftop receivers.

LTE-Advanced supports multimedia broadcast and multicast services eMBMS (LTE Broadcast), which enables MNOs to respond to the increasing demand for mobile video data using point-to-multipoint (P2MP) transmission. A broadcast solution supported over the same frequency carrier as unicast services, eMBMS deployment costs are significantly lower than other broadcast alternatives. However, it also requires the use of dense networks compared to terrestrial broadcast networks and reduces system capacity for unicast services, making business model difficult. The main eMBMS advantages are the end-to-end IP architecture that enables the coexistence of unicast and broadcast services with high capacity, high bandwidth, and high scalability. Furthermore, deployment cost is significantly lower than other broadcast alternatives due to its easy integration with LTE infrastructure and mobile chipsets.

5G Broadcast system has evolved from enTV Release 14 (2017) to Release 16 (2020). 3GPP extends eMBMS to address all broadcast requirements in TS 36.976 description of LTE-based 5G broadcast (April 2020). This provided the foundation for system specification published as ETSI JTC Broadcast TS 103720 that profiles and restricts existing 3GPP 5G specifications in order to enable the deployment of linear TV and radio services (December 2020). Several 3GPP specifications have been extended or newly developed over several releases to address the use cases and requirements for 5G dedicated broadcast networks. ETSI TS 103720 summarizes the basic features of a 5G broadcast system for the carriage of linear television and radio services, and documents these as an implementation profile of a subset of 3GPP specifications. Several functions and reference points are defined. Receiver categories are defined that address implementation profiles to deploy linear television and radio services.

Based on requirements, 3GPP specifications have gradually evolved to meet the use cases and requirements in order to support broadcasting of linear television and radio services. With the completion of the R16, a comprehensive set of 3GPP specifications is available that fulfills the use cases and requirements for a 5G broadcast system:

- support of free-to-air (FTA) service.
- broadcast-only service for UEs without an MNO broadcast subscription.
- decoupling of content, user service, and transport functions.
- exposure of broadcast service and transport capabilities to third parties.
- support for client APIs for simplified access to broadcast services.
- support for mobility scenarios including speeds of up to 250 km/h to support receivers in moving vehicles, with external omnidirectional antennas.
- support for receive-only mode (ROM) services and devices.
- support for user service announcement through broadcast.

5G Advanced Mobile Broadband

- support for common streaming distribution formats such as dynamic adaptive streaming over HTTP (DASH), HLS, and CMAF.
- support for IP-based services such as IPTV or ABR multicast.
- support for different file delivery services such as scheduled delivery or file carousels.

The general architecture for a 5G broadcast system is provided in Figure 2.9. The principal actors in the system are as follows:

- 5G broadcast DTV content service provider runs a head-end providing linear television and radio services.
- 5G broadcast DTV service application runs on devices that include a 5G broadcast receiver.
- Operator runs a 5G broadcast system with transmitters for use by devices including receivers.
- 5G broadcast DTV content service provider makes services available using the 5G broadcast system.
- 5G broadcast DTV service application is able to consume the service by communicating with the receiver through a dedicated set of 5G broadcast client APIs.

The 5G broadcast service consists of a bearer service and a user service. The latter provides the announcement of broadcast user services and also provides information about how to discover and access them. The former provides the distribution means for broadcast user services, including a radio bearer.

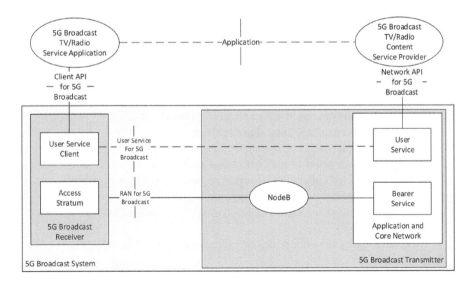

FIGURE 2.9 Reference architecture for 5G broadcast system (transmitter–receiver) for linear DTV services.

The currently implemented broadcast designs are DL only. This is because in normal use cases, different user devices do not have the same content to send to the network. The currently specified broadcast/multicast systems do not use any feedback. As a consequence, there is no opportunity for retransmission and, therefore, the packet losses due to temporal face become a significant performance-limiting factor. While the specification focuses on broadcast-only distribution, a richer application service may be provided to a UE that also supports unicast.

2.2.2.2 5G MBS Multicast and Broadcast Services

3GPP Release 17 (March 2022) has some extensive efforts to add multicast and broadcast capabilities to the 5GS, NR, and 5G MS under the umbrella of multicast broadcast services 5G MBS. Among others, the following aspects are addressed:

- support of multicast services with autonomous RAN-based switching between point-to-point and point-to-multipoint transmission modes
- reuse of PHY channels and signals, without new numerologies
- support broadcast services, always using point-to-multipoint transmission
- only single-carrier point-to-multipoint (SC-PTM) supported in NR RAN, i.e., no support from single frequency networks (SFN)
- delivery and service layer aspects
- APIs on the network and client sides.

The standardization has been conducted for overall 5G system architecture from NR and CN perspectives. Adoption and evolution of multicast and broadcast services in standards posed new requirements and challenges for MBS. New emerging services such as mission-critical delay-sensitive signaling and high-resolution IPTV are required to achieve the same levels of high reliability and low latency as available with unicast services. Therefore, demand exists for an involved protocol stack design with layers and functionalities reinforcing the reliability and latency aspects in RAN protocol, physical layer, and service continuity. In this regard, key features of the standardized 5G MBS are as follows:

- group scheduling mechanism to allow UEs to receive MBS service including simultaneous operation with unicast reception.
- multicast delivery in 5GC, i.e., shared delivery.
- reliability enhancements by dynamic change of multicast/broadcast service delivery between point-to-multipoint and point-to-point, automatic repeat request (ARQ), hybrid automatic repeat request (HARQ), and so forth.
- supporting mobility and lossless handover.
- reception of broadcast data irrespective of UE's radio resource control (RRC) states.
- MBS over legacy network node, for example, Release 15/16 network.

The network architecture of the 5G system has been improved to support MBS, in a way that reuses the existing system as much as possible. Multicast is defined as a service and specific content data provided simultaneously to a dedicated set of UEs

5G Advanced Mobile Broadband

authorized in a service area. Broadcasting is defined as a service and data about specific content that is simultaneously provided to all UEs in the service area. Which cast type is actually used depends on actual service type. To support MBS in 5G system, some network functions are newly introduced as depicted in Figure 2.10.

- Multicast broadcast user plane function MB-UPF is an ingress point to 5GS and works as a session anchor to 5GS.
- Multicast broadcast session management function (MBSMF) manages MBS session and configures a user plane function MB-UPF based on the policy rules for multicast and broadcast services.
- Multicast broadcast service function (MBSF) has service-level functionality to interact with application function/application server (AF/AS) and MB-SMF for MBS session operations. Further, it determines transport parameters and session transport, and control MBSTF if used, which can be implemented in the NEF.

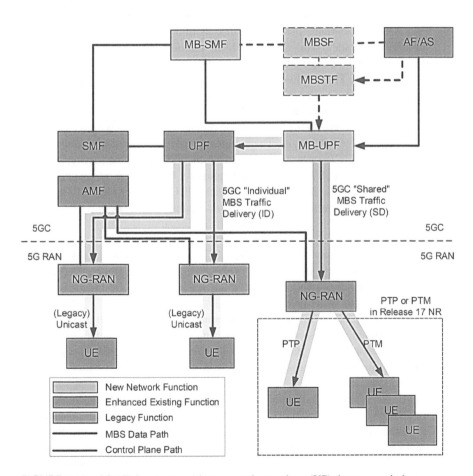

FIGURE 2.10 5G MBS system architecture and user plane (UP) data transmission.

- Multicast broadcast service transport function (MBSTF) has generic packet transport functionalities available to any IP multicast-enabled application such as framing, multiple flows, packet FEC encoding, and therefore, works as a media anchor for MBS data traffic.

NR MBS supports two delivery modes, namely, multicast mode (delivery mode 1) and broadcast mode (delivery mode 2). Each mode has its own characteristics and target services.

- Target services of multicast mode have particular QoS requirements, which the network should guarantee as in case of unicast. Therefore, UEs receiving multicast data are required to stay in CONNECTED state and a dedicated RRC signaling provides the radio resource configuration including MRB configuration, physical-layer configurations, and so on, which can be optimally configured based on the interaction between UE and gNB. Moreover, in case that reliable transmission is required for cell edge user with bad channel quality, the transmission can be switched to PTP leg, i.e., PTP RLC and ARQ can enhance the performance as in unicast transmission. When no multicast data arrival is expected, the multicast session can be deactivated and UEs belonging to the multicast group can transit to INACTIVE or IDLE state. These UEs have to re-enter CONNECTED state when the multicast session is about to be reactivated. In this case, group paging with the corresponding MBS Session ID is used to wake-up these UEs.
- On the contrary, broadcast mode can be provided to all UEs within a coverage regardless of RRC states. The broadcast mode is a similar mechanism to SC-PTM in LTE. In order for UEs out of CONNECTED state to receive the broadcast data, radio resource configuration for broadcast mode is periodically transmitted via MBS control channel (MCCH) from which UEs apply the received configuration for MBS traffic channel (MTCH).

The broadcast mode does not require any interaction between UE and gNB. The network does not have any feedback from UE side on transmission status (ACK/NACK) but it transmits the data only in the best-effort manner. Hence, QoS cannot be guaranteed and only low-QoS services are feasible. Also, the broadcast mode does not mandate RRC state transition.

NR MBS radio protocol architecture is presented in Figure 2.11. It is designed to reuse existing functionalities of unicast, including 4 sublayers, namely, service data adaptation protocol (SDAP), packet data convergence protocol (PDCP), radio link control (RLC), and medium access control (MAC).

- SDAP has a one-to-one mapping with MBS session serving multiple QoS flows (QFs) received from 5GC. SDAP allocates MBS radio bearer (MRB) transmitted for QFs based on required QoS and the network's policy. Since MBS is only for downlink transmission, uplink-related SDAP functions, e.g., reflective QoS and network-initiated QF remapping, are not supported.

5G Advanced Mobile Broadband

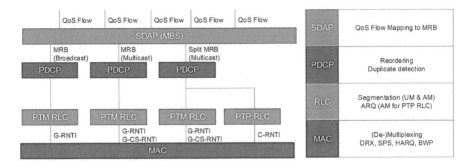

FIGURE 2.11 5G NR MBS protocol stack.

- PDCP is used to support reordering function under an assumption that a UE may receive packets out of order due to ARQ or HARQ retransmissions of the lower layers. A PDCP entity can be linked to multiple RLC entities for point-to-multipoint/point-to-point (PTM/PTP) switching where gNB may send packets via either PTP RLC, PTM RLC, or both. If the gNB sends the same sequenced PDCP packet via both RLCs, duplicate detection function in PDCP discards the later arrived one.
- RLC is responsible for segmentation and ARQ. gNB's segmentation function fragments a large-sized packet into several small-sized subpackets by considering available resource size. UE's RLC entity reassembles the original packet and delivers to upper layer (PDCP). There are two types of RLC entities, namely, PTM RLC and PTP RLC.

5G MBS architecture not only includes new functions on radio and CN layers, but also on user service layer. Similar to the design goals for 5G MS and the lower layers of 5G MBS, the user service definition follows similar principles according to the findings in a feasibility study documented in 3GPP TR 26.802 and the resulting work item on the 5MBS user service architecture as documented in 3GPP TS 26.502 (Figure 2.12).

An MBS application provider may communicate with the service function MBSF to establish user service. The function deals with all internal logic and communicates with the 5GS system to establish QoS and multicast/broadcast delivery. The user service is announced to the UE and the MBSF client discovers the announced user services and sets up the relevant delivery functions in order to receive the data.

It is expected that R17 version of MBS may not be able to address many aspects due to lack of time. Further, MBS feature will continue evolving in future releases of 5G and there are certain important areas and aspects that would be potential candidates:

- multicast service reception and continuity across IDLE/INACTIVE state to cater large user base and critical services
- MBS for localized services and its integration to private networks
- energy efficiency for MBS transmission

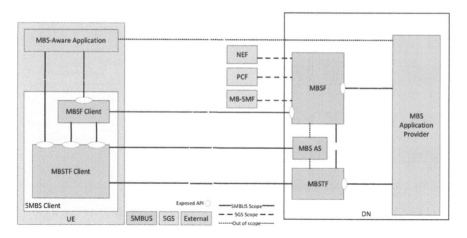

FIGURE 2.12 5G MBS architecture.

- extended service coverage for MBS services
- new service types, including support for enTV and receive-only-mode ROM services that are excluded from R17.

Service requirements for different use cases of MBS diverge in their requirements for reliability, latency, QoS handling, service area coverage, service continuity, and security aspects. Therefore, it becomes imperative to build a comprehensive mechanism for MBS to address these diverse needs' reliability and latency, QoS handling, service area coverage, service continuity, and security.

2.2.3 Multimedia Production and Distribution Ecosystem

5G technology supports the building of mass market ecosystem, which requires collaboration of standards organizations and industry consortia with their own expertise and requirements to develop a healthy ecosystem of unified standards (Table 2.2). 3GPP as the core standard development organization for 5G communication technology is at the center of industry efforts and has evolved to become more open and flexible.

Mobile 5G broadband is becoming an important platform for multimedia distribution. It is important to study technical, operational, and regulatory issues, as well as conduct research into what types of use cases and usage patterns are feasible and beneficial, taking into account current and future requirements and constraints. It starts with a 5G vision and foresight to identify compelling challenges and issues. The implementation of the vision is based on research and innovation activities. Technological innovations are transforming the network into a secure, reliable, and flexible orchestration platform using multiple technologies. The peak of proof-of-concepts (PoCs) and trials targeting verticals occurs with multimedia communication.

TABLE 2.2
A System Approach in Making 5G Technology Ecosystems

Vision	Identify a problem or need, and establish system requirements. It is envisioned in a more efficient way to deliver mass media over cellular networks.
Innovation	Invent new technologies and end-to-end system architecture.
Proof-of-concept	Deliver end-to-end prototypes and impactful demonstrations.
Standardization	Drive ecosystem toward new projects and through standards process.
System trials	Collaborate on field trials that track standards development, preparing for commercialization. It is worked with mobile operators, device manufacturer, and content provider on trials.
Commercialization	Engage with global ecosystems to deploy new products and services.

5G-MAG multimedia action group is a cross-industry organization promoting the commercial adoption of 5G broadcast. The group evaluates the benefits, impacts, and challenges of adopting 5G technologies across the media distribution value chain and positions the media industry as a key vertical for the future development of 3GPP technologies and the 5G ecosystem. The work of the action group is organized around topics proposed by content and service providers, network operators, technology solution providers, equipment manufacturers, R&D organizations, regulators, and policy makers. MAG becomes a 3GPP market partner in 2021. As a market representation partner in 3GPP, the group supports the transition from industry requirements to technical specifications, as well as the transition of standardized technologies for use in new media services.

Digital video broadcasting (DVB) project maintains a set of international open DTV standards published by ETSI, CENELEC, and EBU. The document 5G DVB-I (BlueBook C100, 2021) presents the key elements of 5G networks and systems related to media distribution, including 5G broadcasting, 5G streaming media, and other ongoing activities in 3GPP. A key focus is on creating interfaces and APIs for the distribution of DVB-Internet services over 5G, allowing broadcasters to continue to support the existing service and application layer. DVB-I is service layer that enables the distribution of DTV services over the Internet regardless of access. Broadcasters are now able to expand their traditional broadcast offering with additional event-based services delivered only over the Internet, presenting an integrated DTV offering in a coherent and organized list of services.

EBU 5GCP content production group is a project group of the European Broadcasting Union strategic program distribution. The group studies the technical development within the organizations for the development of 5G standards regarding the production of digital audiovisual content. 5GCP formulates and submits member requirements, monitors and coordinates appropriate support to ensure that these are met. Standardized 5G solutions are expected to reduce costs and increase production flexibility. To this end, solutions of the challenges of standardization, technology availability, regulations, and business models are necessary. The EBU, in cooperation with 3GPP, produced the feasibility study TR 22.827 and the technical specification TS 22.263, which contain the key requirements for professional media production.

2.3 IMMERSIVE COMMUNICATION

The increased bandwidth capacity and reduced latency of 5G communication enable access to complex media experiences over mobile networks. Immersive media and 5G are a combination that supports new services and capabilities within the mobile ecosystem for new verticals such as industrial services, public safety, and automotive. Significant performance such as higher data rates in both directions, ultralow latency and reliability, and greater density of connections.

Immersion is the specific feeling of being surrounded in a virtual space, as well as the feeling of physically located presence. The mechanism of immersion is cognitive or perceptive. 3D interactive media opens up the possibility for users to experience a sense of presence and immersion in an authored environment. While in traditional media, significant steps toward passive consumption of more immersive media include video technology of ultrahigh definition (UHD) and high dynamic range (HDR) for a more realistic presentation of content in terms of contrast and color, full immersion is supported by three key elements:

- visual quality including high fidelity, spherical coordinates and stereoscopic rendering and depth information
- sound quality including high-resolution and 3D audio, positioned such that the directivity of the sound sources can be rendered
- intuitive interactions with the content using natural user interfaces, precise tracking of the motion, and imperceptible latency to avoid lag and motion sickness.

Extended reality (XR) includes all real-and-virtual combined environments and human–machine interactions generated by computer technology and wearable devices. The umbrella term covers different types of reality such as augmented reality (AR), mixed reality (MR), and virtual reality (VR), and the areas within the continuum among them. Levels of virtuality range from partially sensory inputs to full immersion. A key aspect of XR is the extension of human experiences, particularly in relation to feelings of presence (VR) and acquisition of cognition (AR) when a user is provided with additional information or artificially generated items or content overlaid upon their real environment. MR is an advanced form of AR where certain virtual elements are composed into a physical scene creating the illusion that these elements are part of the real scene.

Augmented reality (AR) imposes significant minimum performance requirements on technologies such as tracking, latency, persistence, resolution, and optics. XR covers a wide range of technologies and end devices. The main features are higher data transfer speeds in both directions, ultralow latency and reliability, higher connection density, as well as huge computing resources near the user. Media content resources come in a wide variety of formats and types: 2D/3D, natural or synthetic, compressed or uncompressed, provided by a content provider or recorded locally.

Immersive content is displayed based on spatial tracking and the estimated position/orientation of a XR viewer. Actions and interactions include movements, gestures, and body reactions. In that way, the degrees of freedom (DoF) describes the number of independent parameters used to define the movement of a viewport in the 3D space. The following different types of DoF are described in order of increasing complexity and QoE:

- 3DoF is three rotational and un-limited movements around the 3D (x, y, z) axes (respectively pitch, yaw, and roll). The XR reference space is limited to a single position.
- 3DoF+ is with additional limited translational movements (typically, head movements) along 3D (x, y, z) axes.
- 6DoF is 3DoF with full translational movements along 3D (x, y, z) axes. Beyond the 3DoF experience, (i) it adds moving up and down (elevating/heaving), (ii) moving left and right (strafing/swaying), and (iii) moving forward and backward (walking/surging).
- Constrained 6DoF has constrained translational movements along 3D (x, y, z) axes (typically, a couple of steps walking distance). A typical use case is a user freely walking through VR content displayed on an HMD but within a constrained walking area.

To structure the work on XR and 5G, 3GPP has launched feasibility studies TR 26.918 and TR 26.928 to identify use cases, technologies, and possible gaps that need specifications for interoperable services. Key issues for interoperability expected to be solved include:

- Definition of the spatial environment, the space in which the presentation is valid and can be consumed. Typically, for example what is referred to as windowed 6DoF is the limited amount of possible movements within the 3D space.
- Presentation timeline management. The different resources may have an internal timeline for their presentation.
- Positioning and rendering of the media sources in the 6DoF scene appropriately.
- Interacting with the scene based on sensor and/or user input. The rendered viewport can be dependent on simple aspects such as viewing position or may include complex sensor input or captured.

New media formats are continuously developed, taking into account improvements in capturing and display systems. The evolution of compression technologies in line with high-quality requirements and protocol improvements affects the traffic characteristics of media services on networks. 3GPP report TR 26.925 R17 presents typical requirements for different media services, as well as an overview of their impact on typical network traffic characteristics.

Technical report TR 26.949 collects video distribution formats for 3GPP services. Until R12, 3GPP specifications missing detailed definitions of distribution formats (spatial/temporal resolution, aspect ratios, random access points) for which operators and service providers provide guarantees the QoE. The services' specification only defines video codec profiles and levels. Technical report TR 26.955 R17 analyzes currently specified video compression technologies and their suitability for existing and new services in the context of 5G, and identifies gaps and optimization potentials of new video compression technologies. The relevant interoperability requirements, performance characteristics and limitations of video codec implementation in 5G services are presented:

- collects a summary of the video coding capabilities in 3GPP services;
- collects a subset of relevant scenarios for video codecs in 5G-based services and applications, including video formats (resolution, frame rates, color space, etc.), encoding and decoding requirements, adaptive streaming requirements;
- collects relevant and exemplary test conditions and material for such scenarios, including test sequences;
- defines performance metrics for such scenarios with focus on objective performance metrics;
- collects relevant interoperability functionalities and enabling elements for video codecs in different 5G services supporting the identified scenarios;
- collects relevant criteria and key performance indicators for the integration of video codecs in 5G processing platforms, taking into account factors such as encoding and decoding complexity in the context of the defined scenarios.

Due to the increasing consumption of high-resolution video content, the need for more efficient video compression techniques is also increasing. High compression ratio and reduced complexity are still the basic characteristics of efficient video encoders. The most commonly used encoders are HEVC (High Efficiency Video Coding) and AVC (Advanced Video Coding). The operational parameters of the codec are defined in the TV video profile TS 26.116 and represent the basis of the VR video profile TS 26.118. MPEG/ITU is continuously working on video compression enhancement technologies, creating JVET joint team in October 2015. The development of the new VVC (Versatile Video Coding) codec standard was completed in June 2020. Verification test results show that VVC achieves 40% bitrate reduction versus HEVC for 4K/UHD test sequences using objective metrics for target compression rates (Table 2.3).

Application areas specifically targeted for VVC implementation include 4K/8K ultrahigh-definition video (UHD), high dynamic range (HDR), and wide color gamut (WCG) video, 360° omnidirectional video, as well as conventional standard SD and high definition HD formats. In addition to improving coding efficiency, VVC also provides a highly flexible syntax that supports layered scalable coding.

TABLE 2.3
Video Coding Standards Performance and Bitrate Target

Codec	Objective Coding Performance	Targeted Bitrate	
HEVC	−40% vs. AVC	4K UHD	StatMux: 10–13 Mbps
			CBR: 18–25 Mbps
		8K UHD	CBR: 40–56 Mbps
			High quality: 80–90 Mbps
VVC	−30% vs. HEVC	Expected:	
	CfP best: −42% vs. HEVC	4K UHD CBR: 10–15 Mbps	
	Target: −50% vs. HEVC	8K UHD CBR: 25–35 Mbps	

2.3.1 Immersive Media Over 5G

Immersive media will play a major role in the next period. Each generation of mobile networks is primarily designed for a specific set of target services. 5G and 6G are driving immersive services, supported by advances in sensor technologies. However, networks are required to support the extreme latency sensitivity, and compute and bandwidth intensity of immersive services, while still enabling special offers to other industry verticals (V2Xs). The 5G system is expected to offer solutions to the challenges of immersive services, namely lower latency, higher bandwidth, and ubiquitous connectivity.

Immersive media make intensive use of various technological components on a highly integrated mobile computing platform. The components work together to meet massive computational and real-time demands, both individually and through network interfaces. Media decoding, graphics rendering as well as real-time sensor input processing are covered. Aspects related to communication include:

- high throughput in both directions (uplink and downlink), and equivalently optimized compression technologies
- low latency in the media communication to address the service requirements
- consistency and universally high throughput.

3GPP has been standardizing immersive media since the launch of 5G technical specification activities in 2015. The support of 5G immersive media was investigated in the R14 time frame, with the completion of two studies by working group SA1 on requirements and working group SA4 on codecs and media.

- SA1 R14 feasibility study of new services and enabling market technologies describes a set of diverse use cases of identified market segments and verticals of the 3GPP ecosystem. Immersive media use cases as documented in TR 22.891 media distribution report are also covered. The findings of

the study later headed to the development of 5G Stage 1 requirements during the R15 technical specification service requirements for 5G system TS 22.261.
- 3GPP SA4 study TR 26.918 covers VR services. The technical report documented wide range of streaming, broadcast and conversation cases, relevant audio/video technologies and various subjective quality assessments, which formed the basis of the R15 normative work items.

For the purpose of classifying use cases, the following categories are defined:

- Download. An XR experience is downloaded and consumed offline without requiring a connection. All media and experience-related traffic is downlink.
- Passive streaming. The experience is consumed in real-time from a network server. The user does not interact with the XR experience, or the interaction is not triggering any uplink traffic. All media-related traffic is downlink.
- Interactive streaming. The experience is consumed in real-time from a network server. The user (or the device automatically) interacts with the XR experience and the interaction changes the delivered content. The traffic is predominantly downlink, but certain traffic is uplink (XR viewer pose information). Different interactions exist, for example, viewport adaptation, gaming events, etc. Interaction delay requirements are different, ranging from immersive latency requirements to more static selection interactions.
- Conversational. The experience is generated, shared, and consumed in real-time from two or more participants with conversational latency requirements.
- Split compute/rendering. Network functions run an XR engine to support processing and pre-rendering of immersive scenes and the delivery is split into more than one connection (split rendering, edge computing). The latency and interaction requirements again depend on the use case and the architecture implementation.

The representative use cases considered by 3GPP for XR over 5G are:

- Streaming. The typical media streaming experience is enhanced with the capability of 6DoF within a scene. Motion and interaction are allowed in two possible ways: by changing the viewer's angle within the scene, and by head movements with an HMD. Additionally, the viewer's emotional reactions (facial expressions, eye movements, heartbeats, and biometric data) could be collected by means of body sensors during a watching session and a personalized storyline could be created based on the type of emotions. The stream display may occur over AR glasses on a chosen augmented wall, after the spatial configuration has been analyzed. Synchronized playback and interaction with multiple co-located viewers are possible. This use case relies on volumetric video and 6DoF capture systems. New standardized methods for scene composition and description, social interaction, as well

5G Advanced Mobile Broadband

as new formats for storage and cloud access, content delivery and optimized streaming protocols and formats for biometric, emotion, and spatial metadata would need to be defined.
- Real-time 3D communication. Video chats are captured using 3D models of people's heads, which can be rotated by the receiving party. Multi-party VR conferences support the blended representation of the participants into a single 360-video with a pre-recorded office background. Some of the conference participants may also be overlaid on an AR display. Shared presence using depth cameras is one of the features of this use case. In an instance, virtual meeting space could be created, and the participants' avatars could move and interact with other avatars using 6DoF. Remote participants use an HMD and audio is binaural or spatially rendered.
- Industrial services. One of the use cases covers an AR-guided assistant at a remote location for augmented instructions/collaboration. A remote assistant is guiding a local person to perform maintenance on an industrial machine. The remote assistant can see in real-time the local environment and the machine to be repaired. Part of the repairing instructions is sent as overlays to the AR glasses.
- Other use cases. The list of XR use cases that could benefit from 5G also includes others in the areas of training (possibly the largest market), automotive (engineering, design, marketing/sales), location-based entertainment, digital models, 3D holographic shows, passenger entertainment in self-driving cars, health care, data.

3GPP SA4 technical report TR 26.928 has been investigating the relevance of AR/XR in the context of immersive services addressing aspects such as relevant technologies, media formats, metadata, interfaces and delivery procedures, client and network architectures and APIs, and QoS service parameters and other CN and radio functionalities. The primary scope of the report are following aspects:

- Introducing XR by providing definitions, core technology enablers, a summary of devices and form factors, as well as ongoing related work in 3GPP and elsewhere.
- Collecting and documenting core use cases in the context of XR.
- Identifying relevant client and network architectures, APIs and media processing functions that support XR use cases.
- Analyzing and identifying the media formats (including audio and video), metadata, accessibility features, interfaces and delivery procedures between client and network required to offer such an experience.
- Collecting key performance indicators and QoE metrics for relevant XR services and the applied technology components.

2.3.1.1 Architectural Enhancements for Immersive Media Support

The integration of XR applications in the 5G system is approached according to the 5G MS media streaming model as defined in the 3GPP specification TS 26.501.

Media-centric architecture and simplified rendering focus on the processes in which the following main tasks are performed: rendering, tracking and generating poses, rendering viewports, capturing real-world content, media encoding/decoding, media content delivery, 5G system communication, media formats, metadata, and other data.

5G supports a wider range of QoS requirements, including the high bandwidth and low latency needs of interactive XR applications over the NR air interface, as well as flexible QoS on the 5G CN architecture and NS. Furthermore, the ability of the 5G system to utilize network edge computing EC is essential to meet the performance requirements of immersive media to shift some of the complex XR processing to the edge such as decoding, rendering, graphics, stitching, encoding, transcoding, etc., for reasons of reducing the computing load of client devices.

Based on the conclusions of R15 study, 3GPP SA4 initiated normative work item R16 for the 5G MS streaming media architecture in specification TS 26.501, with the aim of developing MNOs architectures with relevant functions and interfaces to support various collaboration scenarios, including immersive media distribution. Various aspects such as session management, QoS framework, network assistance, QoE reporting, accessibility, content replacement, notifications, content rights management are considered. The relevant UE functions and APIs as well as the use of specific 5G features such as NS and EC are within the scope of the intended work on normative specification. One work item R16 defines 3GPP media codec profiles and network-based media processing functions (video stitching, media transcoding and content reformatting) and also recommends new QoS classes that account for trade-off of video quality and delivery delay of immersive media. 3GPP technical report TR 26.929 investigates QoE parameters and user experience evaluation metrics in XR services (Figure 2.13).

3GPP R17 work is ongoing to identify integration of 5G systems with network edge processing. Technical report TR 23.748 defines modifications to the 5GS system architecture to improve EC. In addition, the TR 23.758 report identified new set of application layer interfaces for EC integration. The candidate items for R18 include immersive media and XR media support in working group SA4 multimedia codecs, systems, and services and SA2 architecture (Figure 2.14).

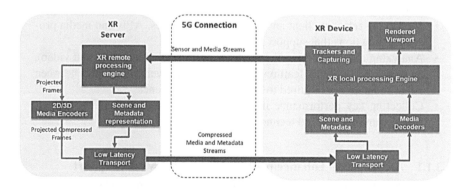

FIGURE 2.13 Generalized immersive XR communication in 5G reference architecture.

5G Advanced Mobile Broadband

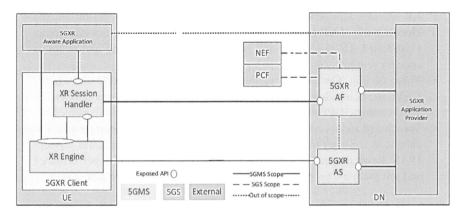

FIGURE 2.14 5G MS immersive XR communication interfaces and architecture.

TABLE 2.4
XR Capacity Evaluation

XR Application	Bit Rate (Mbps)	Packet Delay Bound (ms)	Frame Rate (fps)	XR Capacity (mean/range)
AR/VR DL	30	10	60	7.3
CG DL	30	15	60	9.9
VR/CG UL (Pose)	0.2	10	250	20–225
AR UL (Pose + scene)	10	10	60	4.4

Performance of XR services is evaluated in R17, including VR, AR, and cloud gaming CG in terms of capacity and power requirements (Table 2.4). XR capacity is defined as the number of VR/AR users, 90% of which show satisfactory performance when packet error rate <1%, and delay < packet delay bound (PDB) in dense urban scenario. Potential improvements have been identified, including XR identification in RAN, reduction of UE power consumption, and capacity improvements to better address the nature of traffic specific to XR, such as periodic arrivals, bounded delay/reliability requirements, multiple streams, jitter, and varying sizes package. For these potential improvements, R18 explores areas such as XR traffic characteristics, QoS metrics, application layer attributes, connected mode DRX discontinuous reception, control channel monitoring reduction, and enhancements in semi-persistent, configured, and dynamic scheduling mechanisms.

2.3.1.2 Levels of Immersion and Technological Complexity

Considering the support of latest defined immersive formats in MPEG, it is necessary for 3GPP to develop standardized interfaces that ensure device-complexity dynamic adaptation in the 5G network with mass-scale interoperability. The 3GPP study R17 and normative work until March 2022 are focused on the following areas:

- XR services. Collection of various XR use cases have been performed and identification of XR-based services typical traffic characteristics is ongoing; glass-based AR is studied, and immersive teleconferencing and telepresence for remote terminals is being specified.
- Video coding. Currently specified video codecs H.264/AVC and H.265/HEVC are being characterized over 5G scenarios as a preparation for H.266 VVC coding. 3GPP also plan the addition of 8K VR 360-video profiles.
- Speech and audio coding. New codec selection work is planned for an EVS-enhanced voice services extension for immersive voice and audio services IVAS. Related specification work for terminal audio quality performance and test methods for immersive audio services will follow in a next step.

MPEG experts group organized under ISO/IEC started initiative MPEG-VR to develop a roadmap and coordinate the various activities related to immersive media in June 2016. Currently, MPEG-I project explores standards to digitally represent immersive media and cooperation also with other consortia working on innovative products and services. The first Phase 1A targets the specification of 360-video projection formats OMAF (omnidirectional media application format). The next Phase 1B (Doc. N17069 Requirements on Phase 1B, July 2017) will the extend specification toward 3DoF+ applications. Phase 2 (Doc. N17073 Requirements on 6DoF v1, July 2017) is intended to start from about 2019, aims at addressing 6DoF applications like free viewpoint video. Different levels of experience can be achieved by the user who may freely move his head around three rotational axes 3DoF (yaw, pitch, roll), and along three translational directions 6DoF (left/right, forward/backward, up/down) (Doc. W17285 Visual activities on 6DoF and Light Fields, October 2017).

MPEG initiated a new MPEG-I project on coded representation of immersive media in October 2016. The project was motivated by the lack of common standards that do not enable interoperable services and devices providing immersive, navigable experiences. The MPEG-I project is expected to enable the evolution of interoperable immersive media services. Enabled by the parts of this standard, end users are expected to be able to access interoperable content and services and acquire devices that allow them to consume these. Core technologies as well as additional enablers are implemented in parts of the MPEG-I standard (ISO/IEC 23090-X with X for the part). The MPEG-I project is expected to enable existing services in an interoperable manner and to support the evolution of interoperable immersive media services. After the launch of the project, several phases, activities, and projects have been launched that enable services considered in MPEG-I. The project is divided in tracks that enable different core experiences. Each of the phases is supported by key activities in MPEG, namely in systems, video, audio, and 3D graphics-related technologies. Core technologies as well as additional enablers are implemented in parts of the MPEG-I standard. Currently the following 14(26) parts are under development:

- Part 1 Immersive media architectures. This overview summarizes use cases and architectures that motivate the development of specific components that contribute and support the distribution of immersive media services. Also,

the development of architectures and requirements for the integration of networked and compressed media into 6DoF scenes is underway.
- Part 2 Omnidirectional media format. OMAF addresses a first set of enablers for 3DoF experience based on existing MPEG technologies. It is the first published standard in MPEG (2019) that specifically addresses immersive media by combining and reusing existing MPEG compression (HEVC, MPEG-H) as well as storage and file formats. Currently, the development of the second edition is underway addressing extensions including limited 6DoF with multiple viewpoints, overlays, and improvements to the efficiency of viewport-adaptive streaming.
- Part 3 Versatile video coding. VVC predominantly addresses improved compression efficiency for high-resolution video but is also expected to provide better support for the integration of immersive media signals into the decoding.
- Part 4 Immersive audio coding support rich, immersive and highly interactive audio, MPEG addresses immersive audio in a new part in order to enable fully integrated audiovisual experiences. This specific project is expected to use the existing MPEG-H as a compression engine and will define a rich audio rendering engine with supporting metadata for 6DoF and AR experiences. Integration of low-latency speech codecs is expected to be supported and enables the combination of 3GPP speech codecs (EVS, IVAS), for example to support social VR use cases.
- Part 5/9 Point cloud compression PCC. Point cloud are becoming popular to present immersive volumetric video due to the relative ease of capture and render when compared to other volumetric video representation. Several applications include 6DoF immersive video, VR/AR, immersive real-time communication, autonomous driving, cultural heritage, and a mix of individual point cloud objects with background 2D/360-video. MPEG addresses two ways to compress point clouds, Part 5 video-based V-PCC, and Part 9 geometry-based G-PCC.
- Part 8 Network-based media processing (NBMP). NBMP defines framework that allows content and service providers to describe, deploy, and control media processing for their content in the network/cloud. The NBMP framework provides an abstraction layer on top of existing cloud platforms and is designed to integrate with 5G Core and edge computing EC. A particular aspect on NBMP is the integration of immersive media.
- Part 14 Scene description for MPEG media. The virtual scene composition specifies extensions to existing scene description formats in order to support MPEG immersive media. Extensions include scene description format syntax and semantics and the processing model when using these extensions in combination with a presentation engine.

3GPP adopted the omnidirectional media format MPEG-I Part 2 OMAF architecture and addresses detailed specification for VR360 streaming in TS 26.118 (Figure 2.15).

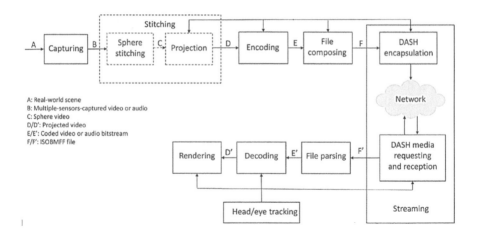

FIGURE 2.15 OMAF (omnidirectional media format architecture).

2.3.2 Extremely Interactive Communication and Low Latency

Unlike the basic 5G enhanced mobile broadband eMBB service, most new applications are time-critical in nature with demanding reliable low latency. The main causes of delays and interruptions are dynamic radio environment, mobility, protocols, network congestion. Critical use cases are real-time digital media and immersive XR communication. Immersive applications are expected to dominate next-generation mesh networks in terms of their popularity and the volume of traffic they generate. XR features in 5G NR have been focused on ways to manage scheduling within the existing framework of the RAN. To unlock the potential of next-generation applications, the challenges posed by the end-to-end (E2E) system need to be addressed.

The key difference for XR applications is that it involves interactive experiences where the entire perspective of the user may be communicated via the network (Figure 2.16). If multiple users are involved, the E2E experience (i.e., user-to-user latency) is the most relevant key performance indicator (KPI) to consider. Several domain experts have identified that the necessary user-to-user latency for a seamless user experience is <50 ms as shown in Figure 2.17.

User-to-user latency includes the access link, gateway, backhauls, Internet service provider (ISP) links, and finally downlink to the other users. Variations in one or more dimensions can be used to maintain a consistent end-user experience. It is time to think of cross-technology optimizations across multiple nodes in the network. It is expected that XR applications will be compute-intensive.

2.3.2.1 Immersive XR Communication

The major components in XR processing include simultaneous localization, mapping, and map optimization (SLAM) with immersive media formats, hand gesture and pose estimation, object detection and tracking, and multimedia processing and transport. Examples of multimedia processing and transport are rendering, asynchronous time warp and video, audio and sensor encoding. 5G XR architecture options in splitting the processing between XR device and network are showed in Figure 2.18:

5G Advanced Mobile Broadband

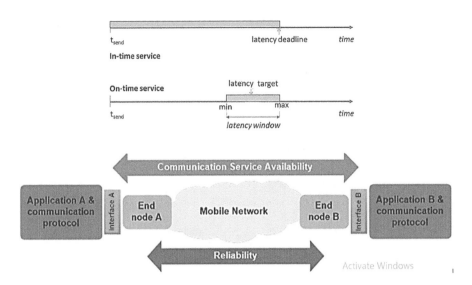

FIGURE 2.16 Illustration of the concepts: (a) in-time vs. on-time services and (b) communication reliability vs. service availability.

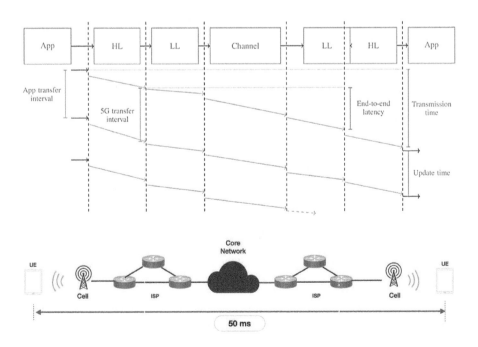

FIGURE 2.17 General timing model from 3GPP and end-to-end user latency for immersive XR communication.

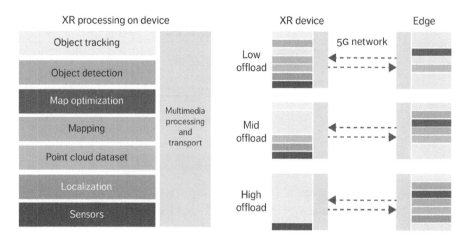

FIGURE 2.18 5G XR architecture options.

- In low-offload architecture, almost all processing is done on the device. Immersive media format, spatial map generation and localization are processed on the device as well as object detection and tracking. Rendering is done locally on the device or at the edge of the network.
- In mid-offload case, the device performs localization and object tracking functions. Spatial map and immersive format are generated in the network, merged with global data at the edge of the network. The overlay rendering occurs at the network edge.
- In high-offload case, only the sensor data is sent over UL connection. Sensor data contains camera data. Many AR/VR devices are equipped with multiple cameras, including IR infrared and RGB cameras. Sensor data also includes LiDAR (light detection and ranging) scanner and IMU (inertial measurement unit) sensor. Image data is encoded using video compression such as the MPEG VVC codec.

5G XR traffic characteristics and connectivity requirements are characterized by a mix of poses and video from/to the same XR device, variable video frame size over time, and quasi-periodic packet reception with application jitter after IP segmentation. The traffic arrival time in the RAN is periodic with non-negligible jitter due to uncertainty of application processing time. Video frame sizes are an order of magnitude larger and, at the same time, not fixed over time compared to packets in industrial control communications. The segmentation of each frame is expected, which implies that packets arrive in bursts that must be handled together to meet stringent bounded latency requirements.

The requirements for XR connectivity depend on the level of shared architecture (Figure 2.18) and the target QoE, resulting in a wide range of transmission rates and bounded latency requirements. The requirements of AR and VR 5G connections based on the development of the 3GPP ecosystem are shown in Table 2.5. A local processing technique is assumed in the split architecture to alleviate the low-latency

5G Advanced Mobile Broadband

TABLE 2.5
XR Use Case Requirements for 5G Networks

Use Cases	DL Bitrates (Mbps)	UL Bitrates (Mbps)	One-Way Latency (ms)	Frame Reliability (%)
AR	2–60	2–20	5–50	>99
VR	30–100	<2	5–20	>99

requirements. Latency and reliability requirements are on a video frame (or file) level excluding application error and delay.

If the XR application is hosted in a central national data center (DC), transport network round-trip is the order of 10–40 ms, depending on the distance to the DC and how well the transport network is built out. Transport latency can be reduced to 5–20 ms by relocate applications in regional DC or reduced to 1–5 ms for edge sites. For a local network deployment with network functions and locally hosted applications, the transport delay becomes negligible.

2.3.2.2 Holographic Communication

Holographic communication refers to real-time capturing, encoding, transporting, and rendering of 3D representations, anchored in space, of remote persons shown as 3D video in XR headsets that deliver a visual effect similar to a hologram. The research area divided into professional-quality or consumer-friendly digital representations of users. Professional-quality digital representations are created through the use of multiple cameras and real-time studio recordings. Consumer-friendly digital representations are generated with the help of an AI-enabled, 3D-capturing setup on consumer-grade phones or tablets.

Holographic communications require the processing and transmission of various formats of immersive visual media. The formats present more realistic and interactive visual representations of people and/or environments than the current 2D video formats used in traditional video conferencing. A point cloud (PC) is a set of points that represent the enclosed volume. Each 3D point contains location information, color voxel, and intensity values in a particular frame. A mesh unites those points with triangles, disregarding redundant points and filling any holes. Meshes can be cleaned and reduced further by decreasing the number of vertices. Depending on their resolution, they can be significantly smaller than point clouds, increasing the speed of storage, transmission and rendering of meshes in comparison with point clouds.

The application of new immersive media formats in XR communication scenarios greatly improves the attractiveness, usefulness, and efficiency of information transmitted between parties in communication. 3GPP working on challenging task of improving the 5G system in R18 research to offer more efficient XR service support with the aim of reusing as much of the previous technical specifications as possible. Standards development organizations have additionally research on the interoperability challenges of respective component technologies:

- ITU-T ILE international standards on immersive live experience have been adopted in November 2019. ILE captures an event and transports it to remote viewing sites in real time and reproduces it with high realism. ILE captures an event and transports it to remote viewing sites in real time and reproduces it with high realism. It is expected that the series of standards on ILE will facilitate the creation of a world where people can enjoy highly realistic reproduction of events in real time wherever they are.
- IEEE Digital reality working on global standards related to digital reality, AR, VR, human augmentation, and related areas. Under P2048 standard, VR/AR working group is developing 12 standards for virtual reality and augmented reality with participants from device manufacturers, content providers, service providers, technology developers, government agencies, and other relevant parties, constituting an excellent mixture for the standards to be widely adopted.
- ETSI ISG ARF has defined the augmented reality framework for the interoperability of AR components, systems and services that specifies relevant components and interfaces required for an AR solution. Transparent and reliable interworking between different AR components is key to the successful roll-out and wide adoption of AR applications and services.
- DVB started commercial module CM study mission group with to deliver commercial requirements for relevant technical module TM groups that will work on technical specifications to deliver VR contents over DVB networks. DVB will consider work done within other organizations such as MPEG, VRIF, and 3GPP.
- ISO/IEC MPEG experts group is developing metadata for immersive video together with media formats and video/audio codecs for XR. MPEG-I project develops standards for volumetric video. For point cloud compression PCC, MPEG defines two representations. V-PCC decomposes point clouds into two separate video sequences, which capture the texture information and geometry. Traditional 2D codec VVC is then applied. However, G-PCC decomposes the 3D space into a hierarchical structure of cubes; each point is encoded as an index of the cube it belongs to. MPEG Immersive video (MIV) standard reached draft international standard DIS status in the January 2021. In 2020, the MIV standard has been aligned with the V-PCC standard. As a result, the MIV standard references the common part of ISO/IEC 23090-5 2nd edition visual volumetric video-based coding V3C – also with DIS status – with V-PCC an Annex H of that document. V3C provides extension mechanisms for V-PCC and MIV. Finally, SC29/WG03 MPEG Systems has also developed a systems standard based on ISOBMFF for the carriage of V3C data ISO/IEC 23090-10 which reached the FDIS stage in January 2021.
- JPEG Pleno is an upcoming standard from the ISO/IEC JTC1/SC29/WG1 committee. It aims to provide a standard framework for coding new imaging modalities derived from representations inspired by the plenoptic function.

The image modalities addressed by the current standardization activities are light field, holography, and point clouds, where these image modalities describe different sampled representations of the plenoptic function.

2.3.3 AI-Based Multimedia Tools

Multimedia tools based on artificial intelligence have been successfully applied in computer vision tasks, and are now arriving immersive media. The main strategy of using AI is to increase efficiency and reduce costs. Learning-based video compression has received much attention in recent years due to its adaptability to content and parallel computation. Machine learning (ML) techniques are used during video encoding to reduce file sizes and bit rates while maintaining perceptual quality. The technique allows encoders to optimize video encode parameters on a scene-by-scene basis while feeding the results back into the system to enhance future encoding sessions. Learning-based video compression has been widely researched and got remarkable milestones.

The advantages of learning-based video compression are as follows:

- The model is based on learning and adapted to the content of a huge amount of training data, so it is more efficient than modules designed for specific tasks.
- The difference from the conventional codec, the learning-based models usually explore the large receptive field in both spatial and temporal domains, therefore provides a more accurate prediction or latent distribution. This manner also helps the codec to avoid the blocking artifact and become flexible in temporal exploration.
- The direct linkable ability allows the learning-based modules to perform the global optimization that is the potential factor for further improvement on the R–D trade-off and specific human vision task.
- The flexibility of the learning-based method allows them to quickly inherit the newest technology, extend the design and transfer knowledge easily.

There are many ways to apply the learning-based method to video compression.

- Learning-based method can be an integrated or replacement module of the conventional codec, then they further are an outside cooperated or guidance module.
- In learning-based end-to-end (E2E) compression methods all components are learnable and linked to solving a global objective function. In compression, the objective function is usually to represent the rate R (number of bits) and distortion D (quality) relation. Different from the conventional codec based on local optimum, the global objective function allows the

learning-based E2E codec find the global optimized R–D operational point, which reveals a huge potential for further performance improvement and on-demand compression ability.

Conventional cooperative approach involves joint processing of standard hybrid video compression and learning-based methods, where the core component is conventional codec. Learning-based methods can use a conventional codec as an internal module (intra-prediction, inter-prediction, in-loop filter), an external enhancement module (intra-prediction, inter-prediction, in-loop filter), or guidance (super-resolution-based, layered-based). The improvement in compression performance is associated with huge trade-off in complexity and increased storage memory. Also, the local optimization problem is inherited from the conventional module. Thanks to the development of differentiable quantization function, recently, learning-based E2E video coding has been intensively researched.

E2E approach contains only trainable components with a global objective function. However, framework is very flexible and can be further separated into predictive video coding and generative video coding. The predictive video coding is usually designed to reflect conventional compression design based on the prediction and residual calculation. Whereas the generative video coding can be seen as an extension of variable autoencoder (VAE) technique.

Predictive video coding approach completely replaces conventionally designed modules with learning-based models, but mainly processes a complete video frame instead of divided into blocks of pixels. Similar to the conventional codec, the key idea is to compress the residual errors between the predicted frame and the current frame. In specific, the motion estimation of interframe prediction is replaced with the learning-based flow estimation, the motion compensation become warping function, and the learning-based frame synthesis model will do the reconstruction task. The estimated flow and the residual error then are compressed and sent to the decoder. Especially, with the join of the differentiable quantization methods, all modules are linked together to perform a global optimization training process for the rate-distortion loss (Figure 2.19).

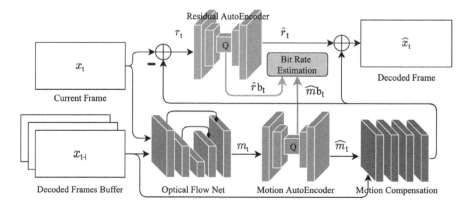

FIGURE 2.19 DVC E2E predictive coding framework (*m* motion flow, *r* estimated residual).

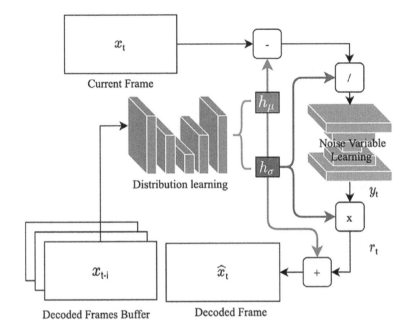

FIGURE 2.20 Basic generative video coding framework.

Generative video coding framework is as an extension of the variable autoencoder (VAE), which learn the parameter of probability distribution for the latent representation. Figure 2.20 shows the pure version of the video coding with only temporal autoregressive transform. Many recent works are studying to improve this module by more complex design, context-adaptive design or including temporal information via sequences-learning technique.

While the conventional-learning-based framework is usually strong on the mean square error (MSE) or peak signal over noise rate (PSNR) because of its conventional baseline, the learning-based E2E framework is surpassed on the structure similarity by avoiding the block artifacts since they process with fully spatial information of the frame. So far, the E2E learning-based video compression methods have surpassed traditional codecs on MS-SSIM structural metric, perceptual quality, or even PSNR on low bitrate. However, E2E compression requires powerful hardware (GPU) to process the compression on both the encoder and decoder sides. The problem becomes more serious since more prior is being leveraged recently required a much more complex network design.

In the trade-off or the compression ratio performance, current learning-based video compression methods are facing many challenges that are required to be further research:

- Complexity and memory requirement. One of the major limitations of the learning-based approach compared to the conventional one is the enormous load of computation and memory requirements.

- Rate-dependent model design. The most common framework design of current learning-based works intuitively assumes that distortion of the predicted frame and rate are in a linear correlation. Hence, the network is designed to produce a better-quality predicted frame that hopefully reduces the compression rate of the residual.
- The generative design for the learning-based model. So far, the generative approach has demonstrated its performance in computer vision tasks. Its advantage becomes clearer in compression where learned distribution can save a lot of sending information.

2.3.3.1 DNNVC Video Codec

With the finalization of VVC video codec standardization, discussion was initiated in MPEG about new codec based on deep neural network (DNN) technology. JVET AdHoc Group 11 started in April 2020 standardization process to create a new AI-based DNNVC codec before the end of the decade, as well as also explores related techniques that can improve existing practices. In the short term, AI presents an opportunity to extend existing conventional coding modules with new tools. Presently primary areas of focus are dynamic frame rate encoding, dynamic resolution encoding, and layering. MPEG created ad hoc group with the following terms of reference:

- evaluate and quantify performance improvement potential of DNN-based video coding technologies (including hybrid video coding system with DNN modules and end-to-end DNN coding systems) compared to existing MPEG VVC standards considering various quality metrics
- study quality metrics for DNN-based video coding
- analyze the encoding and decoding complexity of NN-based video coding technologies by considering software and hardware implementations, including impact on power consumption
- investigate technical aspects specific to NN-based video coding, such as design network representation, operation, tensor, on-the-fly network adaption (updating during encoding).
- building NN environments (small ad hoc deep learning library JVET Doc. W0181).
- From the point of view of codec architecture, two different approaches have been explored in DNNVC:
- Hybrid block-based coding where DNNs replace existing encoding tools or are used as optimization methods, thereby preserving the architecture of a conventional hybrid block-based video codec.
- End-to-end learning-based coding where DNNs play major roles in compression, thus the architectures are highly dependent on the DNN architecture and its usage.

Conventional video codecs are based on hybrid block-based coding. Main encoding tools for hybrid block-based coding include: partitioning frames into blocks, inter/intra prediction of the blocks to remove spatial/temporal redundancies, lossy compression with transform/quantization based on the human visual system, and entropy coding to remove statistical redundancy. In addition, in-loop filtering such as deblocking and sample adaptive offset filters are used to reduce compression artifacts. In hybrid coding approaches, DNNs are used for estimations such as: intra/inter-prediction and compression artifact removal. In entropy coding, DNN is used to predict probability (or possibility) of contexts for context-adaptive binary arithmetic coding CABAC. DNN-based in-loop filtering showed the highest coding enhancements and the deeper network showed a higher enhancement.

One of the hybrid coding approaches DLVC deep learning-based video coding was proposed in 2018. DLVC added two deep convolutional neural network (CNN)-based coding tools on top of HEVC: a CNN-based loop filter, and a CNN-based block adaptive resolution coding.

Compared with the hybrid coding approaches, there is no concrete consensus about a common end-to-end learning-based coding approach until now. But most E2E learning-based coding approaches also try to remove spatial/temporal redundancy and utilize lossy compression with quantization and entropy coding such as the hybrid coding approaches. An E2E neural video coding (NVC) was proposed in 2020. NVC comprises intra, motion, and residual coding networks. In addition, quantization and arithmetic coding is adopted with a context prediction network (named hyper coder).

The first solid performance study on DL-based video codecs JVET NN CTC common test conditions vs. DVC (2019), HLVC (2020), and RLVC (2020) codecs was conducted in 2022.

- DVC (Deep Video Compression) is classical, hybrid video coding architecture, where each module is replaced by a DL-based tool, thus originating one-to-one correspondence between the classical coding modules and the DL-based models. In DVC, motion estimation is performed using a CNN to extract the optical flow, from the previously decoded frame. The resulting residual information is coded using a CNN-based encoder–decoder network and the motion-associated latents are quantized to reduce rate while still preserving a good enough motion representation. The decoded optical flow is used to warp the previous decoded frame to estimate the motion-compensated prediction for the current frame. However, since the motion-compensated frame has artifacts, it is processed by a CNN, together with the previous decoded frame and the decoded optical flow, to obtain a refined prediction frame. Lastly, the prediction residual is compressed into a latent representation using a nonlinear NN, which is quantized to obtain several R–D operational points. The whole DVC architecture is trained with rate-distortion R–D loss function, using as end-to-end distortion MSE metric between the original and decoded frames, while the bitrate is estimated using a rate estimation module.
- HLVC hierarchical learned DL-based video compression solution adopts hierarchical coding architecture inside group of pictures (GOP). At the decoder, the frames are decoded and enhanced using a recurrent network to

increase the decoded frames' quality. The first HLVC layer includes the first and last frames of the GOP, which are coded with a state-of-the-art image coding solution, at the highest quality. The second HLVC layer codes the middle frame of the GOP with medium quality, using a NN for bidirectional prediction, and the frames at the edge of the GOP as anchors. The third and last HLVC layer codes the remaining frames of the GOP with lower quality, using the closest already decoded frame as anchor (from the first or second layers); only unidirectional prediction and a single motion map are used to code two adjacent frames. On the decoder side, after each frame is decoded by the corresponding layer, all frames are processed by an enhancement recurrent NN, which leverages on the frames with better quality to improve the quality of the remaining frames.

- Recurrent learned DL-based video compression (RLVC) solution includes a recurrent autoencoder (RAE) and a recurrent probability model (RPM). Due to its recurrent nature, the RAE allows using temporal information from several frames at a time. In this context, previous inputs are given as hidden states to help generating the latent representation for the current frame, instead of choosing only specific frames as anchors like most DL-based video codecs. The RPM network models the probability mass function of the obtained latent representations, based on previous iterations. In the RLVC codec, the first frame of each GOP is coded with a state-of-the-art image coding solution, either DL-based or conventional. For the remaining frames, motion estimation is performed using a pyramid optical flow network. The RAE codes the estimated motion, considering the hidden states passed on, which express inputs from previous frames; the decoded motion estimation is applied to create a motion-compensated prediction. Next, the residual between the original and motion-compensated frames is obtained and coded using another RAE. The RPM is used to recurrently predict the temporally conditional probability mass function, to reduce the rate when entropy coding the latent representations.

AI-based codecs can be several orders of magnitude more complex than traditional codecs. Neural networks are being used within conventional codecs to replace existing modules, especially where performance is likely to be improved, however the complexity is often exceptionally high. There are other challenges in AI-based video development. Learned models contain millions of parameters, which makes real-time inference on common devices a challenge. Also, it is difficult to interpret learned models or to provide performance bounds on results. Existing video standards offer reference software, common test conditions, and frame sequences, which allow direct comparison of various executions in hardware or software. The video standard is fully described, in detail and verified against agreed metrics. The industry needs common ground to compare AI-based methods. MPEG now have a common data set for AI inferencing which all vendors must use, however the training cannot be cross-checked for validation.

Another active discussion topic was about visual quality metrics. The PSNR and multiscale structural similarity (MS-SSIM) are widely used objective visual quality metrics. But they show inconsistencies with human perceived visual quality, and it is difficult to compare coding performance. In addition, a precise perceptual visual quality metric could help enhancing coding efficiency. Learning-based visual quality metrics such as LPIPS learned perceptual image patch similarity metric is a good candidate.

MPEG VCM video coding for machine is new project with emphasis on exploring video compression solutions for both human perception and machine intelligence. Traditional coding methods aim for the best video under certain bitrate constraints for human consumption. However, with the rise of ML applications, along with the richness of sensors, many intelligent platforms have been implemented with massive data requirements, including scenarios such as connected vehicles, video surveillance, and smart city. The MPEG activity on VCM aims to standardize a bitstream format generated by compressing a previously extracted feature stream and an optional video stream. Deep learning-based coding solutions allow extending the utility of the compressed representations by offering key advantages: reduction of the complexity resources associated to computer vision tasks as already starting from compressed domain features, thus at least partly skipping feature extraction; better analysis accuracy by allowing the computer vision tasks to use the compressed domain features extracted from the original video data instead of extracting them from the lossy decoded video as for conventional coding solutions where feature extraction happens after full decoding.

2.4 CONCLUDING REMARKS

The scope of this chapter is transition of 5G architecture to common media delivery platform based on enhanced broadband mobile access eMBB, 5G Media streaming, 5G Broadcast, and 5G MBS multicast-broadcast services. We point out how is important to study technical issues taking into account current and future requirements and constraints. It starts with a 5G vision and foresight to identify compelling challenges and issues. We address 5G immersive communication aspects such as use cases, relevant technologies, media formats, and delivery procedures, network architectures, and QoS service parameters. Also, initial interest in applying AI to the challenges of video encoding is outlined.

We evaluate the benefits, impacts, and challenges of adopting 5G networks across the media distribution value chain and positions the media industry as a key vertical for the future development of 3GPP technologies. 5G ecosystem promises continued development in future releases, bringing new services and functionalities, better performance and efficiency, as well as rapid replacement cycles. However, there are many compelling future research challenges that still remain to be addressed.

BIBLIOGRAPHY

INTRODUCTION

B. Bendre, E. Gose, M. Thomas, S.P.K. Prasad, U. Mangla, Creating transformational media solutions using 5G edge computing. In *Proc. IBC*, 2021, pp. 1–14.

A. Benjebbour et al., 3GPP defined 5G requirements and evaluation conditions. *NTT DOCOMO Technical Journal*, vol. 10, no. 3, pp. 13–23, 2018.

I. Curcio, S. Gunkel, T. Stockhammer, State of the art of extended reality in 5G networks. In *Proc. IBC*, 2019.

A. DeVita, V. Mignone, D. Milanesio, F.M. Pandolfi, B. Sacco, 5G broadcast and evolution towards new radio: Perspectives for media delivery to mobile devices. In *Proc. IBC*, 2021, pp. 1–11.

N. Finn, Introduction to time-sensitive networking. *IEEE Communications Standards Magazine*, vol. 2, no. 2, pp. 22–28, 2018.

F. Gabin, T. Lohmar, G. Heikkilä, L. D'Acunto, T. Stockhammer, 5G media streaming architecture. In *Proc. IBC*, 2019, pp. 1–9.

M.A. Imran, Y.A. Sambo, Q.H. Abbasi, Enabling the verticals of 5G: Network architecture, design and service optimization (Chapter 1). In *Enabling 5G Communication Systems to Support Vertical Industries*, Wiley, 2019.

H. Kim, *Artificial intelligence for 6G*, Springer, 2022.

Y. Kishiyama, S. Suyama, History of 5G initiatives. *NTT DOCOMO Technical Journal*, vol. 22, no. 2, pp. 4–12, 2020.

T. Stockhammer et al., 5G media distribution - Status and ongoing work. In *Proc. IBC*, 2021, pp. 1–12.

T. Stockhammer, I. Bouazizi, F. Gabin, G. Teniou, Immersive media over 5G – What standards are needed? In *Proc. IBC*, 2018, pp. 1–9.

E. Thomas, E. Potetsianakis, T. Stockhammer, I. Bouazizi, M.-L. Champel, MPEG media enablers for richer XR experiences. In *Proc. IBC*, 2020, pp. 1–12.

EXTENSIONS OF 5G ARCHITECTURE FOR COMMON MEDIA DELIVERY PLATFORM

3GPP TR 22.891, *Feasibility study on new services and markets technology*, Release 14, Sept. 2016.

3GPP TR 26.891, *5G enhanced mobile broadband media distribution*, Release 16, Dec. 2018.

3GPP TR 26.925, *Typical traffic characteristics of media services on 3GPP networks*, Release 17, Sept. 2021.

3GPP TS 22.261, *Service requirements for the 5G system*, Release 18, Dec. 2021.

D. Gomez-Barquero, J.J. Gimenez, G-M. Muntean, Y. Xu, Y. Wu (Eds.), Special issue 5G media production, contribution, and distribution. *IEEE Transactions on Broadcasting*, vol. 68, no. 2, pp. 415–421, 2022.

D. Gomez-Barquero, J.-Y. Lee, S. Ahn, C. Akamine, D. He, J. Montalaban, J. Wang, W. Li, Y. Wu, Special issue on convergence of broadcast and broadband in the 5G era. *IEEE Transactions on Broadcasting*, vol. 66, no. 2, pp. 383–389, 2020.

D. Gomez-Barquero, W. Li, M. Fuentes, J. Xiong, G. Araniti, C. Akamine, J. Wang (Eds.), Special issue on 5G for broadband multimedia systems and broadcasting. *IEEE Transactions on Broadcasting*, vol. 65, no. 2, pp. 351–355, 2019.

W. Zhang et al., Convergence of a terrestrial broadcast network and a mobile broadband network. *IEEE Communications Magazine*, vol. 56. no. 3, pp. 74–81, 2018.

ENHANCED MOBILE BROADBAND EMBB AND MEDIA STREAMING 5G MS

Y. Kim et al., New radio (NR) and its evolution toward 5G-Advanced. *IEEE Wireless Communication*, vol. 26, no. 3, pp. 2–7, Jun. 2019.

3GPP TR 26.949, *Video formats for 3GPP services*, Release 16, Jul. 2020.

3GPP TR 26.955, *Video codec characteristics for 5G-based services and applications*, Release 17, Feb. 2022.

3GPP TS 23.501, *System architecture for the 5G system (5GS)*, Release 17, Dec. 2021.
3GPP TS 23.502, *Procedures for the 5G system (5GS)*, Release 17, Dec. 2021.
3GPP TS 26.238, *Uplink streaming*, Release 17.0.0, Dec. 2021.
3GPP TS 26.501, *5G media streaming (5GMS): General description and architecture*, Release 16/17 2019/2022.
3GPP TS 26.511, *5G media streaming (5GMS): Profiles, codecs and formats*, Release 16/17, 2020/2021.
3GPP TS 26.512, *5G media streaming (5GMS): Protocols*, Release 16, Oct. 2020.
3GPP TSG RAN TS 38.202, *Services provided by the physical layer*, Release 16, Jul. 2020.
3GPP TSG RAN TS 38.300, *NR and NG-RAN overall description*, Release 16, Jan. 2020.
ITU-R Recommendation M.2083, *IMT vision: Framework and overall objectives of the future development of IMT for 2020 and beyond*, Sept. 2015.
B.S. Khan et al., URLLC and eMBB in 5G industrial IoT: A survey. *IEEE Open Journal of the Communications Society*, 2022. (Early access).
X. Lin et al., 5G new radio: Unveiling the essentials of the next generation wireless access technology. *IEEE Communications Standards Magazine*. vol. 3, no. 3, pp. 30–37, Sept. 2019.
P. Popovski et al., A perspective on time towards wireless 6G. *Proceedings of the IEEE*, pp. 1–31, 2022. (Early access).
P. Popovski, K.F. Trillingsgaard, O. Simeone, G. Durisi, 5G wireless network slicing for eMBB, uRLLC, and mMTC: A communication-theoretic view. *IEEE Access*, vol. 6, pp. 5765–5779, 2018.
K. Takeda et al., Understanding the heart of the 5G air interface: An overview of physical downlink control channel for 5G new radio. *IEEE Communications Standards Magazine*, vol. 4, no. 3, pp. 22–29, Sept. 2020.

Evolution of Multimedia Multicast and Broadcast Services 5G MBS

3GPP TR 23.757, *Study on architectural enhancements for 5G multicast-broadcast services* Release 17, Mar. 2021.
3GPP TR 26.802, *Multicast architecture enhancement for 5G media streaming*, Release 17, Jul. 2021.
3GPP TR 36.976, *Overall description of LTE-based 5G broadcast*, Release 16, Apr. 2020.
3GPP TS 23.247, *Architectural enhancements for 5G multicast-broadcast services*, Release 17, Dec. 2021.
3GPP TS 26.346, *Multimedia broadcast/multicast service (MBMS): Protocols and codecs*, Release 16, May 2021.
3GPP TSG RAN TS 36.976, *Overall description of LTE-based 5G broadcast*, Release 16, Apr. 2020.
D. Gómez-Barquero (Ed.), *Next generation mobile broadcasting*, CRC Press, 2013.
DVB Project, *Commercial requirements for DVB-I over 5G*, Bluebook Standard C100, 2021.
ETSI TS 103 720, *5G broadcast system for linear TV and radio services: LTE-based 5G terrestrial broadcast system*, V1.1.1, Dec. 2020.
E. Garro et al., 5G mixed mode: NR multicast-broadcast services. *IEEE Transactions on Broadcasting*, vol. 66, no. 2, Part II, pp. 390–403, Jun. 2020.
D. Gomez-Barquero, J.J. Gimenez, R. Beutler, 3GPP enhancements for television services: LTE-based 5G terrestrial broadcast. In *Encyclopedia of Electrical and Electronics Engineering*, pp. 1–18. Wiley, 2020.
D. He et al., Overview of physical layer enhancement for 5G broadcast in Release 16. *IEEE Transactions on Broadcasting*, vol. 66, no. 2, pp. 471–480, 2020.

J. Lee et al., LTE-advanced in 3GPP Rel-13/14: An evolution toward 5G. *IEEE Communications Magazine*, vol. 54, no. 3, pp. 36–42, Mar. 2016.
D. Mi et al., Demonstrating immersive media delivery on 5G broadcast and multicast testing networks. *IEEE Transactions on Broadcasting*, vol. 66, no. 2, pp. 555–570, Jun. 2020.
J.F. Monserrat et al., Joint delivery of unicast and eMBMS services in LTE networks. *IEEE Transactions on Broadcasting*, vol. 58, no. 2, pp. 157–67, 2012.
A. Sengupta et al., 5G cellular broadcast: Physical layer evolution from 3GPP Release 9 to Release 16. *IEEE Transactions on Broadcasting*, vol. 66, no. 2, pp. 459–470, 2020.
Y. Xu et al., Enhancements on coding and modulation schemes for LTE-based 5G terrestrial broadcast system. *IEEE IEEE Transactions on Broadcasting*, vol. 66, no. 2, pp. 481–489, 2020.
Y. Zhang et al., MBSFN or SC-PTM: How to efficiently multicast/broadcast. *IEEE Transactions on Broadcasting*, vol. 67, no. 3, pp. 582–592, Sept. 2021).

Multimedia Production and Distribution Ecosystem

3GPP TR 22.827, *Study on audio-visual service production*, Release 17, Jan. 2020.
3GPP TR 26.805, *Study on media production over 5G NPN systems*, Release 17 v1.1.0 Feb. 2022.
3GPP TS 22.263, *Service requirements for video, imaging and audio for professional applications (VIAPA)*, Release 17, Jun. 2021.
EBU TR 054, *5G for the distribution of audiovisual media content and services*, May 2020.
F. Gabin, 3GPP multimedia codecs systems and services. *3GPP Highlights Standards for 5G*, no. 2, Mar. 2021.
I. Wagdin et al., Media production over 5G NPN. *3GPP Highlights Standards for 5G*, no. 3, Oct. 2021.

Imersive Communication

S. Schwarz et al., Emerging MPEG standards for point cloud compression. *IEEE Journal on Emerging and Selected Topics in Circuits and Systems*, vol. 9, no. 1, pp. 133–148, 2019.
P. Astola et al., JPEG pleno: Standardizing a coding framework and tools for plenoptic imaging modalities. *ITU Journal: ICT Discoveries*, vol. 3, no. 1, pp. 1–15, 2020.
J.M. Boyce et al., MPEG immersive video coding standard. *Proceedings of the IEEE*, vol. 109, no. 9, pp. 1521–1536, Sept. 2021.
D.B. Graziosi, B. Kroon, Video based coding of volumetric data. In *Proc. ICIP*, 2020, pp. 2706–2710.
M.M. Hannuksela et al., An overview of omnidirectional media format (OMAF). *Proceedings of the IEEE*, vol. 109, no. 9, pp. 1590–1606, Sept. 2021.
M. Kerdranvat et al., The video codec landscape in 2020. *ITU Journal: ICT Discoveries*, vol. 3, no. 1, pp. 1–11, Jun. 2020.
G. Lafruit et al., MPEG-I coding performance in immersive VR/AR applications. In *Proc. IBC*, 2018, pp. 1–9.
G. Lafruit et al., Understanding MPEG-I coding standardization in immersive VR/AR applications. *SMPTE Motion Imaging Journal*, vol. 128, no. 10, pp. 33–39, Nov.–Dec. 2019.
D. Milovanovic, D. Kukolj, An overview of developments and standardization activities in immersive media. *VQEG eLetter*, vol. 3, no. 1, pp. 5–8, Video Quality Experts Group, Immersive Media, Nov. 2017.
D. Milovanovic, D. Kukolj, Emerging levels of immersive experience in MPEG-I video coding. *IEEE COMSOC MMTC Communications – Frontiers*, vol. 13, no. 1, pp. 24–26, Jan. 2018.

D. Milovanovic, D. Kukolj, Recent advances in UHD video coding technology: High dynamic range and wide color gamut. *IEEE COMSOC MMTC Communications – Frontiers*, vol. 11, no. 1, pp. 50–55, Jan. 2016.

D. Milovanovic, D. Kukolj, Standardization activities, section 8 in definitions of immersive media experience (IMEx). (Eds. A. Perkis, C. Timmerer). *arXiv:2007.07032*, 2020.

C. Timmerer, Immersive media delivery: Overview of ongoing standardization activities. *IEEE Communications Standards Magazine*, Dec. 2017.

M. Wien et al., Standardization status of immersive video coding. *IEEE Journal on Emerging and Selected Topics in Circuits and Systems*, vol. 9, no. 1, pp. 5–17, Mar. 2019.

IMMERSIVE MEDIA OVER 5G

3GPP TR 26.928, *Extended reality over 5G*, Release 16, Dec. 2020.

3GPP TR 26.929, *QoE parameters and metrics relevant to the virtual reality (VR) user experience*, Release 17, 2022.

3GPP TR 26.949, *Video formats for 3GPP services*, Release 16, Jul. 2020.

3GPP TR 26.955, *5G video codec characteristics*, Release 17, Feb. 2022.

3GPP TS 26.116, *Television (TV) over 3GPP services: Video profiles*, Release 16, Dec. 2021.

3GPP TS 26.118, *Virtual reality profiles for streaming applications*, Release 16/17, 2020/2021.

3GPP TS 26.918, *Virtual reality media services over 3GPP*, Release 16, Dec. 2018.

E. Bastug et al., Toward interconnected virtual reality: Opportunities, challenges, and enablers. *IEEE Communications Magazine*, vol. 55, no. 6, pp. 11–17, Jun. 2017.

E. Bastug, M. Bennis, M. Medard, M. Debbah, Toward interconnected virtual reality: Opportunities, challenges, and enablers. *IEEE Communications Magazine*, vol. 55, no. 6, pp. 110–117, 2017.

M. Bennis, M. Debbah, H.V. Poor, Ultrareliable and low-latency wireless communication: Tail, risk and scale. *Proceedings of the IEEE*, vol. 106, no. 10, pp. 1834–1853, Oct. 2018.

C.-F. Liu, M. Bennis, Ultra-reliable and low-latency vehicular transmission: An extreme value theory approach. *IEEE Communications Letters*, vol. 22, no. 6, pp. 1292–1295, 2018.

D.G. Morín, P. Perez, A.G. Armada, Toward the distributed implementation of immersive augmented reality architectures on 5G networks. *IEEE Communications Magazine*, vol. 60, no. 2, pp. 46–52, Feb. 2022.

Z. Nadir, T. Taleb, H. Flinck, O. Bouachir, M. Bagaa, Immersive services over 5G and beyond mobile systems. *IEEE Network*, vol. 35, no. 6, pp. 299–306, 2021.

J. Nagao, K. Tanaka, H. Imanaka, Arena-style immersive live experience (ILE) services and systems: Highly realistic sensations for everyone in the world. *ITU Journal: ICT Discoveries*, vol. 3, no. 1, pp. 1–9, May 2020.

LOW-LATENCY AND EXTREMELY INTERACTIVE COMMUNICATION

A. Clemm, M.T. Vega, H.K. Ravuri, T. Wauters, F.D. Turck, Toward truly immersive holographic-type communication: Challenges and solutions. *IEEE Communications Magazine*, vol. 58, no. 1, pp. 93–99, 2020.

ITU-T Focus Group on Technologies for Network 2030 Sub-G2, *New services and capabilities for network 2030: Description, technical gap and performance target analysis*, Oct. 2019.

T. Taleb, Z. Nadir, H. Flinck, J. Song, Extremely interactive and low-latency services in 5G and beyond mobile systems. *IEEE Communications Standards Magazine*, vol. 5, no. 2, pp. 114–119, 2021.

X. Wang, Y. Han, V.C.M. Leung, D. Niyato, X. Yan, X. Chen, Convergence of edge computing and deep learning: A comprehensive survey. *IEEE Communications Surveys & Tutorials*, vol. 22, no. 2, pp. 869–904, 2020.

AI-Based Multimedia Tools

H. Kirchhoffer et al., Overview of the neural network compression and representation (NNR) standard. *IEEE Transactions on Circuits and Systems for Video Technology*, vol. 32, no. 5, pp. 3203–3216, 2022.

A.M. Tekalp, M. Covell, R. Timofte, C. Dong (Eds.), Introduction to the issue on deep learning for image/video restoration and compression. *IEEE Journal of Selected Topics in Signal Processing*, vol. 15, no. 2, Feb. 2021.

M.G. Amankwah, D. Camps, E.W. Bethel, R. VanBeeumen, T. Perciano, Quantum pixel representations and compression for Ndimensional images. *Nature: Scientific Reports*, vol. 12, no. 7712, pp. 1–15, 2022.

N. Anantrasirichai, D. Bull, Artificial intelligence in the creative industries: A review. *Artifcial Intelligence Review*, vol. 55, pp. 589–656, 2022.

J. Ascenso, P. Akyazi, F. Pereira, T. Ebrahimi, Learning-based image coding: Early solutions reviewing and subjective quality evaluation. In *Proc. SPIE Optics, Photonics and Digital Technologies for Imaging Applications VI*, vol. 11353, pp. 1–13, 2020.

J. Ballé et al., Nonlinear transform coding. *IEEE Journal of Selected Topics in Signal Processing*, vol. 15, no. 2, pp. 339–353, Feb. 2021.

J. Ballé, V. Laparra, E.P. Simoncelli, End-to-end optimization of nonlinear transform codes for perceptual quality. In *Proc. Picture Coding Symposium (PCS)*, 2016.

A. Basso et al., AI-based media coding standards. *SMPTE Motion Imaging Journal*, vol. 131, no. 4, pp. 10–20, 2022.

M. Benjak, H. Meuel, T. Laude, J. Ostermann, Enhanced machine learning-based inter coding for VVC. In *Proc. of the IEEE International Conference on Artificial Intelligence in Information and Communication (ICAIIC)*, 2021, pp. 021–025.

J. Cai, L. Zhang, Deep image compression with iterative non-uniform quantization. In *Proc. IEEE International Conference on Image Processing (ICIP)*, 2018, pp. 451–455.

Z. Chen et al., Learning for video compression. *IEEE Transactions on Circuits and Systems for Video Technology*, vol. 30, no. 2, pp. 566–576, Feb. 2020.

T. Chen, H. Liu, Z. Ma, Q. Shen, X. Cao, Y. Wang, End-to-end learnt image compression via non-local attention optimization and improved context modeling. *IEEE Transactions on Image Processing*, vol. 30, pp. 3179–3191, 2021.

L. Chiariglione et al., AI-based media coding and beyond. In *Proc. IBC*, 2021, pp. 1–12.

A.F.R. Guarda, N.M.M. Rodrigues, F. Pereira, Adaptive deep learning-based point cloud geometry coding. *IEEE Journal of Selected Topics in Signal Processing*, vol. 15, no. 2, pp. 415–430, Feb. 2021.

P. Helle, J. Pfaff, M. Schäfer, R. Rischke, H. Schwarz, D. Marpe, T. Wiegand, Intra picture prediction for video coding with neural networks. In *Proc. IEEE Data Compression Conference (DCC)*, 2019.

T.M. Hoang, J. Zhou, Recent trending on learning based video compression: A survey. *Cognitive Robotics*, vol. 1, pp. 145–158, 2021.

Y. Hu, W. Yang, Z. Ma, J. Liu, Learning end-to-end lossy image compression: A benchmark. *IEEE Transactions on Pattern Analysis and Machine Intelligence*, vol. 44, pp. 4194–4211, 2022.

C. Huang et al., Online learning-based multi-stage complexity control for live video coding. *IEEE Transactions in Image Processing*, vol. 30, pp. 641–656, 2020.

ISO/IEC JTC1/SC29/WG1, *Performance evaluation of learning based image coding solutions and quality metrics*. JPEG N85013, Nov. 2019.

ISO/IEC JTC1/SC29/WG5, *Use cases and draft requirements for video coding for machines*. MPEG N0023, Apr. 2020.

ISO/IEC JTC1/SC29 WG7, *EE 5.2 on end-to-end AI PC coding*, MPEG w21259, 2022.

ITU-T SG16/WP3 & ISO/IEC JTC1/SC29, *Common test conditions and evaluation procedures for neural network-based video coding technology.* JVET X2016, Oct. 2021.

ITU-T SG16/WP3 & ISO/IEC JTC1/SC29, *JVET AHG report: Neural network-based video coding.* JVET Y0011, Jan. 2022.

T. Li et al., DeepQTMT: A deep learning approach for fast QTMT-based CU partition of intra-mode VVC. *IEEE Transactions on Image Processing*, vol. 30, pp. 5377–5390, 2021.

G. Lu et al., An end-to-end learning framework for video compression. *IEEE Transactions on Pattern Analysis and Machine Intelligence*, vol. 43, no. 10, pp. 3292–3308, Oct. 2021.

H. Ma et al., End-to-end optimized versatile image compression with wavelet-like transform. *IEEE Transactions on Pattern Analysis and Machine Intelligence*, vol. 44, no. 3, pp. 1247–1263, Mar. 2022.

H. Oliveira, J. Ascenso, F. Pereira, Conventional versus learning-based video coding benchmarking: Where are we? In *Proc. SPIE 12177 International Workshop on Advanced Imaging Technology (IWAIT)*, 2022, pp. 1–6.

X. Pan et al., Analyzing time complexity of practical learned image compression models. In *Proc. IEEE International Conference on Visual Communications and Image Processing (VCIP)*, 2021.

F. Sattlere et al., Trends and advancements in deep neural network communication. *ITU Journal: ICT Discoveries*, vol. 3, no. 1, pp. 1–11, 2020.

I. Shiopu, A. Munteanu, Deep-learning-based lossless image coding. *IEEE Transactions on Circuits and Systems for Video Technology*, vol. 30, no. 7, pp. 1829–1842, Jul. 2020.

D. Silhavy et al., Machine learning for per-title encoding. *SMPTE Motion Imaging Journal*, vol. 131, no. 3, pp. 42–50, 2022.

L. Theis, W. Shi, A. Cunningham, F. Huszár, Lossy image compression with compressive auto-encoders. In *Proc. International Conference on Learning Representations (ICLR)*, 2017.

Y. Wang et al., Multi-scale convolutional neural network-based intra prediction for video coding. *IEEE Transactions on Circuits and Systems for Video Technology*, vol. 30, no. 7, pp. 1803–1815, Jul. 2020.

R. Yang, F. Mentzer, L. Van Gool, R. Timofte, Learning for video compression with hierarchical quality and recurrent enhancement. In *Proc. IEEE/CVF Conference on Computer Vision and Pattern Recognition (CVPR)*, 2020, pp. 6627–6636.

R. Yang, F. Mentzer, L. VanGool, R. Timofte, Learning for video compression with recurrent auto-encoder and recurrent probability model. *IEEE Journal of Selected Topics in Signal Processing*, vol. 15, no. 2, pp. 388–401, Feb. 2021.

S. Zhang, D. Zhu, Towards artificial intelligence enabled 6G: State of the art, challenges, and opportunities. *Computer Networks*, vol. 183 pp. 1–28, Dec. 2020.

3 5G Ultrareliable and Low-Latency Communication in Vertical Domain Expansion

Dragorad A. Milovanovic and Zoran S. Bojkovic
University of Belgrade

CONTENTS

3.1 Introduction 137
3.2 5G Vertical Domain Expansion 138
    3.2.1 uRLLC Services 148
    3.2.2 Key Characteristics of mMTC 151
    3.2.3 5G Private Networks, Energy Efficiency, and AI/ML Support 153
        3.2.3.1 Energy Efficiency 156
        3.2.3.2 AI&ML Support 157
3.3 5G Critical and mMTC Connections 158
    3.3.1 Performance Trade-Offs for uRLLC in Industrial IoT 160
    3.3.2 V2X High-Bandwidth, Low-Latency, and Highly Reliable Communication 165
    3.3.3 AI-Enabled Massive IoT Toward 6G 169
3.4 Concluding Remarks 171
Bibliography 172

3.1 INTRODUCTION

5G wireless networks fundamentally evolve from connectivity-based networks to an intelligent platform for delivering new applications to vertical industries. The 5G advanced network is becoming increasingly capable of supporting many services beyond enhanced mobile broadband (eMBB). Vertical applications impose several stringent performance requirements in terms of high data rates, reliability, latency, coverage, and security. They represent the basic driving force for 5G deployment on network infrastructure adapted to network slicing (NS), private networks (NPN), supported by artificial intelligence (AI), and economical scaling through software

DOI: 10.1201/9781003205494-4

in an open ecosystem. It is expected that the revenue of mobile network operators (MNO) from competitive vertical markets will exceed the revenue from basic mobile broadband communication services by 2025.

The wireless industry and industry verticals have invested significant effort in identifying use cases and requirements for the new 5G platform. Ultrareliable and low-latency communication (uRLLC) has been one area that has attracted a lot of interest. 3GPP initiated R15 research by identifying use cases and specifying corresponding requirements, taking into account input from various industry groups and verticals. In continuation of work on the standardization of technical specifications, the use cases and requirements of industrial Internet of things (IIoT) and vehicle (V2X) communications are documented. Continued strong development is necessary to optimize new services based on the criteria of reduced complexity, energy savings, coverage, and cost.

There is significant interest in continuing to improve the 3GPP 5G new radio (NR) platform based on vertical domain extension requirements. Work on version R16 initiates the expansion of 5G systems to vertical domains from Q3 2018 to Q1 2020 with topics such as uRLLC, IIoT, V2X, non-public/private networks, and unlicensed/shared spectrum. It was further accelerated in R17 from Q2 2020 to Q1 2022 with additional improvements in V2X/uRLLC/IIoT, and NTN satellite networks. The R18 RAN research in the Q2 2022 initiates the development of the transition 5G-Advanced with the goal of being functionally complete by the end of the Q4 2023 in evolution driven by eMBB services and vertical domains. Studies of new areas are not only useful for 5G NR development, they also represent the basis for the next 6G generation. It is expected that the 6G standardization is not only limited to the communication system, it will also represent deep integration of communication, intelligence, and computing. 3GPP is scheduled to launch a comprehensive study of the 6G system at the end of 2025 in version R20, and research on technical specifications will start at the end of 2027.

The major sections of the chapter are an introduction of uRLLC, motivation for 5G vertical domain expansion, and 5G AI IoT (AIoT). In the first part, key characteristics of uRLLC and mMTC services, as well as private networks, energy efficiency, and AI/ML support for verticals are presented. In the second part, performance trade-offs for uRLLC in IT, V2X communication as well as AI-enabled massive IoT toward 6G are outlined. We conclude with remarks on 5G capabilities for industrial IoT and automotive vertical industries.

3.2 5G VERTICAL DOMAIN EXPANSION

5G vertical industries are based on guaranteed high-quality performance, such as throughput, low latency, and reliability provided by new mobile networks. The list of significant vertical industries includes telecommunications, manufacturing, automotive, transportation, energy and utilities, public sector, financial sector, retail, healthcare, critical infrastructure, media, and entertainment. Virtualized end-to-end network functionalities enable network operators and the industry to define distinct and isolated logical networks of individual use cases in real-time, built on a shared physical widely deployed commercial infrastructure.

5G Ultrareliable and Low-Latency Communication

Vertical industries are based on the efficient execution of their core business operations. Therefore, 5G capabilities of key interest to verticals include:

- Differentiated communication services and support services that are aligned with specific use needs (bandwidth, quality of service, availability, cost), enabling the delivery of specific cases and services to industry.
- A service guarantee provides specific guarantees in contrast to the current best-effort approach. Although quality guarantees may not be necessary for all services, supporting this functionality enables new applications.
- Seamless connectivity supports reliable end-to-end (E2E) communication, as well as ubiquitous presence ensures coverage through innovative means and integration with various localized communication systems.

It is useful to view individual vertical use cases as a linear combination of three basic service categories. The 5G concept generalizes the basic characteristics of the use case, harmonizes the requirements, and combines the technology components. Each generic service individually emphasizes a different subset of requirements, but all are relevant to some degree. Generic communication services contain service-specific functions and main drivers contain functions that are common to more than one generic service. Three generic services were implemented:

- uRLLC ultra-reliable and low-latency communication enables connecting network services with extreme demands in terms of availability, latency, and reliability. Typical applications are autonomous vehicle communications (V2X) and industrial IoT (IIoT) applications. Reliability and low latency are prioritized over data transfer speed.
- mMTC machine-type mass communication provides wireless connectivity for a huge number of devices on the network, scalable connectivity for an increasing number of devices, efficient transmission of small data packets, wide coverage, and deep penetration. Requirements are simultaneously and mainly focused on bandwidth, latency, and reliability.
- eMBB enhanced mobile broadband supports extremely high data rates and low-latency communication, as well as extreme coverage in the service area. It includes mobile broadband with bandwidth and availability as basic requirements.

5G services are complex constructions, requiring sophisticated configuration deep in the network, driven by a large number of parameters, in a specific network context, and under limited resource availability. Previous cellular systems have developed extremely efficient architectures in terms of system capacity. However, 5G systems achieve not only high system capacity but also low latency, high connection density and energy efficiency. The key metrics of eMBB, uRLLC, and mMTC core services are system capacity, latency, and connection density. eMBB requires high connection throughput and high network capacity, while uRLLC requires low latency and high reliability for both low and high mobile devices. The relative importance of these requirements is shown in Table 3.1.

TABLE 3.1
Importance of Various Requirements for Three Generic 5G Communication Services

Application	Maximum Connection Throughput	Average Connection Throughput	Spectrum Efficiency	Area Traffic Capacity	Connection Density	Latency	Mobility	Network Energy Efficiency
eMBB	+++	+++	+++	+++	++	++	+++	+++
uRLLC	+	+	+	+	+	+++	+++	+
mMTC	+	+	+	+	+++	+	+	++

Competing different requirements impose a new approach to 5G system design:

- Virtualization of network functions (NFV) and software-defined network (SDN) techniques enable the scalability of 5G networks and faster adaptability to new services. NFV supports the optimization of network resources and network slicing NS, allowing virtual networks to operate on a shared physical infrastructure. SDN enables programmable network management and configuration, abstracts physical network resources (routing, switching), separates network control functions from network forwarding functions, and improves network performance.
- Network slicing (NS) supports the design of virtual E2E networks tailored to different applications. NS enables multiple virtual or logical networks with different E2E performance characteristics on a common physical infrastructure. The performance attributes that can be configured in NS are latency, throughput, reliability, capacity, mobility, security, analytics, and cost profile.
- Distributed network architecture. Distributed computing components are implemented in different locations, networked, and communicate with each other. Distributed systems are more efficient in 5G networks due to their scalability, resilience, fault tolerance, and less network congestion. The distributed architecture of multiaccess edge computing (MEC) integrates network virtualization, cloud computing (CC), software-defined network (SDN), and network slicing (NS). The benefits are reduced network latency and congestion, as well as improved reliability and scalability. The basic idea is that computing is performed at locations near the edges of the network and that more content is cached. Thus, it offloads the core network (CN) traffic, efficiently and dynamically manages traffic flows, adapts network resources for each application, and reduces E2E latency. The distributed architecture supports a new class of cloud-based networks in use cases such as automated vehicles that require ultralow latency and high throughput.
- Latency is the time interval in which a sender receives a response in one-way E2E communication. Two types of delays are considered. The user plane (UP) delay is related to the contribution factor of the radio access network in the period of sending a packet from the source to reception at the destination. It is defined as the one-way time of successful packet delivery (at the application layer) from the ingress point to the egress point of Layers 2 and 3 of the SDU radio protocol. An unloaded network condition is assumed for UL/DL services, the user equipment (UE) is in the active state. UP latency is called transport delay in ms units of measure. The latency of the control plane (CP) is defined as the time difference of the transition of the UE from the inactive state (*Idle state*) to the state of initiation and continuous data transmission (*Active state*). The minimum delay requirement is 20 ms, but less than 10 ms is preferable. Low latency is a fundamental feature of the uRLLC system. Depending on the use cases, the delay requirement can be even more stringent, which is the most difficult 5G system design goal. E2E delay analysis from the access network, aggregation, and CN is the basis for planning the placement of nodes in the network (Figure 3.1).

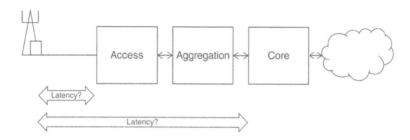

FIGURE 3.1 High-level architecture and latency considerations.

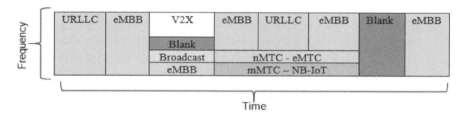

FIGURE 3.2 Frame structure of 5G new radio (NR) with varying length of slots per subframe to enable different applications.

- **New air interfaces.** 5G NR technology is based on scalable OFDM (orthogonal frequency-division multiplexing) radio interface technology. It is designed to be configurable for frequency ranges from sub-GHz to millimeter spectrum up to 100 GHz. The frame structure and radio procedures can be configured to achieve very low latency. 5G NR air interfaces support numerous services, enable the implementation of different network types, and improve system performance. The key components of 5G NR are OFDM-based waveform with cyclic prefix, and windowing, pre-coding, and filtering can be added. The flexible frame structure includes different numerologies and a wide carrier bandwidth as well as high-order modulation. A new channel coding schemes include low-density parity check (LDPC) codes for the UP and polar codes for the CP, massive MIMO, and beamforming. In the downlink (DL), the NR supports a physical control channel (PDCCH) and a physical shared channel (PDSCH). In uplink (UL), uplink control channel (PUCCH) and uplink shared channel (PUSCH) are specified (Figure 3.2).
- The NR radio protocol stack is split into control and user planes. Figure 3.3 shows the radio protocol stack and interfaces for the 5G CN (mobility management function (AMF), user plane function (UPF) from the perspective of UE in case the connections to the base station gNB via NR radio access.
- Reliability is the ability of a network to successfully transmit data over a period of time. It defines as the ability to transmit given amount of traffic in predetermined time with a high probability of successfully transmitting Layers 2/3 packet within the required maximum time. The minimal

5G Ultrareliable and Low-Latency Communication

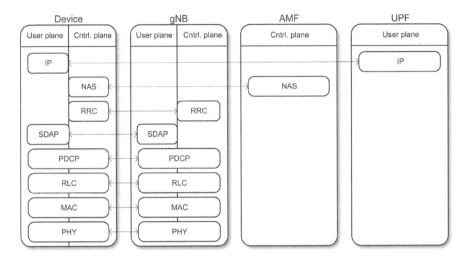

FIGURE 3.3 NR control and user plane protocols and interfaces as seen from the device UE.

requirement defined in report ITU-R M.2410 is 10^{-5} probability of successful transmission of Layer 2 PDU (protocol data unit) of 32 bytes within time limit of 1 ms in urban conditions of uRLLC test environment. Node reliability is defined as the probability of communication between nodes. The probability is calculated by packet error probability, packet drop probability, and queuing delay violation probability. CP reliability represents the probability of successful metadata decoding. Availability is defined as the probability of service available to a user in given coverage area and time, related to channel interference and network implementation.

5G system implementation is based on the 3GPP phased research of multiple architectures specified in the standard. Each phase (Release) usually lasts 1–2 years, and they are sequentially indexed. A release typically develops from study phase (Study Items) to specification phase (Work Items). After each release is completed, the resulting specifications undergo a series of corrections and fixes, often lasting several months or longer. The Working Group (WG) usually has 15–18 months to work on a release and ensure that all agreed relevant topics fit within a strict time frame and are implemented in products on the market within 1 or 2 years after the release is completed (frozen). The biggest advantage of short release phases is the guarantee that only the most relevant topics make onto the release list.

3GPP has successfully completed the first implemented specifications of 5G-NR in releases R15 and R16, which enabled full-scale deployment of 5G-NR. 3GPP continues to advance E2E network functionalities in the network core (5GC), which are crucial for mobile operators to explore advanced opportunities for different vertical market segments. 5GC supports network slicing (NS) and differentiated quality of service (QoS) end-to-end from radio, transport to core, and application server. Vertical markets rely on 5G-NR and 5GC new functionalities for more efficient and reliable services.

The first release of 5G R15 was developed in a short time interval from Q1 2016 to Q1 2017, followed by 9 months of research for the first release focusing on non-standalone (NSA) deployment on LTE EPC core network. This was followed by another 6 months of development for stand-alone deployment (SA) on the new 5GC network core, and another 6 months for the late release specification of additional use cases and architecture options. 5G R15 Phase 1 is focused on eMBB services of enhanced broadband access.

5G R16 Phase 2 supports stand-alone (SA) deployment on the 5GC core network. 5G devices use both UP and CP planes. In addition, the 3GPP standard defines the service requirements and technical goals of the 5G system. R16 extended the 5G standard to new verticals such as NR sidelink (SL) for vehicle V2X communication and public safety, 5G broadcast as an evolution of LTE-based enhanced EnTV, private NPN networks, IIoTs, and shared/unlicensed spectrum (5 GHz and lower 6 GHz bands).

The continued evolution in R17 includes support of satellite communications NTN (non-terrestrial networks) for smartphones and IoT devices, the specification of new user devices UE with reduced capabilities (RedCap) for surveillance cameras and industrial sensors, multicast and broadcast services (MBS), and expansion into FR2 frequency mmWave band. At the same time that 3GPP is technically working on R18 5G-Advanced, new topics for R19 are emerging in parallel (Figure 3.4).

3GPP has supported vertical industries since the first edition of the 5G specifications: uRLLC and mMTC communications. The following work items represent the vertical domain enhancements in R18 5G-Advanced.

Positioning evolution. NR positioning continues its improvements in R18 covering various aspects. Support for sidelink-based positioning and ranging is being explored for V2X, public safety, and commercial and industrial IoT use cases. Integrity improvements for RAT-dependent positioning and precision improvement solutions based on bandwidth aggregation and carrier phase measurements are also explored. In addition, the requirements for low-power, high-precision positioning will be evaluated against current techniques to identify any potential challenges. Finally, the positioning for RedCap devices will be explored with the added challenge of operating in a narrower bandwidth.

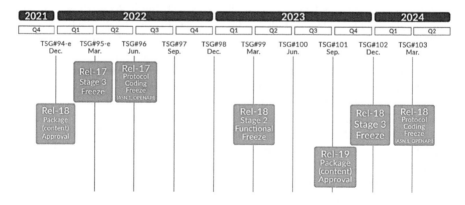

FIGURE 3.4 3GPP R18 5G-Advanced timeline.

5G NR-RedCap (NR-light/lite) evolution. One requirement is emerging for supporting all three primary use cases with a single type of RedCap device in order to reduce market fragmentation and maximize the benefits of economies of scale. Effective support of use cases that are between eMBB, uRLLC, and mMTC impose 3GPP to investigate NR-RedCap devices. UEs with reduced complexity are critical solutions to certain important use cases where requirements are relaxed. The following complexity/cost reduction features are standardized as part of the work item within R17: maximum device bandwidth, minimum UE antenna configuration, minimum number of MIMO layers supported in the DL direction, maximum DL modulation order, and support for half-duplex operation in frequency division duplex (FDD) bands. In addition to the reduced capabilities that mainly involve the physical layer of the radio protocol stack, the following reduced capabilities for multiple layers have been introduced for RedCap devices: maximum number of data radio bearers (DRBs) that the device must mandatorily support from 16 to 8, sequence number (SN) length associated with the packet data unit for packet data convergence protocol, and radio-link control layers in acknowledged mode from 18 to 12 bits, support for automatic neighbor relation (ANR) functionality is not mandatory (Table 3.2).

DECT-2020 (NR+) is the world's first non-cellular 5G IoT technology standard submitted by ETSI DECT Forum and recognized by WP5D ITU-R as an IMT-2020 technology. In the standardization process, each new technology is required to submit a self-evaluation report to ITU-R. Independent evaluation groups (IEGs) evaluate the self-report and proposed radio interface technologies (RITs) and the set of RITs (SRITs) following the guidelines specified by the ITU-R in report M.2412. The results of the evaluation are then submitted to ITU-R and discussed at the 5D working group (WP) meetings. After several such meetings, RIT and SRIT technologies are accepted as IMT-2020 5G standard. The DECT-2020 component technology meets the minimum performance requirements for uRLLC and mMTC usage scenarios. ETSI TS 103.636 specifies MESH topology and relies on multi-hop communication to bring a device onto the network. DECT-2020 is the first 5G technology that supports spectrum sharing and multiple local networks in a mobile system in the 1.9 GHz frequency range. Advantages in IoT application areas are infrastructure-less architecture, autonomous, decentralized, no single point of failure, low cost, and simple to deploy.

NTN IoT evolution. Improvements of mobility, and application of R17 NR NTN mobility enhancements are discussed. The need to support network-verified UE location is also considered in the context of satisfying regulatory requirements

TABLE 3.2
5G-IoT Trade-off Capability vs. Complexity with NR-RedCap

Use Cases	Latency	Throughput	Density	Complexity	Battery Life
Massive IoT	++++++	++	++++++	++	++++++
RedCap NR	++++	++++	++++	++++	++++
Mission-critical IoT	++	++++++	++	++++++	++

(accuracy, privacy, reliability, latency) for potential use cases and services such as emergency calls, lawful interception, and public warning. The aim is to enable the network to verify and confirm the location of the UE derived from its various measurement reports.

Sidelink evolution. R18 extends sidelink operation in multiple directions. Sidelink operation in unlicensed spectrum is focused on FR1 unlicensed band (5 and 6 GHz). Multi-beam operation for sidelink on licensed FR2 bands and definition of carrier aggregation are also explored. R17 defines UE-to-network relays (U2N) for Layers 2 and 3 architectures. In R18, the sidelink relay capability is extended to UE-to-UE relay (L2/L3) assuming one hop and only unidirectional traffic. Improvements to the continuity of service and DRX power down sidelink for the L2 U2N relay are also listed in R18 as a continuation of the work from R17. Finally, multipath operation with direct and indirect paths will be investigated, where either a path is selected or both are simultaneously used. The indirect path is via L2 U2N relay or another UE with ideal inter-UE connection.

Smaller than 5 MHz bandwidth on dedicated spectrum. 5G NR requires minimum system bandwidth of 5 MHz transmitting the signal and channel in initial access. R18 introduces the possibility of deploying NR in a dedicated spectrum with a bandwidth of less than 5 MHz on request of SmartGrid and railways operators. It is intended to support spectrum allocation of 3–5 MHz bandwidth with minimal specification changes, using only 15 kHz subcarrier spacing and normal cyclic prefix. For example, the current synchronization signals will be reused.

3GPP R18 research contains features that explore new areas as well. It includes AI/ML applied to the RAN air interface and to the network-evolved duplex beyond traditional frequency division duplex (FDD), and time-division duplex (TDD) identifies ways to improve network energy efficiency and dramatically reduces UE power consumption.

AI/ML for NR air interface. 3GPP AI/ML research was limited to data collection for various network functions to enable network automation. A new study is launched in R17 to identify use cases that enable AI/ML deployment on the network, leading to R18 standardization work focusing on improvements to data collection and signaling to support AI/ML-based network energy savings, load balancing, and optimization mobility. Research is limited to the definition of data collection metrics and their reporting, allowing the network to run its own AI/ML algorithms. After that, an additional study identifying other new use cases is expected. 3GPP R18 explores how AI/ML techniques solve NR problems with a focus on four areas. The first area concerns the selection of use cases to be studied. An initial set of representative use cases have been identified that include improvements to the channel state information (CSI) reference signals, beam management (BM), and positioning to serve as a pilot for future use cases. AI/ML-based techniques are being explored to improve the performance of selected use cases. The next area studies AI/ML models and their life cycle in relation to different levels of cooperation between network and terminal. A description of model training, inference, testing, and verification is included. The third area evaluates performance and comparison with non-AI/ML-based solutions. Different key performance indicators (KPIs) are being identified for different use cases. AI/ML-based techniques are identified in terms of performance and corresponding complexity. Complexity directly or indirectly relates to significant aspects such as power consumption and memory usage. Finally, the R18 study focuses on the impact of specifications on the implementation and interoperability of AI/ML-based techniques, as well as their appropriate configuration, maintenance, testing, and verification.

Evolution of duplex operation. Wireless networks are deployed in the FDD or TDD spectrum usage techniques. In FDD, the amount of spectrum allocated for DL/UL operations is identical within a carrier. In TDD, the partitioning of resources into DL and UL can be adjusted based on different traffic needs. However, in high-power, macro-cell TDD deployments, special attention has to be paid to avoid severe crosslink interference from not having the same partitioning in the entire network or possibly across operators deployed on adjacent carriers. R16 introduces support for crosslink interference handling for dynamic TDD applications. Notably, converting DL slots to UL is less problematic than converting UL slots to DL for dynamic or flexible TDD operations. Converting UL–DL slots interferes with the reception of neighboring UEs, leading to UE-to-UE interference problems. On the other hand, converting UL–DL slots additionally generates inter-gNB interference, which is typically more significant. Coexistence with legacy operation and inter-operator interference are also challenges to be addressed. Enhanced duplexing in R18 explores subband non-overlapping full duplex (for the gNB) and other potential improvements to dynamic/flexible TDD. Full duplex operation can cause two problems. First, dynamic TDD scenario is created. In addition, full duplex generates self-interference, which stems from the own transmission interfering the own reception at a given time. The assumption that the frequencies do not overlap increases the isolation between transmission and reception because they will not take place on the same frequency resources, creating first level of isolation. Additional levels of isolation are achieved by spatial means, effectively creating different beams for transmission and reception. Finally, the third level of isolation is achieved by nonlinear cancellation techniques on a node that has full duplex capabilities. The goal is to minimize the need for this third level of isolation.

Green networks. Electricity represents 20%–40% of the OPEX costs of mobile network operators, with base stations consuming over 50% of energy in a mobile network. 5G NR is designed with lean underlying structure with very little reliance on always-on signals, so no further standard enhancements are specified. In particular, 5G NR inherently relies on a large number of antenna elements and broadband transmissions. Together with the lower efficiency of the power amplifier as the carrier frequencies increase, it affects the increased power consumption compared to previous generations. Techniques with and without UE support that enable reducing network energy consumption will be explored and specified. Network energy-saving techniques are expected to adjust transmissions in the time, frequency, space, and energy domains to optimize network energy consumption while maintaining adequate QoS for users.

Low-power wake-up signal (LP-WUS). Minimal always-on signals in R15 5G NR are a challenge for UE energy saving, especially due to mobility management, so improvements of UE energy saving in R16, R17, and R18 have been initiated. The common denominator of all previous techniques is the reliance on existing signals and channels, for example, the wake-up signal (WUS) introduced in R16 in the form of DL control channel. In contrast, LP-WUS in R18 is not limited to using existing signals. Instead, wake-up receiver architectures and signal designs for small IoT devices and wearables are being explored, with the goal of achieving significant gains compared to the existing energy-saving mechanisms.

3.2.1 uRLLC Services

A main feature of uRLLC is the ability to transmit data with high reliability within a guaranteed delay limit. However, high reliability at low latency imposes high demands on the 5G system. A very robust connection adaptation is necessary to ensure high reliability. Moreover, it is important to use the diversity as much as possible in the spatial domain, using multiple antennas, or in the frequency domain. To achieve low latency, it is necessary to use appropriate data frame structures. The OFDM system with scalable numerology allows configuring larger spacing between carriers, so that symbol duration and slot duration are reduced. Moreover, the frame structure with early reference symbols and control information is useful. With such configuration, the receiver starts decoding the received packets very early after the initial channel evaluation. For increased robustness during mobility, dual connection is used when data is transmitted over multiple frequency layers.

uRLLC is optimized for ultralow-latency and/or ultrahigh-reliability communication, which is the traffic characteristic of many applications. At the same time, not all applications have large range of required performance. ITU-R specifies user plane delay of 1 ms, control plane delay of 20 ms, reliability of 99.999% with a packet error rate of less than 10^{-5} to 10^{-9}, and near-zero interruption time. Depending on the target applications, latency requirements vary from 10 ms (process automation) to less than 1 ms (factory automation, safety-related V2C communications).

The stringent requirements of uRLLC services impose extreme challenges for wireless networks, the need to optimize for both link-level and system-level efficiency, and it is necessary to coexist with other types of mMTC and eMBB services. Unfortunately, improvements in latency and reliability of wireless communication systems cause certain performance metrics to be sacrificed. System design is complex process. Many design parameters and metrics are in trade-offs relationship because high reliability requires high-latency. In addition, reliability and data transfer speed require trade-off. In coding theory, higher coding rate gives us higher data transfer rate. However, the probability of error is higher and the reliability is lower. Energy efficiency and latency are also in trade-off relationship. Therefore, designers of wireless communication systems strive to find good trade-off points or optimize systems under reasonable design requirements.

3GPP NR did not just support uRLLC as an add-on feature of the existing system design. Since the first version R15, it has been an integral part of the basic NR design. Clearer system design could be done from the scratch up because there was no old system to keep and adapt. Aimed to accommodate a wide range of applications including uRLLC with varying performance requirements, R15 supports high flexibility and configurability at the service provider's choice based on targeted usage requirements.

5G Ultrareliable and Low-Latency Communication

Support of low latency. Some uRLLC use cases limit E2E latency very strictly, on the order of few milliseconds or even less than 1 ms in extreme cases. Taking into account the delay caused by other network components, the delay limit for air interface transmission can be very constrained. Achieving the goal of low latency in NR, it is necessary to optimize every component of the delay in the data delivery procedure. Key features of the NR R15's low-latency design are as follows:

- scalable numerology, with shorter symbol duration for higher subcarrier spacing
- short transmission duration for DL and UL control channels, as short as one symbol
- short transmission duration for DL and UL data channels, as short as one or two symbols
- multiple DL control monitoring occasions within a slot, which reduces the waiting time to schedule a DL or UL data transmission
- grant-free UL data transmissions, which allow the UE to transmit data in configured resources without the need to send scheduling request and wait for UL grant
- flexible TDD frame structure, which allows the gNB to flexibly change the DL and UL direction to accommodate the traffic need
- optimized and significantly reduced UE processing time for both DL and UL.

Support of high reliability. Hybrid automatic repeat request (HARQ) is a very efficient way in wireless systems in order to achieve low block error rate (BLER), or high reliability, after multiple HARQ transmissions. For uRLLC applications, when both high reliability and low latency are required, HARQ may not always be as useful for high reliability because there may be very limited retransmission options within the delay budget. In extreme cases where no retransmission is allowed, the packet is delivered in a single transmission with very high-reliability requirement (10^{-5}). Therefore, it is necessary to design each channel in the system so that it can achieve high-reliability goal in these use cases. There are some basic features in NR that are useful for achieving high reliability, such as:

- **Channel coding design.** In the process of the channel coding design, efficient HARQ support has been one consideration, which is mainly to improve the efficiency. At the same time, low error floor was another consideration factor, which allows very low BLER to be achievable.
- **Time diversity.** This can be achieved by HARQ retransmissions if the latency budget allows, but for uRLLC, the time diversity gain is typically very limited because the HARQ transmissions cannot span over a long duration.
- **Frequency diversity.** This can be achieved by distributed frequency resource allocation (for data and control channel in case of CP-OFDM) or frequency hopping in case of contiguous frequency resource allocation, e.g., for PUSCH and physical uplink control channel (PUCCH).

- **Spatial diversity.** This can be achieved by spatial diversity transmission schemes (e.g., precoder cycling) in R15.
- **High aggregation level for PDCCH.** For PDCCH, the maximum aggregation level of 16 is supported, compared to the maximum aggregation level of 8 in LTE. Higher aggregation level reduces the effective code rate, which allows higher reliability.
- **Slot-based repetitions.** Repetition is a common approach to improve coverage and reliability. Slot-based repetition (transmission in several consecutive slots) is supported for PDSCH, PUSCH, and PUCCH. From reliability perspective, slot-based repetition for data channels is especially useful when the latency budget does not allow the transmitter to wait for the feedback to initiate HARQ retransmission. This is sometimes called blind repetition.

In addition, specific enhancements have been introduced to support uRLLC high-reliability requirements.

- **CSI reporting enhancements.** Traditionally for non-uRLLC traffic, the BLER target for CSI reporting is 10%. For uRLLC, the initial BLER target is likely to be significantly lower than 10%. If the UE only reports CSI with 10% BLER target, it is quite difficult for the gNB to accurately determine the corresponding MCS for a lower BLER target. Therefore, a new BLER target of 10^{-5} for CSI reporting has been introduced and can be configured for a CSI report. Given that lower BLER target naturally means lower spectral efficiency, a new channel quality indicator (CQI) table with lower spectral efficiency entries has been defined accordingly.
- **MCS table enhancements.** Tied together with the CSI reporting enhancements to support lower BLER, a new MCS table with lower spectral efficiency entries was introduced. The lowest spectral efficiency supported becomes 0.0586 bps/Hz instead of 0.2344 bps/Hz in regular MCS tables.
- **PDCP packet duplication.** Reliability can be additionally improved by enabling higher layer packet duplication. In NR, this is done by packet data convergence protocol (PDCP) packet duplication, which enables a packet to get transmitted with two independent radio paths (e.g., in two different carriers) over the air interface.

Basic support for uRLLC was specified in NR R15, while further performance improvements were explored in R16 in two study items followed by two work items. The improvements were targeted for the use cases of the first R15 version and effectively support new use cases such as factory automation, transportation industry, and power distribution.

Design goals in R15 are 1 ms latency and 10^{-5} reliability, and R16 research goals are tightened to 0.5 ms latency and 10^{-6} reliability. At the same time, improvements cover both UE performance and handshake between UEs to improve system performance. All channels carrying control information and user data, including PDCCH, PUCCH, PDSCH, and PUSCH, are enhanced in certain ways to improve latency and/or system reliability or efficiency. Although many new uRLLC features are introduced in R16, there is still space for further improvements in many aspects. A dedicated uRLLC improvement study was conducted in R16. One of the goals of the study is to establish the baseline performance that can be achieved with NR R15, which includes considerations specific to uRLLC and design considerations applicable to both eMBB and uRLLC. Based on the baseline performance, additional improvements to uRLLC were explored and introduced in R16.

A new work item WID R17 on improving support for IIoT and uRLLC contains the following enhancements for physical layer: improvements for intra-UE multiplexing and prioritization, HARQ-ACK improvements, and CSI feedback improvements. The NR specifications already contain large set of tools to meet the requirements of various uRLLC applications. At the same time, it is clear that standards are evolving and supporting new types of applications.

uRLLC applications place strict requirements on E2E performance, so special care is required in every aspect of network design, from system architecture to physical layer design. These include, but are not limited to QoS differentiation in the CN that is transferred to the radio access network (RAN), mobile edge computing that reduces network latency, an always-on session protocol data unit (PDU), a new idle state in the radio control protocol resource allocation (RRC), packet duplication to improve reliability, and various techniques to improve latency and reliability at the physical layer.

3.2.2 KEY CHARACTERISTICS OF mMTC

mMTC connectivity supports numerous devices connected to the Internet or communicating with each other in the so-called massive machine-to-machine (M2M) communication. The connectivity is designed for IoT services. The basic characteristics of systems include interconnectivity, heterogeneity, dynamic changes, and enormous scale. The key challenges are the development of massive communication links for devices that are distributed over wide area and consume ultralow power. The basic requirements can be summarized as massive connectivity, low-cost devices, ubiquitous coverage, low data transfer rates, and ultralow power consumption.

IoT technology is significantly influencing the development of numerous vertical sectors, including factory automation, automotive, healthcare, smart cities, and homes. Depending on targeted applications, the key design considerations of IoT systems are different: connectivity solutions, device requirements (battery life, processing power, memory, bandwidth, throughput, sensing, actuation), network topology, security mechanisms, power management mechanisms, data processing mechanisms, and management via IoT device or servers.

Traditional cellular systems are designed for the transport of large-sized packets, high data rates, and communication links dominated by the DL direction of data transmission. However, mMTC systems require fundamentally changed approach based on small package sizes, low data rates, and communication links dominated by the UL data direction. In addition, the requirement that mMTC systems contain huge number of devices is assumed. The key research challenges and design approaches can be summarized as follows: signaling of unpredictable and sporadic activities of mMTC devices, resource management for massive number of connections, power control and management mechanism, and massive access schemes.

3GPP started MTC research in version R12 with various use cases. The basic characteristics of MTC services are summarized in the following three distinct design goals: low cost/complexity, low power consumption, and extreme coverage. The main drivers for cost savings are reduction of maximum bandwidth, reduction of peak speed, reduction of transmission power, half-duplex (HD) operation, and reduction of supported DL transmission modes. However, the above techniques are in trade-off relations with coverage and power consumption.

MTC research has intensified since R13. The standardization resulted in two branches of support: enhanced MTC (eMTC) and narrowband Internet of things (NB-IoT). The research provided solid foundation for the continuous evolution of MTC in 3GPP, including the R14, R15, R16, and R17 releases. Evolution focuses on further expansion of different MTC types, new services, improved efficiency/latency/power consumption, and applicability. The turning point is the integration of NB-IoT and eMTC into the 5G NR system as de facto mMTC solution, both from the perspective of the radio and from the perspective of the CN. Not only is the duplication of standardization work in 3GPP avoided, but also the potential fragmentation of commercial applications is avoided, thanks to the single continuous direction of MTC standardization shared by 4G LTE. The basic features are summarized as follows: data rate 250 kbps, bandwidth 180 kHz, 50,000 connections per cell, battery life up to 10 years, module price under $5, transmit power 20 dBm, and UL delay less than 10 seconds. NB-IoT includes many advantages such as long battery life, wide coverage, better scalability, QoS and security compared to unlicensed technologies, and coexistence with previous generation cellular systems. However, the disadvantages of NB-IoT are the lack of support for roaming and voice transmission, as well as the relatively low data transfer rate.

NB-IoT overview. To operate in licensed spectrum, narrowband IoT was developed by 3GPP as R13 in June 2016. NB-IoT is subset of the LTE standard, uses limited bandwidth, and relaxes the high-bandwidth configuration. In R14, key improvements have been introduced including a new UE category with higher capabilities, multicast DL transmission based on the agile SC-PTM broadcast/multicast mechanism, and support for multilateral OTDOA positioning. Further improvements were introduced in R15, including reduced system acquisition time by improving cell search and system information collection performance, delay, and power consumption reduction techniques (wake-up signal or channel, relaxed cell reselection tracking, data transfer during random access procedure). In R16 eMTC and NB-IoT are enhanced with group WUS, UL link transfer without authorization, scheduling of multiple transport blocks single grant and coexistence with NR.

eMTC overview. 3GPP initiated research on improved eMTC in R13. In order to reduce device costs, a new UE category (Cat-M1 UE) was introduced, which has an RF bandwidth of 1.4 MHz (compared to 20 MHz for LTE), one receiving antenna chain (compared to two for LTE), the maximum size of the transport block of only 1,000 bits, and usually operates in HD mode. Despite being bandwidth-limited, the Cat-M1 UE can operate within any bandwidth of the LTE system, which is achieved by defining special procedures and channels in cells for bandwidth-limited operation.

In R14, key enhancements were introduced including a new category of higher capability UE (Cat-M2) supporting 5 MHz bandwidth, higher peak data rates for Cat-M1 and Cat-M2, single-cell point-to-point multicast DL transmission multipoint (SC-PTM), and support for multilateral OTDOA positioning. Further improvements were introduced in R15, including reduced system acquisition time by improving cell search and system information collection performance, delay and power consumption reduction techniques, and techniques to improve spectral efficiency (SE) such as 64-QAM support, sub-PRB resource allocation. R16 introduces further enhancements to eMTC, including group wake-up signal, handover of UL link without grant, scheduling of multiple transport blocks using single grant, and coexistence with NR. R17 introduced 5G NR support for satellite communications and is starting project focusing on adapting eMTC/NB-IoT operation to satellite communications.

There are numerous types of MTC devices. Obviously, different use cases may translate into different sets of requirements in terms of cost, complexity, power consumption, and coverage. It is crucial for 3GPP to respond to the needs of MTC communications in an integrated manner with existing and future evolved networks, within the existing standardization framework. Even today, the evolution of standardization continues in a nonstop manner.

3.2.3 5G Private Networks, Energy Efficiency, and AI/ML Support

A private network is a dedicated 5G network with enhanced communication features, unified connectivity, optimized services, and customized security within specific area. Private 5G networks are good solutions for various vertical industries to build dedicated and secure wireless networks in industrial environments. They enable the implementation of services with improved flexibility and device connectivity that encourages Industry 4.0 and the digitalization of the energy and transport sectors.

A private 5G network, non-public network (NPN) in 3GPP terminology, enables the use of technologies to create a dedicated network with unified connections, optimized services, and secure means of communication within a specific area. A private network can be operated by either the company itself or a third party, based on the same or different spectrum owned by the MNO. Unlike public land mobile networks (PLMNs) that offer mobile network services to public subscribers, NPN networks are intended exclusively for use by enterprise customer, such as an industry vertical or a government company. There are two basic deployment options: SNPN that does not rely on PLMN network functions, and PNI-NPN whose deployment is supported by PLMN. Private networks such as SNPN allow the enterprise user to maintain full control over the network. On the other side, the introduction of the PNI-NPN network reduces initial capital CAPEX and operational OPEX costs.

Stand-alone NPN (SNPN) is a private network that operates independently of the public PLMN network, not depending on PLMN network functions. It requires a 5G system separate from the PLMN, and NPN devices must have SNPN subscription to access it. The SNPN is uniquely identified by the combination of the identifier PLMN ID and Network ID (NID). Therefore, the UE is configured with {PLMN ID, NID} to access the private network. The PLMN ID identifier can be the private network ID (based on the MCC 999 mobile country code assigned by the ITU for 3GPP) or the PLMN ID identifier used by the SNPN private network. A NID can be assigned independently (selected by SNPN at implementation time) or coordinately assigned (universally managed NID).

Public network integrated NPN (PNI-NPN) is a private network implemented with the support of the public PLMN. It is necessary that NPN network devices must be subscribed to PLMN in order to access the private network. A publicly integrated private network is supported by the PLMN using dedicated data network name (DNN) or by deploying NS allocated for the private NPN network. 3GPP has defined the basic means for implementing private PNI-NPN networks; however, the implementation of E2E solutions in real 5G systems is still a huge challenge.

Based on these criteria, two scenarios were selected and shown in Figure 3.5 with a private NPN network formed by local data network providing MEC environment for vertical service applications. Edge computing (EC) is an efficient technology that supports private 5G networks. Compared to traditional CC, EC is a decentralized

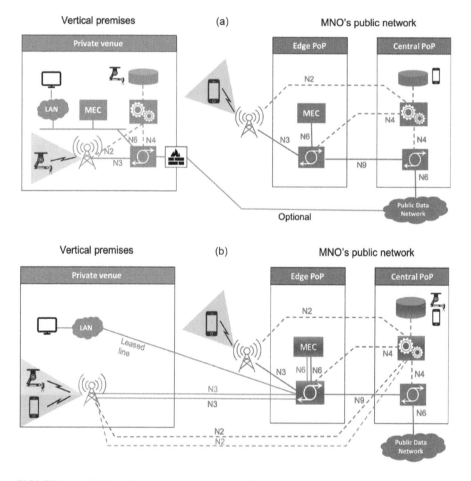

FIGURE 3.5 NPN deployment scenarios: (a) SNPN without RAN sharing and (b) PNI-NPN with full sharing.

computing paradigm, in which computing-intensive tasks and data storage are performed at the edges of network. Due to the close proximity of UE and edge servers, EC enhances local context, improves privacy security, supports cloud storage, reduces energy consumption, and shortens response time. It is necessary to note that EC and CC can cooperate with each other. Specifically, based on high computing and storage capabilities, CC processes non-real-time big data, while EC performs real-time tasks and makes real-time decisions. In general, the reference architecture of EC can be divided into device layer, edge layer, and cloud application layer. Private network computing further enhances EC benefits by adding secure and private services for local requirements and network settings.

The key requirements of private 5G NPN networks from the viewpoint of industry verticals are:

- Guaranteed QoS parameters for NPN networks include throughput, delay, jitter, and availability. The performance requirements in certain use cases of NPN networks are stringent than those imposed by public services in PLMN networks.
- Customization refers to the need for flexibility to include (and configure) additional features for the 5G system to meet user needs in terms of functionality and performance. Unlike PLMN networks, where 5G system is built with components and configuration settings that allow for traffic adaptation/subscriber growth from user-centric services, in NPN networks, 5G systems are designed to handle the specific use cases.
- Network control represents the intention of individual customers of the enterprise to maintain control over their networks, e.g., managing the configuration of certain network functions and deciding on traffic flow policies. PLMN networks are categorized as critical infrastructure, and therefore it is not acceptable for the network operator to allow third parties to freely access the operational and service subsystem (OSS) and network assets.
- Data protection represents the need for customers to ensure that unauthorized entities do not have access to sensitive data, including operational data (configuration information, logs, tracking data), subscriber data, and business-related data (billing information). Proper data security involves implementing appropriate security mechanisms (encryption, secondary authentication), implementing some network functions on-site (unified data management (UDM), user plane function (UPF) and ensuring certain level of redundancy.
- A targeted coverage zone in specific geographic area guarantees that radio signals are locally limited to avoid interference with public subscribers and ensure private communications. It is important to note that QoS is only guaranteed in areas where the enterprise requires coverage. Furthermore, coverage of NPN networks outside the target area is undesirable for the reasons previously stated.
- Backward compatibility in many private use cases requires integration of 5G NPN networks with existing legacy technologies (Wi-Fi, Industrial Ethernet). In this way, the entry barrier is reduced because enterprises

gradually apply the NPN network, while at the same time keeping certain parts of the existing private network unchanged.
- Key solutions enabling 5G NPN networks include spectrum access options, deterministic transport networks, NS, integration with existing private networks, positioning, Open RAN, local edge computing EC, and security and privacy features. These solutions play a significant role in fulfilling the above requirements and complement the capabilities defined in 3GPP standards to support 5G NPN networks.

Basic challenges and future directions of research arising from the realization of 5G NPN networks are NS network intersection, ZTM network management practices, enabling and validating 5G NPNS networks with E2E deterministic QoS support, as well as the availability of PNI-NPN capabilities. NS is a key technology enabling the deployment of PNI-NPN public integrated private 5G networks. NS divides physical network into multiple logical, or network slices, each of which is specialized to provide specific network capabilities and features for specific use case. NFV and SDN are the two core enabling technologies for NS. Network slice architectures consist of an infrastructure layer, an NS instance layer, and a service instance layer. The life cycle of a NS includes four phases: preparation, commissioning, operation, and decommissioning.

3.2.3.1 Energy Efficiency

3GPP Technical Specification Group SA5 researched the energy efficiency (EE) and energy saving (ES) of mobile networks, so in R17 it expanded the scope from the RAN access network to the entire 5G system. EE key performance indicators (KPIs) are defined for the core 5GC network and network components. RAN energy efficienc is defined by the ratio of performance to electrical energy consumption (EC), where the specification of performance depends on the type of network entity to which it refers. Based on this, SA5 defines the best metrics for each of them and their measurement method. The performance of network slices is defined based on the type of NS, i.e., for eMBB, uRLLC, mMTC, and assumption that the volume of traffic at the user level defines the performance of the core of the 5G CN.

A bottom-up approach is adopted for measure energy consumption, starting from the electrical consumption estimation of network functions up to 5GC and NS. Measurement EC of physical network functions PNF is defined by ETSI EE. 3GPP SA5 has defined a method for estimating based on the estimated energy consumption of the underlying instance of virtual computing resource VM (virtual machine). In R17, only the use of virtual CPUs in the VM, obtained by the OSS operator support system ETSI NFV MANO, was taken into account to estimate the electrical consumption. 3GPP SA5 defined mechanisms for collecting, through the operations, administration, and maintenance of OAM standardized API programming interfaces, measurements from 5G network functions, regardless of the type of measurement, including measurements related to performance and energy consumption, over a common OAM channel.

3GPP SA5 working group has also adopted in its network resource model attributes that allow users of the NS network interface to express their requirements regarding the EE of the network slice. The order is based on the GSMA NG.116 generic network interface template, and be periodically informed about the actual EE of the network interface efficiency of the network slice they obtained.

During the work on R18, the SA5 group focused its efforts in various directions, including:

- Work with ETSI NFV to obtain more accurate virtual CPU usage measurements from ETSI NFV MANO, as currently defined measurements lack accuracy.
- Introduce additional metrics when estimating the energy consumption of virtual machines, e.g., their virtual disk or link usage.
- Extend our existing solution to container-based network functions (CNF), in addition to VM-based virtualized network functions.
- Investigate new use cases for energy saving, applied to NGRAN, 5GC, and NS. AI/ML-assisted energy-saving scenarios will be studied, including those based on analytics provided by the management data analytics function (MDAF) or NWDAF.
- Considering that the amount of energy consumed by network functions has dependency on data, signaling, and OA&M traffic volumes to be processed/transported/stored by network functions, we'll introduce such considerations when designing future OA&M solutions.

3.2.3.2 AI&ML Support

3GGP technical specification group RAN plenary approved in September 2020 a new study to explore AI/ML support in 5G RAN architectures. It covered the functional framework for RAN intelligence and identified use cases, based on the current 5G RAN architecture, where the application of AI/ML techniques brings significant benefits. The RAN3 working group study, concluded in March 2022, was motivated by strong 3GPP interest in taking RAN automation in new directions. In order to focus efforts on real solutions and ensure convergence with well-defined guidelines for normative work, RAN3 has identified three core use cases for which AI/ML-based solutions are being explored:

- Network ES where the energy consumption improvements for the whole RAN may be achieved by actions such as traffic offloading, coverage modification, and cell deactivation.
- Load balancing where the objective is to distribute load effectively among cells or areas of cells in a multifrequency/multi-RAT deployment to improve network performance based on load predictions.
- Mobility optimization where satisfactory network performance during mobility events is preserved while optimal mobility targets are selected based on predictions of how UEs may be served.

One of the first steps taken by the RAN3 working group is to identify the basic principles for developing technical solutions for each use case. Introducing AI and ML while maintaining the current 5G architecture is of particular challenge. The RAN3

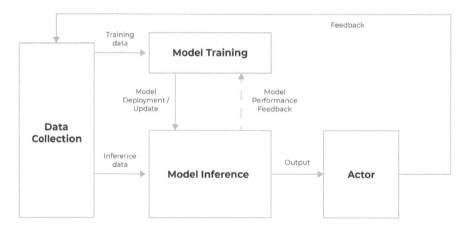

FIGURE 3.6 Functional framework for RAN intelligence (TR 37.817).

group has defined a functional framework for the interworking of the various AI/ML functions, as shown in Figure 3.6.

For every use case, RAN3 working group identified the most relevant sets of inputs, outputs, and feedbacks for implementing and evaluating AI/ML-based processes. Different inputs are described that can be generated by different entities such as UEs, neighboring RAN nodes, and RAN nodes hosting the AI/ML inference process. Examples of input information are UE location, RAN energy efficiency metrics, and RAN resource status measurements. Identified AI/ML outputs consist of various intended metrics and actions. Examples are estimated levels, RAN resource status metrics, and mobility decisions to improve energy efficiency or optimize load balancing. For each use case, a set of feedback is identified that indicates how the system performance is affected by AI/ML-based operations. It can be used, for example, to trigger model retraining. Feedback is case-based and measured after AI/ML-based decisions are made. They include indications of QoS and UE performance, measurements, and RAN resource status metrics. The identified use cases are the focus of the research, and the solutions developed during the study are taken as baseline.

3.3 5G CRITICAL AND MMTC CONNECTIONS

The basic feature of 5G-IoT connection is the separation of connection support from the actual realization of the IoT service. A significant advantage is a reliable long-term solution. 5G-IoT is based on global standards with very strong industry support from a large number of vendors, networks, and service providers. 5G-IoT solutions are embedded in the basic infrastructure of the cellular communication network. Deployment plans are made over decades and systems are built to be highly reliable to high availability standards. 5G-IoT systems are designed for the global market and enable roaming over the networks of multiple operators. IoT services are expected to have a long lifetime, for example, a decade.

One extremely significant benefit of 5G-IoT connectivity is supporting reliable and predictable service performance for future planned operations as well. 5G-IoT uses dedicated spectrum. Radio resources are managed, interference is coordinated, and full QoS is supported. 5G-IoT also follows the continuous evolution of mobile network technologies, where new capabilities and features are continuously added to networks. This evolution is designed to work backward compatible, so that devices that cannot be upgraded to new functionality can continue to work long-term according to the initial capabilities, while new services and devices can simultaneously benefit from newer features.

The 3GPP R16 and R17 standardization supports the necessary capabilities for mission-critical services, uRLLC, mMTC, and eMBB services. Core services can be used for various and heterogeneous industrial IoT use cases and in automotive industry with a unique 5G connectivity solution.

Critical IoT encompasses time-critical industrial use cases with a demanding set of latency and reliability requirements. ITU has set a minimum latency target of 1 ms with packet error rate of 10^{-3}. Covered use are collaboration and control of machines, robots and processes, real-time human–machine interaction (HMI), automated guided vehicles, autonomous cars, and AR/VR applications in the vertical domains of manufacturing, automotive sector. 5G NR introduces several features that enable uRLLC to support critical data streams, grouped into low-latency and ultrareliable communications. The most important options of low-latency communication (LLC) are scalable transmission interval transmission time interval (TTI) corresponding to slots of duration 62.5 μs to 1 ms, fast processing with priority (preemption). A select list of features that improve 5G reliability is ultra-robust modulations and coding, various diversity schemes, and media access/scheduling. The IIoT-oriented 5G CN features are packet duplication, NS, and MEC.

Massive IoT (mIoT) targets huge number of connected low- and mid-range industrial devices (meters, sensors, trackers), mostly battery-powered and sending/receiving a small amount of sensory information (temperature, humidity, location). Although the latency and reliability requirements of mIoT devices are not critical, it is necessary to support extreme device densities (up to $10^6/km^2$) in various industries (manufacturing and automotive).

Broadband-IoT extends eMBB high-data-rate services to data-intensive IIoT use cases. Compared to traditional service, data-intensive use cases pose different requirements and traffic patterns. This is why broadband-IoT supports additional features such as low latency and improved battery savings, coverage, and data rates. 5G NR enables data rates in the tens of gigabits per second (Gbps) through higher-order modulations, multi-antenna transmission, carrier aggregation, UL dynamic TDD, and the deployment of 5G NR in new and existing spectrum bands.

3GPP R16 and R17 research targets new verticals with more stringent requirements improving the capacity and operation of existing functions. R16 expanded uRLLC to new verticals such as factory automation and transportation industry. The stringent uRLLC requirements have focus on reliability improvements in the transmission of control messages and latency. The potential benefits of DL device prioritization and multiplexing are explored, including DL priority and improved power control. It also supported multiple active configurations for configured-grant enhancements adaptation to different service traffic and quicker synchronization for uRLLC UL transmissions. R17 brings more new use cases for 5G system evolution from 2020 onward. uRLLC studies in R17 mainly focus on new emerging verticals and E2E performance of various applications as enhancements to R16 features.

Research has focused on the physical and link layers, but it is becoming increasingly clear that broader redesign at the network level is also essential to meet the specified requirements. NS, SDN, NFV, orchestration, and self-organizing networks (SONs) are covered. 3GPP private NPN networks support deterministic services for industrial use cases, while leaving full control over every aspect (privacy, security adaptation, management, control) of the network.

The digital enterprise of Industry 4.0 is rebuilding business models in different verticals through connectivity. Industries have different requirements and use cases, as it is necessary to transform connectivity from a one-size-fits-all network to a network that supports specialized and stringent requirements. The drivers for 5G NR in industrial networks are increased productivity, digital transformation through wireless technology, and the use of private networks. Digital transformation in industrial enterprises uses the latest 5G technologies including wireless communication, IoT devices, cloud computing, and analytics and automation tools. A private network is considered ideal for maintaining network control (network resources and operational view/tools) and security (traffic within enterprise systems). Dedicated 5G delivers the bandwidth, latency, reliability, wide coverage, and mobility requirements of critical business operations while supporting high security as the convergence of information and operational technologies brings greater vulnerability to cyber attacks. A private 5G network enables customization of the service to fit the needs of the industry and isolates threats typical of public networks.

3.3.1 Performance Trade-Offs for uRLLC in Industrial IoT

5G supports efficient services for industrial IoT applications and their respective use cases for different QoS requirements. However, uRLLC use cases suffers eMBB service degradations. On the contrary, if eMBB is achieved uRLLC would be degraded. Therefore, the trade-off between two services is of particular importance, to simultaneously achieving uRLLC and eMBB with one method of efficient use of resources in order to meet the QoS requirement defined by the ITU. In this way, the coexistence of URLLC and eMBB services initiates new technologies and research directions. The basic trade-offs identified by the researchers are finite versus large block length, SE spectral efficiency versus latency, EE energy efficiency versus latency, reliability versus latency versus data rate. Many existing technologies support trade-offs between uRLLC and eMBB services. However, it is an open research question to obtain the optimal compromise point: 5G NR scalable TTI corresponding to the duration of slots in the frame structure, multiplexed eMBB and uRLLC, NS, edge caching, and grant-free access.

uRLLC and mMTC are two basic types of services in 5G communication systems. Most mMTC use cases are characterized by scenario in which a large number of devices distributed over wide area communicate sporadically without specific latency requirements, ignoring reliability guarantees. Scalable uRLLC supports increased connections of MTC devices with different uRLLC requirements.

This type of traffic cannot be separated and therefore cannot be supported using RAN NS methods or prioritizing used for mixed uRLLC–mMTC services.

uRLLC–mMTC mixture is use cases where uRLLC and mMTC services are supported by separate physical resources within the same network. This heterogeneous system is supported by redesign of cellular resource configuration (numerology), and mini-slot resources are optimized using the NS RAN method in orthogonal or non-orthogonal resource allocation. A simple mixture of conventional mMTC and uRLLC (NS RAN) cannot simultaneously guarantee reliability, latency, and scalability over wireless links at the same level as supposed from wired factory environment (<4 ms latency, 10^6 devices/km^2). These two extreme services are predicted to shortly agglomerate into critical mMTCs with new use cases (factory automation, wide area disaster monitoring), creating new challenges for designing wireless systems beyond 5G.

Critical mMTC supports enhanced uRLLC requirements for a fraction of massively connected devices. It is possible to apply some of the techniques developed for the uRLLC–mMTC mixture to the critical mMTC. However, when the device is sometimes uRLLC or non-uRLLC device, the cost of resource configuration overhead can be enormous. Nevertheless, it cannot be the same, especially for the large number of mission-critical MTC devices applied to the scalable uRLLC. Generic solutions for scalable uRLLC (ML-assisted uRLLC for autonomous vehicles) are not optimized for critical mMTCs. Devices generally lack powerful computing capabilities and sufficient energy, so ML-based solutions for scalable uRLLC are not suitable unless simple and energy-efficient ML architectures and algorithms are deployed. In summary, critical mMTC is still a unique type of service that has not been fully explored.

The main design objectives for critical IIoT applications are energy, latency, throughput, scalability, topology, security as well as reliability, cost-effectiveness, standardization, device maintenance, network monitoring, system configuration and management, scalability and integration, heterogeneity, and interoperability. The architecture of 5G-IIoT is primarily divided into five layers: application layer, architecture layer, communication layer, network layer, and sensor layer.

The Industrial IoT network is scalable in connectivity and number of devices as well as designed to support optimal performance for all industrial applications. In the manufacturing industry, productivity is improved while increasing safety. Processes are automated, available production time is maximized to minimize business interruptions and achieve greater efficiency, competitiveness, and investment value. The technological drivers for the industrial IoT revolution are:

- 5G wireless converged automation protocols will eliminate wires and support time synchronous operations
- private edge cloud will be scalable and provides secure local computing
- slicing for IIoT networks can support multiple stakeholders on one common infrastructure
- machine learning-enabled automated operations will support expertless monitoring, prediction, and optimization.

In summary, IIoT connects industrial assets, such as machines and control systems, with IT and business processes. Integration results in massive data generation and collection. The data can be used to develop analytical solutions to improve industrial operations. Smart manufacturing, on the other hand, primarily concentrates on the production phase of a smart product's life cycle, with the aim of responding quickly and dynamically to changes in demand. As a result, IIoT significantly impacts the industrial value chain and meets the standards for smart manufacturing.

The 5G standard supports use cases that make industrial systems more flexible, useful, and autonomous, while simultaneously meeting the QoS requirements of both 5G and IIoT. Industry stakeholders have identified a variety of potential use cases for the IIoT. The following are general IIoT requirements and goals:

- QoS requirement with low latency, and ultrahigh reliability is vital
- low-cost scalable network with esteemed security and privacy is desired
- the emerging standards should be smoothly implemented and integrated on the IIoT devices in a flexible way
- inter- and intraconnection of networks and IoT devices should be possible frequently.

IIoT is a significant vertical focus of 3GPP NR R16. It includes latency and reliability improvements on the already existing R15 research high reliability and low latency air interface. The version is intended for industrial automation use cases. The 5G-uRLLC foundation has been improved in R16 to ensure more reliable communication up to 99.99%. Along with previously established use cases and deployment options, R17 includes intelligent networking technologies and supports a number of new use cases. One of the more important components of 5G is the application of AI based on machine learning techniques to solve multidimensional problems of online/offline network optimization, enabling the introduction of intelligent network management.

5G offered solutions for multiple vertical markets using uRLLC. R15 established a strong foundation, and R16 added improvements to the 3GPP system architecture and RAN groups to improve support for vertical industries such as factory automation, transportation, and power distribution. Several redundancy schemes have been added to the UP as part of these enhancements, as well as improvements in reliability, latency reduction, and support for time-sensitive communications (TSC). The improvements in R17 aim to increase SE and system capacity, while also supporting uRLLC in unlicensed spectrum environments and strengthening the framework to support TSC communication. Additionally, enhancements to the hybrid automatic HARQ repeat request, CSI channel status, intra-UE multiplexing, and service survival times are included in the assistance information provided by the TSC. R17 introduces mechanisms to detect edge application servers, for example, the application server edge detection feature (EASDF) was defined with the primary purpose of simplifying the session breakout connectivity model.

Industrialists are currently unable to manage the huge amounts of data generated in the 5G-IIoT ecosystem as a whole due to a lack of reliable tools for productive use of big data. AI is capable of autonomously managing itself and will be able

5G Ultrareliable and Low-Latency Communication

to overcome limitations, allowing for optimal utilization and optimization. A connected IIoT ecosystem of devices with AI-based analytical models improves not only manufacturing operations but also the entire industrial process. The combination of AI and IIoT contains many advantages to be overlooked in terms of dependability and reliability.

In summary, the following 3GPP features will be required for IIoT:

- enhancements for latency and reliability in radio and E2E (uRLLC)
- support for wireless industrial ethernet and deterministic communications (uRLLC)
- 5G private networks in licensed ad unlicensed spectrum (uRLLC and eMBB)
- positioning for 5G and IIoT (eMBB, uRLLC, mMTC)
- cellular IoT evolution, e.g., connecting eMTC/NB-IoT to 5G Core (mMTC).

Some scenarios that require very low latency and very high availability of communication services are described below:

- Process automation is characterized by high requirements of the communication system regarding the availability of communication services. Systems that support process automation are implemented in geographically restricted areas, and access is usually limited to authorized users in private networks.
- Discrete automation is characterized by high requirements for the communication system in terms of reliability and availability. Systems that support discrete automation are typically deployed in geographically restricted areas, access is restricted to authorized users, and may be isolated from networks or network resources used by other mobile users.
- Motion control is characterized by high demands on the communication system in terms of delay, reliability, and availability. Systems that support motion control are deployed in geographically limited areas, but may also be deployed in wider areas (networks across city or country), access is limited to authorized users, and may be isolated from networks or network resources used by other mobile users.

The performance requirements in terms of latency and reliability for individual IIoT use cases are shown in Table 3.3. The availability of communication services refers to the availability of a communication service. Reliability is a 3GPP term and refers to the availability of a communication network.

Factory automation encompasses all types of production that result in individual products. Automation of flow control is called process automation. Discrete automation requires open-loop communication of supervisory and control applications, as well as process monitoring and monitoring of operations within an industrial plant. In these applications, a large number of sensors, which are distributed throughout the plant, transmit measurement data to the process controller periodically or when an event occurs. Traditionally, wire-bus technologies have been used to interconnect

TABLE 3.3
uRLLC Requirements for Various IIoT Use Cases

Scenario	E2E Latency	Service Availability	Reliability	User Experienced Data Rate	Payload Size	Traffic Density	Connection Density	Service Area
Process automation monitoring and remote control	50 ms	99.9%	99.9%	1 Mbps	Small	10 Gbps/km^2	10,000/km^2	300×300×50
	50 ms	99.9999%	99.9999%	1–100 Mbps	Small–big	100 Gbps/km^2	1,000/km^2	300×300×50 m
Discrete automation and motion control	10 ms	99.99%	99.99%	10 Mbps	Small–big	1 Tbps/km^2	100,000/km^2	1,000×1,000×30 m
	1 ms	99.9999%	99.9999%	1–10 Mbps	Small	1 Tbps/km^2	100,000/km^2	100×100×30 m
Intelligent transport systems	30 ms	99.9999%	99.999%	10 Mbps	Small–big	10 Gbps/km^2	1,000/km^2	2 km along road

sensors and control equipment. Due to the significant expansion of the plant (up to 10 km^2), a large number of sensors, and the high complexity of the wired infrastructure, wireless solutions have entered the automation of industrial processes. Related use cases require the support of large number of sensor devices per plant, as well as high availability of communication services (99.99%). Furthermore, power consumption is relevant, some sensor devices are powered by batteries with targeted battery life of several years (while providing measurement updates every few seconds). Range also becomes a critical factor due to the low transmission power level of the sensor, the large size of the plant, and the high-reliability requirements in transport. E2E latency requirements range between 10 ms and 1 s. Data transfer speeds to users can be quite slow as each transaction typically contains less than 256 bytes. However, there has been a shift from field buses featuring somewhat modest data rates (~2 Mbps) to higher data rates (~10 Mbps) due to the increasing number of distributed and data-driven applications. An example is the visual control of production processes. For this application, the data rate to the user is usually around 10 Mbps, and the transmitted packets are much larger than what was previously stated.

Process automation shares a lot in common with factory automation. Instead of discrete products (cars), process automation produces bulk products such as gasoline and reactive gases. Unlike factory automation, motion control is of limited or no importance. Typical E2E latency is 50 ms. Data transfer speeds to users, availability of communication services, and connection density vary noticeably between applications.

Factory automation–motion control requires communications for closed-loop control applications. Examples for such applications are motion control of robots, machine tools, as well as packaging and printing machines. In motion control applications, a controller interacts with a large number of sensors and actuators (up to 100) that are integrated in a manufacturing unit. The resulting sensor/actuator density is often very high (up to 1 m^{-3}). Many such manufacturing units may have to be supported within close proximity within a factory (up to 100 in automobile assembly line production).

3.3.2 V2X High-Bandwidth, Low-Latency, and Highly Reliable Communication

The automotive vertical market is undergoing key technological transformations, with a focus on autonomous vehicles that connect and collaborate with each other, with roadside infrastructure elements, with pedestrians and other vulnerable road users (VRUs), and with cloud V2X connections. The collective perception of the vehicle's environment enables decision-making based on shared information, local views, and planned maneuvers of nearby vehicles, instead of relying on the local awareness based only on the vehicle's sensors (radar, LIDAR, cameras). However, the complexity of this environment poses unprecedented challenges. V2X applications such as cooperative sensing and maneuvering pose high computational and communication requirements. Most V2X security applications require communication with ultralow latency (below 10 ms), ultrahigh reliability (close to 100%), and high data rate (on the order of Gbps). In addition, the inherent dynamics in automotive environments associated with rapidly changing network topology, fast-varying wireless channel, and possible sporadic connectivity further increase the complexity of system design and overall require a comprehensive end-to-end approach.

The great interest of the relevant industrial sectors, governments, and organizations in enabling traffic communications is noticeable. The work in the field includes research and industrial work, as well as regulations and standards. Some organizations and governmental bodies are engaged in publishing standards and regulations for communication in vehicles (ASTM, IEEE, ETSI, SAE, 3GPP, ARIB, TTC, TTA, CCSA, ITU, 5GAA, 5G PPP, ITS America, ERTICO, ITS Asia-Pacific). 3GPP is working on standards and technical specifications for cell-based V2X communications, while IEEE is working through the NGV study group to publish the 802.11bd standard for dedicated short-range communications (DSRCs). On the other hand, certain governmental agencies, in cooperation with manufacturer of automobiles, suppliers, consultants, and academic institutions, support research projects.

V2X applications cover wide range of use cases, which can be classified based on their purpose and requirements. 5GAA grouped V2X use cases into four categories: safety aimed at reducing the frequency and severity of vehicle collisions, convenience vehicle status management offering services such as diagnostics and software updates, vulnerable traffic users VRU aimed at safe interactions between vehicles and non-vehicle road users, as well as advanced driving assistance (ADAS).

The use cases of V2X communications can be mainly divided into security, non-security, and infotainment services. Safety services minimize accidents and risks for passengers and road users. Intelligent transportation systems (ITSs) use non-safety services to improve traffic management in order to maximize the efficiency of the existing road network and minimize the negative impacts of traffic such as congestion and its subsequent impacts on economic productivity and environmental quality. Infotainment services provide a range of services to vehicle users, including Internet access, comfort services, video streaming, and content sharing.

3GPP NR-V2X is based on technology designed for high-speed mobile applications and developed specifically for delay-sensitive V2X use cases. Vehicular networks are also part of the IoT scenario. Vehicles themselves can be nodes of a vehicle sensor network. C-V2Xs enable vehicles to support a significant set of functions, such as collaboration, coordination, and sharing of information collected by sensors (comparable to a wireless sensor network for IoT) in advanced driver assistance systems, as well as connected autonomous driving. 5G eMBB, uRLLC, and mMTC services support faster connections, lower latency, higher reliability, higher capacity, and wider coverage for V2X use cases (Figure 3.7).

Despite the diversity, the use cases eventually converge in terms of requirements and lead to the identification of vehicle communication needs, but from different perspectives. In its latest technical report R16 TS 22.186, 3GPP identifies 25 use cases, categorized into four basic groups in addition to a general use case groups and vehicle QoS (Table 3.4). 3GPP defines six levels of automation (LoA SAE J3016) for each use case: no automation $LoA = 0$, driver assistance $LoA = 1$, partial driving automation $LoA = 2$, conditional automation $LoA = 3$, high automation $LoA = 4$, and full automation $LoA = 5$.

5G Ultrareliable and Low-Latency Communication 167

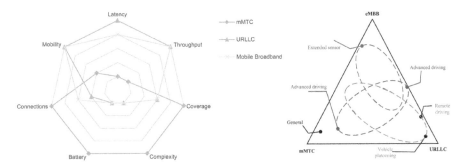

FIGURE 3.7 5G URLLC, eMBB, and mMTC services: (a) requirements and (b) mapping V2X use cases.

Vehicle platooning supports the formation of a group of vehicles that are interconnected in a virtual chain. Vehicles share information, resulting in shorter distances between vehicles, fuel savings, and reducing the number of drivers. A group of use cases enable the dynamic formation of a vehicle unit to travel together.

Advanced driving allows vehicles to share local sensor data and driving plans with nearby vehicles, thus coordinating trajectories and maneuvers. It enables high LoA-level automation assuming longer distance between vehicles. Each vehicle and/or roadside unit RSU shares data obtained from its local sensors with nearby vehicles, thus allowing the vehicles to coordinate their trajectories or maneuvers. In addition, each vehicle shares its driving intention with nearby vehicles. The benefits of this use case groups are safer travel, collision avoidance, and improved traffic efficiency. The group contains seven use cases: collaborative collision avoidance, information sharing for limited automated driving, information sharing for fully automated driving, path alignment to correct relative positions for emergency, intersection safety information for city driving, shared automated vehicle lane change, and 3D video composition for V2X scenario.

Extended sensor group of use cases enables the exchange of raw or processed data acquired from local sensors, video cameras, RSUs, pedestrian devices, and V2X application servers. The group includes three use cases: sensor and state map sharing, collective perception of the environment, and video data sharing for VaD automated driving. Vehicles can increase their awareness of their surroundings beyond what their sensors can detect.

Remote driving use case group aims to control vehicle driving remotely, either by the driver or by cloud/edge computing (CC/EC) applications. Remote driving is necessary in case the vehicle is unable to drive autonomously due to unexpected road situations in dangerous or uncomfortable environments.

3GPP cellular technologies were extended to the automotive vertical in R14 November 2014. The first V2X radio study was approved at the 3GPP RAN plenary meeting in June 2015. Since then, 3GPP has continuously explored V2X and through multiple releases. Many use cases for the automotive vertical (fleet management, infotainment, remote diagnostics) are supported in R15. However, NR

TABLE 3.4
Performance Requirements for Main Group of V2X Use Cases

Scenario	MAX E2E Latency	Reliability	Data Rate	Tx Rate	Payload Size	MIN Service Area
Vehicles platooning: - Cooperative driving	10–25 ms	90%–99.99%	65 Mbps	30–50 message/seconds	50–6,500 bytes	80–350 m
- Information sharing	20 ms	–	50 Mbps	50 message/seconds	6,000 bytes	–
Advanced driving: - Cooperative collision	10 ms	99.99%	10 Mbps	100 message/seconds	2,000 bytes	–
- Emergency trajectory	3 ms	99.999%	30 Mbps	–	2,000 bytes	500 m
Extended sensors: - Sensor information sharing	3–100 ms	99%–99.999%	10–1,000 Mbps	10 message/seconds	1,600 bytes	50–1,000 m
- Video sharing	10–50 ms	90%–99.99%	10–700 Mbps	–	–	400 m

currently does not support direct communication over the SL between a vehicle and nearby vehicles, infrastructure nodes, or pedestrians. Direct communication is useful for transmitting information such as position, speed, and direction without having to transport it through network. R16 research topics were approved in June 2018. Use cases impose different requirements, with some relying on direct vehicle-to-vehicle (V2V) communication, while others take advantage of vehicle-to-infrastructure (V2I) communication.

- Sidelink design including signals, channels, bandwidth part, and resource pools. A sidelink is a direct link between NR devices without having to go through the base station.
- Resource allocation for sidelink. In Mode 1 the base station schedules the resource for sidelink transmission and Mode 2 is where the device autonomously selects the resource for sidelink transmission from a set of resources configured by the network.
- Sidelink synchronization mechanism including synchronization procedure when the devices are out of coverage of the base station.
- Sidelink physical layer procedures including HARQ, channel state acquisition and reporting, and power control.
- Sidelink L2/L3 protocols, signaling, and congestion control.

3GPP is currently working on 5G-based V2X communication specification to meet advanced requirements. For example, the technical specification TS 22.186 defines the performance requirements for remote driving as a maximum E2E delay of 5 ms with a reliability of 99.999% and at speeds up to 250 km/h, for the exchange of information between the UE supporting the V2X application and the V2X application server.

In all V2X use cases, the importance of the scenario with and without mobile network coverage is emphasized. Two scenarios are considered, one where given frequency carriers for V2X communications are dedicated to cell-based V2X services (subject to regional regulation and operator policy), and another scenario where communications are in licensed spectrum used for regular mobile operations. There is extensive documentation in 3GPP with the results of all 5G-V2X studies conducted.

3.3.3 AI-Enabled Massive IoT Toward 6G

Previous generations of wireless networks mainly supported voice, text, video, and Internet capabilities, with increased data rates, while 5G and 6G are becoming increasingly capable of supporting many other services. 6G is expected to expand capabilities to higher levels where millions of connected devices and applications would work seamlessly with confidence, low latency, and high bandwidth. Industries and academia have allocated huge funds and other resources to 6G research and development.

6G is significant enabler of future IoT networks and applications, providing a full dimension of wireless coverage and integrating all functionalities, including sensing, transmission, computation, cognition, and fully automated management.

In fact, compared to 5G systems, next-generation 6G networks are expected to achieve massive coverage and improved adaptability to support IoT connectivity and service delivery. The main focus is expected on the implementation of the massive IoTs (mIoT), interoperability issues, system architecture, energy-efficient technologies, and the application of AI and other novel techniques to improve network performance, security, and privacy.

The evolution of mobile networks is based on inheriting the advantages of previous network architectures and adding conveniences that effectively meet the requirements of the latest era. Similarly, the 6G network adopts the advantages of the 5G architecture and at the same time novel technologies to support the requirements beyond 5G one of the significant components of self-sufficient 6G networks is intelligence, which is a relatively new technology that is being integrated into networks using AI. Obviously, AI could not be applied to previous versions including 4G, since in 5G networks it is applied partially or to a limited extent. It is notable that for 6G real-time communications, the advancement of ML/AI procedures leads to the development of highly intelligent networks. AI techniques possess numerous advantages such as increasing efficiency, reducing delays within communication steps, and effectively solving complex problems. However, examples such as meta-materials, intelligent structures, intelligent networks, intelligent devices, intelligent cognitive radio, self-sustaining wireless networks, and ML provide support for communication systems based on AI. Therefore, the application of AI-based technologies supports the fulfillment of the objectives of numerous 6G services.

AI and ML are used in a range of application domains across all industry sectors. In mobile communication systems, mobile devices (smartphones, cars, robots) are increasingly replacing conventional algorithms (speech recognition, image recognition, video processing) with AI/ML models. The 5G system supports at least three types of AI/ML operations:

- **AI/ML operation splitting between AI/ML endpoints.** The operation/model is divided into multiple parts based on the current task and environment. The intention is to offload the computationally and energy-intensive parts to the network endpoints, while keeping the privacy-sensitive and latency-sensitive parts at the end device. The device executes the operation/model to a specific part/layer and then sends the intermediate data to the network endpoint. The network endpoint executes the remaining parts/layers and sends the inference results back to the device.
- **AI/ML model/data distribution and sharing over 5G system.** Multifunctional mobile terminals might need to switch the AI/ML model in response to task and environment variations. The condition of adaptive model selection is that the models to be selected are available for the mobile device. However, due to the fact that AI/ML models are becoming more and more diverse, and with limited storage resources in the UE, not all AI/ML model candidates are pre-loaded on-board. Online model distribution (new model download) is required, in which an AI/ML model can be distributed from a network endpoint to the devices when they need it to adapt to the

changed AI/ML tasks and environments. It is necessary to continuously monitor the performance of the UE model.
- **Distributed/federated learning over 5G system.** The cloud server trains a global model by aggregating local models partially trained by each end device. Within each training iteration, the UE performs training based on the model downloaded from the AI cloud server using the local training data. Then the UE reports the interim training results to the cloud server via the 5G UL channel. The server aggregates the interim training results from the UE and updates the global model. The updated global model is then distributed back to the UE where training is performed for the next iteration.

6G-IoT impose more stringent requirements for the full realization of massively connected devices and coverage, data-driven services, and autonomous systems by 2030. System development requires stringent standard specifications that call for the cooperation of network operators, service providers, and users. Furthermore, the introduction of vertical 6G-IoT use cases in future intelligent networks imposes significant architectural changes on current mobile networks to simultaneously support a number of different stringent requirements.

3.4 CONCLUDING REMARKS

The development of 5G mobile networks is driven by constant research and innovation. 5G networks are evolving from connectivity-based networks to an intelligent service delivery platform. The applications and services that the 5G system supports are referred to as vertical industries or verticals. There is constant demand for flexible bandwidth, latency, and better coverage from the industry. To accelerate their implementation, verticals classify use cases as a combination of three basic service categories (eMBB, mMTC, URLLC). Although mMTC is specifically categorized and designed to meet IoT requirements, uRLLC contains key features for efficient IoT operations. It is quite challenging to manage mission-critical IoT devices in real-time over wireless. Extremely reliable LLC uRLLC supports delay-sensitive services with a high level of key performance indicators: latency, reliability, and availability. However, many design parameters and metrics are in trade-offs relations, such as reliability vs. latency, reliability vs. data rate, EE vs. latency. Improving one performance indicator affects other metrics. The 3GPP solution is a new service-oriented 5G architecture design. The challenges of uRLLC implementation are QoS, coexistence with eMBB service, packet design, instant and reservation-based scheduling, error handling, and energy efficiency.

This chapter presents an up-to-date overview of vertical requirements as well as technical and architectural concepts critical to the delivery of URLLC services in key 5G verticals (industrial IoT, V2X low latency, and highly reliable communication). The process of standardization of private 5G networks, EE, and AI/ML support for vertical services were analyzed.

The development of new communication networks depends on the emergence of globally accepted standards that support interoperability, economy of scale with affordable cost for system implementation, and end users. From a 3GPP perspective, the research is initiated by identifying use cases and specifying the corresponding requirements from different industry groups and verticals. Later on, use cases and requirements of industrial IoT and vehicle applications on V2X connections were developed and documented. Therefore, there is a strong motivation to further develop standardization to optimize for these new services, especially in terms of cost, complexity, energy savings, and coverage. Furthermore, the intelligent integration of uRLLC and mMTC has the potential to drive massive IoT and industrial automation based on 6G AI.

BIBLIOGRAPHY

INTRODUCTION

3GPP TSG RAN Chair, *Summary for RAN Rel-18 package*. RP-213469, Dec. 2021.
N. Bhus et al., 5G air interface system design principles. *IEEE Wireless Communications*, vol. 24, no. 5, pp. 6–8, Oct. 2017.
A. Centonza, 3GPP WG RAN3 completes AI&ML support study. *3GPP Highlights News Letter*, no. 4, pp. 6–7, May 2022.
W. Chen, P. Gaal, J. Montojo, H. Zisimopoulos, *Fundamentals of 5G communications: Connectivity for enhanced mobile broadband and beyond*. McGraw Hill, 2021.
W. Chen, J. Montojo, J. Lee, M. Shafi, Y. Kim, The standardization of 5G-Advanced in 3GPP. *IEEE Communications Magazine*, pp. 1–7, 2022. (Early access).
J.-M. Cornily, 3GPP SA5 energy efficiency (EE) work and results. *3GPP Highlights News Letter*, no. 4, p. 16, May 2022.
A. Ghosh, A. Maeder, M. Baker, D. Chandramouli, 5G evolution: A view on 5G cellular technology beyond 3GPP Release 15. *IEEE Access*, vol. 7, pp. 127639–127651, 2019.
Y. Kim et al., New radio (NR) and its evolution toward 5G-advanced. *IEEE Wireless Communications*, vol. 26, no. 3, pp. 2–7, Jun. 2019.
H. Kim, *Design and optimization for 5G wireless communications*. Wiley, 2020.
O. Liberg, M. Sundberg, E. Wang, J. Bergman, J. Sachs, G. Wikström, *Cellular internet of things: From massive deployments to critical 5G applications*, 2nd ed. Elsevier, 2020.
O. Liberg, M. Sundberg, E. Wang, J. Bergman, J. Sachs, *Cellular internet of things: Technologies, standards, and performance*, 1st ed. Elsevier, 2018.
J. Park et al., Extreme ultra-reliable and low-latency communication. *Nature Electronics*, vol. 5, no. 3, pp. 1–9, 2022.
B. Sanou, Setting the scene for 5G: Opportunities and challenges. *ITU Discussion Paper*, pp. 1–56, 2018.
R. Vannithamby, A.C.K. Soong, *5G verticals: Customizing applications, technologies and deployment techniques*. Wiley – IEEE Press, 2020.
X. Lin, N. Lee (Eds.), *5G and beyond: Fundamentals and standards*. Springer, 2021.

5G VERTICAL DOMAIN EXPANSION

3GPP RAN 1, *Overall description of radio access network (RAN) aspects for vehicle-to-everything (V2X) based on LTE and NR*. TR 38.985 v17.1.1, Apr. 2022.
3GPP RAN 1, *Study on evaluation methodology of new vehicle-to-everything (V2X) use cases for LTE and NR*. TR 37.885 v15.3.0, Jun. 2019.
3GPP RAN 1, *Study on NR vehicle-to-everything (V2X)*. TR 38.885 v16.0.0, Mar. 2019.

3GPP RAN 4, *V2X services based on NR: User equipment (UE) radio transmission and reception.* TR 38.886 v16.3.0, Apr. 2021.

3GPP RP-201677, *Revised SID on study on support of reduced capability NR devices*, Jul. 2020.

3GPP RP-202933, *New WID on support of reduced capability NR devices*, Dec. 2020.

3GPP Technical Specification Group Services and System Aspects, *Enhancement of 3GPP support for V2X scenarios.* TS 22.186 v17.0.0, Jan. 2022.

3GPP Technical Specification Group Services and System Aspects, *Study on communication for automation in vertical domains.* TR 22.804 v16.0.0, May 2018.

3GPP TR 22.891, *Feasibility study on new services and markets technology.* Release 14, Sept. 2016.

3GPP TR 38.913, *Study on scenarios and requirements for next generation access technologies.* v15.0.0 Jul. 2018, v16.0.0 Jul. 2020, v17.0.0 Apr. 2022.

3GPP TS 22.261, *Service requirements for next generation new services and markets.* Release 18, Dec. 2020.

3GPP TS 38.300 NR, *NR and NG-RAN overall description.* Stage-2. v15.0.0 Jan. 2018, v16.0.0 Jan. 2020, 17.0.0 Apr. 2022.

3GPP TSG Services and System Aspects, *Service requirements for cyber-physical control applications in vertical domains.* TS 22.104 v18.0.0, Feb. 2021.

3GPP TSG Services and System Aspects, *Service requirements for the 5G system.* TS 22.261 v18.0.0, Apr. 2020.

V. Dhanwani et al., Assessment of candidate technology ETSI: DECT-2020 new radio. In *Proc. IEEE 3rd 5G World Forum 5GWF*, 2020, pp. 1–6.

O.O. Erunkulu et al., 5G mobile communication applications: A survey and comparison of use cases. *IEEE Access*, vol. 9, pp. 97251–97295, 2021.

M.A. Imran, Y.A. Sambo, Q.H. Abbasi, Enabling the verticals of 5G: Network architecture, design and service optimization. In *Enabling 5G communication systems to support vertical industries* (Eds. M.A. Imran, Y.A. Sambo, Q.H. Abbasi). Wiley – IEEE Press, 2019.

ITU-R Recommendation M.2083, *IMT vision: Framework and overall objectives of the future development of IMT for 2020 and beyond.* Sept. 2015.

S. Moloudi et al., Coverage evaluation for 5G reduced capability new radio (NR-RedCap). *IEEE Access*, vol. 9, pp. 45055–45067, 2021.

T.W. Nowak et al., Verticals in 5G MEC-use cases and security challenges. *IEEE Access*, vol. 9, pp. 87251–87298, 2021.

S.R. Pokhrel et al., Towards enabling critical mMTC: A review of URLLC within mMTC. *IEEE Access*, vol. 8, pp. 131796–131813, 2020.

A.P.K. Reddy et al., 5G new radio key performance indicators evaluation for IMT-2020 radio interface technology. *IEEE Access*, vol. 9, pp. 112290–112311, 2021.

N. Varsier et al., A 5G new radio for balanced and mixed IoT use cases: Challenges and key enablers in FR1 band. *IEEE Communications Magazine*, vol. 59, no. 4, pp. 82–87, 2021.

uRLLC Services

3GPP SA1 TR 22.186, *Service requirements for enhanced V2X scenarios.* Release 16 v16.2.0, Jun. 2019.

3GPP SA1 TR 22.261, *Service requirements for the 5G system.* Release 18 v18.5.0, Dec. 2021.

3GPP SA1 TR 22.804, *Study on communication for automation in vertical domains (CAV).* Release 16 v16.3.0, Nov. 2020.

3GPP SA4 TR 26.928, *Extended reality (XR) in 5G.* Release 16 v16.1.0, Dec. 2020.

3GPP TSG RAN TR 38.802, *Study on new radio access technology physical layer aspects.* Release 14 v14.2.0, Sept. 2017.

3GPP TSG RAN2 TR 38.804, *Study on new radio access technology: Radio interface protocol aspects.* Release 16 v16.0.0, Mar. 2017.

3GPP TSG RAN1 TR 38.824, *Study on physical layer enhancements for NR ultra-reliable and low latency case (URLLC).* Release 16 v16.0.0, Mar. 2019.

3GPP TSG RAN TR 38.913, *Study on scenarios and requirements for next generation access technologies.* Release 16 v16.0.0, Jul. 2020.

3GPP TSG RAN#86 RP-193241, *New SID on XR evaluations for NR.* Dec. 2019.

3GPP TSG RAN1 TS 38.202, *Services provided by the physical layer.* Release 16 v16.1.0, Jul. 2020.

F. Alriksson et al., XR and 5G: Extended reality at scale with time-critical communication. *Ericsson Technology Review*, no. 8, pp. 1–14, 2021.

A. Avranas, M. Kountouris, P. Ciblat, Energy-latency tradeoff in ultra-reliable low-latency communication with retransmissions. *IEEE Journal on Selected Areas in Communications*, vol. 36, no. 11, pp. 2475–2485, 2018.

A.K. Bairag et al., Coexistence mechanism between eMBB and uRLLC in 5G wireless networks. *IEEE Transactions on Communications*, vol. 69, no. 3, pp. 1736–1749, Mar. 2021.

E. Bastug et al., Toward interconnected virtual reality: Opportunities, challenges, and enablers. *IEEE Communications Magazine*, vol. 55, no. 6, pp. 11–17, Jun. 2017.

M. Bennis, M. Debbah, H.V. Poor, Ultrareliable and low-latency wireless communication: Tail, risk and scale. *Proceedings of the IEEE*, vol. 106, no. 10, pp. 1834–1853, Oct. 2018.

C. Benzaid, T. Taleb, AI-driven zero touch network and service management in 5G and beyond: Challenges and research directions. *IEEE Network*, vol. 34, no. 2, pp. 186–194, Mar. 2020.

T. Braud, F.H. Bijarbooneh, D. Chatzopoulos, P. Hui, Future networking challenges: The case of mobile augmented reality. In *Proc. IEEE International Conference on Distributed Computing Systems (ICDCS)*, 2017, pp. 1796–1807.

H. Chen, R. Abbas, P. Cheng, M. Shirvanimoghaddam, W. Hardjawana, W. Bao, Y. Li, B. Vucetic, Ultra-reliable low latency cellular networks: Use cases, challenges and approaches. *IEEE Communication Magazine*, vol. 56, no. 12, pp. 119–125, Dec. 2018.

G. Durisi, T. Koch, P. Popovski, Toward massive, ultrareliable, and low-latency wireless communication with short packets. *Proceedings of the IEEE*, vol. 104, no. 9, pp. 1711–1726, Sept. 2016.

M.S. Elbamby, C. Perfecto, M. Bennis, K. Doppler, Toward low-latency and ultra-reliable virtual reality. *IEEE Network*, vol. 32, no. 2, pp. 78–84, Mar.–Apr. 2018.

D. Feng, L. Lai, J. Luo, Y. Zhong, C. Zheng, K. Ying, Ultra-reliable and low-latency communications: Applications, opportunities and challenges. *Science China Information Sciences*, vol. 64, no. 120301, pp. 1–12, 2021.

D. Feng, C. She, K. Ying, L. Lai, Z. Hou, T.Q.S. Quek, Y. Li, B. Vucetic, Toward ultrareliable low-latency communications: Typical scenarios, possible solutions, and open issues. *IEEE Vehicular Technology Magazine*, vol. 14, no. 2, pp. 94–102, Jun. 2019.

D. He, C. Westphal, J.J. Garcia-Luna-Aceves, Network support for AR/VR and immersive video application: A survey. In *Proc. International Conference on Signal Processing and Multimedia Applications*, 2018, pp. 359–369.

A. Hoglund, X. Lin, O. Liberg, A. Behravan, E.A. Yavuz, M. VanDerZee, Y. Sui, T. Tirronen, A. Ratilainen, D. Eriksson, Overview of 3GPP Release 14 enhanced NB-IoT. *IEEE Network*, vol. 31, no. 6, pp. 16–22, Nov. 2017.

R. Hong et al., Resource allocation for secure URLLC in mission-critical IoT scenarios. *IEEE Transactions on Communications*, vol. 68, no. 9, pp. 5793–5807, 2020.

S. Jun, Y. Kang, J. Kim, C. Kim, Ultra-low-latency services in 5G systems: A perspective from 3GPP standards. *Wiley ETRI Journal*, pp. 1–13, 2020.

X. Lin et al., 5G new radio: Unveiling the essentials of the next generation wireless access technology. *IEEE Communications Standards Magazine*, vol. 3, no. 3, pp. 30–37, Sept. 2019.

X. Lin, An overview of 5G advanced evolution in 3GPP Release 18. *arXiv:2201.01358v1*, 2022.

C.-F. Liu, M. Bennis, Ultra-reliable and low-latency vehicular transmission: An extreme value theory approach. *IEEE Communications Letters*, vol. 22, no. 6, pp. 1292–1295, 2018.

D.G. Morín, P. Perez, A.G. Armada, Toward the distributed implementation of immersive augmented reality architectures on 5G networks. *IEEE Communications Magazine*, vol. 60, no. 2, pp. 46–52, Feb. 2022.

S. Parkvall et al., NR: The new 5G radio access technology. *IEEE Communications Standards Magazine*, vol. 1, no. 4, pp. 24–30, Dec. 2017.

I. Parvez, A. Rahmati, I. Guvenc, A.I. Sarwat, H. Dai, A survey on low latency towards 5G: RAN, core network and caching solutions. *IEEE Communication Surveys & Tutorials*, vol. 20, no. 4, pp. 3098–3130, 2018.

G. Pocovi, T. Kolding, K.I. Pedersen, On the cost of achieving downlink ultra-reliable low-latency communications in 5G networks. *IEEE Access*, vol. 10, pp. 29506–29513, 2022.

R. Ratasuk et al., Enhancements of narrowband IoT in 3GPP Rel-14 and Rel-15. In *Proc. IEEE Conference on Standards for Communications and Networking (CSCN)*, 2017, pp. 1–6.

R. Ratasuk et al., LTE-M evolution towards 5G massive MTC. In *Proc. IEEE Globecom Workshops*, 2017, pp. 1–6.

R. Ratasuk et al., NB-IoT system for M2M communication. In *Proc. Workshop on Device to Device Communications for 5G Networks WD5G*, 2016, pp. 1–5.

J. Sachs, G. Wikstrom, T. Dudda, R. Baldemair, K. Kittichokechai, 5G radio network design for ultra-reliable low-latency communication. *IEEE Network*, vol. 32, no. 2, pp. 24–31, Mar. 2018.

P. Schulz et al., Latency critical IoT applications in 5G: Perspective on the design of radio interface and network architecture. *IEEE Communication Magazine*, vol. 55, no. 2, pp. 70–78, Feb. 2017.

H. Shariatmadari et al., Machine-type communications: Current status and future perspectives toward 5G systems. *IEEE Communications Magazine*, vol. 53, no. 9, pp. 10–17, 2015.

S.K. Sharma et al., Toward tactile internet in beyond 5G era: Recent advances, current issues, and future directions. *IEEE Access*, vol. 8, pp. 56948–56991, 2020.

Y. Siriwardhana, P. Porambage, M. Liyanage, M. Ylianttila, A survey on mobile augmented reality with 5G mobile edge computing: Architectures, applications, and technical aspects. *IEEE Communications Surveys & Tutorials*, vol. 23, no. 2, pp. 1160–1192, 2021.

K. Takeda et al., Understanding the heart of the 5G air interface: An overview of physical downlink control channel for 5G new radio. *IEEE Communications Standards Magazine*, vol. 4, no. 3, pp. 22–29, Sept. 2020.

Z. Zhu et al., Research and analysis of URLLC technology based on artificial intelligence. *IEEE Communications Standards Magazine*, vol. 5, no. 2, pp. 37–43, 2021.

Key characteristics of massive machine-type communication

3GPP TR 26.558, *Architecture for enabling edge applications*. Release 17 v17.2.0, Dec. 2021.

3GPP TS 23.548, *5G system enhancements for edge computing*. Release 17 v17.1.0, Dec. 2021.

C. Bockelmann, N. Pratas, H. Nikopour, K. Au, T. Svensson, C. Stefanovic, P. Popovski, A. Dekorsy, Massive machine-type communications in 5G: Physical and MAC-layer solutions. *IEEE Communication Magazine*, vol. 54, no. 9, pp. 59–65, Sept. 2016.

X. Chen et al., Massive access for 5G and beyond. *IEEE Journal on Selected Areas in Communications*, vol. 39, no. 3, pp. 615–637, Mar. 2021.

Z. Dawy, W. Saad, A. Ghosh, J.G. Andrews, E. Yaacoub, Toward massive machine type cellular communications. *IEEE Wireless Communications*, vol. 24, no. 1, pp. 120–128, Feb. 2017.

M. Elsaadany, A. Ali, W. Hamouda, Cellular LTE-A technologies for the future internet-of-things: Physical layer features and challenges. *IEEE Communication Surveys & Tutorials*, vol. 19, no. 4, pp. 2544–2572, 2017.

N. Finn, Introduction to time-sensitive networking. *IEEE Communications Standards Magazine*, vol. 2, no. 2, pp. 22–28, 2018.

F. Ghavimi, H.-H. Chen, M2M communications in 3GPP LTE/LTE-A networks: Architectures, service requirements, challenges, and applications. *IEEE Communication Surveys & Tutorials*, vol. 17, no. 2, pp. 525–549, 2015.

N. Hassan, K-L.A. Yau, C. Wu, Edge computing in 5G: A review. *IEEE Access*, vol. 7, pp. 127276–127289, 2019.

R. Li, Z. Zhao, X. Zhou, G. Ding, Y. Chen, Z. Wang, H. Zhang, Intelligent 5G: When cellular networks meet artificial intelligence. *IEEE Wireless Communications*, vol. 24, no. 5, pp. 175–183, Oct. 2017.

A. Rico-Alvarino, M. Vajapeyam, H. Xu, X. Wang, Y. Blankenship, J. Bergman, T. Tirronen, E. Yavuz, An overview of 3GPP enhancements on machine to machine communications. *IEEE Communication Magazine*, vol. 54, no. 6, pp. 14–21, Jun. 2016.

O.B. Sezer, E. Dogdu, A.M. Ozbayoglu, Context-aware computing, learning, and big data in internet of things: A survey. *IEEE Internet of Things Journal*, vol. 5, no. 1, pp. 1–27, Feb. 2018.

S.K. Sharma, X. Wang, Toward massive machine type communications in ultra-dense cellular IoT networks: Current issues and machine learning-assisted solutions. *IEEE Communication Surveys & Tutorials*, vol. 22, no. 1, pp. 426–471, 2020.

X. Wang, X. Li, V.C.M. Leung, Artificial intelligence-based techniques for emerging heterogeneous network: State of the arts, opportunities, and challenges. *IEEE Access*, vol. 3, pp. 1379–1391, 2015.

5G Private Networks, Energy Efficiency, and AI/ML Support

3GPP Management and Orchestration, *5G end to end key performance indicators (KPI)*. TS 28.554 v17.7.0, Jun. 2022.

3GPP Management and Orchestration, *5G network resource model (NRM)*. TS28.541 v18.0.0, Jun. 2022.

3GPP Management and Orchestration, *5G performance measurements*. TS28.552 v17.7.1, Jun. 2022.

3GPP Management and Orchestration, *Energy efficiency of 5G*. TS28.310 v17.0.0, Apr. 2021.

3GPP Management and Orchestration, *Study on new aspects of energy efficiency (EE) for 5G*. TS28.813 v17.0.0, Dec. 2021.

3GPP Telecommunication Management, *Generic radio access network (RAN), network resource model (NRM), integration reference point (IRP), information service (IS)*. TS28.622 v17.2.0, Jun. 2022.

3GPP TR 28.807, *Study on management aspects of non-public networks*. V17.0.0, Dec. 2020.

3GPP TR22.804, *Study on communication for automation in vertical domains*. V16.0.0, Dec. 2018.

3GPP TS 23.501, *System architecture for the 5G system (5GS)*. V16.5.0, Jul. 2020.

A. Adamatzky, Ö. Bulakci, D. Chatterjee, M. Shafik, S.K.S. Tyagi (Eds.), Artificially intelligent green communication networks for 5G and beyond. *Computer Communications*, vol. 169, no. 1, Mar. 2021.

M. Agiwal, A. Roy, N. Saxena, Next generation 5G wireless networks: A comprehensive survey. *IEEE Communication Surveys & Tutorials*, vol. 18, no. 3, pp. 1617–1655, Sept. 2016.

H. Babbar et al., Role of network slicing in software defined networking for 5G: Use cases and future directions. *IEEE Wireless Communications*, vol. 29, no. 1, pp. 112–118, Feb. 2022.

H. Chergui, A. Ksentini, L. Blanco, C. Verikoukis, Toward zero-touch management and orchestration of massive deployment of network slices in 6G. *IEEE Wireless Communications*, vol. 29, no. 1, pp. 86–93, Feb. 2022.

R. Ferrus, O. Sallent, Extending the LTE/LTE-A business case: Mission- and business-critical mobile broadband communications. *IEEE Vehicular Technology Magazine*, vol. 9, no. 3, pp. 47–55, Sept. 2014.

A. Gupta, R.K. Jha, A survey of 5G network: Architecture and emerging technologies. *IEEE Access*, vol. 3, pp. 1206–1232, Sept. 2015.

A. Jerichow et al., 3GPP non-public network security. *Journal ICT Standardization*, vol. 8, no. 1, pp. 57–76, Jan. 2020.

A. Ksentini, N. Nikaein, Toward enforcing network slicing on RAN: Flexibility and resources abstraction. *IEEE Communication Magazine*, vol. 55, no. 6, pp. 102–108, Jun. 2017.

X. Li et al., Multi-domain solutions for the deployment of private 5G networks. *IEEE Access*, vol. 9, pp. 106865–106884, 2021.

Y.-N.R. Li et al., Power saving techniques for 5G and beyond. *IEEE Access*, vol. 8, pp. 108675–108690, 2020.

P.K.R. Maddikunta et al., Green communication in IoT networks using a hybrid optimization algorithm. *Computer Communications*, vol. 159, no. 1, pp. 97–107, Jun. 2020.

J. Ordonez-Lucena, J.F. Chavarria, L.M. Contreras, A. Pastor, The use of 5G non-public networks to support Industry 4.0 scenarios. In *Proc. IEEE Conference on Standards for Communication and Networking (CSCN)*, 2019, pp. 1–7.

T. Panagiotis et al., A cost-efficient 5G non-public network architectural approach: Key concepts and enablers, building blocks and potential use cases. *MDPI Sensors*, vol. 21, no. 16, p. 5578, 2021.

P. Popovski, K.F. Trillingsgaard, O. Simeone, G. Durisi, 5G wireless network slicing for eMBB, URLLC, and mMTC: A communication theoretic view. *IEEE Access*, vol. 6, pp. 55765–55779, 2018.

J. Prados-Garzon et al., 5G non-public networks: Standardization, architectures and challenges. *IEEE Access*, vol. 9, pp. 153893–153908, 2021.

A. Rostami, Private 5G networks for vertical industries: Deployment and operation models. In *Proc. IEEE 2nd 5G World Forum*, 2019, pp. 433–439.

J. Wang, J. Liu, Secure and reliable slicing in 5G and beyond vehicular networks. *IEEE Wireless Communications*, vol. 29, no. 1, pp. 126–133, Feb. 2022.

M. Wen et al., Private 5G networks: Concepts, architectures, and research landscape. *IEEE Journal of Selected Topics in Signal Processing*, vol. 16, no. 1, pp. 7–25, 2022.

S. Wijethilaka, M. Liyanage, Survey on network slicing for internet of things realization in 5G networks. *IEEE Communication Surveys & Tutorials*, vol. 23, no. 2, pp. 957–994, 2021.

5G Critical and Massive Machine-Type Communication

G.A. Akpakwu, B.J. Silva, G.P. Hancke, A.M. Abu-Mahfouz, A survey on 5G networks for the internet of things: Communication technologies and challenges. *IEEE Access*, vol. 6, pp. 3619–3647, Dec. 2018.

I.F. Akyldiz, H. Guo, Wireless communication research challenges for extended reality (XR). *ITU Journal on Future and Evolving Technologies*, vol. 3, no. 1, Apr. 2022.

L. Chettri, R. Bera, A comprehensive survey on internet of things (IoT) toward 5G wireless systems. *IEEE Internet of Things Journal*, vol. 7, no. 1, pp. 16–32, Oct. 2020.

G. Minopoulos, K.E. Psannis, Opportunities and challenges of tangible XR applications for 5G networks and beyond. *IEEE Consumer Electronics Magazine*, 2022. (Early access).

D.G. Morín, P. Perez, A.G. Armada, Toward the distributed implementation of immersive augmented reality architectures on 5G networks. *IEEE Communications Magazine*, vol. 60, no. 2, pp. 46–52, Feb. 2022.

Z. Nadir, T. Taleb, H. Flinck, O. Bouachir, M. Bagaa, Immersive services over 5G and beyond mobile systems. *IEEE Network*, vol. 35, no. 6, pp. 299–306, 2021.

J. Nagao, K. Tanaka, H. Imanaka, Arena-style immersive live experience (ILE) services and systems: Highly realistic sensations for everyone in the world. *ITU Journal ICT Discoveries*, vol. 3, no. 1, pp. 1–9, May 2020.

M. Shafi et al., 5G: A tutorial overview of standards, trials, challenges, deployment, and practice. *IEEE Journal on Selected Areas in Communications*, vol. 35, no. 6, pp. 1201–1221, Jun. 2017.

M.A. Siddiqi, H. Yu, J. Joung, 5G ultra-reliable low-latency communication implementation challenges and operational issues with IoT devices. *MDPI Electronics*, vol. 8, no. 9, Sept. 2019.

Y. Siriwardhana et al., Survey on mobile augmented reality with 5G mobile edge computing: Architectures, applications, and technical aspects. *IEEE Communications Surveys & Tutorials*, vol. 23, no. 2, pp. 1160–1192, 2021.

M.T. Vega et al., Immersive interconnected virtual and augmented reality: A 5G and IoT perspective. *Springer Journal of Network and Systems Management*, vol. 28, no. 6, pp. 1–33, Oct. 2020.

PERFORMANCE TRADE-OFFS FOR URLLC IN INDUSTRIAL IoT

J. Cheng, W. Chen, F. Tao, C.-L. Lin, Industrial IoT in 5G environment towards smart manufacturing. *Journal of Industrial Information Integration*, vol. 10, pp. 10–19, Jun. 2018.

M. Gundall et al., Introduction of a 5G-enabled architecture for the realization of Industry 4.0 use cases. *IEEE Access*, vol. 9, pp. 25508–25521, 2021.

B.S. Khan et al., URLLC and eMBB in 5G industrial IoT: A survey. *IEEE Open Journal of the Communications Society*, pp. 1–25, 2022. (Early access).

A. Mahmood et al., Industrial IoT in 5G-and-beyond networks: Vision, architecture, and design trends. *IEEE Transactions on Industrial Informatics*, vol. 18, no. 6, pp. 4122–4137, Jun. 2022.

A. Narayanan, E. Ramadan, J. Carpenter, Q. Liu, Y. Liu, F. Qian, Z.-L. Zhang, A first look at commercial 5G performance on smartphones. In *Proc. Web Conference*, 2020, pp. 894–905.

C. Paniagua, J. Delsing, Industrial frameworks for internet of things: A survey. *IEEE Systems Journal*, vol. 15, no. 1, pp. 1149–1159, Mar. 2021.

J. Rischke, P. Sossalla, S. Itting, F.H.P. Fitzek, M. Reisslein, 5G campus networks: A first measurement study. *IEEE Access*, vol. 9, pp. 121786–121803, 2021.

E. Sisinni et al., Industrial internet of things: Challenges, opportunities, and directions. *IEEE Transactions on Industrial Informatics*, vol. 14, no. 11, pp. 4724–4734, 2018.

D. Xu, A. Zhou, X. Zhang, G. Wang, X. Liu, C. An, Y. Shi, L. Liu, H. Ma, Understanding operational 5G: A first measurement study on its coverage, performance and energy consumption. In *Proc. Annual Conference of the ACM SIG on Data Communication on the Applications, Technologies, Architectures, and Protocols for Computer Communication*, 2020, pp. 479–494.

V2X High-Bandwidth, Low-Latency and Highly Reliable Communication

3GPP Technical Specification Group Radio Access Network, *NR, study on NR vehicle-to-everything (V2X)*. TR 38.885 Release 16, Mar. 2019.

3GPP Technical Specification Group Services and System Aspects, *Study on enhancement of 3GPP support for 5G V2X services*. TR 22.886 Release 16, Dec. 2018.

A. Alalewi, I. Dayoub, S. Cherkaoui, On 5G-V2X use cases and enabling technologies: A comprehensive survey. *IEEE Access*, vol. 9, pp. 107710–107737, 2021.

S. Gyawali, S. Xu, Y. Qian, R.Q. Hu, Challenges and solutions for cellular based V2X communications. *IEEE Communications Surveys & Tutorials*, vol. 23, no. 1, pp. 222–255, 2021.

S.A.A. Hakeem, A.A. Hady, H.W. Kim, 5G-V2X: Standardization, architecture, use cases, network-slicing, and edge-computing. *Springer Wireless Networks*, vol. 26, pp. 6015–6041, 2020.

M. Harounabadi, D.M. Soleymani, S. Bhadauria, M. Leyh, E. Roth-Mandutz, V2X in 3GPP standardization: NR sidelink in Release-16 and beyond. *IEEE Communications Standards Magazine*, vol. 5, no. 1, pp. 12–21, Mar. 2021.

Z. MacHardy, A. Khan, K. Obana, S. Iwashina, V2X access technologies: Regulation, research, and remaining challenges. *IEEE Communications Surveys & Tutorials*, vol. 20, no. 3, pp. 1858–1877, 2018.

G. Naik, B. Choudhury, J.-M. Park, IEEE 802.11bd & 5G NR V2X: Evolution of radio access technologies for V2X communications. *IEEE Access*, vol. 7, pp. 70169–70184, 2019.

C.R. Storck, F. Duarte-Figueiredo, A survey of 5G technology evolution, standards, and infrastructure associated with vehicle-to-everything communications by internet of vehicles. *IEEE Access*, vol. 8, pp. 117593–117614, 2020.

H. Zhou, W. Xu, J. Chen, W. Wang, Evolutionary V2X technologies toward the internet of vehicles: Challenges and opportunities. *Proceedings of the IEEE*, vol. 108, no. 2, pp. 308–323, Feb. 2020.

AI-Enabled Massive IoT Toward 6G

A. Pouttu (Ed.), Validation and trials for verticals towards 2030's. *6G Research Visions White Paper*, no. 4, pp. 1–36, Jun. 2020.

M.W. Akhtar, S.A. Hassan, R. Ghaffar, H. Jung, S. Garg, M.S. Hossain, The shift to 6G communications: Vision and requirements. *Springer Human-Centric Computing and Information Sciences*, vol. 10, no. 53, pp. 1–27, 2020.

I.F. Akyildiz, A. Kak, S. Nie, 6G and beyond: The future of wireless communications systems. *IEEE Access*, vol. 8, pp. 133995–134030, 2020.

R. Bassoli, F.H.P. Fitzek, E.C. Strinati, Why do we need 6G? *ITU Journal on Future and Evolving Technologies: Wireless Communication Systems in Beyond 5G Era*, vol. 2, no. 6, pp. 1–31, 2021.

E.S. Calvanese et al., 6G: The next frontier – From holographic messaging to artificial intelligence using subterahertz and visible light communication. *IEEE Vehicular Technology Magazine*, vol. 14, no. 3, pp. 42–50, 2019.

A. Clemm, M.T. Vega, H.K. Ravuri, T. Wauters, F.D. Turck, Toward truly immersive holographic-type communication: Challenges and solutions. *IEEE Communications Magazine*, vol. 58, no. 1, pp. 93–99, 2020.

S. Dang, O. Amin, B. Shihada, M.-S. Alouini, From a human-centric perspective: What might 6G be? *Nature Electronics*, vol. 3, pp. 20–9, 2020.

S. Dang, O. Amin, B. Shihada, M.-S. Alouini, What should 6G be? *Nature Electronics*, vol. 3, no. 1, pp. 20–29, 2020.

C. deLima et al., Localization and sensing. *6G Research Visions White Paper*, no. 12, pp. 1–36, Jun. 2020.

A. Dogra et al., A survey on beyond 5G network with the advent of 6G: Architecture and emerging technologies. *IEEE Access*, vol. 9, pp. 67512–67547, 2020.

M. Giordani, M. Polese, M. Mezzavilla, S. Rangan, M. Zorzi, Toward 6G networks: Use cases and technologies. *IEEE Communications Magazine*, vol. 58, no. 3, pp. 55–61, 2020.

F. Guo et al., Enabling massive IoT toward 6G: A comprehensive survey. *IEEE Internet of Things Journal*, vol. 8, no. 15, pp. 11891–11915, 2021.

ITU-T Focus Group on Technologies for Network 2030 Sub-G2, *New services and capabilities for network 2030: Description, technical gap and performance target analysis*, Oct. 2019.

W. Jiang et al., The road towards 6G: A comprehensive survey. *IEEE Open Journal of the Communication Society*, vol. 2, pp. 334–366, 2021.

N. Mahmood et al., Critical and massive machine type communication towards 6G. *6G Research Visions White Paper*, no. 11, pp. 1–36, Jun. 2020.

U.M. Malik et al., Energy efficient fog computing for 6G enabled massive IoT: Recent trends and future opportunities. *IEEE Internet of Things Journal*, 2021. (Early access).

A.R. Medha, M. Gupta, S. Nayak, R. Patgiri, 6G communication: A vision on deep learning in URLLC. In *Proc. Modeling, Simulation and Optimization (CoMSO)*, 2021, pp. 587–598.

S. Mumtaz et al., Guest editorial: 6G the paradigm for future wireless communications. *IEEE Wireless Communications*, vol. 29, no. 1, pp. 14–15, Feb. 2022.

D.C. Nguyen et al., 6G internet of things: A comprehensive survey. *IEEE Internet of Things Journal*, vol. 9, no. 1, pp. 359–383, 2021.

Z. Qadir et al., Towards 6G internet of things: Recent advances, use cases, and open challenges. *ICT Express*, 2022.

T.R. Raddo et al., Transition technologies towards 6G networks. *EURASIP Journal on Wireless Communications and Networking*, no. 100, pp. 1–22, 2021.

N. Rajatheva et al., Broadband connectivity in 6G. *6G Research Visions White Paper*, no. 10, pp. 1–48, Jun. 2020.

T.S. Rappaport et al., Wireless communications and applications above 100 GHz: Opportunities and challenges for 6G and beyond. *IEEE Access*, vol. 7, pp. 78729–3536, 2019.

W. Saad, M. Bennis, M. Chen, A vision of 6G wireless systems: Applications, trends, technologies, and open research problems. *IEEE Network*, vol. 34, no. 3, pp. 134–142, May/Jun. 2020.

S. Sambhwani, Z. Boos, S. Dalmia, A. Fazeli, B. Gunzelmann, A. Ioffe, M. Narasimha, F. Negro, L. Pillutla, J. Zhou, Transitioning to 6G part 1: Radio technologies. *IEEE Wireless Communications*, vol. 29, no. 1, pp. 6–8, Feb. 2022.

H. Sarieddeen, N. Saeed, TY. Al-Naffouri, M.-S. Alouini, Next generation terahertz communications: A rendezvous of sensing, imaging, and localization. *IEEE Communications Magazine*, vol. 58, no. 5, pp. 69–75, May 2020.

T. Taleb, Z. Nadir, H. Flinck, J. Song, Extremely interactive and low-latency services in 5G and beyond mobile systems. *IEEE Communications Standards Magazine*, vol. 5, no. 2, pp. 114–119, 2021.

F. Tariq, M. Khandaker, K.-K. Wong, M. Imran, M. Bennis, M. Debbah, A speculative study on 6G. *IEEE Wireless Communications*, vol. 27, no. 4, pp. 118–125. Aug. 2020.

S. Verma, Toward green communication in 6G-enabled massive internet of things. *IEEE Internet of Things Journal*, vol. 8, no. 7, pp. 5408–5415, 2020.

H. Viswanathan, P.E. Mogensen, Communications in the 6G era. *IEEE Access*, vol. 8, pp. 57063–57074, 2020.

W. Wu et al., AI-native network slicing for 6G networks. *IEEE Wireless Communications*, vol. 29, no. 1, pp. 96–103, 2022.

C. Yeh, G.D. Jo, Y.-J. Ko, H.K. Chung, Perspectives on 6G wireless communications. *ICT Express*, pp. 1–10, Jan. 2022.

X. You et al., Towards 6G wireless communication networks: Vision, enabling technologies, and new paradigm shifts. *Science China Information Sciences*, vol. 64, no. 1, pp. 1–74, 2021.

4 Vehicular Systems for 5G and beyond 5G
Channel Modelling for Performance Evaluation

Caslav Stefanovic and Ana G. Armada
Universidad Carlos III de Madrid

Marco Pratesi and Fortunato Santucci
University of L'Aquila

CONTENTS

4.1 Introduction	184
4.2 Direct V2V Communications	186
4.2.1 Direct V2V Communications over Double-Scattered Fading Channels	187
4.2.2 Performance Measures of Direct V2V Communications	188
4.2.3 Numerical Results of Direct V2V Communications	189
4.3 Relay-Assisted Dual-Hop V2V Communications	189
4.3.1 Relay-Assisted V2V Communications over DSc Fading Channels	190
4.3.2 Performance Measures of Relay-Assisted V2V Communications	192
4.3.3 Numerical Results for Relay-Assisted V2V Communications	195
4.4 V2V Cooperative Dual-Hop Communications	195
4.4.1 V2V Cooperative Dual-Hop Communications over Double-Scattered Fading Channels	195
4.4.2 Performance Measures of V2V Cooperative Dual-Hop Communications	196
4.4.3 Numerical Results for V2V Cooperative Dual-Hop Communications	197
4.5 V2V Cooperative Communications with Direct V2V Link	198
4.5.1 Cooperative Communications with Direct V2V Link over DSc Fading Channels	199
4.5.2 Performance Measures of Cooperative Communications with Direct V2V Link	199
4.5.3 Numerical Results for Cooperative Communications with Direct V2V Link	200

4.6 V2V Mixed RF-FSO Communications 201
    4.6.1 Mixed V2V RF-FSO Communications over Double-Scattered and TF Channels 201
    4.6.2 Performance Measures of Mixed V2V RF-FSO Communications 205
    4.6.3 Numerical Results for Mixed V2V RF-FSO Communications 208
4.7 V2V Cooperative Mixed RF-FSO Communications with Direct and Relay-Assisted RF Links 210
    4.7.1 V2V Cooperative Mixed RF-FSO Communications with Direct and Relay-Assisted RF Links over DSc and TF Channels 211
    4.7.2 Performance Measures of V2V Cooperative Mixed RF-FSO Communications with Direct and Relay-Assisted RF Links............ 211
    4.7.3 Numerical Results for V2V Cooperative Mixed RF-FSO Communications with Direct and Relay-Assisted RF Links............ 212
4.8 Conclusions 214
Acknowledgements 214
References 214

4.1 INTRODUCTION

Vehicular communication (V-Comm) systems are intended to be an essential part of beyond 5G and 6G wireless networks (WNs) (Adhikari et al., 2021; He et al., 2020). The V-Comm systems are usually enabled through direct node-to-node communications without necessity to communicate with the base stations (BSs). Many different types of V-Comm systems can be distinguished, usually termed as vehicle-to-vehicle (V2V), vehicle-to-pedestrian (V2P), vehicle-to-infrastructure (V2I), vehicle-to-everything (V2X), unmanned aerial vehicle (UAV)-assisted communications and others (Fotouhi et al., 2019; MacHardy et al., 2018). Moreover, the V-Comm wireless systems are often under the influence of dynamic channel conditions mainly due to the increased mobility of vehicles and increased traffic density in urban areas or highways. Propagation channels for those types of communications that are in accordance with analytical or empirical results can be modelled as the product of two or more random variables (rvs) (Ai et al., 2018; Bithas et al., 2019, 2020; Dixit et al., 2021; Jaiswal and Purohit, 2019, 2021; Nguyen and Hoang, 2019), which can cause the performance analysis challenging and demanding, especially when higher order statistical (HOS) evaluation is performed (Bithas et al., 2018a; Hajri et al., 2020; Stefanovic et al., 2018a, 2019, 2021a, 2021b). In particular, HOS measures (e.g., average level crossing rate (LCR) and average fade duration (AFD)) can provide valuable data related to time-variant fading channels that can provide better understanding of dynamic propagation conditions involved in V2X communications. The 5G and beyond 5G systems are requiring ultra-reliable low-latency communications (URLLC) for V-Comm systems which can be efficiently addressed through HOS measures. Moreover, LCR and AFD are dependent on maximum Doppler frequency (DF) which is in turn proportional to vehicle's speed and carrier frequency that are other crucial parameters for V2X channel characterization.

In Bithas et al. (2018a), the authors provided performance analysis of double-scattered (DSc) multipath fading channels for V-Comm systems modelled as the product of two α–μ random variables (rvs), known as double α–μ (d–α–μ). In Bithas et al. (2018a), the LCR expression of the d–α–μ distribution for V-Comm systems is calculated as one-folded integral expression. The closed form approximate LCR mathematical formula of d–α–μ distribution has been efficiently derived by Stefanovic et al. (2020). In particular, the general d–α–μ distribution can be reduced to double-Rayleigh (d-Ray), double Nakagami-m (d-Nak-m) and double Weibull (d-Weib) distributions for different values of shaping parameters. The α–μ, d–α–μ and multiple α–μ (m-α-μ) distributions have already been used to address fading in direct V-Comm systems, relay-assisted (R-A) V-Comm systems, reconfigurable intelligent surface (RIS)-assisted V2V communications as well as in UAV communications (Bithas et al., 2018a; Chapala et al., 2022; Dautov et al., 2018; Stefanovic et al., 2018a, 2019). The paper by Badarneh (2016) considers d–α–μ random process as the composite fading model while statistical analysis of the m-α-μ is provided by Leonardo and Yacoub (2015) and Badarneh and Almehmadi (2016). Moreover, the statistical analysis of the ratio of the m-α-μ rvs are given in Leonardo et al. (2016). Moreover, Stefanovic et al. (2020) provided a framework for channel modelling and performance evaluation based on d–α–μ, d–α–μ/gamma and the ratio of two d–α–μ/gamma rvs in order to address multipath in DSc, composite fading in DSc-single shadowed (SS) as well as interference-limited environment in DSc-SS fading conditions, that can appear in 5G propagation scenarios. Indeed, the products and the ratios of rvs are of great importance in channel modelling and performance analysis of WNs (Du et al., 2019; Matovic et al., 2013; Stamenkovic et al., 2014; Stefanovic et al., 2022; Zlatanov et al., 2008).

The R-A communications are often proposed as a useful technique for the system performance improvement and coverage extension in WNs (Di Renzo et al., 2009a, 2009b, 2010). Moreover, R-A communications can be efficiently applied in mobile-to-mobile (M2M) and V-Comm systems (Hajri et al., 2020; Milosevic et al., 2018a, 2018b; Stefanovic et al., 2019; Talha and Pätzold, 2011). The main advantage of cooperative R-A communications is its ability to apply spatial diversity while relying on the existing network nodes. The R-A communications in V-Comm systems can not only be used to extend V2V coverage but also in scenarios without direct V2V line-of-sight (LOS) communications (e.g., cooperative V2V communications at intersections in urban areas) (Belmekki et al., 2019). A cooperative V2X and UAV V-Comm systems with simultaneous wireless information power transport (SWIPT) are studied in Milovanovic and Stefanovic (2021), and Panic et al. (2019). The combining schemes with selection are studied by Al-Hmood and Al-Raweshidy (2017, 2020) and Milic et al. (2016) and due to its relatively lower implementation complexity the selection scheme or switching scheme are of great interest in relay-assisted or distributed antenna systems (DAS) for V-Comm scenarios (Bithas et al., 2018a, 2018b; Stefanovic et al., 2018a, 2018b; Triwinarko et al., 2020; Yoo et al., 2021). Furthermore, optimal relay selection strategy in cooperative V-Comm systems under security constraints is addressed in Poursajadi and Madani (2021).

The main issue in V-Comm networking is whether the direct-short-range communication (DSRC) and cellular (C)-V2X available standards, which usually operate at 5.9 GHz, can provide large enough capacity to support V2X URLLC under all propagation conditions. The performance improvement analysis for C-V2X cooperative

R-A communication under different combining schemes, including selection combining is considered (Pan and Wu, 2021). Ghafoor et al. (2019) considered the co-existence of DSRC and C-V2X standards for efficient V-Comm networking. According to the 5G new-radio (NR) standard, the latest V2X 5G NR releases are considering the application of stochastic models for channel characterization as well as application of frequencies above 52.6 GHz (e.g., V2X 5G NR takes into consideration application of different technologies such as Millimetre-wave, terahertz, and visible light communications-VLC) (Garcia et al., 2021). The Millimetre-wave R-A V2X communications are investigated by Li et al. (2020), Lv et al. (2021), and Tunc and Panwar (2021) while cooperative VLC for V-Comm networking is addressed in Yin et al. (2020). Moreover, mixed radio-frequency (RF)-free-space-optical (FSO) R-A communication systems can be used as redundant links that can improve the capacity of wireless V-Comm systems. Additionally, the FSO links can be distinguished as cost-effective wireless systems. The main disadvantage in FSO-based communications is turbulence fading (TF). The papers by Amer and Al-Dharrab (2019), Balti and Guizani (2018), and Stefanovic et al. (2022) address mixed RF-FSO relay as well as cooperative mixed RF-FSO relay systems. Moreover, FSO dual-hop and multi-hop relaying systems are considered in Stefanović et al. (2021b, 2021e) and Zedini et al. (2016), while RIS-assisted FSO communications are addressed in Boulogeorgos et al. (2022) and Stefanovic et al. (2021a). Park et al. (2013), in their paper, addressed UAVs in RF-FSO relay scenarios while in the work of Dautov et al. (2018) V2V FSO multi-hop UAVs relay-assisted communications is considered. The RF-FSO V2I communication link with UAV as a relay is investigated in Xu and Song (2021) while RF-FSO-RF V2V system is addressed in Stefanovic et al. (2019).

In this chapter, we consider HOS performance measures of V2V communications over d–α–μ fading channels enabled through direct RF links, relay-assisted RF links, cooperative relay-assisted RF links as well as through redundant RF-FSO links. In particular, we derive mathematical expressions for the probability density function (PDF) and the cumulative distribution function (CDF) that are further used for the derivation of outage probability (OP) and HOS measures (e.g., LCR and AFD) for the considered V2V scenarios over d–α–μ fading channels.

The chapter is organized as follows. An introduction is provided in Section 4.1. In Section 4.2, a double-scattered (DSc) fading channel for V2X V-Comm systems modelled as the product of two α–μ rvs is introduced and applied for different types of V2X scenarios, namely (i) direct V2V in Section 4.2, (ii) V2V R-A communications in Section 4.3, (iii) cooperative R-A dual-hop V2V communications in Section 4.4, (iv) direct V2V link assisted by parallel dual-hop V2V link in Section 4.5, (v) direct mixed RF-FSO-RF V2V communications in Section 4.6, and (vi) direct V2V communications assisted by a redundant parallel RF-FSO-RF link in Section 4.7. This chapter ends with conclusions in Section 4.8.

4.2 DIRECT V2V COMMUNICATIONS

Vehicle-to-vehicle (V2V) communications due to latency requirements are established through direct V2V communication links between two communicating vehicles.

Vehicular Systems for 5G and Beyond 5G

FIGURE 4.1 V2V direct communications.

4.2.1 Direct V2V Communications over Double-Scattered Fading Channels

The output signal $z_{d-\alpha-\mu}$ for a direct V2V link between source vehicle (s-v) and destination vehicle (d-v), shown in Figure 4.1, can be modelled with double $\alpha-\mu$ (d–α–μ) distribution (Bithas et al., 2018a) that can appear due to double-scattered (DSc) fading phenomena in vehicular communications as observed in Bithas et al. (2018a, 2019, 2020), Dixit et al. (2021), and Jaiswal and Purohit (2019, 2021).

The $z_{d-\alpha-\mu}$ can be expressed as the product of two $\alpha - \mu$ rvs (Yacoub, 2007):

$$z_{d-\alpha-\mu} = x_{\alpha-\mu} y_{\alpha-\mu} = \underbrace{(x_{N,1} x_{N,2})^{\frac{2}{a}}}_{\text{double-scattered}} \tag{4.1}$$

where a is a parameter of non-homogeneity, x_N and y_N are independent but not identically distributed (i.n.i.d.), and Nak-m rvs is given by Stüber (1996):

$$p_{X_{N,i}}(x_{N,i}) = \frac{2(\mu_i/\vartheta_i)^{\mu_i}}{\Gamma(\mu_i)} x_{N,i}^{2\mu_i-1} e^{-\frac{\mu_i x_{N,i}^2}{\vartheta_i}}, \; i=1,2 \tag{4.2}$$

whose shaping parameters are denoted as μ_i and ϑ_i, $i = 1,2$, respectively. The PDF of $Z_{d-\alpha-\mu}$ by using Stüber (1996) and Gradshteyn and Ryzhik (2000) can be written as:

$$\begin{aligned}
p_{d-\alpha-\mu}(z_{d-\alpha-\mu}) &= \int_0^\infty \left|\frac{dx_{N,1}}{dz_{d-\alpha-\mu}}\right| p_{X_{N,1}}\left(\frac{z_{d-\alpha-\mu}^{\frac{a}{2}}}{x_{N,2}}\right) p_{X_{N,2}}(x_{N,2}) dx_{N,2} \\
&= \frac{2\alpha}{\Gamma(\mu_1)\Gamma(\mu_2)} \left(\frac{\mu_1}{\vartheta_1}\right)^{\mu_1} \left(\frac{\mu_2}{\vartheta_2}\right)^{\mu_2} z_{d-\alpha-\mu}^{a\mu_1-1} \left(\frac{\mu_1 \vartheta_2}{\mu_2 \vartheta_1} z_{d-\alpha-\mu}^a\right)^{\frac{\mu_2-\mu_1}{2}} \\
&\quad K_{\mu_2-\mu_1}\left(2\sqrt{\frac{\mu_1 \mu_2}{\vartheta_1 \vartheta_2} z_{d-\alpha-\mu}^a}\right)
\end{aligned} \tag{4.3}$$

where $K_p(\cdot)$ is a modified Bessel function of the pth order and second kind (Gradshteyn and Ryzhik, 2000). The CDF of $Z_{d-\alpha-\mu}$ can be obtained from $p_{d-\alpha-\mu}(z_{d-\alpha-\mu})$ according to a well-known mathematical formula (Stüber, 1996):

$$F_{d-\alpha-\mu}\left(z_{d-\alpha-\mu}\right) = \int_0^{z_{d-\alpha-\mu}} p_{d-\alpha-\mu}(v)\,dv \tag{4.4}$$

The closed form $F_{d-\alpha-\mu}\left(z_{d-\alpha-\mu}\right)$ is derived by applying (Gradshteyn and Ryzhik, 2000) and for the case where μ_1 is a positive integer given by:

$$F_{d-\alpha-\mu}\left(z_{d-\alpha-\mu}\right) = \frac{1}{\Gamma(\mu_1)}(\mu_1-1)! - \frac{2}{\Gamma(\mu_1)\Gamma(\mu_2)}\left(\frac{\mu_2}{\vartheta_2}\right)^{\mu_2} \cdot (\mu_1-1)!$$

$$\times \sum_{k=0}^{\mu_1-1} \frac{\left(\frac{\mu_1 z_{d-\alpha-\mu}^a}{\vartheta_1}\right)^k}{k!} \left(\frac{\mu_1 \vartheta_2}{\mu_2 \vartheta_1} z_{d-\alpha-\mu}^a\right)^{\frac{\mu_2-k}{2}} K_{\mu_2-k}\left(2\sqrt{\frac{\mu_1 \mu_2}{\vartheta_1 \vartheta_2} z_{d-\alpha-\mu}^a}\right) \tag{4.5}$$

4.2.2 Performance Measures of Direct V2V Communications

The performance measures for direct V-Comm systems such as OP, LCR, and AFD for a given threshold z_{th} are addressed in this section. The OP for a direct V2V communication link can be efficiently obtained from equation (4.5) by using Panic et al. (2013) and Stüber (1996):

$$P_{d-\alpha-\mu}(z_{\text{th}}) = P_{d-\alpha-\mu}\left(z_{d-\alpha-\mu} \le z_{\text{th}}\right) = F_{d-\alpha-\mu}(z_{\text{th}}) \tag{4.6}$$

The LCR for a predetermined threshold z_{th} can be obtained through the integral formula of the joint distribution of $z_{d-\alpha-\mu}$ and its first derivative $\dot{z}_{d-\alpha-\mu}$, given by Stüber (1996):

$$N_{d-\alpha-\mu}(z_{\text{th}}) = \int_0^\infty \dot{z}_{d-\alpha-\mu}\, p_{z_{d-\alpha-\mu}\dot{z}_{d-\alpha-\mu}}\left(z_{\text{th}},\dot{z}_{d-\alpha-\mu}\right)d\dot{z}_{d-\alpha-\mu} \tag{4.7}$$

The LCR of $z_{d-\alpha-\mu}$ provided in equation (4.7) can be evaluated through the mathematical transformations whose detailed derivation is already provided in Stefanovic (2018a) and can be expressed as

$$N_{d-\alpha-\mu}(z_{\text{th}}) = \frac{4\left(\frac{\mu_1}{\vartheta_1}\right)^{\mu_1}\left(\frac{\mu_2}{\vartheta_2}\right)^{\mu_2} \sigma_{\dot{X}_{N,1}}}{\sqrt{2\pi}\,\Gamma(\mu_1)\Gamma(\mu_2)} z_{\text{th}}^{a\left(\mu_1-\frac{1}{2}\right)}$$

$$\times \underbrace{\int_0^\infty \sqrt{1 + \frac{\sigma_{\dot{X}_{N,2}}^2 z_{\text{th}}^a}{\sigma_{\dot{X}_{N,1}}^2 x_{N,2}^4}}\, e^{-\frac{\mu_1}{\vartheta_1}\frac{z_{\text{th}}^a}{x_{N,2}^2} - \frac{\mu_2}{\vartheta_2} x_{N,2}^2 + 2(\mu_2-\mu_1)\ln(x_{N,2})}\, dx_{N,2}}_{E_1} \tag{4.8}$$

Vehicular Systems for 5G and Beyond 5G

where $\sigma^2_{\dot{X}_{N,1}} = \pi^2 (f_{m,1})^2 \vartheta_1/\mu_1$ and $\sigma^2_{\dot{X}_{N,2}} = \pi^2 (f_{m,2})^2 \vartheta_2/\mu_2$ are the variances of $\dot{X}_{N,1}$ and $\dot{X}_{N,2}$, respectively. The maximum Doppler frequencies (DFs) are assumed to be the equal, $f_m = f_{m \cdot 1} = f_{m \cdot 2} = \sqrt{(f_{m \cdot s} - v)^2 + (f_{m \cdot d} - v)^2}, i = 1,2$, where $f_{m,s-v}$ and $f_{m,d-v}$ are maximum DFs for direct RF V2V communications of the s-v and d-v, respectively (Hadzi-Velkov et al., 2009). The expression E_1 is solved by applying the Laplace-based approximation method (LBAM) for one-folded integral (Zlatanov et al., 2008) whose detailed application is also presented in Stefanovic et al. (2020). Here we only re-write previously derived closed form LCR approximation of the $d - \alpha - \mu$ distribution (Stefanovic et al., 2020):

$$N_{d-\alpha-\mu}(z_{th}) \approx \frac{4\left(\frac{\mu_1}{\vartheta_1}\right)^{\mu_1}\left(\frac{\mu_2}{\vartheta_2}\right)^{\mu_2} z_{th}^{\alpha\left(\mu_1-\frac{1}{2}\right)} \sigma_{\dot{X}_{N,1}}}{\sqrt{2\pi}\Gamma(\mu_1)\Gamma(\mu_2)} \left(\frac{2\pi}{\gamma}\right)^{\frac{1}{2}}$$

$$\frac{\left(1 + \frac{\sigma^2_{\dot{X}_{N,2}} z_{th}^a}{\sigma^2_{\dot{X}_{N,1}} x_{N,20}^4}\right)^{1/2} e^{-\gamma\left(\frac{\mu_1 z_{th}^a}{\vartheta_1 x_{N,20}^2} + \frac{\mu_2}{\vartheta_2} x_{N,20}^2 - 2(\mu_2-\mu_1)\ln(x_{N,20})\right)}}{\left(\frac{6\mu_1 z_{th}^a}{\vartheta_1 x_{N,20}^4} + 2\frac{\mu_2}{\vartheta_2} + \frac{2(\mu_2-\mu_1)}{x_{N,20}^2}\right)^{1/2}} \quad (4.9)$$

where $\gamma = 1$, and $x_{N,20} = \left(\frac{\vartheta_1\vartheta_2(\mu_2-\mu_1) + \sqrt{(\vartheta_1\vartheta_2(\mu_2-\mu_1))^2 + 4(\vartheta_1\vartheta_2\mu_1\mu_2 z_{th}^a)}}{2\vartheta_1\mu_2}\right)^{\frac{1}{2}}$

It is important to mention that the approximation of $N_{d-\alpha-\mu}(z_{th})$ provides shorter computing time than its integral form with similar accuracy.

The AFD, $\text{AFD}_{d-\alpha-\mu}$ is obtained as Stüber (1996):

$$\text{AFD}_{d-\alpha-\mu}(z_{th}) = \frac{F_{d-\alpha-\mu}(z_{th})}{N_{d-\alpha-\mu}(z_{th})} \quad (4.10)$$

4.2.3 Numerical Results of Direct V2V Communications

Numerical results for $N_{d-\alpha-\mu}(z_{th})$ and $\text{AFD}_{d-\alpha-\mu}(z_{th})$ are already provided in Stefanovic et al. (2020). However, the presented analytical results will be directly used in the extended cooperative V-Comm scenario when a direct V2V link is supported by relay-assisted (R-A) communications provided in the following sections of this chapter.

4.3 RELAY-ASSISTED DUAL-HOP V2V COMMUNICATIONS

Relay-assisted (R-A) V-Comm systems as well as cooperative V-Comm systems can be used for coverage extension and system performance improvement in V2V communication links.

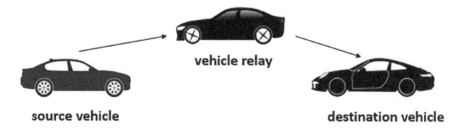

FIGURE 4.2 V2V A–F dual-hop vehicle relay communication.

4.3.1 Relay-Assisted V2V Communications over DSc Fading Channels

The system model for V2V dual-hop amplify-and-forward (A–F) R-A V-Comm systems over $d - \alpha - \mu$ fading channels is shown in Figure 4.2. Although V-Comm systems of R-A as well as cooperative R-A V2V communications over $d - \alpha - \mu$ fading channels are briefly addressed in Stefanovic et al. (2018), in this section as well as in the next section of the chapter we present a comprehensive mathematical framework for their performance evaluation.

We model the signal envelope at the output of a V2V dual-hop A–F relay as a product of two i.n.i.d. $d - \alpha - \mu$ rvs ($z_{sd} = z_{sr} z_{rd}$). The signal envelope from source vehicle (s-v) to vehicle relay (v-r) is denoted with z_{sr} and modelled with $d - \alpha - \mu$ distribution that can be expressed as the product of two $\alpha - \mu$ rvs ($z_{sr} = x_{\alpha-\mu,1} x_{\alpha-\mu,2}$). The signal envelope from v-r to destination vehicle (d-v) is denoted with z_{rd} and also modelled with $d - \alpha - \mu$ distribution that can be further expressed as the product of two other $\alpha - \mu$ rvs ($z_{rd} = y_{d-\alpha-\mu,1} y_{d-\alpha-\mu,2}$). Furthermore, $d - \alpha - \mu$ rvs are expressed through Nakagami-m (Nak-m) rvs, $x_{N,1}$, $x_{N,2}$, $y_{N,1}$, and $y_{N,2}$ and non-homogeneity parameter α (Yacoub, 2007):

$$z_{sd} = z_{sr} z_{rd} = x_{d-\alpha-\mu,1} x_{d-\alpha-\mu,2} y_{d-\alpha-\mu,1} y_{d-\alpha-\mu,2} = \left(x_{N,1} x_{N,2} y_{N,1} y_{N,2} \right)^{\frac{2}{\alpha}} \quad (4.11)$$

The Nak-m is given by the following distributions, respectively (Stüber, 1996):

$$p_{x_{N,i}}(x_{N,i}) = \frac{2}{\Gamma(\mu_{xi})} \left(\frac{\mu_{xi}}{\vartheta_{xi}} \right)^{\mu_{xi}} x_{N,i}^{2\mu_{xi}-1} e^{-\frac{\mu_{xi} x_{N,i}^2}{\vartheta_{xi}}}, \quad i = 1, 2 \quad (4.12)$$

$$p_{y_{N,i}}(y_{N,i}) = \frac{2}{\Gamma(\mu_{yi})} \left(\frac{\mu_{yi}}{\vartheta_{yi}} \right)^{\mu_{yi}} y_{N,i}^{2\mu_{yi}-1} e^{-\frac{\mu_{yi} y_{N,i}^2}{\vartheta_{yi}}}, \quad i = 1, 2 \quad (4.13)$$

where μ_{x1} and μ_{x2} are s-r DSc severity parameters while μ_{Y1} and μ_{Y2} are r-d DSc severity parameters. Moreover, s-r as well as r-d scaling parameters of $x_{N,i}$ and $y_{N,i}$ are denoted as ϑ_{xi} and ϑ_{yi}, respectively. $\Gamma(\cdot)$ presents the gamma function (Gradshteyn and Ryzhik, 2000).

Vehicular Systems for 5G and Beyond 5G

The PDF at the output of the considered V2V system over $d - \alpha - \mu$ fading channels system can be written as Stüber (1996):

$$p_{sd}(z_{sd}) = \int_0^\infty dx_{N,2} \int_0^\infty dy_{N,1} \int_0^\infty \frac{\frac{a}{2} z_{sd}^{\frac{a}{2}-1}}{x_{N,2} y_{N,1} y_{N,2}}$$

$$p_{x_{N,1}}\left(\frac{z_{sd}^{\frac{a}{2}}}{x_{N,2} y_{N,1} y_{N,2}}\right) \cdot p_{x_{N,2}}(x_{N,2}) p_{y_{N,1}}(y_{N,1}) p_{y_{N,2}}(y_{N,2}) dy_{N,2}$$

$$= \frac{8a \, z_{sd}^{a\mu_{x1}-1}}{\Gamma(\mu_{x1})\Gamma(\mu_{x2})\Gamma(\mu_{y1})\Gamma(\mu_{y2})} \left(\frac{\mu_{x1}}{\vartheta_{x1}}\right)^{\mu_{x1}} \left(\frac{\mu_{x2}}{\vartheta_{x2}}\right)^{\mu_{x2}} \left(\frac{\mu_{y1}}{\vartheta_{y1}}\right)^{\mu_{y1}} \left(\frac{\mu_{y2}}{\vartheta_{y2}}\right)^{\mu_{y2}} \quad (4.14)$$

$$\cdot \int_0^\infty dx_{N,2} \int_0^\infty dy_{N,1} \int_0^\infty x_{N,2}^{2\mu_{x2}-2\mu_{x1}-1} x_{N,1}^{2\mu_{y1}-2\mu_{x1}-1} x_{N,2}^{2\mu_{y2}-2\mu_{x1}-1}$$

$$\cdot e^{-\left(\frac{\mu_{x1}}{\vartheta_{x1}(x_{N,2}y_{N,1}y_{N,2})^2} z_{sd}^a + \frac{\mu_{x2}}{\vartheta_{x2}} x_{N,2}^2 + \frac{\mu_{y1}}{\vartheta_{y1}} y_{N,1}^2 + \frac{\mu_{y2}}{\vartheta_{y2}} y_{N,2}^2\right)} dy_{N,2}$$

The CDF of z_{sd} can be obtained from Stüber (1996) and Gradshteyn and Ryzhik (2000), respectively:

$$F_{sd}(z_{sd}) = \int_0^{z_{sd}} p_{z_{sd}}(t) dt$$

$$= \frac{8}{\Gamma(\mu_{x1})\Gamma(\mu_{x2})\Gamma(\mu_{y1})\Gamma(\mu_{y2})} \left(\frac{\mu_{x1}}{\vartheta_{x1}}\right)^{\mu_{x1}} \left(\frac{\mu_{x2}}{\vartheta_{x2}}\right)^{\mu_{x2}} \left(\frac{\mu_{y1}}{\vartheta_{y1}}\right)^{\mu_{y1}} \left(\frac{\mu_{y2}}{\vartheta_{y2}}\right)^{\mu_{y2}}$$

$$\cdot \int_0^\infty dx_{N,2} \int_0^\infty dy_{N,1} \int_0^\infty x_{N,2}^{2\mu_{x2}-2\mu_{x1}-1} y_{N,1}^{2\mu_{y1}-2\mu_{x1}-1} y_{N,2}^{2\mu_{y2}-2\mu_{x1}-1} \quad (4.15)$$

$$\cdot e^{-\left(\frac{\mu_{x2}}{\vartheta_{x2}} x_{N,2}^2 + \frac{\mu_{y1}}{\vartheta_{y1}} y_{N,1}^2 + \frac{\mu_{y2}}{\vartheta_{y2}} y_{N,2}^2\right)} \left(\frac{\vartheta_{x1} x_{N,2}^2 y_{N,1}^2 y_{N,2}^2}{\mu_{x1}}\right)^{\mu_{x1}}$$

$$\gamma\left(\mu_{x1}, \frac{\mu_{x1} z_{sd}^a}{\vartheta_{x1} x_{N,2}^2 y_{N,1}^2 y_{N,2}^2}\right) dy_{N,2}$$

where $\gamma(b, z)$ is incomplete gamma function (Gradshteyn and Ryzhik, 2000).

4.3.2 Performance Measures of Relay-Assisted V2V Communications

Performance evaluation is addressed in terms of OP, LCR, and AFD of a V2V A–F dual-hop R-A communication system in DSc fading environment. The OP (denoted as $P_{sd}(z_{th})$) of a dual-hop A–F relay-assisted V2V link can be written as:

$$P_{sd}(z_{th}) = P_{sd}(z_{sd} \leq z_{th}) = F_{sd}(z_{th}) \qquad (4.16)$$

The average LCR at the output of the considered V-Comm system can be calculated using the formula (Stüber, 1996):

$$N_{z_{sd}} = \int_0^\infty \dot{z}_{sd} p_{z_{sd}\dot{z}_{sd}}(z_{sd}\dot{z}_{sd}) d\dot{z}_{sd} \qquad (4.17)$$

where $p_{z_{sd}\dot{z}_{sd}}(z_{sd}\dot{z}_{sd})$ is the joint distribution of z_{sd} and its first derivative $\dot{z}_{sd}$. The first derivative of z_{sd} can be expressed as:

$$\dot{z}_{sd} = \frac{1}{\frac{\alpha}{2} z_{sd}^{\frac{\alpha}{2}-1}} \left(x_{N,2} y_{N,1} y_{N,2} \dot{x}_{N,1} + x_{N,1} y_{N,1} y_{N,2} \dot{x}_{N,2} + x_{N,1} x_{N,2} y_{N,2} \dot{y}_{N,1} + x_{N,1} x_{N,2} y_{N,1} \dot{y}_{N,2} \right) \qquad (4.18)$$

The $\dot{x}_{N,1}, \dot{x}_{N,2}, \dot{y}_{N,1}$, and $\dot{y}_{N,2}$ are zero-mean Gaussian (ZM-G) rvs. Because a linear transformation of a ZM-G rvs is a ZM-G rv, thus, $\dot{z}_{sd}$ is also ZM-G rv whose variance can be expressed as,

$$\sigma_{\dot{z}_{sd}}^2 = \frac{4}{\alpha^2 z_{sd}^{\alpha-2}} \left((x_{N,2} y_{N,1} y_{N,2})^2 \sigma_{\dot{x}_{N,1}}^2 + (x_{N,1} y_{N,1} y_{N,2})^2 \sigma_{\dot{x}_{N,2}}^2 \right.$$
$$\left. + (x_{N,1} x_{N,2} y_{N,2})^2 \sigma_{\dot{y}_{N,1}}^2 + (x_{N,1} x_{N,2} y_{N,1})^2 \sigma_{\dot{y}_{N,2}}^2 \right) \qquad (4.19)$$

where Yang et al. (2005), $\sigma_{\dot{x}_{N,1}}^2 = \pi^2 f_{m,x1}^2 \frac{\vartheta_{x1}}{\mu_{x1}}$, $\sigma_{\dot{x}_{N,2}}^2 = \pi^2 f_{m,x2}^2 \frac{\vartheta_{x2}}{\mu_{x2}}$, $\sigma_{\dot{y}_{N,1}}^2 = \pi^2 f_{m,y1}^2 \frac{\vartheta_{y1}}{\mu_{y1}}$, and $\sigma_{\dot{x}_{N,2}}^2 = \pi^2 f_{m,y2}^2 \frac{\vartheta_{y2}}{\mu_{y2}}$.

Similarly, the maximal DFs are assumed to be the same ($f_m = f_{m,x1} = f_{m,x2} = f_{m,y1} = f_{m,y2}$). Accordingly, $\sigma_{\dot{z}_{sd}}^2$ is:

$$\sigma_{\dot{z}_{sd}}^2 = \frac{4\pi^2 f_m^2}{\alpha^2 z_{sd}^{\alpha-2}} \frac{\vartheta_{x1}}{\mu_{x1}} (x_{N,2} y_{N,1} y_{N,2})^2 \left(1 + \frac{\mu_{x1}}{\vartheta_{x1}} \frac{\vartheta_{x2}}{\mu_{x2}} \frac{z_{sd}^\alpha}{x_{N,2}^4 y_{N,1}^2 y_{N,2}^2} \right.$$
$$\left. + \frac{\mu_{x1}}{\vartheta_{x1}} \frac{\vartheta_{y1}}{\mu_{y1}} \frac{z_{sd}^\alpha}{x_{N,2}^2 y_{N,1}^4 y_{N,2}^2} + \frac{\mu_{x1}}{\vartheta_{x1}} \frac{\vartheta_{y2}}{\mu_{y2}} \frac{z_{sd}^\alpha}{x_{N,2}^2 y_{N,1}^2 y_{N,2}^4} \right) \qquad (4.20)$$

Vehicular Systems for 5G and Beyond 5G

The joint probability density function (JPDF) of z_{sd}, $\dot{z}_{sd}$, $x_{N,2}$, $y_{N,1}$ and $y_{N,2}$ is,

$$p_{z_{sd}\dot{z}_{sd}x_{N,2}\,y_{N,1}y_{N,2}}\left(z_{sd}\dot{z}_{sd}x_{N,2}y_{N,1}y_{N,2}\right) \tag{4.21}$$
$$= p_{\dot{z}_{sd}|z_{sd}x_{N,2}\,y_{N,1}y_{N,2}}\left(\dot{z}_{sd} \mid z_{sd}x_{N,2}\,y_{N,1}y_{N,2}\right) p_{z_{sd}x_{N,2}\,y_{N,1}y_{N,2}}\left(z_{sd}x_{N,2}y_{N,1}y_{N,2}\right)$$

where

$$p_{z_{sd}x_{N,2}\,y_{N,1}y_{N,2}}\left(z_{sd}x_{N,2}y_{N,1}y_{N,2}\right) = p_{z_{sd}|x_{N,2}y_{N,1}y_{N,2}}\left(z_{sd} \mid x_{N,2}y_{N,1}y_{N,2}\right)$$
$$p_{x_{N,2}}\left(x_{N,2}\right) p_{y_{N,1}}\left(y_{N,1}\right) p_{y_{N,2}}\left(y_{N,2}\right) \tag{4.22}$$

The $p_{x_{N,2}}(x_{N,2})$ is given by equation (4.12), while $p_{y_{N,1}}(y_{N,1})$ and $p_{y_{N,2}}(y_{N,2})$ are given by equation (4.13). The conditional PDF of z_{sd} can be expressed as,

$$p_{z_{sd}|x_{N,2}y_{N,1}y_{N,2}}\left(z_{sd} \mid x_{N,2}y_{N,1}y_{N,2}\right) = \left|\frac{dx_{N,1}}{dz_{sd}}\right| p_{x_{N,1}}\left(\frac{z_{sd}^{\frac{\alpha}{2}}}{x_{N,2}y_{N,1}y_{N,2}}\right) \tag{4.23}$$

where $p_{x_{N,1}}$ is given by equation (4.12), and

$$\left|\frac{dx_{N,1}}{dz_{sd}}\right| = \frac{\dfrac{\alpha}{2}z_{sd}^{\frac{\alpha}{2}-1}}{x_{N,2}y_{N,1}y_{N,2}} \tag{4.24}$$

Since Yacoub et al. (1999),

$$\int_0^\infty \dot{z}_{sd}\, p_{\dot{z}_{sd}|z_{sd}x_{N,2}\,y_{N,1}y_{N,2}}\left(\dot{z}_{sd} \mid z_{sd}x_{N,2}\,y_{N,1}y_{N,2}\right) d\dot{z}_{sd} = \frac{1}{\sqrt{2\pi}}\sigma_{\dot{z}_{sd}} \tag{4.25}$$

and by adequate substitutions, the LCR given by equation (4.17) can be expressed as:

$$N_{sd}(z_{th})$$
$$= f_m \frac{8\sqrt{2\pi}}{\Gamma(\mu_{x1})\Gamma(\mu_{x2})\Gamma(\mu_{y1})\Gamma(\mu_{y2})} \left(\frac{\mu_{x1}}{\vartheta_{x1}}\right)^{\mu_{x1}-\frac{1}{2}} \left(\frac{\mu_{x2}}{\vartheta_{x2}}\right)^{\mu_{x2}} \left(\frac{\mu_{y1}}{\vartheta_{y1}}\right)^{\mu_{y1}} \left(\frac{\mu_{y2}}{\vartheta_{y2}}\right)^{\mu_{y2}} z_{th}^{\frac{\alpha}{2}(2\mu_{x1}-1)}$$
$$\cdot \int_0^\infty \int_0^\infty \int_0^\infty \sqrt{1 + \frac{\mu_{x1}}{\vartheta_{x1}}\frac{\vartheta_{x2}}{\mu_{x2}}\frac{z_{th}^\alpha}{x_{N,2}^4 y_{N,1}^2 y_{N,2}^2} + \frac{\mu_{x1}}{\vartheta_{x1}}\frac{\vartheta_{y1}}{\mu_{y1}}\frac{z_{th}^\alpha}{x_{N,2}^4 y_{N,1}^4 y_{N,2}^2} + \frac{\mu_{x1}}{\vartheta_{x1}}\frac{\vartheta_{y2}}{\mu_{y2}}\frac{z_{th}^\alpha}{x_{N,2}^2 y_{N,1}^2 y_{N,2}^4}}$$
$$\cdot e^{\left(-\frac{\mu_{x1}}{\vartheta_{x1}}\frac{z_{th}^\alpha}{x_{N,2}^2 y_{N,1}^2 y_{N,2}^2} - \frac{\mu_{x2}}{\vartheta_{x2}}x_{N,2}^2 - \frac{\mu_{y1}}{\vartheta_{y1}}y_{N,1}^2 - \frac{\mu_{y2}}{\vartheta_{y2}}y_{N,2}^2 + \ln x_{N,2}^{2\mu_{x2}-2\mu_{x1}} + \ln y_{N,1}^{2\mu_{y1}-2\mu_{x1}} + \ln y_{N,2}^{2\mu_{y2}-2\mu_{x1}}\right)} dx_{N,2} dy_{N,1} dy_{N,2}$$

$$\tag{4.26}$$

The analytical closed form approximation for the $N_{sd}(z_{th})$ at the output of a V2V dual-hop A–F relay system can be obtained by LBAM for three-folded integrals (Hadzi-Velkov et al., 2009):

$$\int_0^\infty dx_{N,2} \int_0^\infty dy_{N,1} \int_0^\infty g(x_{N,2}, y_{N,1}, y_{N,2}) e^{-\lambda f(x_{N,2}, y_{N,1}, y_{N,2})} dy_{N,2}$$

$$\approx \left(\frac{2\pi}{\lambda}\right)^{\frac{3}{2}} \frac{g(x_{N,20}, y_{N,10}, y_{N,20})}{\sqrt{\det A}} e^{-\lambda f(x_{N,20}, y_{N,10}, y_{N,20})}$$

(4.27)

where

$$A = \begin{vmatrix} \dfrac{\partial^2 f(x_{N,20}, y_{N,10}, y_{N,20})}{\partial x_{N,20}^2} & \dfrac{\partial^2 f(x_{N,20}, y_{N,10}, y_{N,20})}{\partial x_{N,20} \partial y_{N,10}} & \dfrac{\partial^2 f(x_{N,20}, y_{N,10}, y_{N,20})}{\partial x_{N,20} \partial y_{N,20}} \\ \dfrac{\partial^2 f(x_{N,20}, y_{N,10}, y_{N,20})}{\partial y_{N,10} \partial x_{N,20}} & \dfrac{\partial^2 f(x_{N,20}, y_{N,10}, y_{N,20})}{\partial y_{N,10}^2} & \dfrac{\partial^2 f(x_{N,20}, y_{N,10}, y_{N,20})}{\partial y_{N,10} \partial y_{N,20}} \\ \dfrac{\partial^2 f(x_{N,20}, y_{N,10}, y_{N,20})}{\partial y_{N,20} \partial x_{N,20}} & \dfrac{\partial^2 f(x_{N,20}, y_{N,10}, y_{N,20})}{\partial y_{N,20} \partial y_{N,10}} & \dfrac{\partial^2 f(x_{N,20}, y_{N,10}, y_{N,20})}{\partial y_{N,20}^2} \end{vmatrix}$$

(4.28)

while $x_{N,20}$, $y_{N,10}$, and $y_{N,20}$ are efficiently obtained from the following differential equations:

$$\frac{\partial f(x_{N,20}, y_{N,10}, y_{N,20})}{\partial x_{N,20}} = 0, \quad \frac{\partial f(x_{N,20}, y_{N,10}, y_{N,20})}{\partial y_{N,10}} = 0,$$

$$\text{and } \frac{\partial f(x_{N,20}, y_{N,10}, y_{N,20})}{\partial y_{N,20}} = 0$$

(4.29)

The functions $g(x_{N,2}, y_{N,1}, y_{N,2})$ and $f(x_{N,2}, y_{N,1}, y_{N,2})$ in equation (4.27) for the considered case are, respectively,

$$g(x_{N,2}, y_{N,1}, y_{N,2})$$

$$= \left(1 + \frac{\mu_{x1}}{\vartheta_{x1}} \frac{\vartheta_{x2}}{\mu_{x2}} \frac{z_{th}^\alpha}{x_{N,2}^4 y_{N,1}^2 y_{N,2}^2} + \frac{\mu_{x1}}{\vartheta_{x1}} \frac{\vartheta_{y1}}{\mu_{y1}} \frac{z_{th}^\alpha}{x_{N,2}^2 y_{N,1}^4 y_{N,2}^2} + \frac{\mu_{x1}}{\vartheta_{x1}} \frac{\vartheta_{y2}}{\mu_{y2}} \frac{z_{th}^\alpha}{x_{N,2}^2 y_{N,1}^2 y_{N,2}^4}\right)^{\frac{1}{2}}$$

(4.30)

Vehicular Systems for 5G and Beyond 5G

and

$$f(x_{N,2}, y_{N,1}, y_{N,2}) = \frac{\mu_{x1}}{\vartheta_{x1}} \frac{z_{\text{th}}^{\alpha}}{x_{N,2}^{2} y_{N,1}^{2} y_{N,2}^{2}} + \frac{\mu_{x2}}{\vartheta_{x2}} x_{N,2}^{2} + \frac{\mu_{y1}}{\vartheta_{y1}} y_{N,1}^{2} + \frac{\mu_{y2}}{\vartheta_{y2}} y_{N,2}^{2}$$
$$- \ln x_{N,2}^{2\mu_{x2}-2\mu_{x1}} - \ln y_{N,1}^{2\mu_{y1}-2\mu_{x1}} - \ln y_{N,2}^{2\mu_{y2}-2\mu_{x1}} \quad (4.31)$$

The $\text{AFD}_{z_{\text{sd}}}$ is evaluated as Stüber (1996):

$$\text{AFD}_{z_{\text{sd}}}(z_{\text{th}}) = \frac{F_{z_{\text{sd}}}(z_{\text{th}})}{N_{z_{\text{sd}}}(z_{\text{th}})} \quad (4.32)$$

4.3.3 Numerical Results for Relay-Assisted V2V Communications

Numerical results for LCR and AFD are presented in the subsequent section as a part of a cooperative relay-assisted (R-F) V2V communication system scenario where approximate and integral-form expressions for $N_{z_{\text{sd}}}$ are presented and compared. One can notice that HOS expressions for R-A V2V systems increase in complexity if compared to direct V2V communications. Stochastic modelling and performance evaluation of the realistic V-Comm networks over fading channels for different V2V communication scenarios can become very complex. This can be a motivation for using advanced channel modelling-related approaches for V2X (e.g., machine learning techniques) (Huang et al., 2019, 2020; Tang et al., 2021).

4.4 V2V COOPERATIVE DUAL-HOP COMMUNICATIONS

We extend the proposed model to a cooperative dual-hop amplify-and-forward (A–F) relay network, consisting of N parallel vehicle relays with selection combining-based antenna system (SC-AS) at the destination vehicle (d-v), as shown in Figure 4.3.

4.4.1 V2V Cooperative Dual-Hop Communications over Double-Scattered Fading Channels

The cooperative V2V communications established through N parallel dual-hop v-r with SC-AS at reception over DSc fading channels are considered in order to extend the coverage and to improve the system performance. The PDF and CDF at the output of SC-AS at the destination over $d - \alpha - \mu$ fading channels are, respectively:

$$p_{N,\text{sd}}(z_{\text{sd}}) = N \cdot p_{\text{sd}}(z_{\text{sd}}) \cdot \left(F_{\text{sd}}(z_{\text{sd}})\right)^{N-1} \quad (4.33)$$

$$F_{N,\text{sd}}(z_{\text{sd}}) = \left(F_{\text{sd}}(z_{\text{sd}})\right)^{N} \quad (4.34)$$

where $P_{\text{sd}}(z_{\text{sd}})$ and $F_{\text{sd}}(z_{\text{sd}})$ are the PDF and CDF at the output of the single dual-hop A–F relay V2V system, already obtained in equations (4.14) and (4.15), respectively.

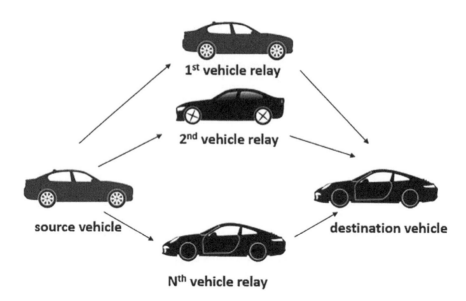

FIGURE 4.3 V2V cooperative A–F dual-hop vehicle relay communication.

4.4.2 Performance Measures of V2V Cooperative Dual-Hop Communications

The performance evaluation of considered cooperative R-A V2V system is addressed in terms of OP, LCR, and AFD. The OP for the predetermined value of the threshold at the output of the proposed model is:

$$P_{N,\,\text{sd}}(z_{\text{th}}) = P_{N,\,\text{sd}}(z_{\text{sd}} \leq z_{\text{th}}) = F_{N,\,\text{sd}}(z_{\text{th}}) \qquad (4.35)$$

where $F_{N,\,\text{sd}}(z_{\text{sd}})$ is calculated in equation (4.34). The LCR for the z_{th} value at the output of the cooperative R-A V2V system is:

$$N_{N,\,\text{sd}}(z_{\text{th}}) = N \cdot N_{\text{sd}}(z_{\text{th}}) \cdot \left(F_{N,\,\text{sd}}(z_{\text{th}})\right)^{N-1} \qquad (4.36)$$

where $N_{\text{sd}}(z_{\text{th}})$ is the LCR at the output of a single V2V dual-hop A–F link obtained in equation (4.27). AFD (denoted as $\text{AFD}_{N,\,\text{sd}}(z_{\text{th}})$) at the output of the V2V A–F dual-hop cooperative R-A system is evaluated by Stüber (1996):

$$\text{AFD}_{N,\,\text{sd}}(z_{\text{th}}) = \frac{F_{N,\,\text{sd}}(z_{\text{th}})}{N_{N,\,\text{sd}}(z_{\text{th}})} = \frac{F_{\text{sd}}(z_{\text{th}})}{N \cdot N_{\text{sd}}(z_{\text{th}})} \qquad (4.37)$$

4.4.3 Numerical Results for V2V Cooperative Dual-Hop Communications

This section presents the numerically evaluated performance measures for a V2V cooperative dual-hop A–F V2V system in DSc fading environment with SC-AS at reception. Due to the motion of s-v, d-v, and v-r, the maximal DF of the single ith path can be expressed as Hadzi-Velkov et al. (2009),

$$f_m = f_{m_i} = \sqrt{f_{ms_i}^2 + 2f_{mr_i}^2 + f_{md_i}^2}, \ i = 1, N \quad (4.38)$$

where f_{ms_i}, f_{mr_i}, and f_{md_i} are the ith path maximal DFs of s-v, r-v, and d-v, respectively.

The considered dual-hop R-A V2V system with r-v for the case when $f_{mr_i} = 0$ can be applied for a dual-hop R-A V2V system with road-side-unit (RSU) as static relay. Further, we assume that $f_m = f_{m_i}$. Normalized LCR ($N_{N,\,sd}(z_{th})/f_m$) and AFD ($AFD_{N,\,sd}(z_{th}) \cdot f_m$) exact analytical expressions as well as approximations derived through LBAM for the various values of fading severity parameters (μ_{x1}, μ_{x2}, μ_{y1} and μ_{y2}), shaping parameters $\left(\Omega_{x1} = \Omega_{x2} = \Omega_{y1} = \Omega_{y2} = 1\right)$, the parameter of non-homogeneity (α), and different number of vehicle relays (N) are graphically shown in Figures 4.4 and 4.5, respectively.

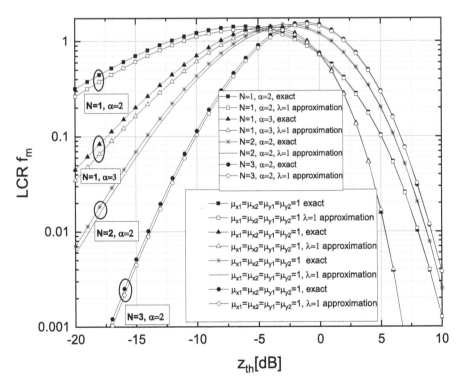

FIGURE 4.4 Normalized LCR of V2V cooperative dual-hop A–F relaying for various system model parameters.

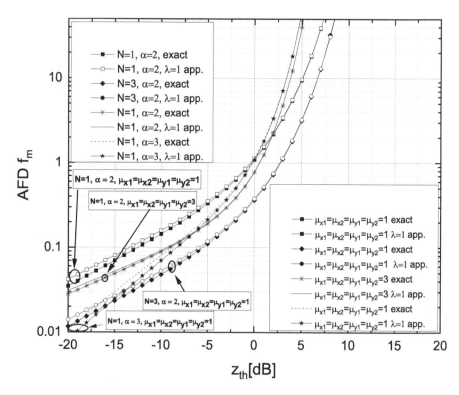

FIGURE 4.5 Normalized AFD of V2V cooperative dual-hop A–F relaying for various system model parameters.

Figure 4.4 shows that by increasing the parameter of non-homogeneity and the number of vehicle relays, $N_{N,\,sd}(z_{th})/f_m$ decreases, evidencing a system performance improvement.

It can be observed that N and α have stronger impact on $N_{N,\,sd}(z_{th})/f_m$ for lower z_{th} values than for the higher values of z_{th}. The system performance improvement (e.g., when normalized $\mathrm{AFD}_{N,\,sd}(z_{th})$ decreases) can be efficiently achieved by increasing the number of parallel relays, what can be observed in Figure 4.5. Moreover, by changing fading severity conditions (e.g., by shifting from $\mu_{x1} = \mu_{x2} = \mu_{y1} = \mu_{y2} = 1$ to $\mu_{x1} = \mu_{x2} = \mu_{y1} = \mu_{y2} = 3$) and nonlinearity parameter values (e.g., by shifting from $\alpha = 2$ to $\alpha = 3$), $\mathrm{AFD}_{N,\,sd}(z_{th}) \cdot f_m$ decreases in lower dB output regime, providing the system performance improvement for lower z_{th} values.

4.5 V2V COOPERATIVE COMMUNICATIONS WITH DIRECT V2V LINK

We examine the V2V communications for a scenario where the direct V2V link is assisted by dual-hop RSU relaying, as shown in Figure 4.6.

Vehicular Systems for 5G and Beyond 5G

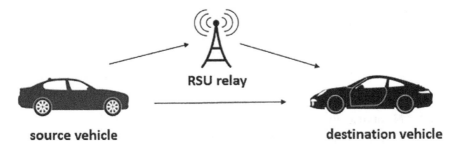

FIGURE 4.6 V2V direct communication assisted by dual-hop A–F relaying link.

4.5.1 Cooperative Communications with Direct V2V Link over DSc Fading Channels

The PDF and CDF for such a model are, respectively:

$$p_{2,\text{sd}}(z_{\text{sd}}) = p_{\text{sd}}(z_{\text{sd}})F_{d-\alpha-\mu}(z_{\text{sd}}) + p_{d-\alpha-\mu}(z_{\text{sd}})F_{\text{sd}}(z_{\text{sd}}) \quad (4.39)$$

$$F_{2,\text{sd}}(z_{\text{sd}}) = F_{d-\alpha-\mu}(z_{\text{sd}})F_{\text{sd}}(z_{\text{sd}}) \quad (4.40)$$

where $p_{d-\alpha-\mu}(z_{\text{sd}})$ and $F_{d-\alpha-\mu}(z_{\text{sd}})$ are PDF and CDF of direct V2V communications over $d-\alpha-\mu$ fading channels obtained in equations (4.3) and (4.5), respectively, while $p_{\text{sd}}(z_{\text{sd}})$ and $F_{\text{sd}}(z_{\text{sd}})$ are PDF and CDF at the output of V2V dual-hop A–F relay obtained in equations (4.14) and (4.15), respectively.

4.5.2 Performance Measures of Cooperative Communications with Direct V2V Link

The performance evaluation of cooperative dual-hop communications with a direct V2V link is considered in terms of OP and HOS measures. The OP is calculated as:

$$P_{2,\text{sd}}(z_{\text{th}}) = P_{2,\text{sd}}(z_{\text{sd}} \leq z_{\text{th}}) = F_{2,\text{sd}}(z_{\text{th}}) \quad (4.41)$$

where $F_{2,\text{sd}}(z_{\text{sd}})$ is given by equation (4.40). The LCR at the output of the proposed model is:

$$N_{2,\text{sd}}(z_{\text{th}}) = N_{\text{sd}}(z_{\text{th}})F_{d-\alpha-\mu}(z_{\text{th}}) + N_{d-\alpha-\mu}(z_{\text{th}})F_{\text{sd}}(z_{\text{th}}) \quad (4.42)$$

where $N_{d-\alpha-\mu}(z_{\text{th}})$ and $N_{\text{sd}}(z_{\text{th}})$ are the LCR expressions at the output of the direct V2V link and R-A dual-hop A–F communication link obtained in equations (4.8) and (4.26), respectively. AFD (denoted as $\text{AFD}_{2,\text{sd}}(z_{\text{th}})$) at the output of the V2V cooperative communications with direct V2V link over DSc fading channels system is evaluated by Stüber (1996):

$$\text{AFD}_{2,\text{sd}}(z_{\text{th}}) = \frac{P_{2,\text{sd}}(z_{\text{th}})}{N_{2,\text{sd}}(z_{\text{th}})}. \qquad (4.43)$$

4.5.3 Numerical Results for Cooperative Communications with Direct V2V Link

This section presents numerical results for a direct V2V link assisted by dual-hop A–F RSU R-A link (Figure 4.6). Due to the motion of only s-v and d-v, the maximal DF can be expressed as Hadzi-Velkov et al. (2009):

$$f_m = f_{m_i} = \sqrt{f_{\text{ms}_i}^2 + f_{\text{md}_i}^2}, \quad i = 1, 2 \qquad (4.44)$$

where f_{ms_i} and f_{md_i} are the ith ($i = 1, 2$) path maximal DFs of s-v and d-v, respectively. Figures 4.7 and 4.8 show $N_{2,\text{sd}}(z_{\text{th}})/f_m$ and $\text{AFD}_{2,\text{sd}}(z_{\text{th}}) \cdot f_m$, exact and approximated numerical results for various values of V2V dual-hop RSU R-A system model parameters ($\mu_{x1}, \mu_{x2}, \mu_{y1}, \mu_{y2}$ and a) and constant values ($\vartheta_{x1} = \vartheta_{x2} = \vartheta_{y1} = \vartheta_{y2} = 1$) as

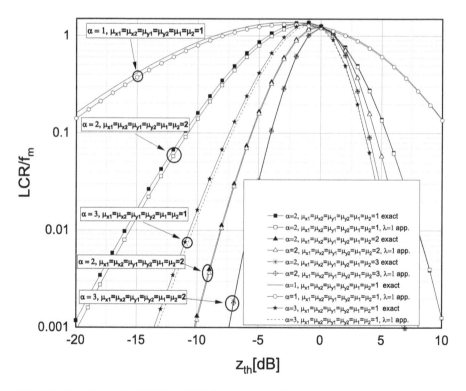

FIGURE 4.7 Normalized LCR of V2V direct communication assisted by an A–F relaying link for various system model parameters.

Vehicular Systems for 5G and Beyond 5G

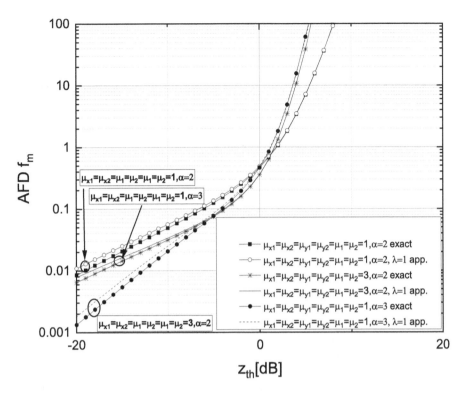

FIGURE 4.8 Normalized AFD of V2V direct communication assisted by an A–F relaying link for various system model parameters.

well as for various values of V2V direct link system model parameters (μ_1, μ_2 and a), and constant values ($\vartheta_1 = \vartheta_2 = 1$), respectively. Similarly, by increasing the set of system model parameters (μ_{x1}, μ_{x2}, μ_{x1}, μ_{x2}, a) and (μ_1, μ_2, a), the normalized $N_{2,\text{sd}}(z_{\text{th}})$ decreases in the whole observed dB output regime while $\text{AFD}_{2,\text{sd}}(z_{\text{th}})$ decreases for lower z_{th} dB values, what can be well observed in Figures 4.7 and 4.8.

4.6 V2V MIXED RF-FSO COMMUNICATIONS

The multi-hop mixed RF-FSO-RF vehicle-to-vehicle (V2V) link consisting of source vehicle (s-v), two A–F RSU-relays, and destination vehicle (d-v) is shown in Figure 4.9.

4.6.1 Mixed V2V RF-FSO Communications over Double-Scattered and TF Channels

We model the output mixed RF-FSO-RF fading signal envelope z_{mix} as the product of the i.n.i.d. random processes:

$$z_{\text{mix}} = \underbrace{\left(z_{N,1} z_{N,2}\right)^{\frac{2}{a}}}_{\text{RF1}} \underbrace{\left(z_{N,3} z_{N,4}\right)^{2}}_{\text{FSO}} \underbrace{\left(z_{N,5} z_{N,6}\right)^{\frac{2}{a}}}_{\text{RF2}} \qquad (4.45)$$

FIGURE 4.9 V2V mixed RF-FSO-RF triple-hop A–F relay link.

- $z_{N,i}$, $i = 1,6$ are i.n.i.d. Nak-m rvs, whose fading severity parameters and shaping parameters are denoted as μ_{mi} and ϑ_{mi} respectively;

$$p_{z_{N,i}}(z_{N,i}) = \frac{2(\mu_{mi}/\vartheta_{mi})^{\mu_{mi}}}{\Gamma(\mu_{mi})} z_{N,i}^{2\mu_{mi}-1} e^{-\mu_{mi} z_{N,i}^2/\vartheta_{mi}} \qquad (4.46)$$

- $(z_{N,i})^{\frac{2}{\alpha}}$ are α–μ rvs expressed through Nak-m rvs, $z_{N,i}$, and non-homogeneity parameter α representing RF fading signals.
- $(z_{N,i})^2$ are gamma (squared Nak-m) rvs, whose product $(z_{N,3} z_{N,4})^2$ represents the FSO fading signal. Moreover, the gamma–gamma process for FSO transmission exposed from strong to moderate TF conditions can be modelled assuming $\vartheta_{m3} = \vartheta_{m4} = 1$ and

$$\mu_{m3} = \alpha^{II} = \left[\exp\left(\frac{0.49\delta^2}{\left(1 + 0.18d^2 + 0.56\delta^{12/5}\right)^{7/6}}\right) - 1\right]^{-1} \qquad (4.47)$$

$$\mu_{m4} = \beta^{II} = \left[\exp\left(\frac{0.51\delta^2\left(1 + 0.69\delta^{12/5}\right)^{-5/6}}{\left(1 + 0.9d^2 + 0.62d^2\delta^{12/5}\right)^{5/6}}\right) - 1\right]^{-1} \qquad (4.48)$$

where $\delta^2 = 0.5 C_l^2 p^{7/6} S^{11/6}$ is the Rytov variance and $d = \sqrt{pD^2/4S}$ is the optical wave number. Further, the gamma–gamma FSO transmission parameters are summarized in Table 4.1.

The PDF of z_{mix} in the case of i.n.i.d. rvs can be transformed through joint and conditional probabilities as:

TABLE 4.1
FSO System Model Parameters

Parameters	Description
α^{II}	Small-scale cells
β^{II}	Large-scale cells
C_i^2	Refractive index (range: 10^{-17} m$^{-2/3}$ to 10^{-13} m$^{-2/3}$)
$p = 2\pi/\lambda$	Wave number
D	Receiver aperture diameter
S	Propagation distance

$$p_{mix}(z_{mix}) = \int_0^\infty dz_{N,1} \int_0^\infty dz_{N,2} \int_0^\infty dz_{N,4} \int_0^\infty dz_{N,5} \int_0^\infty \frac{\frac{1}{2} z_{mix}^{-\frac{1}{2}}}{z_{N,1}^{\frac{1}{a}} z_{N,2}^{\frac{1}{a}} z_{N,4} z_{N,5}^{\frac{1}{a}} z_{N,6}^{\frac{1}{a}}}$$

$$p_{Z_{N,1}}(z_{N,1}) p_{Z_{N,2}}(z_{N,2}) p_{Z_{N,4}}(z_{N,4}) p_{Z_{N,5}}(z_{N,5}) p_{Z_{N,6}}(z_{N,6}) p_{Z_{N,3}}$$

$$\left(\frac{\sqrt{z_{mix}}}{z_{N,1}^{\frac{1}{a}} z_{N,2}^{\frac{1}{a}} z_{N,4} z_{N,5}^{\frac{1}{a}} z_{N,6}^{\frac{1}{a}}} \right) dz_{N,6}$$

(4.49)

The CDF of z_{mix} is derived using Stüber (1996) and Gradshteyn and Ryzhik (2000),

$$F_{mix}(z_{mix}) = \int_0^{z_{mix}} p_{mix}(r) dr$$

$$= \frac{32 \left(\frac{\mu_{m1}}{\vartheta_{m1}} \right)^{\mu_{m1}} \left(\frac{\mu_{m2}}{\vartheta_{m2}} \right)^{\mu_{m2}} (\alpha^{II})^{\alpha^{II}} (\beta^{II})^{\beta^{II}} \left(\frac{\mu_{m5}}{\vartheta_{m5}} \right)^{\mu_{m5}} \left(\frac{\mu_{m6}}{\vartheta_{m6}} \right)^{\mu_{m6}} (\alpha^{II} - 1)!}{\Gamma(\mu_{m1}) \Gamma(\mu_{m2}) \Gamma(\alpha^{II}) \Gamma(\beta^{II}) \Gamma(\mu_{m5}) \Gamma(\mu_{m6})}$$

$$\left(\frac{\Gamma(\mu_{m1}) \Gamma(\mu_{m2}) \Gamma(\alpha^{II}) \Gamma(\beta^{II}) \Gamma(\mu_{m5}) \Gamma(\mu_{m6})}{32 \left(\frac{\mu_{m1}}{\vartheta_{m1}} \right)^{\mu_{m1}} \left(\frac{\mu_{m2}}{\vartheta_{m2}} \right)^{\mu_{m2}} (\beta^{II})^{\beta^{II}} \left(\frac{\mu_{m5}}{\vartheta_{m5}} \right)^{\mu_{m5}} \left(\frac{\mu_{m6}}{\vartheta_{m6}} \right)^{\mu_{m6}}} - \sum_{k=0}^{\alpha^{II}-1} \frac{(\alpha^{II} z_{mix})^k}{k!} E_2 \right)$$

(4.50)

where

$$E_2 = \int_0^\infty dz_{N,1} \int_0^\infty dz_{N,2} \int_0^\infty dz_{N,4} \int_0^\infty dz_{N,5}$$

$$\cdot \int_0^\infty dz_{N,6} e^{-\frac{\mu_{m1}z_{N,1}^2}{\vartheta_{m1}} - \frac{\mu_{m2}z_{N,2}^2}{\vartheta_{m2}} - \frac{a^{II}z_{mix}}{z_{N,1}^{\frac{2}{a}} z_{N,2}^{\frac{2}{a}} z_{N,4}^{\frac{2}{a}} z_{N,5}^{\frac{2}{a}} z_{N,6}^{\frac{2}{a}}} - \beta^{II}z_{N,4}^2 - \frac{\mu_{m5}z_{N,5}^2}{\vartheta_{m5}} - \frac{\mu_{m6}z_{N,6}^2}{\vartheta_{m6}}}$$

$$\cdot e^{\left(2\mu_{m1} - \frac{2}{a}k - 2\right)\ln(z_{N,1}) + \left(2\mu_{m2} - \frac{2}{a}k - 1\right)\ln(z_{N,2}) + \left(2\beta^{II} - 2k - 1\right)\ln(z_{N,4}) + \left(2\mu_{m6} - \frac{2}{a}k - 1\right)\ln(z_{N,5}) + \left(2\mu_{m6} - \frac{2}{a}k - 1\right)\ln(z_{N,6})}$$

(4.51)

for the case where α^{II} is a positive integer. The closed form $F_{mix}(z_{mix})$ expression can be obtained by evaluating E_2 given in equation (4.51) by applying LBAM for three-folded integrals (Stefanovic et al., 2021c; Stefanovic et al., 2021d):

$$\int_0^\infty dz_{N,1} \int_0^\infty dz_{N,2} \int_0^\infty dz_{N,4} \int_0^\infty dz_{N,5} \int_0^\infty g(z_{N,1}, z_{N,2}, z_{N,4}, z_{N,5}, z_{N,6}) e^{-\gamma f(z_{N,1}, z_{N,2}, z_{N,4}, z_{N,5}, z_{N,6})} dz_{N,6}$$

$$\approx \left(\frac{2\pi}{\gamma}\right)^{\frac{5}{2}} \frac{g(z_{N,10}, z_{N,20}, z_{N,40}, z_{N,50}, z_{N,60})}{\sqrt{\det B}} e^{-\gamma f(z_{N,10}, z_{N,20}, z_{N,40}, z_{N,50}, z_{N,60})}$$

(4.52)

where the arguments in equation (4.52) are, respectively, $\gamma = 1$, $g(z_{N,1}, z_{N,2}, z_{N,4}, z_{N,5}, z_{N,6}) = 1$,

$$f(z_{N,1}, z_{N,2}, z_{N,4}, z_{N,5}, z_{N,6})$$

$$= \frac{\mu_{m1}z_{N,1}^2}{\vartheta_{m1}} + \frac{\mu_{m2}z_{N,2}^2}{\vartheta_{m2}} + \frac{a^{II}z_{mix}}{z_{N,1}^{\frac{2}{a}} z_{N,2}^{\frac{2}{a}} z_{N,4}^{\frac{2}{a}} z_{N,5}^{\frac{2}{a}} z_{N,6}^{\frac{2}{a}}} + \beta^{II}z_{N,4}^2$$

$$+ \frac{\mu_{m5}z_{N,5}^2}{\vartheta_{m5}} + \frac{\mu_{m6}z_{N,6}^2}{\vartheta_{m6}} - \left(2\mu_{m1} - \frac{2}{a}k - 2\right)\ln(z_{N,1}) - \left(2\mu_{m2} - \frac{2}{a}k - 1\right)\ln(z_{N,2})$$

$$- \left(2\beta^{II} - 2k - 1\right)\ln(z_{N,4}) - \left(2\mu_{m6} - \frac{2}{a}k - 1\right)\ln(z_{N,5}) - \left(2\mu_{m6} - \frac{2}{a}k - 1\right)\ln(z_{N,6})$$

(4.53)

Furthermore, $z_{N,10}$, $z_{N,20}$, $z_{N,40}$, $z_{N,50}$ and $z_{N,60}$ in equation (4.52) can be calculated when all first-order partial derivatives of $f(z_{N,10}, z_{N,20}, z_{N,40}, z_{N,50}, z_{N,60})$ are 0. The B in equation (4.52) is:

Vehicular Systems for 5G and Beyond 5G

$$B = \begin{vmatrix} \dfrac{\partial^2 f}{\partial z_{N,10}^2} & \dfrac{\partial^2 f}{\partial z_{N,10} \partial z_{N,20}} & \dfrac{\partial^2 f}{\partial z_{N,10} \partial z_{N,40}} & \dfrac{\partial^2 f}{\partial z_{N,10} \partial z_{N,50}} & \dfrac{\partial^2 f}{\partial z_{N,10} \partial z_{N,60}} \\ \dfrac{\partial^2 f}{\partial z_{N,20} \partial z_{N,10}} & \dfrac{\partial^2 f}{\partial z_{N,20}^2} & \dfrac{\partial^2 f}{\partial z_{N,20} \partial z_{N,40}} & \dfrac{\partial^2 f}{\partial z_{N,20} \partial z_{N,50}} & \dfrac{\partial^2 f}{\partial z_{N,20} \partial z_{N,60}} \\ \dfrac{\partial^2 f}{\partial z_{N,40} \partial z_{N,10}} & \dfrac{\partial^2 f}{\partial z_{N,40} \partial z_{N,20}} & \dfrac{\partial^2 f}{\partial z_{N,40}^2} & \dfrac{\partial^2 f}{\partial z_{N,40} \partial z_{N,50}} & \dfrac{\partial^2 f}{\partial z_{N,40} \partial z_{N,60}} \\ \dfrac{\partial^2 f}{\partial z_{N,50} \partial z_{N,10}} & \dfrac{\partial^2 f}{\partial z_{N,50} \partial z_{N,20}} & \dfrac{\partial^2 f}{\partial z_{N,50} \partial z_{N,40}} & \dfrac{\partial^2 f}{\partial z_{N,50}^2} & \dfrac{\partial^2 f}{\partial z_{N,50} \partial z_{N,60}} \\ \dfrac{\partial^2 f}{\partial z_{N,60} \partial z_{N,10}} & \dfrac{\partial^2 f}{\partial z_{N,60} \partial z_{N,20}} & \dfrac{\partial^2 f}{\partial z_{N,60} \partial z_{N,40}} & \dfrac{\partial^2 f}{\partial z_{N,60} \partial z_{N,50}} & \dfrac{\partial^2 f}{\partial z_{N,60}^2} \end{vmatrix}$$

(4.54)

Approximate and exact analytical expressions for $F_{\text{mix}}(z_{\text{mix}})$ are used for derivation of AFD of the considered mixed V2V R-A RF-FSO-RF system.

4.6.2 Performance Measures of Mixed V2V RF-FSO Communications

The performance analysis of a mixed V2V RF-FSO-RF system in terms of OP is:

$$P_{\text{mix}}(z_{\text{th}}) = P_{\text{mix}}(z_{\text{mix}} \leq z_{\text{th}}) = F_{\text{mix}}(z_{\text{th}}) \qquad (4.55)$$

The LCR (denoted as $N_{\text{mix}}(z_{\text{th}})$) at the output of the considered V2V R-A RF-FSO-RF system model can be evaluated by Stüber (1996):

$$N_{\text{mix}}(z_{\text{th}}) = \int_0^\infty \dot{z}_{\text{mix}} p_{Z_{\text{mix}} \dot{Z}_{\text{mix}}}(z_{\text{th}} \dot{z}_{\text{mix}}) \, d\dot{z}_{\text{mix}} \qquad (4.56)$$

The joint PDF of i.n.i.d. rvs Z_{mix}, $\dot{Z}_{\text{mix}}$, $Z_{N,1}$, $Z_{N,2}$, $Z_{N,4}$, and $Z_{N,6}$ can be expressed as:

$$\begin{aligned} & p_{Z_{\text{mix}} \dot{Z}_{\text{mix}} Z_{N,1} Z_{N,2} Z_{N,4} Z_{N,5} Z_{N,6}}(z_{\text{mix}} \dot{z}_{\text{mix}} z_{N,1} z_{N,2} z_{N,4} z_{N,5} z_{N,6}) \\ & = p_{\dot{Z}_{\text{mix}} | Z_{\text{mix}} Z_{N,1} Z_{N,2} Z_{N,4} Z_{N,5} Z_{N,6}}(\dot{z}_{\text{mix}} | z_{\text{mix}} z_{N,1} z_{N,2} z_{N,4} z_{N,5} z_{N,6}) \\ & \quad p_{Z_{\text{mix}} | Z_{N,1} Z_{N,2} Z_{N,4} Z_{N,5} Z_{N,6}}(z_{\text{mix}} | z_{N,1} z_{N,2} z_{N,4} z_{N,5} z_{N,6}) \\ & \quad p_{Z_{N,1}}(z_{N,1}) \, p_{Z_{N,2}}(z_{N,2}) p_{Z_{N,4}}(z_{N,4}) p_{Z_{N,5}}(z_{N,5}) p_{Z_{N,6}}(z_{N,6}) \end{aligned}$$

(4.57)

where after the simple transformation of the conditional probability function $p_{z_{\text{mix}} | z_{N,1} \, z_N, 2z_{N,4} \, z_{N,5} \, z_{N,6}}$ is:

$$p_{z_{\text{mix}} | z_{N,1} \, z_N, 2z_{N,4} \, z_{N,5} \, z_{N,6}} (z_{\text{mix}} | z_{N,1} \, z_N, 2z_{N,4} \, z_{N,5} \, z_{N,6})$$

$$= \left| \frac{dz_{N,3}}{dz_{\text{mix}}} \right| \cdot p_{z_{N,3}} \left(\frac{\sqrt{z_{\text{mix}}}}{z_{N,1}^{\frac{1}{a}} \, z_{N,2}^{\frac{1}{a}} z_{N,4} \, z_{N,5}^{\frac{1}{a}} z_{N,6}^{\frac{1}{a}}} \right) \tag{4.58}$$

$$= \left| \frac{\frac{1}{2} z_{\text{mix}}^{-\frac{1}{2}}}{z_{N,1}^{\frac{1}{a}} \, z_{N,2}^{\frac{1}{a}} z_{N,4} \, z_{N,5}^{\frac{1}{a}} z_{N,6}^{\frac{1}{a}}} \right| p_{z_{N,3}} \left(\frac{\sqrt{z_{\text{mix}}}}{z_{N,1}^{\frac{1}{a}} \, z_{N,2}^{\frac{1}{a}} z_{N,4} \, z_{N,5}^{\frac{1}{a}} z_{N,6}^{\frac{1}{a}}} \right)$$

Based on the facts that the variance of the first derivative of ZM-G rv is ZM-G rv as well as that the linear transformation of the ZM-G rvs is also ZM-G rv, the variance of $\dot{Z}_{\text{out}}$ is ZM-G rv. After simple manipulations, the variance of $\dot{Z}_{\text{out}}$ can be expressed through the variances of $\dot{Z}_{N,1}, \dot{Z}_{N,1}, \dot{Z}_{N,2}, \dot{Z}_{N,3}, \dot{Z}_{N,4}, \dot{Z}_{N,5}$ and , $\dot{Z}_{N,6}$ as:

$$\sigma_{\dot{z}_{\text{mix}}}^2 = \frac{4 z_{\text{mix}}^2}{a^2 z_{N,1}^2} \sigma_{\dot{z}_{N,1}}^2$$

$$\left(1 + \frac{z_{N,1}^2}{z_{N,2}^2} \frac{\sigma_{\dot{z}_{N,2}}^2}{\sigma_{\dot{z}_{N,1}}^2} + a^2 \frac{z_{N,1}^{2\left(\frac{1}{a}+1\right)} z_{N,2}^{\frac{2}{a}} z_{N,4}^2 z_{N,5}^{\frac{2}{a}} z_{N,6}^{\frac{2}{a}}}{z_{\text{mix}}} \frac{\sigma_{\dot{z}_{N3}}^2}{\sigma_{\dot{z}_{N1}}^2} \right.$$

$$\left. + a^2 \frac{z_{N,1}^2}{z_{N,4}^2} \frac{\sigma_{\dot{z}_{N,4}}^2}{\sigma_{\dot{z}_{N,1}}^2} + \frac{z_{N,1}^2}{z_{N,5}^2} \frac{\sigma_{\dot{z}_{N,5}}^2}{\sigma_{\dot{z}_{N,1}}^2} + \frac{z_{N,1}^2}{z_{N,6}^2} \frac{\sigma_{\dot{z}_{N,5}}^2}{\sigma_{\dot{z}_{N,1}}^2} \right) \tag{4.59}$$

Since Yacoub et al. (1999),

$$p_{\dot{z}_{\text{mix}} | z_{\text{mix}} z_{N,1} \, z_N, 2z_{N,4} \, z_{N,5} \, z_{N,6}} (\dot{z}_{\text{mix}} | z_{\text{mix}} z_{N,1} z_N, 2z_{N,4} \, 5z_{N,6}) d\dot{z}_{\text{mix}}$$

$$= \frac{1}{\sqrt{2\pi}} \sigma_{\dot{z}_{\text{mix}}} \tag{4.60}$$

The $N_{\text{mix}}(z_{\text{th}})$ is obtained after some algebra and presented as integral-form expression:

$$N_{\text{mix}}(z_{\text{th}})$$

$$= \frac{64 \left(\frac{\mu_{m1}}{\vartheta_{m1}} \right)^{\mu_{m1}} \left(\frac{\mu_{m2}}{\vartheta_{m2}} \right)^{\mu_{m2}} (\alpha^{II})^{\alpha^{II}} (\beta^{II})^{\beta^{II}} \left(\frac{\mu_{m5}}{\vartheta_{m5}} \right)^{\mu_{m5}} \left(\frac{\mu_{m6}}{\vartheta_{m6}} \right)^{\mu_{m6}} \sigma_{\dot{z}_{N,1}} z_{\text{th}}^{\alpha^{II}}}{a \sqrt{2\pi} \, \Gamma(\mu_{m1}) \Gamma(\mu_{m2}) \Gamma(\alpha^{II}) \Gamma(\beta^{II}) \Gamma(\mu_{m5}) \Gamma(\mu_{m6})} \tag{4.61}$$

Vehicular Systems for 5G and Beyond 5G

where E_3 is:

$$E_3 = \int_0^\infty dz_{N,1} \int_0^\infty dz_{N,2} \int_0^\infty dz_{N,4} \int_0^\infty dz_{N,5}$$

$$\cdot \int_0^\infty dz_{N,6} \sqrt{\left(1 + \frac{z_{N,1}^2}{z_{N,2}^2} \frac{\sigma_{z_{N,2}}^2}{\sigma_{z_{N,1}}^2} + a^2 \frac{z_{N,1}^{2\left(\frac{1}{a}+1\right)}}{z_{th}} \frac{z_{N,2}^{\frac{2}{a}} z_{N,4}^{\frac{2}{a}} z_{N,5}^{\frac{2}{a}} z_{N,6}^{\frac{2}{a}}}{\sigma_{z_{N,1}}^2}\right.}$$

$$\overline{\left. + a^2 \frac{z_{N,1}^2}{z_{N,4}^2} \frac{\sigma_{z_{N,4}}^2}{\sigma_{z_{N,1}}^2} + \frac{z_{N,1}^2}{z_{N,5}^2} \frac{\sigma_{z_{N,5}}^2}{\sigma_{z_{N,1}}^2} + \frac{z_{N,1}^2}{z_{N,6}^2} \frac{\sigma_{z_{N,6}}^2}{\sigma_{z_{N,1}}^2} \right)}$$

$$\cdot \exp\left[-\frac{\mu_{m1} z_{N,1}^2}{\vartheta_{m1}} - \frac{\mu_{m2} z_{N,2}^2}{\vartheta_{m2}} - \frac{\alpha^{II} z_{mix}}{z_{N,1}^{\frac{2}{a}} z_{N,2}^{\frac{2}{a}} z_{N,4}^{\frac{2}{a}} z_{N,5}^{\frac{2}{a}} z_{N,6}^{\frac{2}{a}}} \right.$$

$$\left. - \beta^{II} z_{N,4}^2 - \frac{\mu_{m5} z_{N,5}^2}{\vartheta_{m5}} - \frac{\mu_{m6} z_{N,6}^2}{\vartheta_{m6}} \right]$$

$$\cdot \exp\left[\left(2\mu_{m1} - \frac{2}{a}\alpha^{II} - 2\right)\ln(z_{N,1}) + \left(2\mu_{m2} - \frac{2}{a}\alpha^{II} - 1\right)\ln(z_{N,2})\right.$$

$$+ \left(2\beta^{II} - 2\alpha^{II} - 1\right)\ln(z_{N,4}) + \left(2\mu_{m5} - \frac{2}{a}\alpha^{II} - 1\right)\ln(z_{N,5})$$

$$\left. + \left(2\mu_{m6} - \frac{2}{a}b^{II} - 1\right)\ln(z_{N,6}) \right] \quad (4.62)$$

The closed form expression for $N_{mix}(z_{th})$ can be obtained by evaluating the integrals E_3 given in equation (4.62) by LBAM for five-folded integrals already given in equation (4.52), where $\gamma = 1$, whereas $g(z_{N,1}, z_{N,2}, z_{N,4} z_{N,5}, z_{N,6})$ and $f(z_{N,1}, z_{N,2}, z_{N,4} z_{N,5}, z_{N,6})$ for the considered case are, respectively:

$$g(z_{N,1}, z_{N,2}, z_{N,4} z_{N,5}, z_{N,6})$$

$$= \left(1 + \frac{z_{N,1}^2}{z_{N,2}^2} \frac{\sigma_{z_{N,2}}^2}{\sigma_{z_{N,1}}^2} + \frac{a^2 z_{N,1}^{\frac{2}{a}+2}}{z_{th}} \frac{z_{N,2}^{\frac{2}{a}} z_{N,4}^{\frac{2}{a}} z_{N,5}^{\frac{2}{a}} z_{N,6}^{\frac{2}{a}}}{\sigma_{z_{N,1}}^2} \sigma_{z_{N,3}}^2 \right.$$

$$\left. + \frac{a^2 z_{N,1}^2}{z_{N,4}^2} \frac{\sigma_{z_{N,4}}^2}{\sigma_{z_{N,1}}^2} + \frac{z_{N,1}^2}{z_{N,5}^2} \frac{\sigma_{z_{N,5}}^2}{\sigma_{z_{N,1}}^2} + \frac{z_{N,1}^2}{z_{N,6}^2} \frac{\sigma_{z_{N,6}}^2}{\sigma_{z_{N,1}}^2} \right)^{\frac{1}{2}} \quad (4.63)$$

$$f(z_{N,1}, z_{N,2}, z_{N,4} z_{N,5}, z_{N,6})$$

$$= \frac{\mu_{m1} z_{N,1}^2}{\vartheta_{m1}} + \frac{\mu_{m2} z_{N,2}^2}{\vartheta_{m2}} + \frac{\alpha^{II} z_{mix}}{z_{N,1}^{\frac{2}{a}} z_{N,2}^{\frac{2}{a}} z_{N,4}^{\frac{2}{a}} z_{N,5}^{\frac{2}{a}} z_{N,6}^{\frac{2}{a}}} + \beta^{II} z_{N,4}^2 + \frac{\mu_{m5} z_{N,5}^2}{\vartheta_{m5}}$$

$$+ \frac{\mu_{m6} z_{N,6}^2}{\vartheta_{m6}} - \left(2\mu_{m1} - \frac{2}{a}\alpha^{II} - 2\right) \ln(z_{N,1}) - \left(2\mu_{m2} - \frac{2}{a}\alpha^{II} - 1\right) \ln(z_{N,2})$$

$$- \left(2\beta^{II} - 2\alpha^{II} - 1\right) \ln(z_{N,4}) - \left(2\mu_{m5} - \frac{2}{a}\alpha^{II} - 1\right) \ln(z_{N,5}) - \left(2\mu_{m6} - \frac{2}{a}b^{II} - 1\right) \ln(z_{N,6})$$

(4.64)

The AFD_{mix} is evaluated as:

$$AFD_{mix}(z_{th}) = \frac{F_{mix}(z_{th})}{N_{mix}(z_{th})} \quad (4.65)$$

4.6.3 Numerical Results for Mixed V2V RF-FSO Communications

In Figures 4.10 and 4.11, HOS results for a V2V mixed R-A RF-FSO-RF A–F communication link in mixed TF and F environments are presented. It can be seen that exact

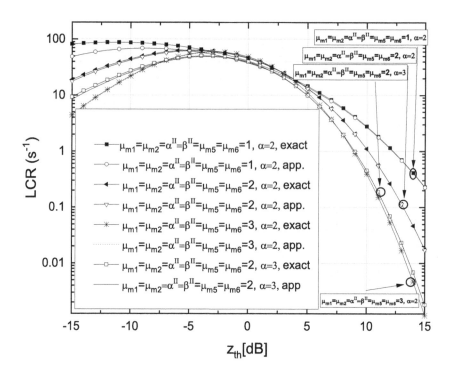

FIGURE 4.10 Normalized $N_{mix}(z_{th})$ for V2V R-A mixed RF-FSO-RF triple-hop A–F relay link for various system model parameters.

Vehicular Systems for 5G and Beyond 5G

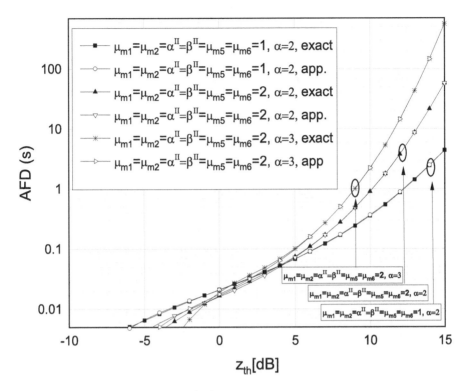

FIGURE 4.11 Normalized $\text{AFD}_{\text{mix}}(z_{\text{th}})$ for V2V R-A mixed RF-FSO-RF triple-hop A–F relay link for various system model parameters.

analytical expressions fit well with approximations (closed form expression approximated by LBAM), especially for higher values of the signal envelope output threshold z_{th}[dB]. The RF sections of R-AV2V RF-FSO-RF links are each modelled with the $d-\alpha-\mu$ distribution and numerically evaluated for various $d-\alpha-\mu$ distribution parameters such as non-homogeneity parameter a, fading severity parameters μ_{m1}, μ_{m2}, μ_{m5}, μ_{m6}, and normalized, $\vartheta_{m1}=\vartheta_{m2}=\vartheta_{m5}=\vartheta_{m6}=1$. The variances in equation (4.59) are expressed as $\sigma^2_{\dot{Z}_{N,i}}=\pi^2(f_{mi})^2\vartheta_{mi}/\mu_{mi}$, $i=1,2,5,6$ where the maximum DFs are assumed to be the same, $f_m=f_{m1}=f_{m2}=f_{m5}=f_{m6}=\sqrt{(f_{ms})^2+(f_{md})^2}$, where f_{ms} and f_{md} are maximum DFs for R-A V2V RF-FSO-RF link of the s-v and d-v, respectively.

The FSO section of a V2V RF-FSO-RF A–F relay link is modelled with gamma–gamma distribution, where numerical results are computed for various optical fading severity parameters (α^{II}, β^{II}) and various gamma–gamma irradiance variance values ($\sigma_{GG}^2=1/\alpha^{II}+1/\beta^{II}+1/(\alpha^{II}\beta^{II})$) (Andrews and Phillips, 2005). The $\sigma^2_{\dot{Z}_{N3}}$ and $\sigma^2_{\dot{Z}_{N4}}$ in equation (4.59) are assumed to take the same values and in the case of the GZM rvs can be expressed as $\sigma^2_{\dot{Z}}=\sigma^2_{\dot{Z}_{N3}}=\sigma^2_{\dot{Z}_{N4}}=f_0^2\pi^2\sigma_{GG}^2 Z$, as obtained in Jurado-Navas et al. (2017), where $Z=1$ for gamma–gamma distribution. f_0 is the frequency of fades and can be further expressed as $f_0=1/(\pi t_0\sqrt{2})$ (Jurado-Navas et al., 2017).

Furthermore, $t_0 = \sqrt{\lambda S}/u$ is the turbulence correlation time, where λ is the optical window, S is the optical distance, and u is the average wind speed (Jurado-Navas et al., 2017).

Figure 4.10 reports the behaviour of $N_{\text{mix}}(z_{\text{th}})$ for various FSO sets of parameters (α^{II}, β^{II}, $\lambda = 1{,}550$ nm, $u = 1$ m/s, $S = 200$ m) and for various RF sets of parameters (μ_{m1}, μ_{m2}, μ_{m5}, μ_{m6}, a, $f_m = 90$ Hz). It can be seen that by increasing RF fading severity as well as FSO optical severity parameters (μ_{m1}, μ_{m2}, μ_{m5}, μ_{m6}, α^{II} and β^{II}), the $N_{\text{mix}}(z_{\text{th}})$ decreases, as expected. A similar trend can be noticed by increasing the non-homogeneity parameter a. Moreover, it is evident that (μ_{m1}, μ_{m2}, μ_{m5}, μ_{m6}, α^{II} and β^{II}) have slightly stronger impact on $N_{\text{mix}}(z_{\text{th}})$ than a for the considered system values.

The behaviour of $\text{AFD}_{\text{mix}}(z_{\text{th}})$ is shown in Figure 4.11. By increasing non-homogeneity parameter, multipath severity, and turbulence severity parameters, $\text{AFD}_{\text{mix}}(z_{\text{th}})$ decreases in the lower dB output regime while $\text{AFD}_{\text{mix}}(z_{\text{th}})$ increases in the higher dB output regime.

4.7 V2V COOPERATIVE MIXED RF-FSO COMMUNICATIONS WITH DIRECT AND RELAY-ASSISTED RF LINKS

The V2V direct link supported by a redundant multi-hop R-A mixed RF-FSO-RF V2V link consisting of source vehicle (s-v), two AF RSU-relays and destination vehicle (d-v) is shown in Figure 4.12, whereas a V2V R-A link supported by redundant multi-hop R-A mixed RF-FSO-RF V2V link is shown on Figure 4.13.

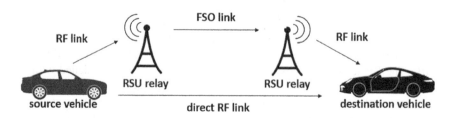

FIGURE 4.12 V2V direct RF link supported by V2V mixed RF-FSO-RF A–F relay link.

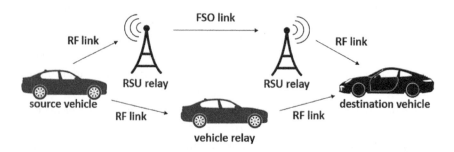

FIGURE 4.13 V2V R-A RF links supported by V2V mixed RF-FSO-RF A–F relay link.

Vehicular Systems for 5G and Beyond 5G 211

4.7.1 V2V Cooperative Mixed RF-FSO Communications with Direct and Relay-Assisted RF Links over DSc and TF Channels

The PDF and CDF of a V2V direct RF link supported by a mixed RF-FSO-RF A–F relay link at the output of the proposed model with SC-AS system at reception (for the scenario presented in Figure 4.10) is:

$$p_{\text{coop},1}(z_{\text{coop},1}) = p_{\text{mix}}(z_{\text{coop},1})F_{d-\alpha-\mu}(z_{\text{coop},1}) + p_{d-\alpha-\mu}(z_{\text{coop},1})F_{\text{mix}}(z_{\text{coop},1}) \quad (4.66)$$

$$F_{\text{coop},1}(z_{\text{coop},1}) = F_{d-\alpha-\mu}(z_{\text{coop},1})F_{\text{mix}}(z_{\text{coop},1}) \quad (4.67)$$

where $p_{d-\alpha-\mu}$ and $F_{d-\alpha-\mu}$ are PDF and CDF of direct V2V communications over $d-\alpha-\mu$ fading channels obtained in equations (4.3) and (4.5), respectively, while p_{mix} and F_{mix} are PDF and CDF at the output of V2V mixed RF-FSO A–F relay system obtained in equations (4.49) and (4.50), respectively.

Similarly, the PDF and CDF of V2V R-A RF links supported by a mixed RF-FSO-RF A–F relay link with SC-AS (for the scenario presented in Figure 4.11) is:

$$p_{\text{coop},2}(z_{\text{coop},2}) = p_{\text{sd}}(z_{\text{coop},2})F_{d-\alpha-\mu}(z_{\text{coop},1}) + F_{\text{sd}}(z_{\text{coop},2})F_{\text{mix}}(z_{\text{coop},2}) \quad (4.68)$$

$$F_{\text{coop},2}(z_{\text{coop},2}) = F_{\text{sd}}(z_{\text{coop},2})F_{\text{mix}}(z_{\text{coop},2}) \quad (4.69)$$

where p_{sd} and F_{sd} are PDF and CDF of R-A V2V communications over $d-\alpha-\mu$ fading channels obtained in equations (4.14) and (4.15), respectively.

4.7.2 Performance Measures of V2V Cooperative Mixed RF-FSO Communications with Direct and Relay-Assisted RF Links

The LCR at the output of the SC-AS with two i.n.i.d. branches for the scenario presented in Figures 4.12 and 4.13 are, respectively:

$$N_{\text{coop},1}(z_{\text{th}}) = N_{\text{mix}}(z_{\text{th}})F_{d-\alpha-\mu}(z_{\text{th}}) + N_{d-\alpha-\mu}(z_{\text{th}})F_{\text{mix}}(z_{\text{th}}) \quad (4.70)$$

$$N_{\text{coop},2}(z_{\text{th}}) = N_{\text{mix}}(z_{\text{th}})F_{\text{sd}}(z_{\text{th}}) + N_{\text{sd}}(z_{\text{th}})F_{\text{mix}}(z_{\text{th}}) \quad (4.71)$$

where $N_{d-\alpha-\mu}(z_{\text{th}})$, $N_{\text{sd}}(z_{\text{th}})$, and $N_{\text{mix}}(z_{\text{th}})$ are the LCR expressions at the output of the direct V2V link, R-A V2V link, and mixed R-A V2V RF-FSO-RF A–F link obtained in equations (4.9), (4.26), and (4.61), respectively. AFD at the output of the direct V2V link and R-A V2V supported by redundant mixed R-A V2V RF-FSO-RF A–F link are, respectively:

$$\text{AFD}_{\text{coop, 1}}(z_{\text{th}}) = \frac{F_{\text{coop, 1}}(z_{\text{th}})}{N_{\text{coop, 1}}(z_{\text{th}})} \tag{4.72}$$

$$\text{AFD}_{\text{coop, 2}}(z_{\text{th}}) = \frac{F_{\text{coop, 2}}(z_{\text{th}})}{N_{\text{coop, 2}}(z_{\text{th}})} \tag{4.73}$$

4.7.3 Numerical Results for V2V Cooperative Mixed RF-FSO Communications with Direct and Relay-Assisted RF Links

The HOS measures of cooperative V2V mixed RF-FSO-RF A–F relay link with a direct RF V2V link (scenario presented in Figure 4.12) evaluated from exact integral-form expressions (equations 4.70 and 4.72) for various sets of RF-FSO-RF link parameters ($\mu_{m1}, \mu_{m2}, \mu_{m5}, \mu_{m6}, \alpha^{II}, \beta^{II}, f_m = 90$ Hz, $\lambda = 1{,}550$ nm, $u = 1$ m/s, $S = 200$ m) and various values of direct RF link parameters ($\mu_1, \mu_2, a, f_m = 90$ Hz) are presented in Figures 4.14 and 4.15, respectively.

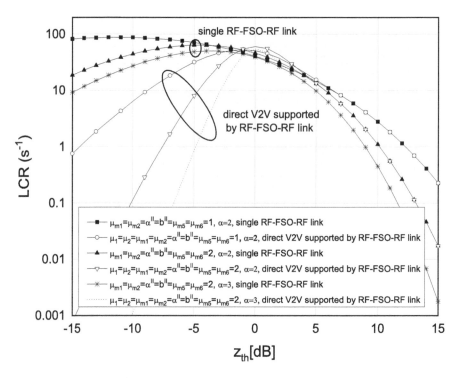

FIGURE 4.14 LCR (s^{-1}) for dual-hop R-A RF V2V link supported by redundant mixed V2V RF-FSO-RF A–F relay link for various system model parameters.

Vehicular Systems for 5G and Beyond 5G

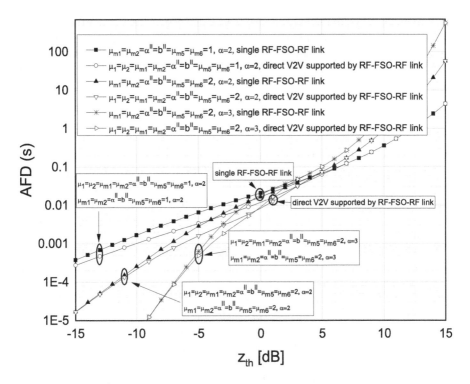

FIGURE 4.15 AFD (s) for dual-hop R-A RF V2V link supported by redundant mixed V2V RF-FSO-RF A–F relay link for various system model parameters.

Since $f_m = vf_c/c$, where v is speed, f_c is the carrier frequency, and c is the speed of light, it can be seen that by increasing f_m (by increasing speed or carrier frequency), the $N_{\text{coop},1}(z_{\text{th}})$ increases, while $\text{AFD}_{\text{coop},1}(z_{\text{th}})$ decreases. Similarly, by increasing the fading severity parameters and non-homogeneity parameter, $N_{\text{coop},1}(z_{\text{th}})$ and $\text{AFD}_{\text{coop},1}(z_{\text{th}})$ decrease for lower z_{th} values. It can be concluded from Figure 4.10 that the direct V2V link assisted by a R-A RF-FSO-RF link provides a significant decrease of $N_{\text{coop},1}(z_{\text{th}})$ values for lower z_{th} if compared with a single RF-FSO-RF link. Interestingly, the direct V2V assisted by RF-FSO-RF link has the same $N_{\text{coop},1}(z_{\text{th}})$ values for higher z_{th} as the single RF-FSO-RF link. Although the addition of the direct V2V link tends to decrease the values of $\text{AFD}_{\text{coop},1}(z_{\text{th}})$ for the same values of the system model parameters, in boundary regions (for small and high thresholds), the direct V2V assisted by a RF-FSO-RF link has the same of $\text{AFD}_{\text{coop},1}(z_{\text{th}})$ values as the single RF-FSO-RF link.

The HOS measures of a cooperative V2V mixed RF-FSO-RF link with a relay-assisted RF V2V link (scenario presented in Figure 4.13) evaluated from exact integral-form expressions (equations 4.71 and 4.73) for various sets of RF-FSO-RF link parameters ($\mu_{m1}, \mu_{m2}, \mu_{m5}, \mu_{m6}, \alpha^{\text{II}}, \beta^{\text{II}}, f_m = 90$ Hz, $\lambda = 1{,}550$ nm, $u = 1$ m/s, $S = 200$ m) and various values of direct RF link parameters ($\mu_{x1}, \mu_{x2}, \mu_{y1}, \mu_{y2}, a, f_m = 90$ Hz) are presented in Figures 4.14 and 4.15, respectively.

In particular, by comparing the mixed V2V RF-FSO-RF link and the dual-hop R-A V2V link assisted by a redundant V2V RF-FSO-RF link in terms of HOS measures presented in Figures 4.14 and 4.15, one can conclude that the addition of a parallel dual-hop R-A V2V RF link can lead to the decrease of the $N_{\text{coop}}(z_{\text{th}})$ and $\text{AFD}_{\text{coop}}(z_{\text{th}})$.

4.8 CONCLUSIONS

This chapter considers HOS performance measures of V2V communications over d–α-μ fading channels enabled through direct RF links, relay-assisted RF links, cooperative relay-assisted RF links as well as through redundant RF-FSO links. The TF in FSO links is modelled with gamma–gamma distribution. Numerical examples in terms of HOS measures show that exact analytical expressions fit well with the approximations, especially in the higher signal envelope dB output regime. The HOS performance measures of the considered V2V configurations are analysed under different propagation conditions. Namely, in less severe fading conditions (e.g., by increasing the values of multipath fading and non-homogeneity parameters) the system performance improvement in terms of HOS measures can be achieved. In the case of mixed RF-FSO-RF V2V scenarios, the less severe RF fading conditions and TF severity conditions improve the system performance.

Moreover, in relation to the considered mixed RF-FSO-RF and cooperative mixed RF-FSO-RF V2V scenarios, the direct V2V communications assisted by a RF-FSO-RF link ensure the lowest LCR values for the lower thresholds under the same propagation conditions. Since the mathematical complexity of exact integral expressions as well as approximate expressions derived by LBAM with addition of V-Comm multi-hop and cooperative multi-hop systems increases, the provided mathematical expressions can be subjected to machine learning techniques in order to additionally decrease the computation time; this will be considered in our future work.

ACKNOWLEDGEMENTS

The authors would like to acknowledge the CONEX-Plus project that has received research funding from UC3M and the European Union's Horizon 2020 program under the Marie Skłodowska–Curie grant agreement No. 801538. This work has been supported by the project IRENE-EARTH (PID2020–115323RB-C33/ AEI/10.13039/501100011033) and the Cost Actions CA19111, CA20120, and CA16220. This work was also supported by the Centre of EXcellence on Connected, Geo-Localized and Cybersecure Vehicles (EX-Emerge), funded by the Italian Government under CIPE resolution n. 70/2017 (August 7, 2017).

REFERENCES

Adhikari, M., Hazra, A., Menon, V.G., Chaurasia, B.K., & Mumtaz, S. (2021). A roadmap of next-generation wireless technology for 6G-enabled vehicular networks. *IEEE Internet of Things Magazine*, 4(4), 79–85.

Ai, Y., Cheffena, M., Mathur, A., & Lei, H. (2018). On physical layer security of double Rayleigh fading channels for vehicular communications. *IEEE Wireless Communications Letters*, 7(6), 1038–1041.

Al-Hmood, H., & Al-Raweshidy, H. (2017). Unified modeling of composite kappa-mu/gamma, eta-mu/gamma, and alpha-mu/gamma fading channels using a mixture gamma distribution with applications to energy detection. *IEEE Antennas and Wireless Propagation Letters*, 16, 104–108.

Al-Hmood, H., & Al-Raweshidy, H.S. (2020). Selection combining scheme over non-identically distributed Fisher–Snedecor F fading channels. *IEEE Wireless Communications Letters*, 10(4), 840–843.

Amer, M.A., & Al-Dharrab, S. (2019). Performance of two-way relaying over α–μ fading channels in hybrid RF/FSO wireless networks. *arXiv preprint arXiv:1911.05959*.

Andrews, L.C., & Phillips, R.L. (2005). *Laser Beam Propagation through Random Media* (2nd ed.) SPIE Press Book.

Badarneh, O.S. (2016). The α-μ/α-μ composite multipath-shadowing distribution and its connection with the extended generalized K distribution. *AEU: International Journal of Electronics and Communications*, 70(9), 1211–1218.

Badarneh, O.S., & Almehmadi, F.S. (2016). Performance of multihop wireless networks in α-μ fading channels perturbed by an additive generalized Gaussian noise. *IEEE Communications Letters*, 20(5), 986–989.

Balti, E., & Guizani, M. (2018). Mixed RF/FSO cooperative relaying systems with co-channel interference. *IEEE Transactions on Communications*, 66(9), 4014–4027.

Belmekki, B.E.Y., Hamza, A., & Escrig, B. (2019). Cooperative vehicular communications at intersections over Nakagami-m fading channels. *Vehicular Communications*, 19, 100165.

Bithas, P.S., Kanatas, A.G., da Costa, D.B., Upadhyay, P.K. & Dias, U.S. (2018a). On the double-generalized gamma statistics and their application to the performance analysis of V2V communications. *IEEE Transactions on Communications*, 66(1), 448–460.

Bithas, P.S., Kanatas, A.G., & Matolak, D.W. (2018b, September). Shadowing-based antenna selection for V2V communications. In *2018 IEEE 29th Annual International Symposium on Personal, Indoor and Mobile Radio Communications (PIMRC)* (pp. 106–110). IEEE.

Bithas, P.S., Nikolaidis, V., & Kanatas, A.G. (2019). A new shadowed double-scattering model with application to UAV-to-ground communications. In *Proceedings of the International Conference Wireless Communications and Networking Conference (WCNC)* (pp. 1–6). IEEE.

Bithas, P.S., Nikolaidis, V., Kanatas, A.G., & Karagiannidis, G.K. (2020). UAV-to-ground communications: Channel modeling and UAV selection. *IEEE Transactions on Communications*, 68(8), 5135–5144.

Boulogeorgos, A.A.A., Chatzidiamantis, N., Sandalidis, H.G., Alexiou, A., & Di Renzo, M. (2022). Cascaded composite turbulence and misalignment: Statistical characterization and applications to reconfigurable intelligent surface-empowered wireless systems. *IEEE Transactions on Vehicular Technology*, 71(4), 3821–3836.

Chapala, V.K., Malik, A., & Zafaruddin, S.M. (2022). RIS-assisted vehicular network with direct transmission over double-generalized gamma fading channels. *arXiv preprint arXiv:2204.09958*.

Dautov, K., Arzykulov, S., Nauryzbayev, G., & Kizilirmak, R.C. (2018). On the performance of UAV-enabled multihop V2V FSO systems over generalized α–μ channels. In *Proceedings of the International Conference on Computing and Network Communications (CoCoNet)* (pp. 69–73). IEEE.

Di Renzo, M., Graziosi, F., & Santucci, F. (2009a). A comprehensive framework for performance analysis of dual-hop cooperative wireless systems with fixed-gain relays over generalized fading channels. *IEEE Transactions on Wireless Communications*, 8(10), 5060–5074.

Di Renzo, M., Graziosi, F., & Santucci, F. (2009b). A unified framework for performance analysis of CSI-assisted cooperative communications over fading channels. *IEEE Transactions on Communications, 57*(9), 2551–2557.

Di Renzo, M., Graziosi, F., & Santucci, F. (2010). A comprehensive framework for performance analysis of cooperative multi-hop wireless systems over log-normal fading channels. *IEEE Transactions on Communications, 58*(2), 531–544.

Dixit, D., Kumar, N., Sharma, S., Bhatia, V., Panic, S., & Stefanovic, C. (2021). On the ASER performance of UAV-based communication systems for QAM schemes. *IEEE Communications Letters, 25*(6), 1835–1838.

Du, H., Zhang, J., Peppas, K.P., Zhao, H., Ai, B., & Zhang, X. (2019). On the distribution of the ratio of products of Fisher–Snedecor $\mathcal{F}$ random variables and its applications. *IEEE Transactions on Vehicular Technology, 69*(2), 1855–1866.

Fotouhi, A., Qiang, H., Ding, M., Hassan, M., Giordano, L.G., Garcia-Rodriguez, A., & Yuan, J. (2019). Survey on UAV cellular communications: Practical aspects, standardization advancements, regulation, and security challenges. *IEEE Communications Surveys & Tutorials, 21*(4), 3417–3442.

Garcia, M.H.C., Molina-Galan, A., Boban, M., Gozalvez, J., Coll-Perales, B., Şahin, T., & Kousaridas, A. (2021). A tutorial on 5G NR V2X communications. *IEEE Communications Surveys & Tutorials, 23*(3), 1972–2026.

Ghafoor, K.Z., Guizani, M., Kong, L., Maghdid, H.S., & Jasim, K.F. (2019). Enabling efficient coexistence of DSRC and C-V2X in vehicular networks. *IEEE Wireless Communications, 27*(2), 134–140.

Gradshteyn, I.S., & Ryzhik, I.M. (2000). *Table of Integrals, Series, and Products* (6th ed.). New York: Academic Press.

Hadzi-Velkov, Z., Zlatanov, N., & Karagiannidis, G.K. (2009). On the second order statistics of the multihop Rayleigh fading channel. *IEEE Transactions on Communications, 57*(6), 1815–1823.

Hajri, N., Khedhiri, R., & Youssef, N. (2020). On selection combining diversity in dual-hop relaying systems over double rice channels: Fade statistics and performance analysis. *IEEE Access, 8*, 72188–72203.

He, J., Yang, K., & Chen, H.H. (2020). 6G cellular networks and connected autonomous vehicles. *IEEE Network, 35*(4), 255–261.

Huang, C., Molisch, A.F., He, R., Wang, R., Tang, P., Ai, B., & Zhong, Z. (2020). Machine learning-enabled LOS/NLOS identification for MIMO systems in dynamic environments. *IEEE Transactions on Wireless Communications, 19*(6), 3643–3657.

Huang, C., Molisch, A.F., He, R., Wang, R., Tang, P., & Zhong, Z. (2019). Machine learning-based data processing techniques for vehicle-to-vehicle channel modeling. *IEEE Communications Magazine, 57*(11), 109–115.

Jaiswal, N., & Purohit, N. (2019, December). Performance evaluation of non-orthogonal multiple access in V2V communications over double-Rayleigh fading channels. In *2019 IEEE Conference on Information and Communication Technology* (pp. 1–5). IEEE.

Jaiswal, N., & Purohit, N. (2021). Performance analysis of NOMA-enabled vehicular communication systems with transmit antenna selection over double Nakagami-m fading. *IEEE Transactions on Vehicular Technology, 70*(12), 12725–12741.

Jurado-Navas, A., Garrido-Balsells, J.M., Castillo-Vázquez, M., Puerta-Notario, A., Monroy, I.T., & Olmos, J.J.V. (2017). Fade statistics of M $\mathcal{M}$-turbulent optical links. *EURASIP Journal on Wireless Communications and Networking, 2017*(1), 1–9.

Leonardo, E.J., & Yacoub, M.D. (2015). Product of α-µ variates. *IEEE Wireless Communications Letters, 4*(6), 637–640.

Leonardo, E.J., Yacoub, M.D., & de Souza, R.A. (2016). Ratio of products of α-µ variates. *IEEE Communications Letters, 20*(5), 1022–1025.

Li, Z., Xiang, L., Ge, X., Mao, G., & Chao, H.C. (2020). Latency and reliability of mmWave multi-hop V2V communications under relay selections. *IEEE Transactions on Vehicular Technology*, *69*(9), 9807–9821.

Lv, J., He, X., & Luo, T. (2021). Blockage avoidance based sensor data dissemination in multi-hop mmWave vehicular networks. *IEEE Transactions on Vehicular Technology*, *70*(9), 8898–8911.

MacHardy, Z., Khan, A., Obana, K., & Iwashina, S. (2018). V2X access technologies: Regulation, research, and remaining challenges. *IEEE Communications Surveys & Tutorials*, *20*(3), 1858–1877.

Matović, A., Mekić, E., Sekulović, N., Stefanović, M., Matović, M., & Stefanović, Č. (2013). The distribution of the ratio of the products of two independent-variates and its application in the performance analysis of relaying communication systems. *Mathematical Problems in Engineering*, *2013*, 1–6.

Milic, D., Djosic, D., Stefanovic, C., Panic, S., & Stefanovic, M. (2016). Second order statistics of the SC receiver over Rician fading channels in the presence of multiple Nakagami-m interferers. *International Journal of Numerical Modelling: Electronic Networks, Devices and Fields*, *29*, 222–229.

Milosevic, N. et al., (2018a). Performance analysis of interference-limited mobile-to-mobile κ–μ fading channel. *Wireless Personal Communications*, *101*(3), 1685–1701.

Milosevic, N., et al. (2018b). First- and second-order statistics of interference-limited mobile-to-mobile Weibull fading channel. *Journal of Circuits, Systems and Computers*, *27*(11), 1850168.

Milovanovic, I., & Stefanovic, C. (2021). Performance analysis of UAV-assisted wireless powered sensor network over shadowed κ–μ fading channels. *Wireless Communications & Mobile Computing* [Online], *2021*, 1–7.

Nguyen, B.C., & Hoang, T.M. (2019). Performance analysis of vehicle-to-vehicle communication with full-duplex amplify-and-forward relay over double-Rayleigh fading channels. *Vehicular Communications*, *19*, 100166.

Pan, B., & Wu, H. (2021). Success probability analysis of cooperative C-V2X communications. *IEEE Transactions on Intelligent Transportation Systems*, *23*(7), 7170–7183.

Panic, S., Perera, T.D.P., Jayakody, D.N.K., Stefanovic, C., & Prlincevic, B. (2019, November). UAV-assisted wireless powered sensor network over Rician shadowed fading channels. In *2019 IEEE International Conference on Microwaves, Antennas, Communications and Electronic Systems (COMCAS)* (pp. 1–5). IEEE.

Panic, S., Stefanovic, M., Anastasov, J., & Spalevic, P. (2013). *Fading and Interference Mitigation in Wireless Communications*. New York: CRC Press.

Park, J., Lee, E., Park, G., Roh, B., & Yoon, G. (2013, November). Performance analysis of asymmetric RF/FSO dual-hop relaying systems for UAV applications. In *MILCOM 2013-2013 IEEE Military Communications Conference* (pp. 1651–1656). IEEE.

Poursajadi, S., & Madani, M.H. (2021). Adaptive optimal relay selection in cooperative vehicular communications under security constraints. *Vehicular Communications*, *31*, 100360.

Stamenovic, G., Panic, S.R., Rancic, D., Stefanović, C., & Stefanović, M. (2014). Performance analysis of wireless communication system in general fading environment subjected to shadowing and interference. *EURASIP Journal on Wireless Communications and Networking*, *2014*(1), 1–8.

Stefanovic, C., Djosic, D., Panic, S., Milic, D., & Stefanovic, M. (2020). A framework for statistical channel modeling in 5G wireless communication systems. In *5G Multimedia Communications: Technology, Multiservices, Deployment* (pp. 31–54). CRC Press.

Stefanovic, C., Milovanovic, I., Panic, S., & Stefanovic, M. (2022). LCR and AFD of the products of Nakagami-m and Nakagami-m squared random variables: Application to wireless communications through relays. *Wireless Personal Communications*, *123*(3), 2665–2678.

Stefanovic, C., Morales-Céspedes, M., & Armada, A.G. (2021a). Performance analysis of RIS-assisted FSO communications over Fisher–Snedecor F turbulence channels. *Applied Sciences, 11*(21), 10149.

Stefanovic, C., Morales-Céspedes, M., Róka, R., & Armada, A.G. (2021b). Performance analysis of N-Fisher–Snedecor F fading and its application to N-hop FSO communications. In *2021 17th International Symposium on Wireless Communication Systems (ISWCS)* (pp. 1–6). IEEE.

Stefanovic, C., Panic, S., Bhatia, V., & Kumar, N. (2021c). On second-order statistics of the composite channel models for UAV-to-ground communications with UAV selection. *IEEE Open Journal of the Communications Society, 2*, 534–544.

Stefanovic, C., Panic, S.R., Bhatia, V., Kumar, N., & Sharma, S. (2021d). On higher-order statistics of the channel model for UAV-to-ground communications. In *2021 IEEE 93rd Vehicular Technology Conference (VTC2021-Spring)* (pp. 1–5). IEEE.

Stefanovic, C., Panic, S., Djosic, D., Milic, D., & Stefanovic, C. (2021e). On the second order statistics of N-hop FSO communications over N-gamma-gamma turbulence induced fading channels. *Physical Communication, 45*, 101289.

Stefanovic, C., Pratesi, M., & Santucci, F. (2018a). Performance evaluation of cooperative communications over fading channels in vehicular networks. In *Proceedings of the International Conference on Atalntic Science Radio Meeting (AT-RASC)* (pp. 1–4). IEEE.

Stefanovic, C., Veljkovic, S., Stefanovic, M., Panic, S., & Jovkovic, S. (2018b). Second order statistics of SIR based macro diversity system for V2I communications over composite fading channels. In *First International Conference on Secure Cyber Computing and Communication (ICSCCC)* (pp. 569–573). IEEE.

Stefanovic, C., Pratesi, M., & Santucci, F. (2019). Second order statistics of mixed RF-FSO relay systems and its application to vehicular networks (pp. 1–6). In *Proceedings of the International Conference on Communications (ICC)*. IEEE.

Stüber, G.L. (1996). *Principles of Mobile Communication* (Vol. 2). Norwell, MA: Kluwer Academic.

Talha, B., & Pätzold, M. (2011). Channel models for mobile-to-mobile cooperative communication systems: A state of the art review. *IEEE Vehicular Technology Magazine, 6*(2), 33–43.

Tang, F., Mao, B., Kato, N., & Gui, G. (2021). Comprehensive survey on machine learning in vehicular network: Technology, applications and challenges. *IEEE Communications Surveys & Tutorials, 23*(3), 2027–2057.

Triwinarko, A., et al. (2020). A PHY/MAC cross-layer design with transmit antenna selection and power adaptation for receiver blocking problem in dense VANETs. *Vehicular Communications, 24*, 100233.

Tunc, C., & Panwar, S.S. (2021). Mitigating the impact of blockages in millimeter-wave vehicular networks through vehicular relays. *IEEE Open Journal of Intelligent Transportation Systems, 2*, 225–239.

Xu, G., & Song, Z. (2021). Performance analysis of a UAV-assisted RF/FSO relaying systems for internet of vehicles. *IEEE Internet of Things Journal, 9*(8), 5730–5741.

Yacoub, M.D. (2007). The α-μ distribution: A physical fading model for the stacy distribution. *IEEE Transactions on Vehicular Technology, 56*(1), 27–34.

Yacoub, M.D., Bautista, J.V., & de Rezende Guedes, L.G. (1999). On higher order statistics of the Nakagami-m distribution. *IEEE Transactions on Vehicular Technology, 48*(3), 790–794.

Yang, L., Hasna, M.O., & Alouini, M.S. (2005). Average outage duration of multihop communication systems with regenerative relays. *IEEE Transactions on Wireless Communications, 4*(4), 1366–1371.

Yin, R.R., Ghassemlooy, Z., Zhao, N., Yuan, H., Raza, M., Eso, E., & Zvanovec, S. (2020, July). A multi-hop relay based routing algorithm for vehicular visible light communication networks. In *2020 12th International Symposium on Communication Systems, Networks and Digital Signal Processing (CSNDSP)* (pp. 1–6). IEEE.

Yoo, S.K., Cotton, S.L., Zhang, L., Doone, M.G., Song, J.S., & Rajbhandari, S. (2021). Evaluation of a switched combining based distributed antenna system (DAS) for pedestrian-to-vehicle communications. *IEEE Transactions on Vehicular Technology*, *70*(10), 11005–11010.

Zedini, E., Soury, H., & Alouini, M.S. (2016). Dual-hop FSO transmission systems over gamma–gamma turbulence with pointing errors. *IEEE Transactions on Wireless Communications*, *16*(2), 784–796.

Zlatanov, N., Hadzi-Velkov, Z., & Karagiannidis, G.K. (2008). Level crossing rate and average fade duration of the double Nakagami-m random process and application in MIMO keyhole fading channels. *IEEE Communications Letters*, *12*(11), 822–824.

5 Distribution of NFV Infrastructure Providing Efficient Edge Computing Architecture for 5G Environments

Gjorgji Ilievski
Makedonski Telekom AD

Pero Latkoski
Ss. Cyril and Methodius University

CONTENTS

5.1 Introduction 221
5.2 Related Work 225
5.3 Analytical Models 227
    5.3.1 Assumptions 227
    5.3.2 Centralized NFV Model 228
    5.3.3 NFV Model with Distributed Data Plane and Central MANO 231
    5.3.4 Fully Distributed NFV Model 233
5.4 Performance Evaluation of the Analytical Models 236
5.5 Experimental Evaluation of a Distributed NFV 241
    5.5.1 Experimental Testbed 241
    5.5.2 Experimental Results 242
5.6 Concluding Remarks 245
References 246

5.1 INTRODUCTION

Telecommunication industry is already deploying 5G standard-based [1] networks, but the process has a lot of challenges, not only from a technological perspective but also from a financial point of view [2]. The process is inevitable as the user density is rising at an incredible rate, but with unpredictable clustering, due to the user mobility. It is expected that the traffic consumption will rise exponentially with the massive machine-type communication (mMTC) [3], which will be imposed by the penetration of Internet of Things (IoT), thus the need for increased traffic capacity, both on the 5G radio access network (RAN) and on the core network. But the challenge expands further into solving power consumption problems as well as providing low data rates for the consumers.

The last decade focused on virtualization technologies that acted as driver for cloud services, both private and public. Most of the services were virtualized using virtual machines and appliances, followed by containerization, which has further improved the usage scenarios, allowing complex end-to-end services, with reduced economic impact driven by the economy of scale. Networking concepts are included this stream. Software-defined networking (SDN) allowed the separation of the control plane and the data plan and has improved the cloud orchestration significantly. Network functions virtualization (NFV) architecture has allowed the virtualization of the network functions (VNFs) on a commodity hardware, improving the financial aspect of the new networking concepts, but also allowing agile approach toward building scalable and adaptable network services. Although different, SDN and NFV complement each other and have become key technology enablers [4] of the 5G networks.

Another 5G requirement is the need for user plane latency of just 1 ms for ultra-reliable low-latency communications (URLLC) [5], which will allow mission-critical and safety-critical applications to run on the network with high reliability. This has to be provided in a situation where there are many connected entities that require high bandwidth and, in many cases, communicate among each other, and in many cases quality of service (QoS) needs to be provided. The SDN and the NFV architecture are the key enablers of the services in such cases, and allow simple and fast service provisioning, network management automation, and innovative service development approach, while at the same time the CAPEX and OPEX are significantly reduced. The ETSI Industry Specification Group for Network Functions Virtualization (ETSI ISG NFV) [6] is responsible for the standardization of the NFV architecture within telecom networks.

The 5G networks have to comply with strict requirements, and yet the services built upon them require increased reliability, scalability, and availability. One option is to bring the compute and storage resources used by the services and the application that drive them, close to the end user and the network edge. This is applicable not only to the virtualized RAN (vRAN) but also to the 5G core network. It can be done using an NFV-based approach, where the NFV elements are distributed geographically in multiple locations, allowing the network to grow toward the edge. The NFV Architecture Framework [7] is depicted in Figure 5.1.

The OpenFlow (OF) communication protocol is one of the first standards defined in SDN. It allows the network flow among the control plane and the data plane. The controllers communicate with the networking elements through the southbound APIs, but also with the business logic through the northbound API. OpenFlow is used by the controllers to push updates into the flow tables of the OF-capable switches and routers in the data plane. The path of the individual packets is defined through the flow tables [8], which are updated every time a new path is needed. The packets come directly into the data plane of the SDN network where its source and destination addresses are matched to an entry into the flow table. If such an entry

Distribution of NFV Infrastructure

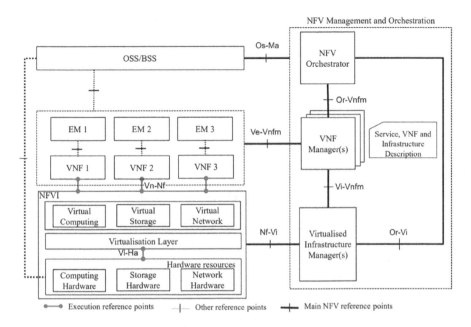

FIGURE 5.1 NFV reference architectural framework [7].

is found, the packet continues its journey, but if an entry isn't found, the packet is directed into the control plane, which calculates its optimal path [9–11]. The controller updates the flow table accordingly and any consecutive packet is directed to its destination immediately without the need to interact with the control plane. This concept complements the concept of the NFV, and the possibility for distribution of the SDN control plane has been researched and analyzed [12–14].

In this work, we are investigating the possibilities for distribution of the NFV architecture and the impact that such a distribution has on the network latency. We have created analytical models for three scenarios: centralized infrastructure, an environment with distributed data plane and central management and orchestration (MANO), and finally fully distributed infrastructure in which every location has a data plane and a MANO environment. Our motive is to evaluate the impact that distribution of the network architecture has on the packet sojourn time when an end-to-end service is built in a monolith architecture and when the service spans across multiple locations, thus bringing it closer to the end consumer. This is a significant prediction in a 5G-based environment. For the analytical models, we are using the queuing theory and we model the network elements: the switches, the controllers, the VNFs, and the VNF manager as M/M/1 queues. The system is conceptualized as a Jackson network. In the distributed environment, the communication among the location is on both the data plane level and the MANO level. We consider the network packet delay caused within one location due to the OF concept, but also the delay caused by the distribution and the links that connect the geo-locations. The modes are tested in different scenarios of MATLAB [15] simulation that can help to pinpoint the main factors that have an effect on the network latency.

In the next phase, we are introducing an experimental environment based on an NFV infrastructure (NVFI) with two geographically separated datacenters. We conduct experiments to analyze and to compare the latencies within single-location services and distributed services. It is important to consider the proximity of the service to the end user. The availability and the reliability of the service must be on a highest possible level; thus, this scenario is valid for disaster recovery planning. At the end we compare the analytical results with the experimental results in order to evaluate the validity of the analytical models.

Most of the works toward distribution of the NFV-based network are concentrated on specific parts of the architecture. Some authors are investigating the SDN approach to evaluate the network performance [16,17]. Software algorithms for SDN elements placements are often proposed [18]. Others are evaluating only the control plane and the latency it adds to the system [19] or only the data plane and its elements [20]. Our work distinguishes in the approach that combines SDN and NFV, as they are inevitably connected in practice. ETSI has presented this connection as given in Figure 5.2 [21].

The analytical models of systems as complex as this have multiple degrees of freedom. We are making some simplifications and generalizations in the models, but the use of queuing theory in the modeling comes naturally when OF elements are modeled. Other authors have used M/M/1 queues [17,22], but other types of queues such as M/M/m and G/G/m are also used [23,24]. Another stream of exploration is the network slicing in 5G researches [25], which divide the network architecture

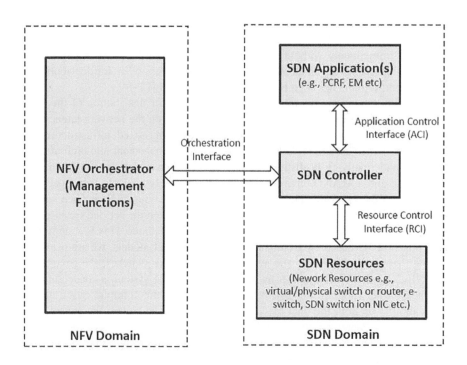

FIGURE 5.2 NFV and SDN domains interaction according to ETSI [21].

Distribution of NFV Infrastructure

horizontally, on multiple slices, each with its own purpose. NFV architecture is a basis for it also.

The objective of our work to help researchers and network architects to build cost-effective, secure, easily manageable, and innovative network environments that will form a basis for new generation services, especially in the 5G era. The conclusions that we take are usable when building the RAN network, but also in the later stages when services are planned upon it, especially for the latency in an NFV-based environment.

In the next sections, we will go through the related work on this subject and then we will present the analytical models that are proposed. The simulations conducted in MATLAB are shown next. Latter we present the experimental environment and the results from it. In the end, we draw the conclusions.

5.2 RELATED WORK

The academic community, but also the related industries, are researching the possibilities for building optimized services by using NFV architecture that will fully utilize the potentials that 5G mobile networks are providing. It is a broad field and many aspects are being investigated. Some authors have chosen a theoretical path by constructing mathematical models to simulate different scenarios and predict the behavior of the system [17,22,23,26,27]. Markovian queuing models are the most used approximation of the VNFs and the network elements that constitute the MANO. Other authors are proposing algorithms, applications, and frameworks that are optimizing some process within the NFV environment [28,29]. Experimental researches are also being performed to confirm or to improve the features of the services that run within the system [30,31].

In our work, we are both proposing analytical models that represent different network topologies in which we introduce an NFV geo-distribution and testing the efficiency of the proposed models in terms of network packet delay. We are also introducing a novel experimental testbed in which we calculate the delay in the service chains that span across separated geo-locations, and we compare the results with monolithic service chains on a centralized NFV environment.

The authors of [17] modeled an SDN environment with a single SDN controller and a data plane with multiple nodes. Every node has a single switch. The controller is modeled as an M/M/1 queue, and the data plane is modeled as an open Jackson network, similar to our work. They performed simulations to evaluate key system metrics such as the average time a packet spends in the network. Our work distinguishes from theirs because we consider an NFV-based system taking into consideration the NFV MANO that consisted of a controller but also an VNF manager. In the analytical models we consider, the MANO can be also distributed in multiple locations, with multiple controllers and VNF managers that communicate among each other.

Jarschel et al. [26] modeled an SDN-based system in which the controller and the switch were modeled as Markovian servers with an M/M/1 forward model and an M/M/1-S for the feedback model. Their model was based on a single switch and a single controller and their accent was on the probability of packet loss in such a system. The authors of [22] worked on an OpenFlow SDN environment that had a single controller and multiple switches. They analyzed how the probability of packet-in messages impacted the performance of the switches and the controllers. The controller was modeled as an M/M/1 queue, while the switches were modeled as M/H2/1 queues. They concluded that the latency in the network rises exponentially with a linear rise of the probability for OF packet-in messages. In our analytical models we are also using M/M/1 queues to model the switches and the controllers. But our main accent is on the distribution of the NFV-based model and the impact that it has on the packet sojourn time.

In the work of [23], the subject of the analytical model was the network chains made of VNFs that form the data layer of the system. The VNFs were modeled as G/G/m queues in a three-tiered architecture that represented an LTE virtualized Mobility Management Entity.

Sarkar et al. [27] explored an SDN environment and they also used M/M/1 queues. They propose two scenarios, one in which they explore the limitation of the controller regarding the number of OF switches it can handle based on the flow rate, and a second one in which they explored the tolerance of the OF switches based on packet sojourn time to decide when to shift from one to another controller.

Sadaf et al. [28] used software modeling of an NFVI. An integrated process modeling and enactment environment called MAPLE was introduced. They generated traceability information and used analysis support built into Eclipse Papyrus to review the automation of network service management in NFV. The goal was to define an automated process for the design, deployment, and management of network services that form a VNF chain.

A review of NFV and service function chain (SFC) implementation frameworks is done in [29]. Based on the primary objectives of each of the surveyed frameworks they categorized them into three: resource allocation and service orchestration, performance tuning, and resilience and fault recovery. Their accent was also on the NFV slicing, which is related to the NFV distribution that we are investigating in our research.

A novel experimental testbed of an NFV-based system is proposed by Vergara-Reyes et al. [30]. Their main objective was to benchmark a selected set of supervised machine learning (ML) algorithms to efficiently classify the traffic within the system. Such an analysis was important for performing various tasks within an NFV environment, such as QoS, deep packet inspection (DPI), establishing network security, and controlling the virtual network. As the traffic mostly goes in the virtual layer in the east–west direction, such an analysis was a challenge. Our experimental testbed uses similar tools, but the experiments and the research go in another direction.

The authors of [31] implemented a novel experimental testbed, which they named 5GIIK. They focused on the management and orchestration of network slices. They both identified design criteria that are a superset of the features present in other testbeds and determined appropriate open-source tools for implementing them. Their testbed was about slice provision dynamicity, real-time monitoring of VMs, and VNF onboarding to different virtual infrastructure managers (VIM). We are also introducing an experimental NFV testbed, but the concept of network slices is more focused on the virtual network overlays, which divides the network horizontally, while our investigation is on the architecture distribution and the vertical network division.

In [32], a framework that uses network slicing is proposed to provide seamless and isolated access to corporate-based content while moving through heterogeneous networks. This solution allows mobile network operators to dynamically instantiate isolated network slices for corporate users, and handover them between 3GPP and non-3GPP networks while users move away from the corporate network. The authors investigated remote working in a corporate environment in which multiple locations were used, which is related to our network distribution research.

5.3 ANALYTICAL MODELS

To create the models for a system built upon NFV architecture we are using queuing theory principles. We are using SDN principle and the interaction between the SDN and the NFV is given in Figure 5.2. There are three models that are examined [33]:

- A centralized system where all the elements are in a single location (usually a single data center). The system has one controller, one VNF manager, and one switch.
- A system that has a distributed data plane, meaning that it has NFVI elements on multiple locations. The VNF chains can span on multiple datacenters and different geographical locations. Every location has one switch. The MANO environment has one controller and one VNF manager that serve all locations.
- Fully distributed system, where every location has all the elements of the NFV. Every location has a data plane made upon NFVI with a single switch, and MANO with a controller and a VNF manager. The controllers communicate with each other and every controller is aware of the entire system.

5.3.1 Assumptions

In practice, it is normal that a single location has multiple switches. In our models, we are simplifying the system with a single switch on every location. The approximation does not have a high impact on the overall system generalization as multiple serially connected switches that are connected with high bandwidth links on a small distance can easily be replaced with a single more powerful switch. We are modeling the data plane as a Jackson network, similar to [24]. The switches are Jackson's servers. The arrival of the data at every location is assumed to be a Poisson process [34,35].

We assume that the arrival of network packets at the nodes and the service times of the nodes are independent from each other. The system is OF-based. The packet service time has an exponential distribution [32,36]. The network queues are in balanced state and every component can serve the required traffic, which means that the network queues have reached a balanced state with utilization less than one.

Under these assumptions, we can model the switches, the controllers, and the VNF managers as M/M/1 queues that have infinite queue sizes. We assume that all the packets that arrive in the system are processed and there is no packet drop-out.

In the models with a single controller, it is responsible for all the changes in the OF tables, in all locations. In the models with multiple controllers, every controller is responsible for changes in the OF tables in the switches. The controllers that have received the packet-in message are responsible for the packet-out message. The controllers communicate with each other and every controller is completely aware of the entire setup and the architecture in all locations in order to calculate the optimal packet flow. The SDN service chain can span on multiple locations.

The VNF managers are connected to the element managers (EM) and the VNFs and control their creation, destruction, and scaling. The VNF managers are connected to the controllers and react in the NFVI when the controller cannot find an available VNF for the service.

The VIM is connected to the NFVI, but we assume that it controls the infrastructure independently. The infrastructure always has available resources, which can be assigned to the VNFs when a certain threshold is reached, thus it does not contribute to the network packets sojourn time. Due to this, we do not put the VIM into the analytical models.

5.3.2 CENTRALIZED NFV MODEL

A graphical representation of the centralized NFV model can be seen in Figure 5.3. The arrival rate at the switch is λ_0. The probability of packet-in message is denoted as q_0. While the probability of packet going from the controller to the VNF manager is denoted as r_m. A full explanation of the notation used is given in Table 5.1.

The utilization of the controller, the switch, and the VNF manager is calculated as:

$$\rho = \frac{\Lambda}{\mu} \tag{5.1}$$

We calculate the total arrival rate at the switch Λ_0 as a sum of the external arrival rates and the packets sent to the switch by the controller:

$$\Lambda_0 = \lambda_0 + q_0 \cdot \lambda_0 \tag{5.2}$$

Similarly, we calculate the total arrival rate at the controller Λ_c and the total arrival rate at the VNF manager Λ_v as:

$$\Lambda_c = q_0 \cdot \lambda_0 + r_V \cdot q_0 \cdot \lambda_0 \tag{5.3}$$

$$\Lambda_V = r_V \cdot q_0 \cdot \lambda_0 \tag{5.4}$$

The average time the packets spend in the switch, the controller, and the VNF manager are calculated as:

$$M[T_0] = \frac{1}{\mu_0 - \Lambda_0} \tag{5.5}$$

Distribution of NFV Infrastructure

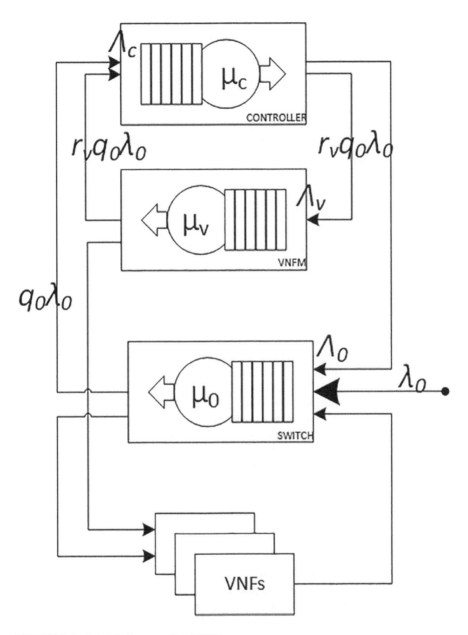

FIGURE 5.3 Model of a centralized NFV.

$$M[T_c] = \frac{1}{\mu_c - \Lambda_c} \tag{5.6}$$

$$M[T_V] = \frac{1}{\mu_V - \Lambda_V} \tag{5.7}$$

TABLE 5.1
Notation Used

Symbol	Parameter
λ_i	Arrival rate at switch i
q_i	Probability of packet going from switch i to controller
p_{ij}	Probability of packet going from switch i to switch j
r_V	Probability of packet going from controller to VNF manager
μ_i	Service rate of the switch i
μ_c	Service rate of the controller
μ_V	Service rate of the VNF manager
B	Binary decision, 0 or 1
Λ_i	Total arrival rate at the switch i
Λ_c	Total arrival rate at the controller
Λ_V	Total arrival rate at the VNF manager
T_i	Sojourn time at the switch i
T_c	Sojourn time at the controller
T_V	Sojourn time at the VNF manager
$M[T]$	Mean value for the sojourn time of packets in M/M/1 queue
L_T	Transmission latency
L_P	Propagation latency
L_{Tij}	Transmission latency from i to j
L_{Pij}	Propagation latency from i to j
T_N	Time spent in network
T_{NC}	Time spent in the network for packets going to the controller
T_{NS}	Time spent in the network for packets going between switches
$M[T_{NC}]$	Mean value for the sojourn time for packets going to the controller
$M[T_{NS}]$	Mean time for the sojourn time for packets going between switches
B_{ij}	Number of packets on link from i to j
Bw_{ij}	Bandwidth on link from i to j
d_{ij}	Distance from i to j
s	Speed of data in link

In this centralized scenario, we assume that all network traffic goes within a single datacenter. The transmission and the propagation latency of the connections among elements are minimal and do not produce additional latency. The average time the packet spends in the system $M[T_{\text{AVG}}]$ is a sum of the average times given in equations (5.5–5.7).

$$M[T_{\text{AVG}}] = M[T_0] + M[T_c] + M[T_V] \qquad (5.8)$$

$$M[T_{\text{AVG}}] = \frac{1}{\mu_0 - \lambda_0 \cdot (1+q_0)} + \frac{1}{\mu_c - q_0 \cdot \lambda_0 \cdot (1+r_V)} + \frac{1}{\mu_V - r_V \cdot q_0 \cdot \lambda_0} \qquad (5.9)$$

Distribution of NFV Infrastructure 231

5.3.3 NFV Model with Distributed Data Plane and Central MANO

The following model shows a scenario in which the data plane is distributed among multiple physical location, meaning multiple data centers, which can be in different geographical regions. These regions are used by almost all public cloud providers and are defined as a set of datacenters deployed within a latency-defined perimeter and connected through a dedicated regional low-latency network [37,38]. The MANO is central and manages the entire data plane. This scenario can be used for services that are logically near the end consumer. One such example is multimedia services, such as IPTV, where the streaming services are close to the viewers, while the backend can be in completely different location, together with the MANO elements. The MANO is used to start and stop streaming containers on the distributed locations. The model is graphically represented in Figure 5.4.

The number of locations is k. The arrival rate at switch i is denoted as λ_i where:

$$i \in \{1, 2, 3, \ldots, k\}.$$

The utilization of the switch, the controller, and the VNF manager are given in equation (5.1). The switches are in balanced state and the total arrival rate for the switch i is a sum of the arrival rate from outside and the arrival rates from the other switches.

$$\sigma_i = \lambda_i + \sum_{j=1,\, j \neq i}^{k} (p_{ij} \cdot b_{ij}) \sigma_j \tag{5.10}$$

As in the VNF chains, there can be a situation where there is no connection between some switches, we set b_{ij} at value 0 when there is no connection between switches i and j, or value 1 when there is no connection between switches i and j.

$$b_{ij} = \begin{cases} 1 & \text{if update from } j \text{ involves } i \\ 0 & \text{if update from } j \text{ doesn't involve } i \end{cases} \tag{5.11}$$

The total arrival rates for the switch i, the controller c, and the VNF manager v are:

$$\Lambda_i = \sigma_i + q_i \cdot \lambda_i + \sum_{j=1,\, j \neq i}^{k} (q_j \cdot b_{ij}) \sigma_j \tag{5.12}$$

$$\Lambda_c = \sum_{i=1}^{k} q_i \cdot \lambda_i + \sum_{i=1}^{k} q_i \cdot \lambda_i \cdot r_i \tag{5.13}$$

$$\Lambda_V = \sum_{i=1}^{k} q_i \cdot \lambda_i \cdot r_i \tag{5.14}$$

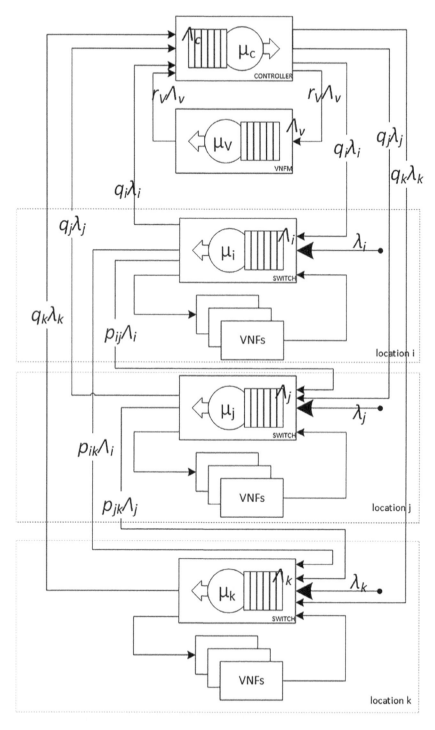

FIGURE 5.4 NFV model with distributed data plane and central MANO.

Distribution of NFV Infrastructure

The average time the packets spend in the switch, the controller, and the VNF manager are:

$$M[T_i] = \frac{1}{\mu_i - \Lambda_i} \quad (5.15)$$

$$M[T_c] = \frac{1}{\mu_c - \Lambda_c} \quad (5.16)$$

$$M[T_V] = \frac{1}{\mu_V - \Lambda_V} \quad (5.17)$$

With this we can calculate the average time the packet spends in the system, without the impact of the links between the distributed locations:

$$M[T_{\text{SYS}}] = \frac{1}{\sum_{i=1}^{k} \lambda_i} \cdot \left[\sum_{i=1}^{k} M[T_i] + M[T_c] + M[T_V] \right] \quad (5.18)$$

To calculate the impact of the links to the packet sojourn time, we calculate the transmission and the propagation latency between i and j as:

$$L_{Tij} = \frac{B_{ij}}{Bw_{ij}} \quad (5.19)$$

$$L_{Pij} = \frac{d_{ij}}{s} \quad (5.20)$$

The total latency is made of the latencies of the communication between the switches and the controller, and between the switches, given with:

$$M[T_N] = \sum_{i=1}^{k} q_i \cdot (L_{Tic} + L_{Pic}) + \sum_{i=1}^{k} \sum_{j=1, j \neq i}^{k} p_{ij} \cdot (L_{Tij} + L_{Pij}) \cdot b_{ij} \quad (5.21)$$

With this, the overall packet latency in the system is a sum of the latency caused by the system $M[T_{\text{SYS}}]$ and the latency caused by the network links $M[T_N]$:

$$M[T_{\text{AVG}}] = M[T_{\text{SYS}}] + M[T_N] \quad (5.22)$$

5.3.4 Fully Distributed NFV Model

The fully distributed environment is set up with a separate data plane and MANO elements, including a controller and a VNF manager in every location. In real situation, every location must have a VIM manager that manages the underlying physical infrastructure, but we choose not to model it as the assumption are that the infrastructure has sufficient resources and the VIM manager does not influence

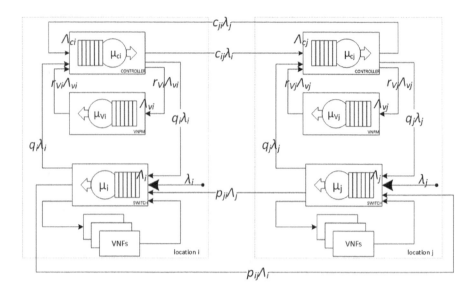

FIGURE 5.5 Fully distributed NFV model.

the packet sojourn time. As we noted, the controllers communicate with each other synchronously and every controller is fully aware of the total network infrastructure. This process goes parallel with the main controller function, to update the OF tables, so it also does not influence the packet sojourn time. The model is presented in Figure 5.5.

The fully distributed architecture is mostly used when it is needed that the services are close to the end consumer, but the VNF chain is such that all the VNFs are in a single location. If the chain is spread across multiple locations, the traffic will go through the links multiple times, and the overall network latency will be dependable from the latency caused by the links among the datacenters or the regions. If the chain spreads across three or more regions, the links have a decisive role in the overall latency, so such a scenario is used only if necessary for the purpose it is built for. This architecture is beneficial when disaster recovery is planned, and it can apply zero downtime if a certain region fails.

The number of locations is k. The locations are noted as i, j where:

$$i, j \in \{1, 2, 3, \ldots, k\}.$$

Similar to equation (5.1), the utilization is given as:

$$\rho_i = \frac{\Lambda_i}{\mu_i} \qquad (5.23)$$

The switches, controllers, and the VNF managers are in balanced state. The net input σ_i is calculated in equation (5.24), while the total arrival rate of the switch is calculated with equation (5.25).

Distribution of NFV Infrastructure

$$\sigma_i = \lambda_i + \sum_{j=1,\, j \neq i}^{k} (p_{ij} \cdot b_{ij}) \sigma_j \tag{5.24}$$

$$\Lambda_i = \sigma_i + q_i \cdot \lambda_i + \sum_{j=1,\, j \neq i}^{k} (q_j \cdot b_{ij}) \sigma_j \tag{5.25}$$

The total arrival rates of the VNF managers and the controllers on location i are calculated as:

$$\Lambda_{vi} = q_i \cdot \lambda_i \cdot r_i + \sum_{j=1, j \neq i}^{k} b_{ij} \cdot \lambda_j \cdot r_j \tag{5.26}$$

$$\Lambda_{ci} = q_i \cdot \lambda_i + q_i \cdot \lambda_i \cdot r_i + \sum_{j=1, j \neq i}^{k} b_{ji} \cdot \lambda_j \tag{5.27}$$

With these formulas we can calculate the average sojourn time that the packets spend in the elements:

$$M[T_i] = \frac{1}{\mu_i - \Lambda_i} \tag{5.28}$$

$$M[T_{ci}] = \frac{1}{\mu_{ci} - \Lambda_{ci}} \tag{5.29}$$

$$M[T_{Vi}] = \frac{1}{\mu_{Vi} - \Lambda_{Vi}} \tag{5.30}$$

Now, the average time the packet spends in the system $M[T_{SYS}]$, without the impact of the links between the distributed locations is given in equation (5.31):

$$M[T_{SYS}] = \frac{1}{\sum_{i=1}^{k} \lambda_i} \cdot \left[\sum_{i=1}^{k} M[T_i] + \sum_{i=1}^{k} M[T_{ci}] + \sum_{i=1}^{k} M[T_{Vi}] \right] \tag{5.31}$$

The transmission and propagation latency are calculated as in equations (5.19 and 5.20). The total latency produced from the inter-location links for the communications between the switches and the controllers is given in equations (5.32 and 5.33).

$$M[T_{Ni}] = \sum_{i=1}^{k} \sum_{j=1, j \neq i}^{k} p_{ij} \cdot \left(L_{Tij} + L_{Pij} \right) \cdot b_{ij} \tag{5.32}$$

$$M[T_{NC}] = \sum_{i=1}^{k} \sum_{j=1, j \neq i}^{k} c_{ij} \cdot \left(L_{Tij} + L_{Pij} \right) \cdot b_{ij} \tag{5.33}$$

With this, the overall latency produced by the links that connect the distributed locations is equation (5.34):

$$M[T_N] = \sum_{i=1}^{k}\sum_{j=1,j\neq i}^{k} c_{ij} \cdot (L_{Tij}+L_{Pij}) \cdot b_{ij} + \sum_{i=1}^{k}\sum_{j=1,j\neq i}^{k} p_{ij} \cdot (L_{Tij}+L_{Pij}) \cdot b_{ij} \quad (5.34)$$

The average packet sojourn time in the system is equation (5.35):

$$M[T_{\text{AVG}}] = M[T_{\text{SYS}}] + M[T_N] \quad (5.35)$$

5.4 PERFORMANCE EVALUATION OF THE ANALYTICAL MODELS

Simulations of the proposed models are done using MATLAB programs. In the simulations, we are going with an assumption that 4% of the packets need to go from the switch to the controller [16,17,39,40], which is a OF parameter that is widely measured and established. We also assume that 4% of the packets that go through the controller need to go to the VNF manager that has to react in the VNFs and create new VNF instances.

We also define that the controller's service rate is 90,000 pkts/seconds and the VNF manager's service rate is 95,000 pkts/seconds. For the links, we have taken that the distance between the distributed locations is 100 km and that the bandwidth of the links is 10 Gbps.

The first simulation that we've done is to check the packet sojourn time when there is a change in the number of packets (as a percentage) that are sent from the switch to the controller. As expected, the simulations have shown that in all the scenarios, the packet sojourn time rises with the rise of the packets sent to the controller. The rise is exponential with linear rise of the probability for packets sent to the controllers. Also, as expected, with scenarios with multiple controllers, the rise is more visible.

Another simulation done to check the dependency of the packet sojourn time from the possibility of packets sent to the VNF manager. Again, as expected, the packet sojourn time rises with the rise of the possibility for packets sent to the VNF manager, which means that the VNF manager needed to intervene more often in the VNF chains by creating/starting or removing new VNFs. There are multiple studies [37,38,41] that are researching this part of the NFV-based system. The packet sojourn time, and the overall system performance can be optimized by using "smart VNF managers" that are using artificial intelligence, pattern learning, and ML methods to predict the need for new VNFs and to create them in a timely manner.

The number of distributed locations is one of the main factors that impact the latency of a VNF chain that spans multiple locations. We compare the two scenarios when the arrival rate at the switches is steady at $\lambda = 50,000$ pkts/seconds. Figure 5.6 shows how the packet sojourn time rises as the number of locations rises. It can be concluded that for two locations, the systems behave almost the same, but the scenario with central controller behaves much worse as the number of locations rises. This happens due to the packet-in messages that always have to travel through a network link, while in the fully distributed scenario the controller is on-site, within a same datacenter, which significantly reduces the latency.

Distribution of NFV Infrastructure

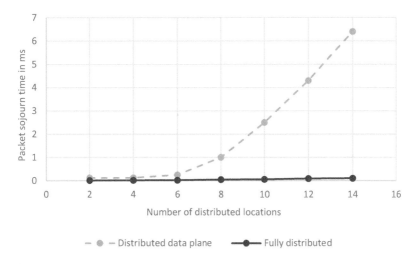

FIGURE 5.6 Packet sojourn time relative to the number of locations.

In the next simulation, we were investigating the impact that probability of packets sent to the controller has on the system latency. First, we work on the distributed system with central MANO. If more packets are sent to the controller the latency gets higher. The latency rises faster as the number of switches gets larger. The simulations are done with service rates: $\lambda = 10{,}000$ pkts/seconds, $\lambda = 11{,}000$ pkts/seconds, and $\lambda = 12{,}000$ pkts/seconds. The results are graphically presented in Figures 5.7–5.9. We were working with situations in which there are 5, 10, and 15 distributed locations.

The fully distributed system behaves similar to the data plane distributed system when we simulate the service time relative to the probably of Openflow table-miss

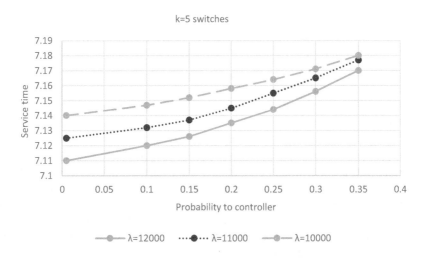

FIGURE 5.7 Service time relative to the probability of packet sent to the controller with five distributed data planes.

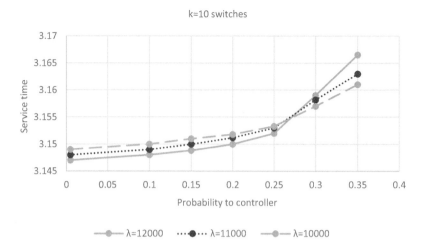

FIGURE 5.8 Service time relative to the probability of packet sent to the controller with ten distributed data planes.

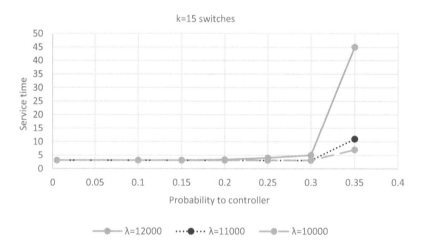

FIGURE 5.9 Service time relative to the probability of packet sent to the controller with 15 distributed data planes.

and the packet is forwarded to the controller. From Figures 5.10–5.12 we can see that the service time is better when we simulate five and ten locations. But if we see the 15 locations scenario the service time of the system with central MANO is much worse.

We also compare the two distributed scenarios in a situation where we investigate the dependency of the service time relative to the probability of packets going from one distributed location to another. We again have three simulations, with 5, 10, and 15 locations (each with one switch as).

If we analyze Figures 5.13–5.15, we can draw the following conclusions:

Distribution of NFV Infrastructure

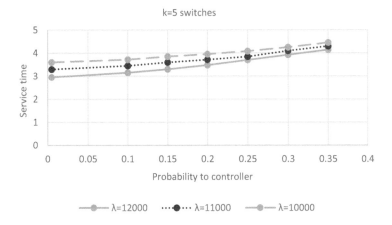

FIGURE 5.10 Service time relative to the probability of packet sent to the controller with five fully distributed locations.

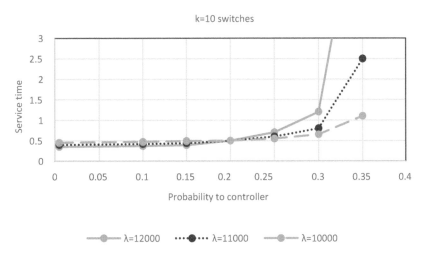

FIGURE 5.11 Service time relative to the probability of packet sent to the controller with ten fully distributed locations.

When the number of locations is small, the architecture with central MANO and the fully distributed environment behave similarly. But in the same situation, when the probability of packets sent to other distributed location is higher, the fully distributed environment is better.

In the ten locations scenario, again when the probability of packets sent to other location is small, the central MANO architecture is better. But this changes rapidly as the probability for packets sent to other locations gets higher. So, if the VNF chain spans across multiple locations, the fully distributed environment performs better.

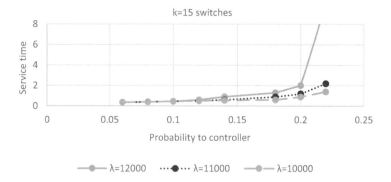

FIGURE 5.12 Service time relative to the probability of packet sent to the controller with 15 fully distributed locations.

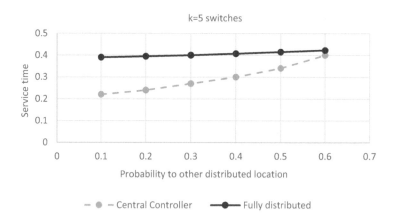

FIGURE 5.13 Comparing two scenarios with five locations.

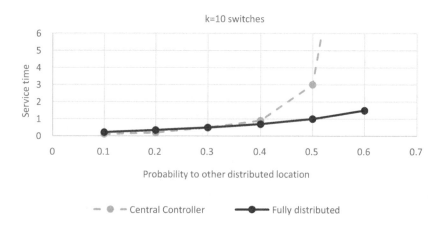

FIGURE 5.14 Comparing two scenarios with ten locations.

Distribution of NFV Infrastructure

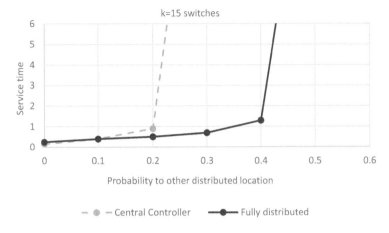

FIGURE 5.15 Comparing two scenarios with 15 locations.

When we simulate 15 locations, the fully distributed scenario is much more effective.

5.5 EXPERIMENTAL EVALUATION OF A DISTRIBUTED NFV

The previous simulations are purely theoretical. We have laid down the assumptions and we have made analytical models that were then tested using MATLAB simulations. Another point of view over the distribution of the NFV architecture, in order to analyze the packet latency, is the experimental approach. We have made a novel experimental testbed in order to compare the network packet latencies in a centralized network architecture versus a distributed network architecture. The conclusion that we draw are of a high importance, especially for the 5G networks and the possibility to bring the services close to the consumer, which will then use the 5G RAN to connect to the network. The service has to be stable and reliable. The scenarios that we simulate can also be used for designing disaster recovery sites, to predict the influence that the migration of the service chain to a different geo-location will have when the primary location fails. This research is important for the network architects when they plan the services that will run on the network platform. The most cost-effective scenario can be used, which will allow the services to have the smallest possible latency.

5.5.1 EXPERIMENTAL TESTBED

In order to simulate a distributed NFV environment we have laid down our infrastructure on two geographically separated datacenters [42]. They are 160 km apart and are connected with a 10 Gbps link. There is no QoS implemented. Two hypervisors were used, one on each side. The hypervisors were based on Ubuntu 18.04 LTS and Oracle VirtualBox [43] for the machines. Open vSwitch [44] was used for network connection. Each hypervisor has a network card with 1 Gbps bandwidth.

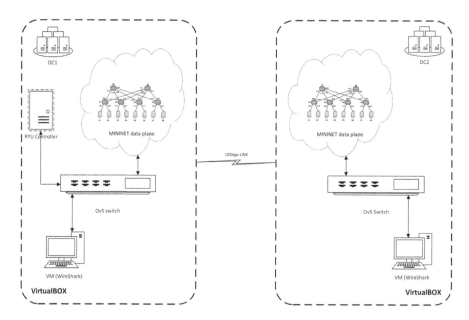

FIGURE 5.16 Experimental testbed of a distributed NFV.

To simulate the data plane, we used Mininet [45] on which we have built virtual network. Each location has 200 virtual VNFs that are interconnected on Mininet switches and links. GRE tunneling was used to connect the Mininet infrastructure to the Open vSwitch so that the VNF chains have outside access.

The MANO part of the NFV was simulated with Mininet network generator as a VNF manager and Ryu Controller [46] was used as a controller. It is installed on a separate VM that is connected to Mininet instances on both locations.

We used Distributed Internet Traffic Generator (D-ITG) [47] to generate both TCP- and UDP-based traffic. The generator is designed to produce traffic at packet level, replicating appropriate stochastic processes for both IDT (inter-departure time) and PS (packet size) random variables. We used it to create multiple simultaneous flows with standard payload information. IDT was set with normal distribution.

The traffic was captured inside specially installed VMs that were used as a destination machine for the flows. Wireshark [48] was used for packet capturing.

For each experiment, we were conducting 30 iterations, which lasted from 30 seconds to 10 minutes. The time intervals on which traffic was generated follow the Poisson distribution. We were using 4, 8, 12, and 16 parallel dataflows in the experiments. Using the captured data, we were calculating the mean packet delivery time. An average result from all the iterations within an experiment was taken as a final result.

The experimental environment is illustrated in Figure 5.16.

5.5.2 Experimental Results

First, we will present the results when the VNF service chain is on a single location, in our case the primary location, meaning that the end consumer is very close to

Distribution of NFV Infrastructure

TABLE 5.2
Mean Packet Delivery Time – Centralized System

Flows	4	8	12	16
Mean packet delivery time in ms	0.434	1.629	2.971	5.86

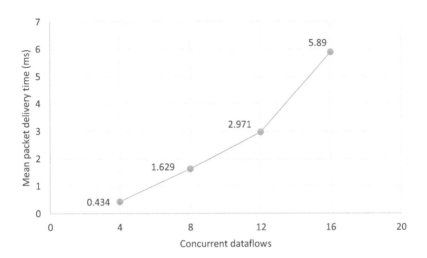

FIGURE 5.17 Packet delivery time in single location.

the network services. We are only chaining the number of concurrent data flows in the network, with a destination, the VM on the primary location. As expected, the mean packet delivery time that we conclude from the iterations rises as the number of concurrent flows is bigger, but it can be noted that the overall latency is just a few milliseconds. The results can be seen in Table 5.2 and are graphically represented in Figure 5.17.

When the VNF chain is distributed in two locations, the network packets can pass through the link multiple times, depending on the type of VNFs needed in the chain. To test this scenario, we have made experiments when every flow goes through the link 1, 3, 5, and 7 times. As the number of passes is odd and the source is always in the primary location, the destination is always in the secondary location. Again, we are using 4, 8, 12, and 16 simultaneous dataflows, and 30 measurement iterations for every situation. The data results are given in Table 5.3 and Figure 5.18. It can be seen that the link has a significant impact on the overall latency. The mean packet delivery time rises exponentially as the number of passes through the link rises. Now the latency is not just a few milliseconds, but it is much larger. From deduction, we can conclude that the switches that are connecting the two datacenters and are used for link termination add a lot of the transport latency. In our case, the switches work in traffic-shaping mode.

TABLE 5.3
Mean Packet Delivery Time – Two Locations

No. of Passes Through Link	Concurrent Flow			
	4	8	12	16
1	5.121 ms	5.67 ms	6.195 ms	6.634 ms
3	5.72 ms	11.21 ms	31.461 ms	48.389 ms
5	8.432 ms	31.961 ms	61.31 ms	89.455 ms
7	19.651 ms	58.12 ms	112.942 ms	193.82 ms

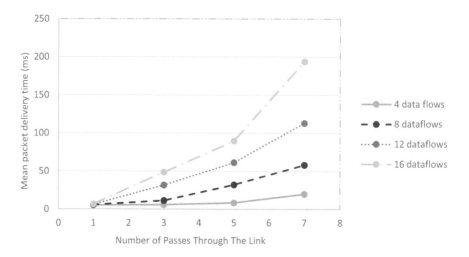

FIGURE 5.18 Mean packet delivery time – two locations.

We have made another experiment to see the dependence of the mean packet delivery time from the percentage of traffic flows that pass through the link. In this case, the destination of the flows can be in the primary, or in the secondary location, but we manage the percentage of flows that go to the secondary location. Now we have used 40 concurrent dataflows. We can mention that when 100% of the dataflows were directed through the link, the network interfaces of the hypervisors were 97% utilized, which means that we have reached the limit of the setup. We made 30 iterations for every case (for 10%, 20%, 30%, etc.) As it can be seen in Figure 5.19 and Table 5.4, the mean packet delivery time rises exponentially as the percentage of the dataflows through the link rises linearly. The time is from 5 ms when only 10% of the traffic goes to the second location, up to almost 200 ms when all the traffic goes through the link. This experiment can be viewed from a disaster recovery perspective. If the primary location fails and the end service consumer is close to it, all services will run on the second location, meaning that all the traffic will go through the link. Of course, we assume that the link itself is stable and is not affected by the disaster.

Distribution of NFV Infrastructure

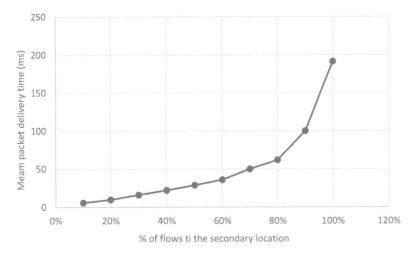

FIGURE 5.19 Mean packet delivery time relative to the % of flows through the link.

TABLE 5.4
Mean Packet Delivery Time Relative to the % of Flows through the Link

%	10%	20%	30%	40%	50%	60%	70%	80%	90%	100%
Mean packet delivery time in ms	5.86	9.82	16.13	22.07	28.91	35.87	5005	61.81	100.12	191.58

The presented experiments are beneficial for network architects that can draw conclusions for VNF chains that span across multiple locations and to plan development of the network, including disaster recovery (DR) scenarios. It can be seen that the proximity of the VNF chain (from both geographical and network point of view) can be crucial for the stability of the service that is offered. If regulatory, legal, or contractual obligations are taken, it is important to be able to predict the system behavior in different situations.

5.6 CONCLUDING REMARKS

The SDN technology and the NFV architecture are providers of advanced services that must be brought close to the end user in order to comply with existing standards and increase the service to the end entities, especially now, when the 5G enters the market with high speed. Distributing the NFV architecture and deploying virtual network features in separate locations close to the end user significantly helps to that end. This conclusion is proven by the research done in this work.

To draw the conclusions, the first approach is to make analytical models of systems using VNF service chains that are operating on a centralized and distributed NFV architecture. The models are made using assumptions that are derived by analyzing, synthesizing, and comparing the data used in our research, with other researches that work on similar problems, as well as with the classical physical networks. MATLAB programs are used to simulate the analytical models. The results have proved the initial hypothesis that NFV distribution can be of a great value when the network latency is a key factor for a successful end-to-end service. But, the distribution has to be handled carefully, as spanning the service chain across multiple locations can have an opposite effect. Using the models, network architects can assess the network setup and design an optimal solution from a point of technical efficiency, but also from a point of financial impact of the setup.

Further, an experimental scenario is built, using two geographically distant location, to evaluate the network packets sojourn time (and thus the latency) in an SDN service chain that is built on top of an NFV architecture. Different scenarios of service placement were explored, which also confirmed the initial hypothesis. This scenario is also feasible when designing a highly available DR system with two redundant locations. It allows a researcher to predict the latency that will be caused by a DR scenario when one of the two locations becomes unavailable.

REFERENCES

1. ETSI. 5G standard [Online], cited: 2022.05.15. Available at: https://www.etsi.org/technologies/mobile/5g.
2. J. Reis, M. Amorim, N. Melao, P. Matos. Digital transformation: A literature review and guidelines for future research. In *Trends and Advances in Information Systems and Technologies. WorldCIST'18*, vol. 745, pp. 411–421, Mar. 2018. doi:10.1007/978-3-319-77703-0_41.
3. C. Bockelmann, et al. Massive machine-type communications in 5G: Physical and MAC-layer solutions. *IEEE Communications Magazine*, vol. 54, no. 9, pp. 59–65, Sept. 2016. doi:10.1109/MCOM.2016.7565189.
4. F. Yousaf, M. Bredel, S. Schaller, F. Schneider. NFV and SDN – Key technology enablers for 5G networks. *IEEE Journal on Selected Areas in Communications*, vol. 35, no. 11, pp. 2468–2478, Nov. 2017. doi:10.1109/JSAC.2017.2760418.
5. M. Eiman. Minimum technical performance requirements for IMT-2020 radio interface(s). Presentation 2018 [Online], cited: 2022.05.15. Available at: https://www.itu.int/en/ITU-R/studygroups/rsg5/rwp5d/imt-2020/Documents/S01-1_Requirements%20for%20IMT-2020_Rev.pdf.
6. https://www.etsi.org/technologies/nfv.
7. https://www.etsi.org/images/files/ETSIWhitePapers/ETSI_White_Paper_Network_Transformation_2019_N32.pdf.
8. Open Networking. OpenFlow switch specification (Version 1.3.1). Tech. Republic, 2012.
9. W. Braun, M. Menth. Software-defined networking using OpenFlow: Protocols, applications and architectural design choices. *Future Internet*, vol. 6, no. 2, pp. 302–336, 2014. doi:10.3390/fi6020302.
10. A. Kalyaev, I. Korovin, M. Khisamutdinov, G. Schaefer, R.A. Md. Atiqur. A hardware approach for organization of software defined network switches based on FPGA. In *2015 4th International Conference on Informatics, Electronics and Vision (ICIEV 2015)*, pp. 1–4, 2015. doi:10.1109/ICIEV.2015.7334067.
11. S. Panev, P. Latkoski. Handover analysis of openflow-based mobile networks with distributed control plane. *Computers & Electrical Engineering*, vol. 81, p. 106546, 2020. doi:10.1016/j.compeleceng.2019.106546.

12. T. Issa, Z. Raoul, A. Konaté, J.C. Adepo, B. Cousin, A. Olivier. Analytical load balancing model in distributed open flow controller system. *Scientific Research Engineering*, vol. 10, no. 12, pp. 863–875, 2018.
13. F. Bannour, S. Souihi, A. Mellouk. Distributed SDN control: Survey, taxonomy and challenges. *IEEE Communications Surveys & Tutorials*, vol. 20, no. 1, pp. 333–354, 2017. doi:10.1109/COMST.2017.2782482.
14. A. Aissioui, A. Ksentini, A. Gueroui. An efficient elastic distributed SDN controller for follow-me cloud. In *2015 IEEE 11th International Conference on Wireless and Mobile Computing, Networking and Communications (WiMob)*, Abu Dhabi, pp. 876–881, 2015. doi:10.1109/WiMOB.2015.7348054.
15. MATLAB:R. Natick, Massachusetts, The Mathworks, Inc. MATLAB (Version 9.6.0.1072779), 2019.
16. M. Jarschel, S. Oechsner, D. Schlosser, R. Pries, S. Goll, P. Tran-Gia. Modeling and performance evaluation of an OpenFlow architecture. In *2011 23rd International Teletraffic Congress (ITC)*, San Francisco, CA, 2011.
17. M. Jarschel, O. Østerbø, A. Chilwan, K. Mahmood. Modelling of OpenFlow-based software-defined networks: The multiple node case. *Special Issue: Software Defined Networking*, vol. 4, no. 5, pp. 278–284, 2015. doi:10.1049/iet-net.2014.0091.
18. M. Iushchenko, V. Shuvalov, B. Zelentsov. Modeling the software dependability for an SDN/NFV controller. In *2019 International Multi-Conference on Engineering, Computer and Information Sciences (SIBIRCON)*, Novosibirsk, Russia, 2019. doi:10.1109/SIBIRCON48586.2019.8958337.
19. B. Xiong, X. Peng, J. Zhao. A Concise queuing model for controller performance in software-defined networks. *Journal of Computers*, vol. 11, pp. 232–237, 2016. doi:10.17706/jcp.11.3.232-237.
20. W. Saied, N. Ben Youssef, A. Saadaoui, A. Bouhoula. Deep and automated SDN data plane analysis. In *2019 International Conference on Software, Telecommunications and Computer Networks (SoftCOM)*, Split, Croatia, 2019. doi:10.23919/SOFTCOM.2019.8903846.
21. ETSI NFV ISG. GS NFV-EVE 005 V1.1.1 network function virtualization (NFV); ecosystem; report on SDN usage in NFV architectural framework, Dec. 2015.
22. S. Zhihao, K. Wolter. Delay evaluation of OpenFlow network based on queueing model. In 12th European Dependable Computing Conference (EDCC 2016), Gothenburg, Sweden, 2016. doi:10.48550/arXiv.1608.06491
23. K. Sood, S. Yu, Y. Xiang, H. Cheng. A general QoS aware flow-balancing and resource management scheme in distributed software-defined networks. *IEEE Access*, vol. 4, pp. 7176–7185, 2016.
24. J. Prados, P. Ameigeiras, J. Ramos, P. Andres-Maldonado, J. Lopez-Soler. Analytical modeling for virtualized network functions. In *2017 IEEE International Conference on Communications Workshops (ICC Workshops)*, pp. 979–985, 2017. doi:10.1109/ICCW.2017.7962786.
25. P. Subedi, A. Alsadoon, P.W.C. Prasad, et al. Network slicing: A next generation 5G perspective. *Journal of Wireless Communications and Networking*, vol. 102, 2021. doi:10.1186/s13638-021-01983-7.
26. G. Honghao, Y. Yuyu, H. Guangjie, Z. Wenbing. Edge computing: Enabling technologies, applications, and services. *Transactions on Emerging Telecommunications Technologies*, vol. 32, no. 6, p. e4197, 2021. doi:10.1002/ett.4197.
27. C. Sarkar, S.K. Setua. Analytical model for openflow-based software-defined network. In Progress in Computer, Analytics and Networking, vol. 710, pp. 583–592, 2018.
28. M. Sadaf, H. Omar, D. Guillaume, K. Ferhat, T. Maria. Model-driven process enactment for NFV systems with MAPLE.In Software and Systems Modeling 19, pp. 1263–1282, 2020. doi:10.1007/s10270-020-00783-9.

29. H. Adoga, D. Pezaros. Network function virtualization and service function chaining frameworks: A comprehensive review of requirements, objectives, implementations, and open research challenges. *Future Internet*, vol. 14, no. 2, p. 59, 2022. doi:10.3390/fi14020059.
30. J. Vergara-Reyes, M.C. Martinez-Ordonez, A. Ordonezy, O.M.C. Rendon. IP traffic classification in NFV: a benchmarking of supervised Machine Learning algorithms. In *IEEE Colombian Conference on Communications and Computing*, 2017.
31. A. Esmaeily, K. Kralevska, D. Gligoroski. A cloud-based SDN/NFV testbed for end-to-end network slicing in 4G/5G. In *2020 6th IEEE Conference on Network Softwarization (NetSoft)*, pp. 29–35, 2020. doi:10.1109/NetSoft48620.2020.9165419.
32. E. Modiano. Introduction to queueing theory: Lectures 5 and 6. Massachusetts Institute of Technology. Available at: https://web.mit.edu/modiano/www/6.263/lec5-6.pdf.
33. G. Ilievski, P. Latkoski. Analytical modeling of distributed network functions virtualization architectures and the impact of the distribution on the packet sojourn time. *In Transactions on Emerging Telecommunications* Technologies. 33, 2022. doi:10.1002/ett.4559.
34. T.C. Brown, P.K. Pollet. Poisson approximations for telecommunications networks. *Journal of the Australian Mathematical Society – Series B-Applied Mathematics*, vol. 32, no. 3, pp. 348–364, 1990.
35. J. Zhang, J. Tang, X. Zhang, W. Ouyang, D. Wang. A survey of network traffic generation. In *Third International Conference on Cyberspace Technology (CCT 2015)*, pp. 1–6, 2015. doi:10.1049/cp.2015.0862.
36. B. Bellalta, S. Oechsner. Analysis of packet queueing in telecommunication networks. In *Network Engineering Course Notes*, Universitat Pompeu Fabra, Barcelona, Spain. 2020. Available at: https://www.upf.edu/documents/221712924/231130239/NetworkEngBook-2020/46de69a2-8cb7-4618-a9c4-4d52d572c149.
37. A. Bikakis, X. Zheng. Multi-disciplinary trends in artificial intelligence. In *9th International Workshop, MIWAI 2015*, Fuzhou, China, November 13–15, pp. 390–403, 2015.
38. P. Gope, I. Pedone, A. Lioy, F. Valenza. Towards an efficient management and orchestration framework for virtual network security functions. *Security and Communication Networks*, vol. 2019, p. 2425983, 2019. doi:10.1155/2019/2425983.
39. A. Fahmin, Y. Lai, M.S. Hossain, Y.D. Lin, D. Saha. Performance modeling of SDN with NFV under or aside the controller. In *5th IEEE International Conference on Future Internet of Things and Cloud Workshops, W-FiCloud 2017*, Prague, Czech Republic, pp. 211–216, 2017. doi:10.1109/FiCloudW.2017.76.
40. J.R. Jackson. Networks of waiting lines. *Operations Research*, vol. 5, no. 4, pp. 518–521, 1957.
41. T. Subramanya, R. Riggio. Machine learning-driven scaling and placement of virtual network functions at the network edges. In *2019 IEEE Conference on Network Softwarization (NetSoft)*, Paris, France, pp. 414–422, 2019. doi:10.1109/NETSOFT.2019.8806631.
42. G. Ilievski, P. Latkoski. (2021). Experimental evaluation of network packet latency within a distributed NFV Infrastructure. In *TELFOR 2021*, doi:10.1109/TELFOR52709.2021.9653395.
43. Oracle VirtualBox. [Online], 2019. Available at: https://www.virtualbox.org.
44. M.V. Bernal, I. Cerrato, F. Risso, D. Verbeiren. Transparent optimization of inter-virtual network function communication in open vSwitch. In *IEEE Cloudnet*, Pisa, Italy, pp. 76–82, 2016.
45. M. Team. Mininet: An instant virtual network on your laptop (or other PC) – Mininet [Online], 2017. Available at: http://mininet.org.

46. S. Asadollahi, B. Goswami, M. Sameer. Ryu controller's scalability experiment on software defined networks. In *2018 IEEE International Conference on Current Trends in Advanced Computing (ICCTAC)*, pp. 1–5, 2018. doi:10.1109/ICCTAC.2018.8370397.
47. A. Botta, A. Dainotti, A. Pescapè. A tool for the generation of realistic network workload for emerging networking scenarios. *Computer Networks*, vol. 56, no. 15, pp. 3531–3547, 2012.
48. Wireshark. [Online], 2006. Available at: https://www.wireshark.org/.

Part 2

Machine Learning-Based Communication and Network Automation

6 5G-AIoT Artificial Intelligence of Things
Opportunity and Challenges

Dragorad A. Milovanovic
University of Belgrade

Vladan Pantovic
University Union – Nikola Tesla

CONTENTS

6.1 Introduction 253
6.2 5G for Smart IoT Connectivity 254
    6.2.1 Co-Desing of AI and 5G Network Technology 256
    6.2.2 Intelligent Connectivity in Complex Use Cases 260
6.3 5G Artificial Intelligence IoT 262
    6.3.1 IIoT Framework 263
    6.3.2 DT Models and Application Scenarios 265
        6.3.2.1 5G-AIoT Initiative 267
        6.3.2.2 5G Network Digital Twin 269
        6.3.2.3 Manufacturing Digital Twin 269
6.4 Concluding Remarks 270
References 271

6.1 INTRODUCTION

Artificial intelligence of things (AIoT) represents combination of artificial intelligence (AI) technologies and Internet of things (IoT) infrastructure with the goal of achieving more efficient services, improving human–machine interaction, as well as improving data management and analytics [1–7]. The convergence of AI and IoT is an obvious logical progression in the evolution of these technologies. Data collected by IoT devices and sensors can now be effectively analyzed and contextualized using AI models. These two technological concepts absolutely complement each other, developing rapidly and have vital role in industry and society [8–11]. AI is the real driving force behind the full potential of IoT. However, in reality there are still many challenges. AI and IoT encompass different design principles, industry standards, and operate on heterogeneous computing platforms and network topologies. A successful AIoT environment requires standardization that includes interoperability, compatibility, reliability, and efficiency of new devices that can be self-configured and self-adapted [12–19].

Mobile network systems have evolved from communication infrastructure to critical and necessary industrial and social infrastructure. The concept of ICon intelligent connections is based on the combination of the new 5th-generation (5G) networks and AI, in order to accelerate technological development and digital transformation. With the continuous development, new opportunities and challenges arise. The technical success of 5G depends on providing a wide range of data rates for much wider set of devices and users. The capacity of the 5G network has been greatly increased and supports the connection for huge number of IoT devices. Increased data rates, reduced end-to-end (E2E) latency, improved coverage, as well as network slicing (NS) and mobile edge computing (MEC) support even the most demanding and sophisticated industrial IoT (IIoT) applications. The convergence of 5G and AI enables the development of flexible and adaptive intelligent networks that support scalability by using NS techniques and orchestration, decentralized intelligence by moving AI computing capabilities from central nodes to the edge, improved operational efficiency by reducing human intervention and simpler technological complexity, as well as improved network security by big data (BD) analysis. The convergence is expected to take center stage in key vertical domains: industrial manufacturing, transportation and logistics, healthcare, security, and the entertainment industry. However, there are numerous 5G-AIoT challenges that need to be addressed in the near future. For AIoT systems with a high level of complexity, it is possible to apply the concept of digital twins (DT) to create a digital representation of physical entities. The DT solution has been developed over the years under different names such as virtual space, digital mirror, digital copy, and then the term digital twins. The 5G systems and the DT support each other: the new use cases are based on DT, and on the other hand, DT allows us to better understand how 5G fits into our connectivity ecosystem.

The major sections of this chapter introduce 5G, IoT and AI, and smart connectivity for critical IoT. In the first part, the key characteristics of co-design of 5G and AI as well as ICon intelligent connectivity in complex use cases are presented. In the second part, IIoT framework as well as DT models and application scenarios are outlined. We conclude with remarks on opportunity and challenges of 5G-AIoT.

6.2 5G FOR SMART IoT CONNECTIVITY

The past few years have seen the rapid growth and convergence of mobile networks, devices, and applications. AI and IoT have advanced with a new innovative combination called artificial intelligence of things (AIoT). The 5G-AIoT paradigm represents the convergence of networking, intelligence, and things (Figure 6.1). The new intelligent network, based on 5G communication, is designed to connect the sensing regions (sensors) and the processing center (cloud). Cloud computing (CC) is the concept of sharing the resources of remote servers located on the Internet for data storage, management, and processing. Data is stored on the Internet platform instead of on local devices. The 5G-AIoT is a new field of research into the various capabilities of mobile networks and AI algorithms [20–28].

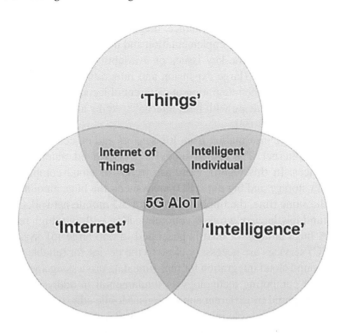

FIGURE 6.1 5G-AIoT convergence of networking, intelligence, and smart objects.

The IoT ecosystem is highly complex, fragmented, and rapidly evolving. The IoT ecosystem is not fully developed in terms of technology and application. It is in the early stage of implementation, and different applications have been developed for different services. Internet facilities have a huge potential for the transformation of industry, transport, healthcare, smart home, and smart city. However, there are several key challenges such as standardization and interoperability. IoT supports various system functions, such as detection, identification, activation, communication, and management. Before moving to the application layer, it is necessary to store the data and process it efficiently. AI and machine learning (ML) algorithms and analytics are essential for the development of AIoT systems. There are three main components in IoT: data collection, communication, and computing infrastructure and analytics. In the implementation and development of IoT analytics, there are the following life-cycle stages. The acquisition unit contains smart devices, sensors, production systems, and other objects, from which it collects raw data. It is necessary to check data integrity, accuracy, and consistency, and prepare for transmission. The Internet is the backbone of the communication infrastructure. It connects the acquisition to the computerized data center (DC), which processes and analyzes the data. The processed data is delivered to smart devices in real time. Therefore, these three components also contain numerous challenges.

A typical IoT architecture connects objects to the network at any time via a communication infrastructure. The IoT has great potential for the development of smart applications and as a key provider activates new business models based on the IoT services. The deployed applications can be classified into four groups: monitoring (control devices, environmental status, notifications, alerts), control (device function control), optimization (device performance, diagnostics, repair, etc.), and autonomy (autonomous operations). In the implementation and development of IoT applications it is necessary to consider the key issues of availability, management, reliability, interoperability, scalability (large expansion and integration), security, and privacy. The IoT network enables the integration of different objects with embedded technology, so that they communicate with both internal systems and the external environment at the same time [29–33].

Applications utilize information from IoT sensors and objects that can also communicate on the Internet. The system supports personalized, automated, and more intelligent services. In this regard, there are trends of cloud computing (CC) for large-scale data storage and big data (BD) analysis on the huge amount of collected IoT data. At the same time, the implementation of 5G mobile networks enables new speed ranges and low-latency wireless communication with ultrahigh reliability and availability. A huge amount of data is generated in one large IoT system, so edge computing (EC) services are necessary close to the device for reliable local unloading and background cloud integration for real-time data processing and content localization. EC edge computing technology is fundamental in addressing the various network demands posed by different applications. Mobile edge computing (MEC) as platform plays a significant role in this architecture. The three core functionalities it enables are real-time and low-latency analysis, local content/caching, close collaboration, and complementarity with CC. EC is suitable for local, real-time, short-term data processing and analysis, while CC focuses on global, long-term, non-real-time processing and analysis of BD.

The deep integration of IoT technology with BD and AI enables solutions for many platforms. AI provides algorithms that analyze data collected from IoT devices, identify different modes of operation, and make intelligent predictions and decisions. When using AI-based applications, any added value comes from the collected data. BD analytics can then be used to create predictions, gain insights, and trigger actions using defined workflows. In other words, information becomes action in AIoT.

The 5G network technology is becoming the key infrastructure for intelligent business and social transformation. The development of the digital economy is promoted through the industrial Internet and intelligent upgrades of traditional industries. Despite supporting innovative solutions, there are various challenges such as integration, connectivity, scalability, adaptability as well as economic issues. The emergence of new technologies has forced competitors to adapt in order to maintain relevant position in the market. As technology continues to grow, there are both best practices and risks that are associated with it [34–41].

6.2.1 CO-DESING OF AI AND 5G NETWORK TECHNOLOGY

The focus of mobile system development is communication aspects, while other services were introduced in the final design phase with low priority. The consequence is that the achieved performance is not optimal or the system capabilities are not fully utilized. In the 5G systems, the co-design of AI and IoT technologies is not only desirable, but also critical to achieve high performance of future network services [42–45].

AI is not a new concept. Over the last few decades, there was several cycles of development, alternating with phases of disappointment and funding cuts (*AI Winter*). Today's huge investment in AI has significantly fueled the progress and many practical applications that are now being implemented. Many modern AI methods are based on advanced statistical methods. Traditional AI research problems include reasoning, knowledge representation, planning, learning, natural language processing, perception, and the ability to move and manipulate objects. Probably the most relevant method of AI at the moment is ML. Machine learning refers to a set of algorithms that automatically improves through experience and data usage. Within ML, a significant category is deep learning (DL), which uses multilayer neural networks (DNNs) and convolutional neural networks (CNNs). The three most common ML methods are supervised (SL), unsupervised (USL), and reinforcement learning (RFL).

- Supervised learning (SL) method relies on manually labeled data samples, which are used to train the model so that it can then be applied to similar but new and unlabeled data. The two basic types of supervised models are regression and classification. Some widely used examples of supervised ML algorithms include linear regression (regression problems), random forest (classification and regression problems), and support vector machines (classification problems).
- Unsupervised (USL) method attempts to automatically discover structures and patterns from unlabeled data. The main goal is to discover previously unknown patterns in the data. Unsupervised ML is used when there is no data on the desired outcomes. Typical applications of USL include clustering (automatically divide a data set into groups based on similarity), anomaly detection (automatically detect unusual data points in a data set), association mining (identify clusters items that frequently occur together in a data set), and latent variable models (data pre-processing, reducing the number of features in a data set – dimensionality reduction).
- Reinforcement learning (RFL) combines trial-and-error approach with rewards or penalties. An agent learns to achieve its goals in an uncertain, potentially complex environment. The goal of the agent is to maximize the total reward for the actions it performs in the simulation. One of the fundamental challenges in RFL is creating an appropriate simulation environment. For example, an RFL environment for training autonomous driving algorithms should realistically simulate road situations. The advantage is that it is usually much more convenient to train model in simulated environment, rather than risk damaging real physical objects by using model in development. However, challenge is then to transfer the model out of the training environment and into the real world.

A specialized area within ML is artificial neural networks (ANNs), ambiguously inspired by the neural networks that constitute biological brains. An ANN is represented by a collection of connected nodes called neurons. Each connection can transmit signals to other neurons similar to synapses in the human brain. The receiving neuron processes the incoming signal and then signals to other connected neurons.

Signals are numbers, which are computed using statistical functions. Neuron connections are weighted, increasing or decreasing signal strength. Weights can be iteratively adjusted in learning process. Neurons are grouped into layers, where different layers perform different transformations on their input signals. Signals propagate through these layers, potentially multiple times. The adjective deep in DL refers to the use of multiple layers in these networks. A popular implementation of ANNs is convolutional neural networks (CNNs), which are often used for processing visual and other two-dimensional data. Another example is generative adversarial networks (GANs), where multiple networks compete with each other.

Understanding the basic categories of AIoT data and their corresponding AI methods is key to the success of R&D project. In ML, we need datasets for training and testing models. A dataset is a collection of data (set of files or a specific table in a database). The rows in the table correspond to the members of the data set, while each column of the table represents specific variable. The dataset is usually divided into training (approximately 60%), validation (approximately 20%), and testing datasets (approximately 20%). The training data set is used to train the model. Validation sets are used to select and tune the final ML model by evaluating the tuned model compared to other models. The test data set is used to evaluate how well the model is trained.

The use cases of AI technologies in the 5G ecosystem can be grouped into service management (operations support systems, resource provisioning, fault localization, fault root analysis, business support systems, security), network and cloud resource management (flexible feature deployment, network function virtualization (NFV) orchestration, network cutting, green operation), and radio link management (air interface coordination, site collaboration, user mobility).

In the first step, ML is applied to optimize and configure the network and devices. The mobile network, as well as the devices, contain hundreds or even thousands of operational parameters that can be configured. Algorithms for radio resource management are used, for example, for efficiently allocating radio resources, performing handovers, and assigning devices to different frequency bands. The use of ML has the potential to improve network configuration and optimize radio resource management algorithms, in particular when multiple algorithms interact. As a second step, 5G networks are being developed toward the wider application of ML. ML essentially depends on the availability of data for training and analysis. New 5G systems will provide improved reporting mechanisms to support machine learning-based algorithms.

In 2017 the International Telecommunications Union Radiocommunication sector (ITU-R) defined the fundamental framework for the work on 5G by the publication of the report Minimum requirements related to technical performance for IMT-2020 radio interfaces(s). It presents the uses cases and requirements associated with the so-called International Mobile Telecommunication-2020 (IMT-2020) system. IMT-2020 is what in layman's terms is referred to as 5G. IMT-2020 is intended to support three major categories of use cases, namely mMTC, cMTC, and eMBB. At the end of 2015, 3GPP has started studies to develop channel models for wireless networks operating above 6 GHz and to define 5G requirements. In R14 a study item for a new 5G New Radio (NR) air interface was concluded and R15 work item has been defined for NR specification. R15 will provide a first phase of an NR specification, which will be extended with a second phase in R16 that builds on studies on NR enhancements and significant extensions. The work items R17 approved in December 2019 lead to the introduction of new features for the three main use case families: enhanced mobile broadband (eMBB), URLLC, and massive machine-type communications (mMTC). The 3GPP has plans to publish the release at the end of the first quarter of 2022. The roadmap toward 5G-Advanced begins with R17. Meanwhile, the discussions on the scope of R18 are well underway. The 3GPP RAN standardization team began discussing the scope in June 2021 and aims for approval of the detailed scope by December 2021. 5G-Advanced early planning indicates that it will significantly evolve 5G in the areas of AI and extended XR reality. It improves the user experience in many ways by lowering latency, expanding bandwidth, and improving reliability and energy savings. We expect a final R18 standard by the end of 2023, so the first 5G-Advanced networks won't appear until 2025 [46].

In order to efficiently support such use cases that are in-between eMBB, URLLC, and mMTC, 3GPP has studied reduced-capability NR devices (NR-RedCap), previously known as NR-light and NR-lite, in R17. The RedCap study item has been completed in December 2020 and is continued as a work item. The use cases envisioned for RedCap include industrial wireless sensor network (IWSN), video surveillance cameras, and wearables (smart watches, rings, eHealth-related devices, medical monitoring devices, etc.). The following general requirements are common to all RedCap use cases: lower device cost and complexity as compared to high-end eMBB and URLLC devices of R15/R16, smaller device size or compact form factor, and support deployment in all FR1/FR2 bands for frequency division duplex (FDD) and time division duplex (TDD). In order to meet the above generic requirements, and more specifically the one on device complexity and device size, the following features have been considered in the RedCap study item: reduced number of user equipment (UE) receiver (Rx) and/or transmitter (Tx) branches, UE bandwidth reduction, half-duplex FDD, relaxed UE processing time, relaxed UE processing capability. The complexity reduction features that are expected to have the largest impact on coverage performance are reduced number of UE Rx/Tx branches and UE bandwidth reduction. The NR-RedCap UE is designed to have lower cost, lower complexity (e.g., reduced bandwidth and number of antennas), a longer battery life, and enable a smaller form factor than regular NR UEs. These devices support all frequency range bands (FR1 and FR2) for both FDD and TDD operations [47–49].

Next, as envisioned today, 6G mobile communication networks are expected to provide extreme peak data rates over 1 Tbps. The end-to-end delays will be imperceptible and lie even beneath 0.1 ms. 6G networks will provide access to powerful edge intelligence that has processing delays falling below 10 ns. Network availability and reliability are expected to go beyond 99.99999%. An extremely high connection density of over 10^7 devices/km^2 is expected to be supported to facilitate IoE. The spectrum efficiency of 6G will be over 5× than 5G, while support for extreme mobility up to 1,000 kmph is expected. Many new research works and projects are start ups that concentrated on developing technologies, use cases, applications, and standards. According to presented technology roadmap, collaborative efforts of industry and academia focus on 5G-Advanced evolution for the first five years, while setting the specifications of 6G. The first 6G testbeds can be expected to appear only post-2025 [48–56].

6.2.2 Intelligent Connectivity in Complex Use Cases

In the case of AIoT systems with high level of complexity, it is possible to apply the concept of digital twins (DT) to create digital representation of physical entities. The concept enables efficient management of complexity and the formation of semantic layer on top of technical layers. DT is a key enabler of data-driven decision-making, complex systems monitoring, product validation and simulation, and object life-cycle management. Relatively recent developments in 5G are undoubtedly innovative and based on the experiences of complex systems simulation [57–59].

DT is virtual representation of physical object or process in real time. Continuous upgrading and growing demands of DT technology generate new requirements and trends. In addition, DT can be divided into three categories based on different levels of integration, different degrees of data, and information flow that exist between the physical object and the digital copy (Figure 6.2):

- Digital model (DM) is digital version of an existing or planned physical object. There is no need for automatic data exchange between the physical model and the DM. DM examples are building plans, design, and product development. When a DM is created, a change made to the physical object has no effect on the DM in any way.
- Digital shadow (DS) is a digital representation of an object that contains direct one-way data flow between the physical and digital object. Changing the state of physical object causes a change in the digital object, but not vice versa.
- Digital twin (DT) establishes data flows between an existing physical object and a digital object, so that they are fully bidirectionally integrated. A change made to a physical object automatically leads to a change in the digital object, and vice versa.

Traditional three-dimensional (3D) DT model contains three components: physical entity in physical space, virtual entity in virtual space, and link between data and information that connects the physical and virtual entities. The new modeling

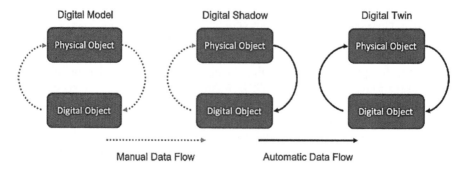

FIGURE 6.2 The concept of digital model (DM), digital shadow (DS), and digital twin (DT).

requirements extend the five-dimensional (5D) DT model by adding DT data and services.

- DT creates virtual models for physical entities digitally and simulates their behavior. The physical world is the basis of DT and includes devices, products, physical systems, activity processes, and organizations. DT conducts activities in accordance with physical laws in uncertain environments. Physical entities can be divided based on functions and structures, which are unit level, system level, and system of systems (SoS).
- Virtual models are faithful replicas of physical entities, so they reproduce physical geometries, properties, behaviors, and rules. The 3D geometric models describe the characteristics of a physical entity (shape, size, tolerance, and structural relation). Based on the physical properties, the model reflects the physical properties of the entity. The behavioral model describes state transition, performance degradation, and coordination, as well as entity response mechanisms to changes in the external environment. The rule model describes DT logical abilities such as reasoning, judgment, evaluation, and autonomous decision-making, following rules based on historical data or domain expert rules.
- DT data is multi-temporal, multidimensional, multisource, and heterogeneous, and is the fundamental driver. Individual data is generated by physical entities, including static attribute data and dynamic condition data. The data also generates virtual models as a result of the simulation. Some data is obtained from the service that describes the calling and execution of the service. Individual data represents the knowledge of experts in the field or generated from existing data. There is also fusion data, which is the result of the integration of all the above data.
- Service is an essential DT component. The model provides users with application services related to simulation, verification, monitoring, optimization, diagnosis, and prediction. In the process of building functional DT, a number of third-party services are needed, such as data services, knowledge services, and algorithm services. Finally, the functioning of DT requires the ongoing support of various service platforms, which can accommodate custom software development, model building, and service delivery.
- Digital representations are dynamically linked to their real-world counterparts, enabling advanced simulation, operations, and analysis. Connections between physical entities, virtual models, services, and data enable the exchange of information and data. The following DT relationships exist between physical entities and virtual models, between physical entities and data, between physical entities and services, between virtual models and data, between virtual models and services, between services and data. Connections enable the cooperation of four parts: physical entity, virtual entity, data, and services.

DT digital twins establish semantic layer on top multiple technical layers, thus supporting the realization of business goals and the implementation of AI/ML solutions.

A key advantage of the DT concept is managing complexity through abstraction. Especially for complex, heterogeneous collections of physical assets, the concept allows complexity to be managed through well-defined interfaces and relationships between different DT instances.

6.3 5G ARTIFICIAL INTELLIGENCE IoT

The 5G network provides a great opportunity for IoT growth with huge bandwidth, better coverage, and overall higher speeds compared to previous mobile networks. IoT is developing globally, so its ecosystems are built on key elements:

- collect data generated from the connections of devices and information
- connect heterogeneous devices and information
- cache involving stored information in the distributed IoT computing environment
- compute with advanced processing and computation of data and information
- cognize information analytics, insights, extractions, real-time AI processing
- create new interactions, services, experiences, business models, and solutions.

Next-generation AIoT applications embrace AI technology in a dynamic IoT environment. Localized IoT environments are formed by heterogeneous devices (edge computers as well as resource-constrained devices) that execute semi-autonomous IoT applications, which include functions for sensing, action, reasoning, and management functions. Next-generation applications require a paradigm shift from classic ML to distributed, low-latency, and reliable machine learning at the wireless network edge [60–63].

Intelligent IoT environments require an efficient, reliable, and secure communication infrastructure, by which resource-constrained IoT devices and more powerful edge assets must be efficiently managed. Computing and communication form a closed-loop system through which the infrastructure is integrally optimized. The infrastructure enables URLLC ultra-reliable low-latency communication through dynamic network management and heterogeneous network technologies (5G NR, NB-IoT). The wireless front-end is specially designed to support the communication requirements of advanced techniques.

AIoT applications are moving from the cloud to the network edge, so that computing is performed in the local environment of data producers and consumers. While IoT/edge devices can provide the computing side of the infrastructure, the communication side is driven by advanced networking technologies, such as 5G-NR extensions to private networks and industrial IoT. The solutions enable efficient use of spectrum in the downlink (DL) direction, with massive sensory feedback on the uplink (UL) link. Ultra-reliable, low-latency communication for industrial IoT requires 5G mmWave radio link (with comparable fiber-optic data rate, real-time reactivity, and massive sensing capacity).

EC enables services to leverage local devices by providing computing resources closer to the end nodes, thus enabling ultra-low latency and high data rate communication. At the same time, EC supports managing and limiting the propagation of sensitive data. Multiaccess edge computing is an European Telecommunications Standards Institute (ETSI) standard for 5G networks, among other things, to offload processing and data storage from mobile and IoT devices to the edge of mobile networks instead of transporting all data and computing to data centers (DCs). Fog computing (FC) is closely related to both EC and CC. FC mainly refers to the logical architecture that covers caching, data processing, and analytics that occur near the data source, such as improved performance at the network edges, reduced load on DC data centers and core networks, as well as improved resilience against networking problems [64–76].

Multiple committees, working groups, and standardization bodies around the world have been created. The most important ones are MEC industry specification group (ISG) within the ETSI, and MEC in 5G networks within the 3rd Generation Partnership Project (3GPP). Technical specification 3GPP TS 23.501, on the architecture for 5G systems, introduces new functional enablers for integration of MEC in 5G networks as an application function (AF). On the other hand, ETSI MEC introduces a reference architecture and technical requirements enabling efficient and seamless execution as well as interoperability and deployment of a wide range of EC scenarios that include IIoT. Important aspects such as latency, energy efficiency, system resource utilization, network throughput, and quality of service are constantly emphasized.

6.3.1 IIoT Framework

IIoT is a specific segment that enables high level of resilience, communication availability, security, precision, automation, and Industry 4.0 compatibility. There is a constant trend of the need for high production efficiency, increasing product customization, shortening the production cycle, and dynamic global supply chain. Industries are extremely incorporating IoT and key technologies. The combination of IoT and AI provides the basis for communication, management, and automated data acquisition, while data analysis processes expand data into meaningful information [77–82].

IIoT applies IoT approaches in an industrial context to support the digital transformation of sectors, such as manufacturing, process industry, energy distribution, and so forth. IIoT is part of critical system operation. The consequences of incorrect operation are, in extreme cases, may at the worst be production stops, power outage, and the like. As result, extremely high reliability and availability, resilience, as well as cybersecurity are prerequisites for successful IIoT adoption. Many IIoT use cases also contain real-time operation requirements [83–89].

Next-generation industrial automation adopts AI technologies, leading to 5G, AI, and IoT convergence with noticeable improvements in efficiency, modularity, management, automation, and usability of future smart factories in all layers of manufacturing processes. Real-time processing of collected data from industrial IoT devices continuously brings intelligence to 5G orchestration mechanisms. ML mechanisms are designed and applied to support various industrial processes, such as time-critical operations, mobile robotics, and predictive maintenance functionalities.

The first step in this direction is based on accurate profiling of IIoT applications in terms of resource usage, capacity constraints, and reliability characteristics. Profiling vertical industry applications in terms of resource consumption, resilience efficiency, ability to adapt to dynamic network conditions and failures, as well as identifying patterns in overall application behavior are extremely useful.

IIoT is a network of intelligent devices connected into systems that monitor, collect, exchange, and analyze data. IIoT applications, on the other hand, connect machines and devices in industries such as utilities and manufacturing. System failures and downtime in IIoT implementation can result in high-risk situations or life-threatening situations. IIoT applications are aimed at improving efficiency by improving security, in relation to the user-centric nature of IoT applications.

In the industrial IoT sector, many devices require extremely low-latency to achieve maximum operational efficiency, depending on the application. The digital transformation in the 5G vertical industry and the huge number of users promote the rapid development of Industry 4.0. Smart factories not only need low-latency and high stability, they need to adopt customized NS technology based on user needs. Smart manufacturing requires devices connected to the cloud via a network, a platform based on ultrahigh computing power, and real-time operation and control of the manufacturing process through BD and AI. Therefore, wireless communication networks with extremely low-latency and high reliability are necessary in the intelligent manufacturing process [90–95].

Factory equipment typically requires different networking based on use, purpose, and connected factories. 5G NS E2E technology enables different quality of services in the core network and flexible adjustment on demand. Solutions applicable in various production scenarios are supported, as well as real-time efficiency and low energy consumption.

- NS technology provides on-demand allocation of network resources to meet network demands in various production scenarios. NS capabilities combined with new technologies support flexible and dynamic resource allocation. In the process of creating network slices, it is necessary to schedule infrastructure resources, including reception, transmission, and CC resources.
- In addition to critical NS, 5G smart factories also require mobile broadband NS connections and massive NS connections. Different NSs share the same infrastructure within the management system planning and do not interfere with each other. Instead, they support the independence of their services.
- 5G optimizes network connections and offloads local traffic to meet low-latency requirements. The optimization of each NS for business needs is not only reflected in different network functions and characteristics, but also in flexible deployment schemes.
- In addition, it adopts distributed CC technology to deploy industrial applications and main network functions based on NFV technology in local or centralized data centers in a flexible manner.
- The high-bandwidth and low-latency characteristics of 5G networks have greatly improved the capabilities of intelligent processing by migrating to the cloud and directed the improvement of intelligence.

6.3.2 DT Models and Application Scenarios

The DT technology enables a simple and economical approach to 5G with highly flexible and repeatable development approaches. DT enables proactive modeling of data traffic and security risks for testing/validation purposes, promotes operational and energy efficiency, and accelerates research and time to market of new services [96].

However, DT occurrence in networks is not common, despite the obvious potential in the development and deployment of complex 5G systems. The key difference between DT and conventional simulation methods is the two-way data connection and updating process. 5G DT would handle how data is generated during network maintenance, operations, design, development, testing, and validation, and how it is routed and used at the destination object. It is important that the 5G DT architecture allows the virtual system to start with a simple form and then, using AI mechanisms, evolve to a more comprehensive model with high precision through data updating.

The coherence relations of different 5G DT components are shown in Figure 6.3. ML supports DT for continuous prototyping, testing, assuring, and self-optimizing a functional 5G network for different use cases. 5G has enabled cloud-native core and virtualized radio access network (Cloud-RAN) since the R15 standard. Thus, AI/cloud-based DT technology has the potential to accelerate processes.

DT usage is evolving from the manufacturing environment and later in the IoT and cyber–physical systems (CPS). It captures the properties, conditions, and behaviors of a real object through models and data. DT is a set of realistic models that can simulate the behavior of an object in the deployed environment. DT represents its physical twin and remains its virtual counterpart throughout the object's life cycle. DT has a dual role in IoT: implemented and recognized as the basic approach for creating IoT applications, and at the same time, DT is naturally associated with the ability to discover and activate IoT technologies.

DT objects can be deployed either on end devices, at the edge of the network, or in cloud. Depending on the application, the DT architecture can be divided into three categories: digital twin at the network edge, digital twin on the cloud, and edge-cloud-based collaborative digital twin. Edge-based twin objects are more suitable for applications with strict latency constraints (bulk URLLC) due to their location near the end devices. Cloud-based DTs are used for latency-tolerant and computationally demanding applications. In general, the available computing power in the cloud is greater than that at the edge of the network, but with higher latency and higher communication costs. Edge-cloud-based twins take advantage of both cloud-based twins (high computing power) and edge-based twins (instant analytics with low communication costs).

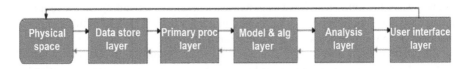

FIGURE 6.3 Digital twin for 5G.

TABLE 6.1
Comparison of Edge-, Cloud-, and Edge-Cloud-Based Twins

Metrics	Description	Edge-Based Twin	Cloud-Based Twin	Edge-Cloud-Based Twin
Scalability	Scalability refers to fulfilling latency requirements for massive number of 6G devices. Furthermore, the addition of new nodes should not significantly degrade the system performance in terms of latency.	High	Lowest	Low
Latency	This metric represents the overall delay that accounts for latency from service request until service provision in providing 6G services.	Low	High	Medium
Geo-distribution	This metric tells us about the geographical distribution of twin objects for enabling a 6G service.	Distributed	Centralized	Hybrid
Elasticity	This metric refers to on-demand dynamic resource allocation for digital twins operation in an elastic way in response to highly dynamic requirements.	High	Low	High
Context awareness	Context awareness is the function that deals with the knowledge about the end devices location and network traffic.	High	Low	Medium
Mobility support	Mobility support deals with the ability of digital twins to seamlessly serve mobile end devices.	High	Low	Medium
Twins' robustness (reliability)	Robustness refers to seamless operation of digital-twin-enabled 6G application in case of failure of twin objects.	Highest (for multiple edge-based twins)	Lowest	Medium

Various DT performance metrics are shown in Table 6.1. Considering scalability, DT at the edge of the network has the most value because it has the lowest latency compared to its cloud-based twins. It is possible to add endpoints in the case of twins at the edge of the network, up to their maximum serving limit without significantly increasing latency. However, the cloud twin has low scalability due to the increase in latency when the number of devices increases. Therefore, depending on the requirements of the 5G application, it is necessary to deploy the DT object on the appropriate location in the selected system architecture [97–100].

Since 2002, interest in the DT concept has grown exponentially, both in industry and academia. In recent years several standard development organizations (SDOs) have been working on standardizing the definition of DTs to facilitate common understanding, align stakeholder requirements and expectations, and improve clarity on the topic. There are various activities regarding standardization of DT, even if not directly termed as digital twin. IEC 62832 is a well-established standard, which defines a digital factory framework with the representation of digital factory assets in its center, although it is not called digital twin. ISO/IEC JTC1 provided a technology trend report by its joint advisory group on emerging technology and innovation (JETI). In the report, DT was identified as the number one area needing in-depth analysis, where JETI is also looking at how cooperation with the open source community can be established. SC 41 IoTs and DT subcommittees prepare standards for the IoT and has widened its scope and terms of reference to include DT, building on the exploratory work of JTC 1 Advisory Group. In addition to the working group WG 6 Digital Twin, SC 41 established an advisory group AG 27 related to the technology, which is expected to identify synergies with existing SC 41 activities and relationships as well as elaborate standardization strategy. The group will especially address life-cycle issues, standardization opportunities in virtual systems, devices, and sensors. The key standards to track from ISO/IEC include:

- ISO/IEC AWI 30172 Digital twin use cases are currently under development and at Stage 10.99 (Proposal) – now approved as a new project with a working draft under development.
- ISO/IEC AWI 30173 Digital twin concepts and terminology are currently under development and at Stage 20 (Preparatory) with a working draft already prepared, comments received, and approved for registration as a Committee Draft.
- ISO/FDIS AWI 23247 series defines a framework to support the creation of DTs of observable manufacturing elements including personnel, equipment, materials, manufacturing processes, facilities, environment, products, and supporting documents. Part 1 overview and general principles, Part 2 reference architecture, Part 3 Digital representation of manufacturing elements, and Part 4 Information exchange between entities within the reference architecture. This framework enables plug and play for twin elements, focusing mainly on the interfaces and functions of DTs.

In addition, in 2019, the IEEE Standards Association initiated a project IEEE P2806 that aims to define the system architecture of digital representation for physical objects in factory environments. The German Plattform Industrie 4.0 launched Asset Administration Shell as the implementation of the DT for smart manufacturing, IEC PAS 63088. This was deepened by partnerships between France, Italy, and Germany.

6.3.2.1 5G-AIoT Initiative
In the AIoT initiative, the DT concept has been assigned a significant role in providing a semantic abstraction layer. The IoT provides a connectivity service. AI is an important tool for the reconstruction process in which a virtual representation is created based on the input data from the sensors. After the DT is reconstructed, another AI algorithm is applied on the semantically rich DT representation [101–106].

DT is a very good synergy example of IoT, BD, and AI. A DT can be defined as a virtual representation of either physical objects, workflows, or systems in general. The technology is enabled by a huge number of IoT sensors. DT is a virtual representation of a physical asset, machine, vehicle, or device on an IoT platform. The DT represents the data, processes, operating states, and life cycle of the asset. The DT design requires inputs from massive sensors that collect relevant characteristics (in the form of BD) of the physical twin. In this case, AI is an effective tool to discover the underlying characteristics of a physical entity, offer recommendations and insight to performance validation. AI also effectively reacts to dynamic DT contexts and enables real-time improvements.

It is necessary for the DT system to continuously learn and change its mode of operation based on inputs and updates using methods such as AI and ML. Dynamic DT models enable the projection of technical objects into the digital platform.

DT technology includes the creation of virtual simulation models of technical and physical objects that are maintained and changed by information inside the physical object. The models are quite dynamic and accept real-time information readings from sensors of physical objects, sensors of control devices, and the environment. Models also include historical information to adjust the parameters of the physical asset. It is necessary that the models are dynamically updated in the DT concept in order to function correctly. The systems update measurements in real time and are regularly refreshed with new and old data using ML algorithms. The goal of the system is to achieve optimal operational capabilities of the physical facility through regular and dynamic updates.

DT integrates seamlessly into IoT and data analytics by connecting the physical and virtual twin. The data collected by IoT sensors is huge. Due to the transfer of large data between virtual and physical assets in DT, it is necessary to implement a high-performance network. The combination of 5G networks, IoT, and AI enables improved services and new technology applications. The DT applies analytical algorithms to manage this data. The IoT is particularly dependent on AI to help protect facilities, support analytics, reduce fraud, and enable automated data analysis decisions.

An obvious DT synergy of AI and IoT technologies is emerging, which also results in common challenges. The first step in facing challenges is to identify them. Some of the challenges in the field of data analytics are general IT infrastructure, data quality, privacy and security, trust, and expectations. Challenges in the IoT area are infrastructure, connectivity, data, privacy, security and trust, and expectations. The next challenges within all forms of DT development are caused by the lack of a standardized modeling approach. A standard approach from DT design initiation to simulation, whether physics-based or design-based, is essential. Standardized approaches ensure domain and user understanding, while ensuring the flow of information within each phase of DT development and implementation. Another challenge is to transfer information related to each of the development and functional stages of the modeling of a DT. Domain compatibility is ensured, allowing successful use of DT in the future.

6.3.2.2 5G Network Digital Twin

The adoption of digital twins to support network operation and maintenance, as well as network planning and design, is the first step toward 5G DT. The new approach to testing and provisioning provides an emulated, software-based replica of a 5G physical network that enables continuous prototyping, testing, provisioning, and self-optimization of the living network [102–107].

A simulation behaves similar to the system being modeled and gives us insight into how something works. Emulation behaves exactly like a system being modeled. It follows all the rules of the system being emulated, creating an exact, always-updated replica that functions in a completely identical manner and exhibits features and results accurately. The ability to continuously monitor and deliver reliable results, emulation provides significant advantages (simplicity, cost-effectiveness, repeatability, and predictability) in testing complicated real-world network conditions. The benefits further add to greater flexibility, comfort, confidence, cost savings, and a momentum for research and development.

Simulation may not be sufficient to operate and manage the complex 5G system, real network emulation is necessary, a task well suited for DT. Emulation of the core network, base stations (gNodeB), and channels (interference evaluation) is suggested. Each of these system components can be connected to the DT. Examples of specific use cases are a private 5G network in a factory, NS operations and management, and 5G-based V2X communication between vehicles and infrastructure for virtual vehicle testing.

Emulators are used to test the performance of a functional network, as well as network functions and services that are too remote, complex, and expensive to easily configure and access. A software-emulated replica of the 5G physical network is DT and enables continuous prototyping, testing, assuring, and self-optimizing a living network.

6.3.2.3 Manufacturing Digital Twin

DT is one of the most promising technologies driving digitization in the industry. DT is a digital replica or model of any physical object (physical twin). What differentiates DT from digital or CAD models is the automatic two-way exchange of data between the digital and physical twins in real time. The benefits of implementing DT in the sector include reduced operating costs and time, increased productivity, better decision-making, improved predictive/preventive maintenance, and so forth. DT technology finds its application in basic industries undergoing digital transformation: aerospace, manufacturing, healthcare, energy, automotive, public sector (education), mining, maritime, and agriculture [108–110].

Manufacturing is widely seen as the leading vertical for DT. Digital transformation is shaping Industry 4.0 with new technologies such as AI, IoT, and EC. New technologies are capable of optimizing processes and creating value, while 5G is becoming key to connectivity and digital transformation. It includes a range of capabilities to transform industry use cases: low latency, high bandwidth, high capacity,

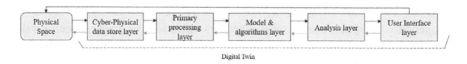

FIGURE 6.4 5D cross-industry DT application model.

strong reliability, advanced mobility, and longer battery life. In the context of DT, 5G is critical for use cases that depend on data being transported from one location to another quickly, securely, and reliably.

Digitization in manufacturing opens new options for significant improvements in the productivity and effectiveness of complex systems. The integration of cutting-edge technologies such as IoT, ML, data mining, and BD enable the smart production of Industry 4.0. A significant feature of smart manufacturing is the interaction of both physical and cyber systems. The CPS consists of a set of physical devices and products that communicate in virtual cyberspace through a communication network. DT technology is rapidly developing and Industry 4.0 has been marked as the beginning of a new era of production, leading to a complete transformation of existing production systems and their management. Industry 4.0 brings together technologies that remove the boundaries of the physical and digital realms. Each physical object is represented by a digital prototype that extends data, information, and knowledge about the physical object.

Uncertainty, imperfection, and unknown information are inherent in production processes, so the DT approach is a more suitable method for optimizing the entire process than simulation. An accurate DT model improves safety, reduces costs, accelerates manufacturing of new products, and implements new processes. Much research describes the DT on the basis of its three components: the physical model, the collected data and information, and the virtual model. Comprehensive, accurate, and extensive virtual model is important for a variety of fields of manufacturing and production.

The five-dimensional structure of DT is shown in Figure 6.4. The basic data storage layer contains collected data from the physical model, historical data, and data from the enterprise's production systems. In the next primary processing layer, appropriate processes on the collected data are executed. The primary processing layer performs operations such as data-to-information conversion to enable the flow of information between different levels of production. Mathematical, statistical, and CAD models are stored in the models and algorithms layer. The fourth layer is the analysis layer for prediction, optimization, reconfiguration, monitoring, and management. The last layer is the visualization and user interface layer, which provides users with a graphical interface to access the DT.

6.4 CONCLUDING REMARKS

The IoT ecosystem is highly complex, fragmented, and rapidly evolving. AI is the real driving force behind the IoT's full potential. However, there are several key challenges such as standardization and interoperability. The 5G network is emerging as an excellent opportunity for further IoT growth, with massive bandwidth, better coverage, and overall higher speeds. The convergence of 5G-IoT and AI represents a breakthrough in the evolution of these technologies. As technology continues to grow, there are both opportunities and challenges that are associated with it. AI and IoT encompass different design principles, industry standards, and developments. They function on heterogeneous computer platforms and network topologies. A successful AIoT environment requires standardization that supports the interoperability, compatibility, reliability, and efficiency of new devices. The technical success of 5G depends on bringing a wider range of data rates to a much wider set of devices and users. 5G-AIoT is expected to play a fundamental role in key vertical industry domains.

For AIoT systems with a high level of complexity, it is justified to apply the concept of digital twin DT to create a digital representation of physical entities. A key advantage of DT is managing complexity through abstraction. DT and 5G support each other in a virtuous circle – leverage digital twins to build the use case for 5G and understand how 5G fits within your connectivity ecosystem and existing connectivity solutions.

An obvious DT synergy of AI and IoT technologies is emerging, which also results in common challenges. The first step in facing challenges is to identify them. The numerous challenges within all forms of DT development are caused by the lack of a standardized modeling approach. The standardization process ensures domain and user understanding, development, and implementation, enabling successful DT use in the future.

REFERENCES

1. Y. Wu, H. Huang, C.-X. Wang, Y. Pan, *5G-Enabled Internet of Things*, CRC Press, 2019.
2. A.M. Al-Sartawi, A. Razzaque, M.M. Kamal (Eds.), *Artificial Intelligence Systems and the Internet of Things in the Digital Era*, Springer, 2021.
3. K.G. Manoharan, J.A. Nehru, S, Balasubramanian, *Artificial Intelligence and IoT*, Springer, 2021.
4. R. Ashfaq, Study and analysis of 5G enabling technologies, their feasibility and the development of the Internet of Things (Chapter 5). In *Intelligence of Things: AI-IoT Based Critical-Applications and Innovations* (Eds. F. Al-Turjman, A. Nayyar, A. Devi, P.K. Shukla), Springer, 2021.
5. S. Mathur, et al., AIIoT: Emerging IoT with AI technologies (Chapter 15). In *A Fusion of Artificial Intelligence and Internet of Things for Emerging Cyber Systems* (Eds. P. Kumar, et al.), Springer, 2022.
6. N.P. Gupta, M.A. Alam, Challenges in the adaptation of IoT technology. (Chapter 19). In *A Fusion of Artificial Intelligence and Internet of Things for Emerging Cyber Systems* (Eds. P. Kumar, et al.), Springer, 2022.
7. A. Yarali, Artificial Intelligence, Big Data analytics, and IoT. (Chapter 11). In *Intelligent Connectivity AI, IoT, and 5G*, Wiley, 2022.
8. J.P. Lemayian, F. Al-Turjman, Intelligent IoT communication in smart environments: An overview. (Chapter 10). In *Artificial Intelligence in IoT* (Ed. F. Al-Turjman), Springer, 2019.

9. A. Bröring, V. Kulkarni, A. Zirkler, P. Buschmann, K. Fysarakis, M. Konstantinos, S. Mayer, B. Soret L.D. Nguyen, P. Popovski, S. Samarakoon, M. Bennis, J. Härri, M. Rooker, G. Fritz, A. Bucur, G. Spanoudakis, S. Ioannidis, IntelIIoT: Intelligent IoT environments. In *Proceedings of the Global IoT Summit (GIoTS)*, pp. 1–14, 2022.
10. D. Wang, et al., From IoT to 5G I-IoT: The next generation IoT-based intelligent algorithms and 5G technologies. *IEEE Communications Magazine*, vol. 56, no. 10, pp. 114–120, Oct. 2018.
11. N. Javaid, A. Sher, H. Nasir, N. Guizani, Intelligence in IoT-based 5G networks: Opportunities and challenges. *IEEE Communications Magazine*, vol. 56, no. 10, pp. 94–100, Oct. 2018.
12. H. Shariatmadari, et al., Machine-type communications: Current status and future perspectives toward 5G systems. *IEEE Communications Magazine*, vol. 53, no. 9, pp. 10–17, 2015.
13. D. Milovanovic, V. Pantovic, G. Gardasevic, Converging technologies for the IoT: Standardization activities and frameworks (Chapter 3). In *Emerging Trends and Applications of the Internet of Things* (Eds. P. Kocovic, R. Behringer, M. Ramachandran, R. Mihajlovic), pp. 71–103, IGI Global, 2017.
14. D. Milovanovic, V. Pantovic, Interoperability in Internet of Media Things and integration Big Media: Conceptual model and frameworks (Chapter 4). In *Emerging Trends in IoT and Integration with Data Science, Cloud Computing, and Big Data Analytics* (Ed. P.Y. Taşer), pp. 59–79, IGI Global, 2021.
15. Y. Wu, et al., Convergence and interoperability for the energy Internet: From ubiquitous connection to distributed automation. *IEEE Industrial Electronics Magazine*, vol. 14, no. 4, pp. 91–105, Dec. 2020.
16. C.E. Palau, et al. (Eds.), *Interoperability of Heterogeneous IoT Platforms: A Layered Approach*, Springer, 2021.
17. M. Roy, P. Kar, S. Datta (Eds.), *Interoperability in IoT for Smart Systems*, CRC Press, 2021.
18. M. Noura, M. Atiquzzaman, M. Gaedke, Interoperability in Internet of Things: Taxonomies and open challenges. *Mobile Networks and Applications*, pp. 1–14, 2018.
19. Z. Alansari, et al., Challenges of Internet of Things and Big Data integration. In *Proceedings of the International Conference (iCETiC)*, 2018.
20. D. Milovanovic, Z. Bojkovic, D. Kukolj, Machine learning in 5G multimedia communications: Open research challenges and applications (Chapter 17). In *Handbook of Research on Emerging Trends and Application of Machine Learning* (Eds. A. Solanki, S. Kumar, A. Nayyar), pp. 361–382, IGI Global, 2020.
21. J. Ding, M. Nemati, C. Ranaweera, J. Choi, IoT connectivity technologies and applications: A survey. *IEEE Access*, vol. 8, pp. 67646–67673, 2020.
22. K. Shafique, B.A. Khawaja, F. Sabir, S. Qazi, M. Mustaqim, Internet of things (IoT) for next-generation smart systems: A review of current challenges, future trends and prospects for emerging 5G-IoT scenarios. *IEEE Access*, vol. 8, pp. 23022–23040, 2020.
23. T.Q. Duong, V.-P. Hoang, C.-K. Pham (Eds.), Convergence of 5G technologies, artificial intelligence and cybersecurity of networked societies for the cities of tomorrow. *Mobile Networks and Applications*, vol. 26, pp. 1747–1749, 2021.
24. L.J. Poncha, et al., 5G in a convergent Internet of Things era: An overview. In *Proceedings of the IEEE International Conference on Communications Workshops (ICC Workshops)*, pp. 1–6, 2018.
25. J. Prieto, R.C. Vara (Eds.), The convergence of 5G and IoT in a smart city context. *MDPI Smart Cities*, Sept. 2022.
26. S.K. Sharma, X. Wang, Towards massive machine type communications in ultra-dense cellular IoT networks: Current issues and machine learning-assisted solutions. *IEEE Communications Surveys & Tutorials*, vol. 22, no. 1, pp. 426–471, 2020.

27. F. Guo, et al., Enabling massive IoT toward 6G: A comprehensive survey. *IEEE Internet of Things Journal*, vol. 8, no. 15, pp. 11891–11915, 2021.
28. S. Verma, S. Kaur, M.A. Khan, P.S. Sehde, Towards green communication in 6G-enabled massive Internet of Things. *IEEE Internet of Things Journal*, vol. 8, no. 7, pp. 5408–5415, 2021.
29. H. Tataria, M. Shafi, A.F. Molisch, M. Dohler, H. Sjöland, F. Tufvesson, 6G wireless systems: Vision, requirements, challenges, insights, and opportunities. *Proceedings of the IEEE*, vol. 109, no. 7, pp. 1166–1199, Jul. 2021.
30. W. Jiang, et al., The road towards 6G: A comprehensive survey. *IEEE Open Journal of the Communication Society*, vol. 2, pp. 334–366, 2021.
31. R. Bassoli, et al., Why do we need 6G? *ITU Journal on Future and Evolving Technologies*, vol. 2, no. 6, pp. 1–31, Sept. 2021.
32. C. DeAlwis, et al., Survey on 6G frontiers: Trends, applications, requirements, technologies and future research. *IEEE Open Journal of the Communications Society*, vol. 2, pp. 836–886, 2021.
33. J.H. Kim, 6G and Internet of Things: A survey. *Journal of Management Analytics*, vol. 8, no. 2, pp. 316–332, 2021.
34. Y. Chen, P. Zhu, G. He, X. Yan X, H. Baligh, J. Wu, From connected people, connected things, to connected intelligence. In *Proceedings of the 2nd 6G Wireless Summit*, pp. 1–7, 2020.
35. F. Tariq, M. Khandaker, K.-K. Wong, M. Imran, M. Bennis, M.A. Debbah, Speculative study on 6G. *IEEE Wireless Communication*, vol. 27, no. 4, pp. 118–125, Aug. 2020.
36. K. David, H. Berndt, 6G vision and requirements: Is there any need for beyond 5G? *IEEE Vehicular Technology Magazine*, vol. 13, no. 3, pp. 72–80, 2018.
37. K.B. Letaief, W. Chen, Y. Shi, J. Zhang, Y.-J.A. Zhang, The roadmap to 6G: AI empowered wireless networks. *IEEE Communications Magazine*, vol. 57, pp. 84–90, 2019.
38. N.H. Mahmood, et al., Machine type communications: Key drivers and enablers towards the 6G era. *EURASIP Journal on Wireless Communications and Networking*, no. 1, Art. 134, 2021.
39. J.R. Bhat, S.A. Alqahtani, 6G ecosystem: Current status and future perspective. *IEEE Access*, vol. 9, pp. 43134–43167, 2021.
40. A. Dogra, et al., A survey on beyond 5G network with the advent of 6G: Architecture and emerging technologies. *IEEE Access*, vol. 9, pp. 67512–67547, 2021.
41. P. Porambage, G. Gür, D.P.M. Osorio, M. Liyanage, A. Gurtov, M. Ylianttila, The roadmap to 6G security and privacy. *IEEE Open Journal of the Communications Society*, vol. 2, pp. 1094–1122, 2021.
42. X. You, C. Zhang, X. Tan, S. Jin, H. Wu, AI for 5G: Research directions and paradigms, *Science China Information Sciences*, vol. 62, pp. 1–13, Feb. 2019.
43. N. Janbi, I. Katib, A. Albeshri, R. Mehmood, Distributed artificial intelligence-as-a-service (DAIaaS) for smarter IoE and 6G environments. *MDPI Sensors*, vol. 20, 2020.
44. J.K. Nurminen, H. Mfula, A unified framework for 5G network management tools. In *Proceedings of the IEEE Conference on Service-Oriented Computing and Applications*, pp. 41–44, 2018.
45. G. Zhu, J. Zan, Y. Yang, X. Qi, A supervised learning based QoS assurance architecture for 5G networks. *IEEE Access*, vol. 7, pp. 43598–43606, 2019.
46. M. Agiwal, A, Roy, N. Saxena, Next generation 5G wireless networks: A comprehensive survey. *IEEE Communications Surveys & Tutorials*, vol. 18, no. 3, pp. 1617–1655, 2016.
47. S. Moloudi, et al., Coverage evaluation for 5G reduced capability new radio (NR-RedCap). *IEEE Access*, vol. 9, pp. 45055–45067, 2021.
48. 3GPP, *Revised SID on Study on Support of Reduced Capability NR Devices*. RP-201677, Jul. 2020.

49. 3GPP, *New WID on Support of Reduced Capability NR Devices*. RP-202933, Dec. 2020.
50. M. Suryanegara, A.S. Arifin, M. Asvial, The IoT-based transition strategy towards 5G. In *Proceedings of the International Conference on Big Data and Internet of Thing*, pp. 186–190, 2017.
51. A. Luntovskyy, L. Globa, Performance, reliability and scalability for IoT. In *Proceedings of the International Conference on Information and Digital Technologies*, pp. 316–321, 2019.
52. G. Su, M. Moh, Improving energy efficiency and scalability for IoT communications in 5G networks. In *Proceedings of the International Conference on Ubiquitous Information Management and Communication*, pp. 1–8, 2018.
53. D. Loghin, et al., The disruptions of 5G on data-driven technologies and applications. *IEEE Transactions on Knowledge Data Engineering*, vol. 32, pp. 1179–1198, 2020.
54. L.U. Khan, et al., Network slicing: Recent advances, taxonomy, requirements, and open research challenges. *IEEE Access*, vol. 8, pp. 36009–36028, 2020.
55. Q. Liu, T. Han, When network slicing meets deep reinforcement learning. In *Proceedings of the International Conference on Emerging Networking Experiments and Technologies*, pp. 29–30, 2019.
56. H.X. Nguyen, et al., Digital twin for 5G and beyond. *IEEE Communications Magazine*, vol. 59, no. 2, pp. 10–15, 2021.
57. 5G-ACIA, *Using Digital Twins to Integrate 5G into Production Networks*. 5G Alliance for Connected Industries and Automation [White paper], pp. 1–48, 2021.
58. L.U. Khan, W. Saad, D. Niyato, Z. Han, C.S. Hong, Digital-twin-enabled 6G: Vision, architectural trends, and future directions. *IEEE Communications Magazine*, vol. 60, no. 1, pp. 74–80, 2022.
59. D.C. Nguyen, et al., 6G Internet of Things: A comprehensive survey. *IEEE Internet of Things Journal* vol. 9, no. 1, pp. 359–383, 2022.
60. T.-W. Sung, P.-W. Tsai, T. Gaber, C.-Y. Lee (Eds.), Artificial Intelligence of Things (AIoT) technologies and applications. *Wireless Communications and Mobile Computing*, ID: 9781271, pp. 1–2, 2021.
61. J. Jagannatha, N. Poloskya, A. Jagannatha, F. Restuccia, T. Melodia, Machine learning for wireless communications in the Internet of Things: A comprehensive survey. *Ad Hoc Networks*, vol. 93, Oct. 2019.
62. M.S. Mahdavinejad, et al., Machine learning for Internet of Things data analysis: A survey. *Digital Communications and Networks*, vol. 4, no. 3, pp. 161–175, Aug. 2018.
63. A. Ghosh, D. Chakraborty, A. Law, Artificial intelligence in Internet of Things. *CAAI Transactions on Intelligence Technology*, pp. 208–218, Oct. 2018.
64. Y. Zhang, *Mobile Edge Computing*, Springer, 2022.
65. P. Mach, Z. Becvar, Mobile edge computing: A survey on architecture and computation offloading. *IEEE Communications Surveys & Tutorials*, vol. 19, no. 3, pp. 1628–1656, 2017.
66. X. Wang, et al., Convergence of edge computing and deep learning: A comprehensive survey. *IEEE Communications Surveys & Tutorials*, vol. 22, no. 2, pp. 869–904, 2020.
67. F. Wang, et al., Deep learning for edge computing applications: A state-of-the-art survey. *IEEE Access*, vol. 8, pp. 58322–58336, 2020.
68. M.T. Beck, M. Werner, S. Feld, S. Schimper, Mobile edge computing: A taxonomy. In *Proceedings of the International Conference on Advances in Future Internet*, Lisbon, Portugal, pp. 48–55, 2014.
69. S. Wang, et al., A survey on mobile edge networks: Convergence of computing, caching and communications. *IEEE Access*, vol. 5, pp. 6757–6779, 2017.
70. C. Mouradian, et al., A comprehensive survey on fog computing: State-of-the-art and research challenges. *IEEE Communications Surveys & Tutorials*, vol. 20, no. 1, pp. 416–464, 2018.

71. L. Gao, T.H. Luan, B. Liu, W. Zhou, S. Yu, Fog computing and its applications in 5G. In *5G Mobile Communications* (Eds. W. Xiang, K. Zheng, X. Shen), Springer, 2017.
72. U.M. Malik, M.A. Javed, S. Zeadally, S. Islam, Energy efficient fog computing for 6G enabled massive IoT: Recent trends and future opportunities. *IEEE Internet of Things Journal*, 2021 (Early access).
73. J. Mour, D. Hutchison, Fog computing systems: State of the art, research issues and future trends, with a focus on resilience. *Journal of Network and Computer Applications*, pp. 1–39, Jul. 2020.
74. A.H. Jafari, H.S. Shahhoseini, IoT integration with MEC. In *Mobile Edge Computing* (Eds. A. Mukherjee, D. De, S.K. Ghosh, R. Buyya), pp. 111–144, Springer, 2021.
75. G. Kaur, M. Moh, Cloud computing meets 5G networks: Efficient cache management in cloud radio access networks. In *Proceedings of the ACMSE*, pp. 1–8, 2018.
76. B. Liu, C. Liu, M. Peng, Resource allocation for energy-efficient MEC in NOMA-enabled massive IoT networks. *IEEE Journal on Selected Areas in Communications*, vol. 39, no. 4, pp. 1015–1027, 2021.
77. X. Wang, L. Gao, *When 5G Meets Industry 4.0*, Springer, 2020.
78. B.S. Khan, et al., URLLC and eMBB in 5G industrial IoT: A survey. *IEEE Open Journal of the Communications Society*, 2022 (Early access).
79. A. Mahmood, et al., Industrial IoT in 5G-and-beyond networks: Vision, architecture, and design trends. *IEEE Transactions on Industrial Informatics*, vol. 18, no. 6, pp. 4122–4137, Jun. 2022.
80. J. Prados-Garzon, P. Ameigeiras, J. Ordonez-Lucena, P. Munoz, O. Adamuz-Hinojosa, D. Camps-Mur, 5G non-public networks: Standardization, architectures and challenges. *IEEE Access*, vol. 9, pp. 153893–153908, Nov. 2021.
81. J. Cheng, W. Chen, F. Tao, C.-L. Lin, Industrial IoT in 5G environment towards smart manufacturing. *Journal of Industrial Information Integration*, vol. 10, pp. 1019, 2018.
82. L.P.I. Ledwaba, G.P. Hancke, Security challenges for industrial IoT. In *Wireless Networks and Industrial IoT: Applications, Challenges and Enablers* (Eds. N.H. Mahmood, N. Marchenko, M. Gidlund, P. Popovski), Springer, 2021.
83. R. Pethuru, T. Poongodi, B. Balusamy, M. Khari, *The Internet of Things and Big Data Analytics Integrated Platforms and Industry Use Cases*, CRC Press, 2020.
84. E. Harjula, A. Artemenko, S. Forsström, Edge computing for industrial IoT: Challenges and solutions. In *Wireless Networks and Industrial IoT: Applications, Challenges and Enablers* (Eds. N.H. Mahmood, N. Marchenko, M. Gidlund, P. Popovski), Springer, 2021.
85. Y. Wang, M. Nekovee, E.J. Khatib, R. Barco, Machine learning/AI as IoT enablers. In *Wireless Networks and Industrial IoT: Applications, Challenges and Enablers* (Eds. N.H. Mahmood, N. Marchenko, M. Gidlund, P. Popovski), Springer, 2021.
86. A. Zafeiropoulos, et al., Benchmarking and profiling 5G verticals applications: An industrial IoT use case. In *Proceedings of the IEEE Conference on Network Softwarization (NetSoft)*, pp. 1–9, 2020.
87. S.K. Rao, R. Prasad, Impact of 5G technologies on industry 4.0. *Wireless Personal Communications*, vol. 100, pp. 145–159, 2018.
88. P. Varga, et al., 5G support for industrial IoT applications – Challenges, solutions, and research gaps. *MDPI Sensors*, vol. 20, no. 828, pp. 1–43, 2020.
89. M. Singh, et al., Applications of digital twin across industries: A review. *MDPI Applied Sciences*, vol. 12, pp. 1–28, 2022.
90. A. Shahraki, M. Abbasi, A comprehensive survey on 6G networks: Applications, core services, enabling technologies, and future challenges. *IEEE Transactions on Network and Service Management*, Jun. 2021.
91. M. Alsharif, A.H. Kelechi, M. Albreem, S.A. Chaudhry, M.S. Zia, S. Kim, Sixth generation (6G) wireless networks: Vision, research activities, challenges and potential solutions. *MDPI Symmetry*, vol. 12, no. 4, pp. 1–21, 2020.

92. X. Yang, Z. Zho, B. Huang, URLLC key technologies and standardization for 6G power internet of things. *IEEE Communications Magazine*, vol. 5, no. 2, pp. 52–59, Jun. 2021.
93. A. Mukherjee, P. Goswami, M.A. Khan, L. Manman, L. Yang, P. Pillai, Energy efficient resource allocation strategy in massive IoT for industrial 6G applications, *IEEE Internet of Things Journal*, vol. 8, no. 7, pp. 5194–5201, 2021.
94. K. Pedersen, T. Kolding, Overview of 3GPP new radio industrial IoT solutions. In *Wireless Networks and Industrial IoT: Applications, Challenges and Enablers* (Eds. N.H. Mahmood, N. Marchenko, M. Gidlund, P. Popovski), Springer, 2021.
95. B. Ji, et al., A survey of computational intelligence for 6G: Key technologies, applications and trends. *IEEE Transactions on Industrial Informatics*, vol. 17, no. 10, pp. 7145–7154, 2021.
96. R. Minerva, G.M. Lee, N. Crespi, Digital twin in the IoT context: A survey on technical features, scenarios, and architectural models. *Proceedings of the IEEE*, vol. 108, no. 10, pp. 1785–1824, 2020.
97. R.S. Mendonça, et al., Digital twin applications: A survey of recent advances and challenges. *MDPI Processes*, vol. 10, pp. 1–12, 2022.
98. W. Kritzinger, et al., Digital twin in manufacturing: A categorical literature review and classification. *IFAC-PapersOnLine*, vol. 51, no. 11, pp. 1016–22, 2018.
99. Q. Qia, et al., Enabling technologies and tools for digital twin. *Journal of Manufacturing Systems*, vol. 58, Part B, pp. 3–21, Jan. 2021.
100. J. Jicheng, Z. Xitong, Five-dimension digital twin model and its ten applications. *Proceedings of the Computer Integrated Manufacturing Systems CIMS*, vol. 25, no. 1, pp. 1–18, 2019.
101. S.M. Bazaza, M. Lohtandera, J. Varis, 5-Dimensional definition for a manufacturing digital twin. *Procedia Manufacturing*, vol. 38, pp. 1705–1712, 2019.
102. S.R. Newryella, D.W. Franklin, S. Haider, 5-dimension cross-industry digital twin applications model and analysis of digital twin classification terms and models. *IEEE Access*, vol. 9, pp. 131306–131321, 2021.
103. M. Pernoa, L. Hvama, A. Haug, Implementation of digital twins in the process industry: A systematic literature review of enablers and barriers. *Computers in Industry*, vol. 134, Jan. 2022.
104. M. Singh, E. Fuenmayor, E.P. Hinchy, Y. Qiao, N. Murray, D. Devine, Digital twin: Origin to future. *MDPI Applied System Innovation*, vol. 4, no. 2, 2021.
105. D.M. Botín-Sanabria, et al., Digital twin technology challenges and applications: A comprehensive review. *MDPI Remote Sensing*, vol. 14, 2022.
106. T. Wenhu, et al., Technologies and applications of digital twin for developing smart energy systems. *Strategic Study of CAE*, 2020, vol. 22, no. 4, pp. 1–13.
107. R. Saracco, Digital twins: Bridging physical space and cyberspace. *IEEE Computer*, vol. 52, no. 12, pp. 58–64, 2019.
108. A. Rasheed, et al., Digital twin: Values, challenges and enablers from a modeling perspective. *IEEE Access*, vol. 8, pp. 21980–22012, 2020.
109. ISO/DIS, 23247-1, *Automation Systems and Integration: Digital Twin Framework for Manufacturing, Part 1 — Overview and General Principles*, 2020.
110. S. Malakuti, et al., *Digital Twins for Industrial Applications: Definition, Business Values, Design Aspects, Standards and Use Cases*. Industrial Internet Consortium [White paper], pp. 1–19, 2020.

7 Machine Learning-Based Scheduling in 5G/6G Communication Systems

M.I. Sheik Mamode and Tulsi Pawan Fowdur
University of Mauritius

CONTENTS

7.1	Overview of 5G/6G	278
7.2	Definition and Importance of Scheduling	278
7.3	Scheduling Schemes in 5G/6G	280
7.4	Machine Learning Techniques Used in 5G/6G	285
	7.4.1 Supervised Learning	285
	7.4.1.1 Classification	285
	7.4.1.2 Regression	287
	7.4.1.3 Decision Trees	288
	7.4.1.4 Forecasting	288
	7.4.2 Semi-Supervised Learning	288
	7.4.3 Unsupervised Learning	289
	7.4.3.1 K-Means Algorithm	289
	7.4.3.2 Hidden Markov Model	290
	7.4.3.3 Deep Learning	291
	7.4.4 Reinforcement Learning	291
7.5	Overview of Previous Works Using Machine Learning for Scheduling in 5G/6G Systems	292
	7.5.1 Reinforcement Learning-Based Scheduling for Data Traffic Management	293
	7.5.2 Reinforcement Learning for 5G Scheduling Parameter Optimization	295
	7.5.3 Knowledge-Assisted Deep Reinforcement Learning in 5G Scheduler Design	296
	7.5.4 Deep Reinforcement Learning for Radio Resource Scheduling in 5G MAC Layer	297
	7.5.5 Intelligent Resource Scheduling for 5G Radio Access Network Slicing	299
	7.5.6 Delay-Aware Cellular Traffic Scheduling with DRL	300
	7.5.7 A Fairness-Oriented Scheduler Using Multiagent RL	300
	7.5.8 Deep Learning-Based User Scheduling for Massive MIMO Downlink System	302

DOI: 10.1201/9781003205494-9

    7.5.9   Scheduling Based on DRL Model for UAV Video 302
    7.5.10 Reinforcement Learning Algorithms in Fairness-Oriented
           OFDMA Schedulers 304
7.6 Conclusion 305
References... 305

7.1 OVERVIEW OF 5G/6G

5G is the fifth generation of wireless networks. It is a combination of two radio technologies namely New Radio (NR) and Long-Term Evolution (LTE) [1]. In the early stages of 5G development, the 5G network was a non-standalone NR architecture, which was finalized in 2017. LTE also formed part of this architecture to enable initial access to the network as well as mobility handling. In 2018, the standalone architecture was completed and in 2019, the final version of 5G was released.

5G uses OFDM (orthogonal frequency-division multiplexing) technology as the modulation method and is able to provide enhanced flexibility and scalability compared to LTE. 5G technology caters for greater bandwidth by using a broader range of frequencies in both lower bands below 1 GHz to mid bands (from 1 to 6 GHz) and higher bands known as millimeter wave (mmWave). The peak data rate achievable by 5G is 20 Gbps and the average data rate is 100 Mbps [2].

5G is designed to support a 100-fold more traffic than LTE and has significantly low latency of 1 ms. It is used for three main types of services, namely massive IoT, mission-critical communications, and more importantly, enhanced broadband through new experiences such as virtual reality (VR) and augmented reality (AR) [2]. Works pertaining to 5G advanced Evolution have started in 2022 and the latter will build on works completed in Releases 15, 16, and 17.

Sixth-generation (6G) wireless technology is the successor to 5G technology and will enable the use of even greater frequencies compared to those used by 5G networks. The use of terahertz spectrum will increase capacity, improve spectrum sharing, and lower latency. One of the objectives of 6G is to support 1 µs latency as compared to 1 ms in 5G. The incorporation of artificial intelligence (AI) into the 6G infrastructure will enable huge improvements in areas of imaging, presence technology, and location awareness [3]. 6G will have the capacity to support ten times more devices per area than 5G, thus enabling the collection of a great amount of data from countless devices as well as its storage [4]. The peak data rate achievable by 6G is estimated to be around 1 Terabyte per second and it will support sub-mm Waves (wavelengths less than 1 mm) and frequency selectivity. 6G networks will have built-in Mobile Edge computing while currently it is a separate entity from the 5G networks. This will enable better access to AI technology and support for sophisticated mobile devices and systems. 6G is targeted for commercial launch in 2030.

7.2 DEFINITION AND IMPORTANCE OF SCHEDULING

In 5G, scheduling can be described as the method of allocating resources for transmission of data. There are a number of variables that can influence how and when resources are given to a certain user. Figure 7.1 depicts some factors considered for

Machine Learning-Based Scheduling in 5G/6G Systems 279

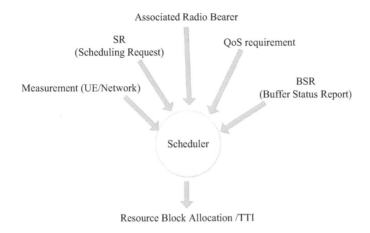

FIGURE 7.1   5G scheduler (5G/NR – scheduling) [5].

scheduling in 5G. 5G uses OFDM technology as the modulation method and is able to provide enhanced flexibility and scalability compared to LTE. 5G technology caters for greater bandwidth by using a broader range of frequencies in both lower bands below 1 GHz to mid bands (from 1 to 6 GHz) and higher bands known as millimeter wave (mmWave). The peak data rate achievable by 5G is 20 Gbps and the average data rate is 100 Mbps [2].

When allocating resources among UEs for the scheduler operation, each UE's QoS needs and those of the radio carriers it is connected to, are considered, together with the condition of the UE buffers. The scheduler's resource allocation may also be impacted by the radio circumstances at the UE, which are determined by measurements effected at the base station and/or reported by the UE. The distribution of radio resources is made using slots (e.g., one mini-slot, one slot, or several slots), and they consist of resource blocks. After a scheduling request, the UE will receive a scheduling channel and will subsequently learn the resources allotted. The reports for the uplink buffer status, which assess the data stored in the logical channel queues of the UE, are one of the measurements used to determine scheduler operation. They are used to facilitate QoS-aware packet scheduling. In addition, power-aware packet scheduling makes use of power headroom reports, which assess the disparity between the UE's peak transmit power and the approximated power for uplink transmission [5].

Frequency domain scheduling and time-domain scheduling are the two main scheduling categories used in 5G [6]. The fundamental time-frequency unit of resource, which can be employed for uplink or downlink transmission, is called a resource element. It can as well be defined a single subcarrier over a single OFDM symbol [7]. Twelve subcarriers that are continuous in frequency and span one-time slot constitute a resource block (RB), in which i is the smallest radio resource unit that can be allotted to an individual user. The four types of radio resource categories are radio frames, subframes, slots, and mini-slots. Ten subframes, each of duration 1 ms, make up the radio frame, which has a duration of 10 ms. One or more adjacent

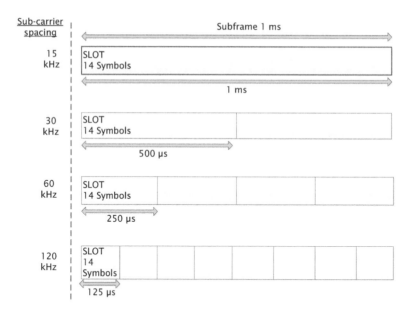

FIGURE 7.2 Structure of frame for 5G [8].

slots with 14 OFDM symbols make up each subframe. As illustrated in Figure 7.2, in Release 15, a mini-slot consists of 2, 4, and 7 OFDM symbols, and the length for one slot is dependent on the spacing of the subcarrier [8].

To accommodate the considerably high number of users and sophisticated functionalities, 5G has introduced numerous novel scheduling methods. Massive MIMO is one of the key features that 5G has introduced. It comprises utilizing a large number of antennas and terminals. Multi-User MIMO (MU MIMO) is essential for Massive MIMO [9]. The transmission of multiple data streams would cause interference with a single antenna, but in MU MIMO, the signals are sent over separate pathways, allowing the receiving antenna, with the appropriate encoding, to reconstruct the initial signal. One more aspect that has gained in viability is dynamic TDD, which mainly alters the frame configuration of the cell to adapt to the fluctuating traffic so as to maximize the throughput of the system. With the ability to adapt the pattern of TDD to the uplink or downlink transmission of a specific user, the use of dynamic TDD has turned out to be more feasible in small cell scenarios. The end-to-end latency of 5G was also designed to be ten times lower than that of LTE. This covers the frame size, round-trip time, HARQ processing time, transmit time interval, and intermittent reception [10].

7.3 SCHEDULING SCHEMES IN 5G/6G

Figure 7.3 depicts the block diagram of a 5G communication system.

A transport channel is included in the downlink shared channel (DL-SCH), which is used, among other things, to transmit user data. To obtain transport and control resources via the radio transmission link, control and data are encoded and decoded

Machine Learning-Based Scheduling in 5G/6G Systems 281

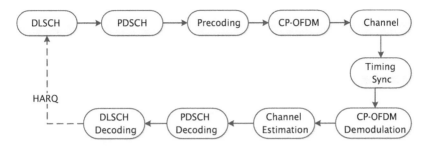

FIGURE 7.3 5G system model [11].

FIGURE 7.4 DL-SCH processing blocks [12].

to and from the media access control (MAC) layer [11]. Error detection, error correction, rate matching, interleaving, and transport channel, or control information mapping onto/splitting from physical channels are the building components that make up a channel coding system.

Figure 7.4 illustrates the building blocks of the DL-SCH [12].

Errors on transport blocks are found using the cyclic redundancy check (CRC). The block of transport is taken into account when generating the CRC parity bits, and the results are then connected to the transport block. The type of channel coding used determines how code blocks are segmented. More often, LPDC coding is utilized with code rate $R:D = K/R$ where the parity check matrix H uses $D-K$ to encode the K filler bits, where D denotes the quantity of encoded bits and $D-K$ is the amount of parity check bits.

Each coded bit stream identifies the rate matching for LDPC transport channels. The outputs from the rate matching block are consecutively concatenated in order to perform code block concatenation. The DL-SCH coded data is transported over the physical downlink shared channel (PDSCH). Figure 7.5 illustrates the building blocks that the PDSCH is comprised of [13].

The following procedures describe the simplified steps:

- Scrambling of the coded bits is performed for transmission
- Complex-valued modulation symbols are generated by modulating the output from the scrambling block
- The output of the modulation phase is mapped onto a single or several layers of transmission.

FIGURE 7.5 PDSCH processing blocks [13].

The precoding block in Figure 7.3 is then used to precode the modulation symbols before transmission to the antenna ports. The symbols of modulation are assigned to resource elements at each antenna port. The CP-OFDM (Cyclic Prefix-OFDM) generates complex-valued time-domain OFDM signals for each antenna port. After that, a channel is used to send the CP-OFDM block's output. To compare the received waveform at the timing sync block, the PDSCH demodulation reference signal (DM-RS) is utilized. OFDM demodulation is then performed on the synchronized signal [11].

The PDSCH DM-RS is used at the channel estimation block. The precoding matrix for the subsequent transmission is produced using singular value decomposition (SVD). By averaging the channel conditions, a singular matrix is produced for the entire assignment. Hence, there was a possibility for the precoding matrix to be less accurate for higher bandwidths assigned in cases where there is frequency selectivity in the channel [11].

Then, using the demodulation and descrambling of the recovered PDSCH symbols along with an estimation of the noise, a codeword estimate is created. The DL-SCH decoder block further decodes the vector of decoded soft bits and the block CRC error is acquired. The system's throughput is generated using the error. For a hybrid automatic repeat request (HARQ) procedure, the block CRC error is also recorded.

Some common time-domain packet scheduling techniques used in 5G are studied. By analyzing variable channel conditions, the maximum rate scheduling method takes high capacity and maximum throughput into account. Users with better channel conditions are preferred by the algorithm, and UEs having serious channel degradation are not scheduled. Hence, the allocation of resources for users is not fair. The scheduling technique chooses a UE maximizing the subsequent algorithm at each TTI:

$$x_i = y_i(t) \tag{7.1}$$

where $y_i(t)$ shows the user i's instantaneous data rate while he or she is utilizing the entire bandwidth at time t.

In LTE mobile systems, the round robin (RR) scheduling method was developed to distribute resources equally across users. As opposed to the maximum rate algorithm, the RR algorithm permits users to transmit packets in turn with equal opportunity. Hence, the RR technique hugely enhances fairness but it also leads to throughput degradation because of the fact that channel quality is not taken into consideration [14]. The proportional fair (PF) scheduling technique was first created for CDMA networks serving non-guaranteed bit rate (GBR) services. By improving the throughput of UEs with better momentary attainable data rates than average throughput, it sought to accomplish a fair trade-off between fairness and throughput. Owing to the fact that the PF algorithm does not consider the buffer status of the UE, it is not the most appropriate algorithm for real-time services. PF technique can be represented by the following equations.

$$a_i(t) = \frac{b_i(t)}{B_i(t)} \tag{7.2}$$

$$B_i(t+1) = \left(1 - \frac{1}{t_c}\right) B_i(t) + C_i(t+1) * \frac{1}{t_c} * b_i(t+1) \qquad (7.3)$$

$$C_i(t+1) = \left\{ \begin{array}{l} 1 \text{ when packets of user } i \text{ are allocated at interval } t+1 \\ 0 \text{ when packets of user } i \text{ are not allocated at interval } t+1 \end{array} \right\} \qquad (7.4)$$

where
- $b_i(t)$ indicates the instantaneous rate of data for user i generated through time interval t,
- $B_i(t)$ represents the average throughput of user i for time interval t,
- $C_i(t+1)$ denotes the choice of packet for transmission for time interval $t+1$,
- t_c indicates a time constant that could be utilized to make the most of throughput and fairness with the PF algorithm.

The blind equal throughput (BET) scheduling technique has been used mostly in LTE systems and it does not consider channel conditions to allocate resources. On the other hand, the past instance mean throughput of each UE is considered in order to reach a fair assignment of resources among UEs. The BET technique can be represented as follows:

$$d_i(t) = \frac{1}{B_i(t)} \qquad (7.5)$$

where $d_i(t)$ indicates the what the UE prefers at a particular time interval and $B_i(t)$ denotes the mean throughput for user i during time interval t.

Due to the fact that the BET algorithm does not consider channel conditions, it is not appropriate to generate high throughput as opposed to PF and maximum rate algorithms. The delay prioritized scheduling (DPS) technique comprises packet delay information. The DPS algorithm prioritizes UEs in downlink LTE networks that have delays, which are more than a threshold so as to fulfill the QoS requirements for GBR services. The following equation represents the DPS algorithm [10].

$$x_i(t) = T_i - F_i(t) \qquad (7.6)$$

where
- $F_i(t)$ indicates the head of line (HoL) packet delay for user i during time interval t,
- T_i represents the threshold for the delay of the buffer, which relies on the category of service,
- $x_i(t)$ denotes the real time of the HoL packet for user i during time interval t.

Increasing the QoS of real-time UEs is the goal of the modified-largest weighted delay first (M-LWDF) scheduling strategy. The M-LWDF method takes into account a number of variables, including packet delay, mean throughput, instantaneous data rate, and bandwidth. Systems with CDMA-high data rate (HDR) have employed this algorithm. The following equation represents the M-LWDF algorithm [10].

$$g_i(t) = h_i * F_i(t) * \frac{b_i(t)}{B_i(t)} \qquad (7.7)$$

$$h_i = -\frac{\log \delta_i}{T_i} \qquad (7.8)$$

where
  g_i denotes the QoS requirement of user i,
  $F_i(t)$ indicates user i's HoL packet delay through time interval t,
  $b_i(t)$ denotes the instantaneous data rate,
  $B_i(t)$ indicates the mean throughput for user i for time interval t,
  δ_i denotes the packet loss ratio and T_i denotes the threshold of the buffer delay for user i.

The HDR/CDMA framework used the exponential rule (EXP) scheduling mechanism primarily for real-time and non-real-time services. The EXP scheduling technique can be illustrated by the following equations.

$$k_i(t) = \alpha_i * F_i(t) * \frac{b_i(t)}{B_i(t)} * \exp\left(\frac{\alpha_i * F_i(t) - \alpha F_{-avg}}{1 + \sqrt{\alpha F_{-avg}}} \right) \qquad (7.9)$$

$$\alpha F_{-avg} = \frac{1}{N} \sum_{i=1}^{i=N} \alpha_i * F_i(t) \qquad (7.10)$$

where
  $k_i(t)$ indicates the priority for user i to obtain packets for time interval t,
  α_i represents user i's QoS requirement,
  $F_i(t)$ denotes user i's HoL packet delay for time interval t,
  $b(t)$ indicates the instantaneous data rate,
  $B_i(t)$ represents the mean throughput for user i for time interval t
  N denotes the overall quantity of users.

Channel-dependent earliest due deadline (CD-EDD) scheduling technique was created in order to handle mobile systems with sensitive traffic. While displaying a similarity to the M-LWDF and EXP algorithms, the CD-EDD scheduling algorithm considers the mean throughput, instantaneous data rate, and information regarding packet delay when allocating resources. In the event that a user's mean throughput and instantaneous data rate are similar, the CD-EDD will prioritize transmission for the user with the more pressing HoL delay. On the other hand, the M-LWDF and EXP algorithms give priority to the base station's longest buffer delay. The following formula represents the CD-EDD algorithm.

$$m(t) = \alpha_i * \frac{b_i(t)}{B_i(t)} * \frac{F_i(t)}{T_i - F_i(t)} \qquad (7.11)$$

Machine Learning-Based Scheduling in 5G/6G Systems 285

where
  $m_i(t)$ represents the precedence for user I for time interval t,
  α_i indicates user i's QoS requirement,
  $b_i(t)$ denotes the momentary data rate,
  $B_i(t)$ denotes the average throughput for user i for time interval t,
  $F_i(t)$ denotes user i's HoL packet delay for time interval t,
  T_i denotes user i's threshold of the buffer delay.

The user with the highest channel quality indicator (CQI) is chosen by the Best CQI algorithm. Essentially, this algorithm allocates resources according to feedback from the UE regarding the radio channel quality, including BER, CQI, and SINR. Since the allocation of resources largely depends on the channel's condition or the strength of the radio signal, fairness is not a concern for this method. CQI vs. MCS tables are already specified in both the 5G-NR and LTE specifications. To transmit data, one of the available transport block sizes is chosen based on the CQI value supplied by the UE. When a high CQI value is provided, data transmission uses a bigger transport block size [15]. The Best CQI algorithm can be demonstrated using the following equation.

$$n = \max_{t=1-N} \left(P_i(t) \right) \tag{7.12}$$

where N is the overall quantity of active users, $P_i(t)$ is the data rate that a specific UE i can maintain at time t, and n is the user.

7.4 MACHINE LEARNING TECHNIQUES USED IN 5G/6G

Machine learning forms part of AI and has undergone many advances recently. It enables a particular device to perform tasks without giving specific instructions on the procedures involved in solving them. It creates a learning process, whereby the device identifies patterns and makes decisions in accordance with a user-defined goal [16].

7.4.1 SUPERVISED LEARNING

The system is taught by example in this learning approach. When given a set of desired inputs and outputs, the machine learning algorithm creates a function to map the inputs to the outputs. An operator monitors results and is able to intervene to correct predictions made by the algorithm. The algorithm undergoes a training process until the required level of accuracy is achieved. Supervised learning is further divided into three categories namely classification, regression, and forecasting [16].

7.4.1.1 Classification
In classification duties, the machine learning program has to reach a decision from values observed and classify the values into categories. An example would be a program that filters emails as "spam" or "not spam" by considering the existing observational data and filtering emails accordingly [17].

Some classifications of machine learning algorithms are elaborated upon.

The Naïve Bayes classifier. The Naïve Bayes classifier is called so because it presumes that the occurrence of a specific feature is independent of the presence of other features [18]. For example, if a device is identified on the basis of size, shape, and capacity then a small, rectangular, and 32 Gb device can be generally identified as a pen drive. The Naïve Bayes algorithm depends on the principle of the Bayes theorem. The latter is a formula utilized for generating conditional probabilities. A conditional probability is the gauge of the probability of an incident happening provided that a previous incident has happened. The formula is given as follows [19]:

$$P(A|B) = \frac{P(B|A) \cdot P(A)}{P(B)} \qquad (7.13)$$

where $P(A|B)$ is the probability of event A occurring provided that event B has already happened, also called the posterior probability.

$P(B|A)$ denotes the probability of event B happening provided that A has already happened, also called the likelihood probability.

$P(A)$ indicates the probability of A occurring before the event happens, also called the prior probability

$P(B)$ is the probability of B occurring also known as the marginal probability.

Thus, the Bayes theorem indicates a means of discovering a probability when other probabilities are known. It considers each feature as independent and equally contributing toward the results.

K-nearest neighbors algorithm. This algorithm is utilized for both classification and regression problems but it is more often utilized for classification. It collects all cases available and classifies the new case or new data according to a similarity measure. It considers the classification of a neighbor data point in order to classify a new data point [20]. An example would be the case of determining whether a particular device is a pen drive or hard disk. Using the K-nearest neighbor (KNN) algorithm, the latter will identify the similarities between the device and either a pen drive or a hard disk and based on the most similar features, the algorithm will classify it as pen drive or hard disk. The KNN mainly forms a majority vote based on the most similar instances to a particular observation [20]. The similarity can be defined as a distance metric between two data points. The Euclidean distance method forms part of the distance metric method that is used in the KNN algorithm [20].

$$d(x, y) = \sqrt{\sum_{i=1}^{k}(x_i - y_i)^2} \qquad (7.14)$$

where:

$d(x, y)$ indicates the distance between two points in Euclidean k-space,
x_i and y_i are Euclidean vectors, starting from the original point,
k indicates k-space.

Other distance metric methods that are used include Manhattan, Minkowski, and Hamming distance methods.

Machine Learning-Based Scheduling in 5G/6G Systems

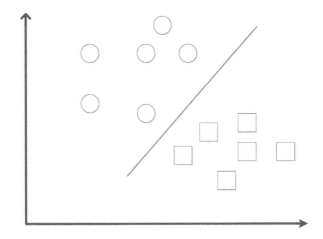

FIGURE 7.6 Example for SVM [21].

Support vector machine. The support vector machine (SVM) algorithm is a classification technique in which raw data is plotted as points in an *n*-dimensional space (where *n* indicates the quantity of instances present). The value of a particular instance is attached to a specific coordinate, thus enabling easier classification of data [18]. Basically, it is an algorithm, which sorts data into one of two categories [21].

A simple example would be a 2D model with a set of data points. The objective is to separate those data points into different categories by using a line called the decision boundary, which is the line between the two closest points that keep the other data points separated (Figure 7.6).

7.4.1.2 Regression

Regression machine learning algorithm is mainly used for prediction and forecasting as the machine learning program needs to estimate and realize how different variables are related. This technique focuses on one dependent variable while other variables are changing. Linear regression is used for the prediction of the value of a variable based on another variable. Thus, the value to be predicted is the dependent variable while the other variable is independent. A relationship is determined between those variables in the form of a line known as the regression line that is represented by the following equation [18].

$$Y = a*X + b \tag{7.15}$$

where
    y is the dependent variable, *x* is the independent variable,
    a is the slope and *b* is the intercept,
    a and *b* are obtained by reducing the sum of squared difference of distance between the data points and the regression line.

A simple example to explain the linear regression algorithm would be arranging devices in increasing order of their capacity but the capacity of each device is unknown to the operator. In this case the operator will have to guess the capacity by performing a check for the size, shape, and manufacturing date of the devices. Thus, a combination of those parameters will give an estimate of the capacity of the device.

7.4.1.3 Decision Trees

Decision trees can be used for classification or regression models. In classification models, the decision tree is finite while in regression models, it is continuous. The dataset is split into smaller subsets and the tree is developed into decision nodes and leaf nodes [22]. A decision node is made up of two or more branches and it represents values for the attribute tested. Lead node indicates the decision on the numerical target. The root note is the uppermost decision node in a tree, which correlates with the best forecaster. Both categorical and numerical data can be managed by a decision tree [22]. The main algorithm for constructing decision trees is called ID3 by J. Quinlan that consists of a top-down, greedy search without backpedaling through the space of possible branches.

An example of a decision tree is illustrated in Figure 7.7.

7.4.1.4 Forecasting

Forecasting algorithm takes into account past and present data to make future predictions and this method is used to identify trends.

7.4.2 Semi-Supervised Learning

The difference between this learning method and supervised learning is the utilization of both labeled and unlabeled data. Labeled data indicates data that has specific labels in order for the algorithm to interpret the data correctly while unlabeled data do not have any specific label. Thus, by using both types of data, machine learning algorithms can manage to label the unlabeled data [17].

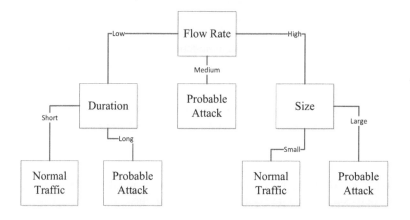

FIGURE 7.7 Example of decision tree for classifying network traffic [22].

Machine Learning-Based Scheduling in 5G/6G Systems

The procedures for semi-supervised learning are as follows [23]:

a. The machine is first trained with a small amount of training data as in supervised learning until target results are obtained.
b. The unlabeled training data set are then used with the model to predict pseudo labels that are not as accurate.
c. The labels from the original training data are linked with the pseudo labels.
d. The data inputs in the labeled training data are matched with the inputs of the unlabeled set.
e. The model is then trained in a similar manner as with the labeled data set in order to get more accurate results.

A simple example would be a text document classifier where there is a significant number of unlabeled documents.

7.4.3 Unsupervised Learning

In unsupervised learning, there is no operator to monitor the process. The machine learning algorithm identifies patterns and correlations by studying the data available. In this process, the algorithm mainly organizes the data into groups and as more data is being assessed, the algorithm improves its capability of making decisions on that data.

There are two categories of unsupervised learning namely clustering and dimension reduction. Clustering consists of gathering similar sets of data according to a particular criteria and carrying out an analysis on the data sets to investigate correlations. Dimension reduction consists of reducing the quantity of variables that is taken into consideration, to output the required data [17].

Some clustering algorithms are illustrated subsequently.

7.4.3.1 K-Means Algorithm

In this algorithm, the data is arranged into a number of clusters (identified as K) such that all the components of the cluster are homogenous and heterogenous from the components of other clusters [24].

The procedure involved in this algorithm is as follows:

a. K number of data points called centroids are selected for every cluster.
b. Data points are gathered into a cluster such that the sum of the squared distance between the data points and the centroid of a particular cluster is at a minimum value.
c. New centroids are recalculated
d. The shortest distance for each data point is generated for the new centroids and this procedure is repeated until a constant value is obtained

The equation used by the K-means algorithm is called the squared error function and is as follows [24].

$$J(V) = \sum_{I=1}^{C}\sum_{j=1}^{C_i}\left(\|x_i - v_j\|\right)^2 \qquad (7.16)$$

where
$\|x_i - v_j\|$ represents the Euclidean distance between x_i and v_j,
c_i denotes the quantity of data points in the ith cluster,
C indicates the quantity of cluster centers,
V is the set of data centers.

The new centroids are calculated based on the following equation

$$v_i = \left(\frac{1}{c_i}\right)\sum_{j=1}^{c_i} x_i \qquad (7.17)$$

where x_i is the ith data point.

A simple example would be the installation of new access points. The *K*-means algorithm can be used based on some information, for example, at which location there are more clients being connected, how many clients are being connected at any given time, and how to keep the distance between the access point and the server to a minimum.

7.4.3.2 Hidden Markov Model

The hidden Markov model designates a probabilistic model used to generate the probabilistic characteristic of any random process [25]. In this model, an observed event is thus linked to a set of probability distributions. If the event is believed to be a Markov chain and there are some hidden states, those hidden states are assumed to form part of the main Markov process. The main objective of HMM is to gather information about a Markov chain by studying the hidden states. An example of an HMM is if an agent B sends the results of data analysis to agent A on a daily basis. Agent A assumes two states from the data analysis report. First, that the system is generating the data correctly and that B is accessing the data for that particular day. On the day A does not receive a report from B, A again assumes two states, either B has not accessed the data or the system is faulty. Thus, A can deduce hidden states based on whether or not the report is received.

The HMM is a process whereby a symbol is generated from some alphabet Σ at every time step in accordance with emission probability depending on state.

$$M = (Q, \Sigma, a, e) \qquad (7.18)$$

where Q denotes a finite set of n states,
$A = n*n$ transition probability.

Matrix $a(i,j) = \Pr[q_{t+1} = j \mid q_{t+1} = i]$ (q_t is the state in position t in the sequence)

$$\Sigma = \{\sigma_1, \ldots, \sigma_k\}$$

$e(i, j)$ is the probability of generating symbol σ_j in state $q_i = P[a_t = \sigma_j \mid q_t = i]$; where a_t is the tth element of the generated sequence.

7.4.3.3 Deep Learning

Deep learning consists of a category of machine learning that is established on the parameterization of multiplayer (deep) neural networks, which can manage to grasp representations of the data [16].

Deep learning is widely used in image classification, speech recognition, and language processing. The neural network strives to reproduce the behavior of the human brain, allowing it to learn from a consequent amount of data [16]. Deep learning differs from other machine learning algorithms according to the type of data used as well as the learning techniques. While machine learning mainly use structured data that are first pre-processed, deep learning does not require data to be pre-processed. Deep learning algorithms can process unstructured data and feature recognition is automated [26]. An example using deep learning is a set of pictures of different devices and the objective is to categorize the devices as "storage," "accessories," and so forth. The deep learning algorithm can establish which features are the most essential in order to correctly differentiate one device from another. In other machine learning algorithms, this priority order of features is determined by a human operator. Thus, through a process of gradient descent and backpropagation, the deep learning algorithm evolves and becomes more accurate, enabling it to predict a new picture more precisely. Gradient descent is an algorithm used to calculate errors in predictions and afterwards alters the function by going through the layers in order to train the model. Two main types of deep learning algorithms are utilized namely convolutional neural networks (CNNs) and recurrent neural network (RNNs). CNNs are used mainly in image classification applications, and RNNs are used for speech recognition [26].

7.4.4 REINFORCEMENT LEARNING

A set of parameters, actions, and end values are provided to the machine learning algorithm in reinforcement learning (RL). The algorithm aims to discover several options and possibilities while assessing each outcome to determine which is the best after a set of rules has been defined [17]. It is a trial-and-error method whereby the machine learns from past experience and uses the knowledge to make accurate future decisions.

In RL, decisions have to be made or actions taken so that the idea of cumulative reward is maximized [27]. As opposed to supervised learning, labeled inputs/outputs are not required and correction of suboptimal actions is not needed. The main goal is achieving an equilibrium between exploration and exploitation. The Markov decision process (MDP) is usually utilized to model the problem in RL applications. However, the RL algorithms used do not follow the explicit mathematical model of the MDP [28]. RL is also described as approximate dynamic programming or neurodynamic programming. RL uses function approximation and samples to handle large environments and to optimize performance [27].

The setup of a RL model is given in Figure 7.8 [28].

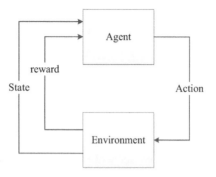

FIGURE 7.8 Interaction of agent–environment [28].

The agent denotes the learner and decision maker.
The environment is where the agent learns and makes decisions to act.
Action refers to the set of actions that the agent is able to do.
State refers to the state of the agent in the particular environment.
Reward refers to the feedback provided by the environment to the agent for every action the agent performs.

An example is rewarding a driver for changing state. In this case, the driver is the agent in the environment, which is the city. Two states are defined namely the driver sitting and driving. The agent (driver) performs an action by driving, thus changing its current state. The policy is defined as a technique to choose an action based on a specific state, in expectation of a better outcome is the specific state. Thus, the agents receive a reward or penalty as per its current state [29].

There are three methods for implementing RL algorithms:

a. **Value-based:** in value based, a value function $V(s)$ is maximized. In this technique, the agent expects a long-term return of the current states.
b. **Policy-based:** in this method, an action is performed to yield maximum reward. There are two types of policy-based methods namely deterministic and stochastic. In deterministic, the same action is output by the policy for any state while in stochastic, every action has a probability determined by a specific function.
c. **Model-based:** in this method a virtual model is created for each environment.

The two learning models mainly used are the MDP and Q-learning. MDP has been illustrated in the previously discussed cat example.

In Q-learning, a value-based method is used to inform the agent which action it should take.

7.5 OVERVIEW OF PREVIOUS WORKS USING MACHINE LEARNING FOR SCHEDULING IN 5G/6G SYSTEMS

In this section a review of state-of-the-art machine learning techniques used to enhance the conventional scheduling algorithms in 5G systems is provided.

7.5.1 Reinforcement Learning-Based Scheduling for Data Traffic Management

In this chapter, the authors have developed a new scheduling program to choose various scheduling algorithms according to the momentary scheduler states to reduce delays and drop rates of packets for applications with harsh QoS requirements. Realtime scheduling is made possible by using RL rules to map the scheduling algorithms to every state and determine when to execute each algorithm. To deal with the complexity of the RL model, neural networks are also utilized as function approximations and to represent the scheduler space state [30]. The proposed RL framework is shown in Figure 7.9. The optimal action-value functions are given as follows:

$$K*(v) = h(\theta_t, \psi(v)) \quad (7.19)$$

$$J*(v, d) = h^d\left(\theta_t^d, \psi(v)\right) \quad (7.20)$$

where $\{h, h^1, h^2, \ldots, h^D\}$ represent the neural networks serving as a rough guide for the value and action-value functions, $\psi(v)$ indicates the feature vector, and $\{\theta, \theta^1, \theta^2, \ldots, \theta^D\}$ represent the set of weights that should be tuned.

Five RL algorithms were implemented and evaluated, namely QV-learning, QV2-learning, QVMAX-learning, QVMAX2-learning, and actorcritic learning automata (ACLA). Variable window size, traffic type, objective metrics, and ever-changing network conditions were all taken into consideration when evaluating their performance. The scheduling algorithms used were exponential 1 (EXP 1), exponential 2 (EXP 2), logarithmic rule (LOG), and earliest deadline first (EDF) rule.

The system was implemented using a RRMScheduler C/C++ object-oriented utility tool that includes the LTESim.

Other scheduler parameters are depicted in Table 7.1.

Simulation results showed that the average percentage of possible TTIs remains relatively constant for the delay target, when the windowing factor is varied.

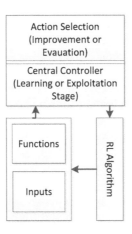

FIGURE 7.9 Proposed RL framework [30].

TABLE 7.1
Parameters for the Packet Scheduler [30]

Parameters	Value
Bandwidth/radius of cell	20 MHz (100 RBs)/1,000 m
Speed of user/mobility model	120 kmph/random direction
Channel	Jakes model
Path loss/penetration Loss	Macro cell model/10 dB
Interfered cells/shadowing standard deviation	6/8 dB
Frequency of carrier/power for DL	2 GHz/43 dBm
Structure of frame	FDD
Reporting mode for CQI	Full-band, periodic at each TTI
PUCCH model	Errorless
Type of traffic	CBR, VBR
Number of users that can be scheduled	10 each TTI
RLC ARQ	Acknowledged model (five retransmissions)
Levels for AMC	QPSK, 16-QAM, 64-QAM
BLER target	10%
Quantity of users	Variable: 15–120
Exploration/exploitation duration	500/95 seconds
Factor for windowing, ρ	{5.5, 100, 200, 200}

However, it was discovered that QVMAX and QVMAX2 outperformed over a range of $\rho = \{5.5, 100\}$ while ACLA outperformed both for a range of $\rho = \{200, 400\}$. When the PDR target is taken into account, for $\rho = \{5.5\}$, the QV policy has the fewest penalties and most reasonable rewards, while the QVMAX2 algorithms perform better for values of $\rho = \{100, 200, 400\}$. However, ACLA and QVMAX2 produced the lowest average percentage when $\rho = \{5.5, 100, 200\}$ when both delay and PDR targets are taken into account. In cases where there are wide windows used in PDR simulations, the policies are limited and apply the scheduling algorithms to satisfy PDR requirements. For VBR traffic, it is noticed that QV, QVMAX2, and ACLA display better performance for $\rho = \{5.5\}$ and when the windowing factor is enhanced to $\rho = \{200, 400\}$. The following policies produce the best outcomes when merging the delay and PDR targets: QV, QVMAX2, ACLA for $\rho = 5.5$, ACLA, QVMAX, QVMAX2 for $\rho = 100$, and QVMAX for $\rho = \{200, 400\}$. The simulations suggest that RL algorithms learn more effectively under VBR traffic compared to CBR traffic [30].

The RL-based framework was then compared with the four baseline scheduling algorithms namely EXP 1, EXP 2, LOG, and EDF. It was observed that the novel technique outperformed the scheduling rules in terms of average percentage of TTIs, q for both CBR and VBR traffic. When CBR traffic is scheduled, the suggested framework obtains more than 10% of viable TTIs across all windowing factor settings. At every TTI, the most appropriate scheduling technique is invoked, enabling all active users to adhere to the lower limit of the delay parameter. When the VBR traffic is scheduled with $\rho = 5.5$ for $q [\%] = \{90, 92, 94\}$, a degradation is noticed.

The proposed system performed best when optimizing the percentages of possible TTIs when all active users met the requirements for both packet delay and PDR. The novel technique outperforms traditional scheduling techniques for the CBR traffic type and the windowing factor of $\rho = \{5.5, 100, 200, 400\}$ by more than 15% by selecting appropriate scheduling strategies for diverse traffic loads, network conditions, and QoS requirements. The system developed showed a gain of about 10% for VBR traffic due to the fact that some VBR packets are larger in size when compared with CBR.

7.5.2 Reinforcement Learning for 5G Scheduling Parameter Optimization

In [31], the authors proposed a model to use RL with the cross-entropy method to learn the best set of parameters for a certain traffic profile. The outcomes are contrasted with those obtained using hand-tuned parameter thresholds by RF subject matter experts (SME) [31].

An environment with three 5G UEs and one 5G cell with no other user was set up. To enter commands and implement the system, a vendor-provided scripting language and a unique operation and maintenance interface were used. The environment created offered fundamental ways to carry out complete steps with the specified actions and return the new state, much as OpenAI's gym [32]. The environment was then connected to an RL algorithm built using the cross-entropy method (CEM). Almost 50 sessions were generated. The architecture of the system is depicted in Figure 7.10.

Simulation results showed that as from 25 iterations, the algorithm outperformed the SME-generated results. The simulations were terminated by the 140th iteration because the overall rewards did not considerably improve with more iterations. The algorithm was able to deliver 25th percentile rewards by epoch 140, which were consistently superior to the baseline.

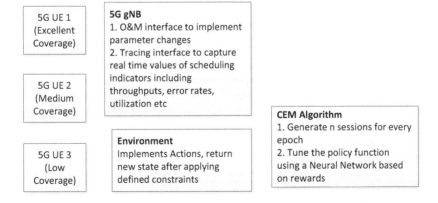

FIGURE 7.10 Proposed RL framework [31].

7.5.3 KNOWLEDGE-ASSISTED DEEP REINFORCEMENT LEARNING IN 5G SCHEDULER DESIGN

In this research, the authors generated a deep RL algorithm for wireless schedulers in 5G with traffic that is sensitive to time. Utilizing a deterministic scheduling policy, the RL algorithm links the channel and queue states to scheduling operations. The approach takes advantage of information about the scheduler design problem, such as the QoS and the target scheduling policy, by adapting a deep deterministic policy gradient (DDPG). A deep RL framework was first established where a MDP process was applied to create the model [33]. A knowledge-assisted (K-DDPG) algorithm was designed. The scheduler was trained and fine-tuned in accordance with feedback from actual networks. In the 5G downlink scheduler considered, K users were served by a single base station (BS). The kth user's packets were waiting in the kth queue in the BS's buffer, and every queue was handled in FIFO (first-in, first-out) order. The structure of the novel algorithm is shown in Figure 7.11.

The simulation setup is shown in Table 7.2.

In the simulation framework, the users' movement were random with a velocity of 5 m/s in a cell of radius 100 m. The path loss model is defined as $45 + 30 \log(l)$ dB where l stands for the measurement in meters separating a user from the BS. At the start, the users are placed at random positions in the cell. The BS is assumed to

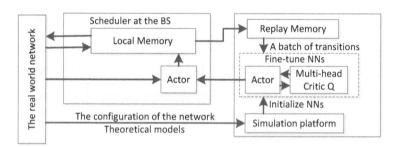

FIGURE 7.11 Proposed online DDPG architecture.

TABLE 7.2
Simulation Setup [33]

Spectrum density for transmit power spectrum	20 dBm/Hz
Spectrum density for noise power	−90 dBm/Hz
Time slot duration (one TTI)	125 μs
Bandwidth of a RB	180 kHz
Packet size	32 bytes
Packet arrival probability	10%
Required decoding error probability	10−5
Timeliness requirement	[5,7]
Maximum SNR	3.8

be in a factory and a Rician distribution is thought to govern the small-scale channel gain [34]. The mean power in the line-of-sight path was set to be 0.6 times greater than that in the non-line-of-sight ways. In the simulation, a discrete-time channel model was taken into account. In the actual slot, the small-scale channel gain slot was maintained in the following slot when the probability is 80%. When the probability is 20%, the channel gain changes depending on the Rician fading. One input layer, one output layer, and two hidden levels were shared by both the actor and the critic.

Every five episodes, the packet loss probabilities were calculated. According to simulation data, DDPG converged in the T-DRL framework after 25 minutes when there are few users. DDPG did not, however, unite to a strategy with little probabilities of packet loss. When compared to DDPG, K-DDPG can further cut convergence time by 50% (in the T-DRL framework). In cases when there are a lot of users, DDPG can scarcely get a scheduler that is satisfactory without help from knowledge. The scheduler learned more quickly using K-DDPG in the T-DRL system than in circumstances where knowledge or theoretical models are lacking. The reward of various DDPG algorithms, including the original DDPG, an extension of DDPG in [35], and the K-DDPG created in this work, is then compared with the T-DRL system. The outcomes show that the method in [35] results in a greater reward at the start of the training phase. This is due to the fact that exploration uses a human-written scheduler, which performs better than the new scheme's randomly initialized actor. Toward the end of the training phase, the K-DDPG, however, outperformed the other two algorithms in terms of learning speed and performance. Simulation results demonstrated that the novel approach reduced the convergence time of DDPG to a large extent and yielded better QoS than the current schedulers (reducing packet losses by 30%–50%). The novel framework obtained superior initial QoS with offline initialization than random initialization, according to experimental data, and online fine-tuning converges quickly.

7.5.4 DEEP REINFORCEMENT LEARNING FOR RADIO RESOURCE SCHEDULING IN 5G MAC LAYER

The authors in [36] have developed a novel deep reinforcement learning model called learn to schedule (LEASCH), which managed to resolve the issue concerning radio resource scheduling pertaining to the MAC layer of 5G networks. LEASCH has been assessed under several numerology settings. In order to increase the adaptability of training the agents and shorten training time, an off-simulator training strategy was proposed. The scheduling problem was converted into an episodic deep reinforcement learning (DRL) problem during the training phase, and LEASCH was trained until convergence. The deployment of LEASCH takes place during the testing phase using a 5G system-level simulator. For a number of episodes, LEASCH was trained. Every episode started with the creation of a random state. The agent was subsequently instructed in a series of episode steps. In each phase, the agent guides the online Q-neural network, and the learnt parameters were then applied to the target critic neural network. The memory R was replayed following the conclusion of an episode, and the acquired weights were then applied to the subsequent episode,

TABLE 7.3
Parameters for Testing of LEASCH in 5G

Parameter	Value
Number of frames	250 frames
Simulation scenario	100 runs with various deployment scenarios
Numerology index	{0, 1, 2}
Bandwidth	{5, 10, 20 MHz}
Number of UEs	4
Subcarrier spacing	{15, 30, 60 kHz}
Number of resource blocks	{25, 24, 24}
Period of scheduling	1 resource block group (RBG)
RBG size	2 resource blocks
Total tested RBGs	250 × 100 × {130, 240, 480} RBGs
Channel	Randomly changes each ¼ s
HARQ	True

and so on. Each episode begins with a reset of the state. Following the training stage, LEASCH is put into a 5G simulator for testing. No retraining was required at this step. The parameters used for simulations are displayed in Table 7.3.

The 5G simulation was implemented using MATLAB 2019b. Simulation results showed that less than 300 episodes were necessary for LEASCH to converge. A stable learning of LEASCH, depicted by a steady increase, was noticed for the theoretical (long-term) reward. Two baseline algorithms, namely PF and RR, have been used to compare with and assess the novel scheduler. Results from the first set of settings, 5 MHz BW and 15 kHz SCS, showed that LEASCH outperformed the baseline in every KPI. As opposed to PF and RR, LEASCH boosted throughput by 2.4% and 18%, respectively. In terms of goodput, LEASCH performed better than PF and RR by 3% and 20%, respectively, suggesting greater stability in LEASCH performance as opposed to the benchmark. In comparison to PF and RR, LEASCH increased throughput for the other set of settings, 10 MHz BW and 30 kHz SCS, by ≈ 3% and 19%, respectively. LEASCH outperformed PF and RR with regard to goodput by ≈3.3% and 21%, respectively. The third set of parameters, 20 MHz BW and 60 kHz SCS, likewise demonstrated comparable performance, with LEASCH increasing throughput by ≈3% and 18%, respectively, above PF and RR. In terms of goodput, LEASCH fared better than PF and RR by ≈4% and 20%, respectively. Additionally, LEASCH was able to scale effectively and enhance the performance when choosing a setting with a greater potential throughput (for instance, 10 MHz with 30 kHz SCS as opposed to 5 MHz with 15 kHz SCS). Additionally, to retest all approaches, the quantity of UEs was doubled, and the second settings were chosen. The findings demonstrated that even with a wider number of UEs, the suggested model could still produce accurate results. It outperformed PF and RR in terms of throughput by 5% and 13%, respectively. In terms of goodput, it outperformed PF and RR by 7% and 14%, respectively.

7.5.5 Intelligent Resource Scheduling for 5G Radio Access Network Slicing

In this study, the authors have presented an intelligent resource scheduling method (iRSS) for 5G RAN slicing [37]. The basic objective of iRSS was to create a framework for collaborative learning that combines deep learning (DL) and RL. While RL is utilized for online resource scheduling to address small-scale network dynamics, comprising erroneous prediction and unanticipated network states, DL is primarily employed for large-scale resource allocation. iRSS can alter the importance between prediction and online decision modules with flexibility and this depends on the capacity of historical traffic data and thus assist the RAN to make resource scheduling decisions. Long short-term memory (LSTM) was utilized to investigate the consistency of data traffic and produce allocation of RAN slices' resources over a long period of time. Moreover, the distributed architecture-based asynchronous advantage actor-critic (A3C) method was used to carry out online resource scheduling of RAN slices in order to deal with faulty prediction and unanticipated network conditions in brief timescales. In the large timescale, LSTM was employed as a DL technique to anticipate traffic volume for resource assignment, and in the small timescale, the parallel computing-based A3C algorithm was used for resource scheduling. When iRSS was compared with other benchmark algorithms, significant performance improvements were seen in the cumulated reward and resource usage. The proposed iRSS was evaluated through extensive simulations. The Gaussian distribution was used to simulate the data traffic at the start of the training process of iRSS. In the simulation trials, the quantity of slices, M, was fixed at 10. To guarantee the convergence of the method, the step-size in the actor process and the step-size in the critic process were specified as constants at extremely low values.

According to the simulation results, the MSE can ultimately approach a minimum value in about 18 epochs. Thus, the LSTM can merge at the conclusion of a prediction window and deliver the best prediction outcomes. Additionally, for almost all cases at that specific MSE point, the errors between the targets and outputs were mostly on either side of 0, demonstrating that the LSTM algorithm could be used to predict the traffic volume. Different numbers of network slices and resource blocks were used to contrast the mean network system rewards. To analyze the system average cost, the simulation included a range of 1–20 network slices and 1–200 RBs, respectively. When the quantity of RBs was fixed, the simulation results revealed that the system mean cost grew monotonically as the quantity of slices increased. That happened as a result of the need for extra resources to accommodate the growing quantity of slices. The system mean cost, on the other hand, dropped to zero when the number of RBs was raised while the number of slices was fixed. The effectiveness of the proposed iRSS was also evaluated against other cutting-edge machine learning techniques, such as the conventional (or tabular) Q-learning and the standard AC algorithm, in addition to a heuristic resource allocation approach (HRSA). Utilization of resources and the total reward were two of the performance criteria employed. While the other configurations remained the same as in the prior experiment, the resource utilization (RU) of the four methods was examined as the number of DTIs was increased from 0 to 200. It should be noted that the RU of the iRSS algorithm consistently appeared greater than that of the Q-learning, traditional AC, and HRSA techniques, respectively. Additionally, it was shown that the RU of Q-learning, AC, and HRSA changed, although that of iRSS fluctuated just slightly, when the number of DTIs was less than 40. When there were more than 40 DTIs, the four algorithms' RUs progressively stabilized, and the RU of iRSS was greater than those of Q-learning, AC methods, and HRSA by roughly 16.3%~19.8%, 8.3%~13.7%, 30.5%~34.7%, respectively.

7.5.6 Delay-Aware Cellular Traffic Scheduling with DRL

In this research, cellular packet scheduling was designed using DRL. The observed system status was mapped to scheduling choices using a delay-aware cell traffic scheduling method. An RNN was employed to estimate the ideal action-policy function owing to the size of the state space. In contrast to traditional rule-based scheduling techniques, the proposed approach might take into account communications with the environment and determine the optimal scheduling choice for each TTI adaptively. The MDP was utilized to model the packet scheduling issue. To grasp a delay-optimized scheduling solution via communications with the environment, a deep-Q-learning agent built on a RNN was developed [38].

Consideration was given to cellular network downlink transmission in which base stations supplied UEs (BS). The case where 2 UEs were requesting data from the BS was considered whereby one UE followed the Poisson distribution and the other UE followed the uniform distribution. The maximum queue length Qmax was set as 100. The highest quantity of bits that could be transferred on the RB was calculated based on the assumption that the quality of each RB could be gauged at each TTI. In this simulation, it was believed that the most bits that could be evenly broadcast on the RB for a particular UE could only take distinct values from a set of {2, 3, 4, 5}. In the neural network used, between the input and the output, there was three levels: two dense layers and one LSTM layer. Each layer had 50 neurons and the number of episodes was 200. The total number of steps in each episode was 300. One thousand simulations were generated and the results averaged. According to the simulation results, the reward value started to converge after about 100 episodes of training. Additionally, at about training episode 150, the maximum award was earned. To attain the optimum performance, the ideal training episode count could be 150. Moreover, the proposed DRL-based scheduling algorithm was shown to have the lowest delay, 0.4049 seconds. The proposed algorithm had a latency that was nearly 3.5 times as high as max-CQI's, which sought to increase traffic throughput. The max-CQI technique, however, had the longest queue, which was about nine times longer than the novel approach.

7.5.7 A Fairness-Oriented Scheduler Using Multiagent RL

The issue of fairness-oriented user scheduling was looked into in this work, particularly for the distribution of RB groups. To augment the fairness of the communication system, a user scheduler was created utilizing multiagent reinforcement learning (MARL), which carried out distributional optimization. The agents explored the best solution in accordance with a clearly defined reward function designed to optimize

Machine Learning-Based Scheduling in 5G/6G Systems

fairness, using cross-layer information (such as RSRP, Buffer size, etc.) as state and the RBG assignment outcome as action. Additionally, the performance of MARL scheduling was compared with that of PF scheduling and RRF scheduling by running a large number of simulations. The fifth%-tile data rate (5TUDR) of a user was chosen as the key performance indicator (KPI) of fairness. An LTE network with one BS scheduling its RBGs to multiple users in bursty traffic was considered. In MARL, there are multiple agents interacting with the environment. In this work, each agent was responsible for a single RBG and a MARL system was built to learn a scheduling policy, which maximizes the system fairness. All the agents were set to be fully cooperative and accept the specific cross-layer network data as input [39].

The UE arrival in the BS configuration was modeled as a Poisson process and the arrival rate λ was considered. Each UE requested a unique, but limited, quantity of traffic data on arrival. The data was kept in the transmitter's buffer. After its request was completed, the corresponding UE suddenly left the network and this action simulated the bursty traffic mode. Furthermore, the frequency resources were divided into RBGs. The most common network setting was considered in which one RBG could be assigned a maximum of one UE in each TTI. The simulation parameters are illustrated in Table 7.4.

The agents were trained following the parameters listed earlier. One epoch represented one simulation, which lasted 1,000 TTIs. The system was evaluated using four schedulers, namely MARL scheduling, PF1 scheduling, PF2 scheduling, and RRF scheduling, respectively. Experimental results showed that the MARL scheduling outperformed the other three schedulers in most of the simulations considering the 5TUDR. The simulation results showed that MARL scheduling could significantly improve the fairness while maintaining good performance in average user data rate. After a period of scheduling, the number of users changed, and some users left the BS. For the MARL scheduling, it insisted upon following the distributional policy and allocating the RBGs to different users. But PF scheduling and RRF scheduling still assigned all the RBGs to a unique user. Moreover, users with lower buffer size took higher priorities in two instances of MARL scheduling.

TABLE 7.4
The Base Station Parameters

Each RBs transmit power	18 dBm
Quantity of RB in every RBG	3
Bandwidth	10 MHz
Noise power density	−174 dBm/Hz
Minimum MCS	1
Maximum MCS	29
HARQ number	8
Feedback period of HARQ	8
Initial RB CQI value	4
Each RB's transmit power	18 dBm

7.5.8 DEEP LEARNING-BASED USER SCHEDULING FOR MASSIVE MIMO DOWNLINK SYSTEM

In this chapter, the authors have investigated a scheduling technique for users with massive multiple-input multiple-output (MIMO) systems using corelated Rician fading channels. A novel user scheduling algorithm was proposed in order to reduce latency and achieve high throughput. Only statistical channel state information (CSI) was used by the algorithm [40]. Through supervised learning, the scheduling network developed was taught to comprehend the mapping from the statistical signal and interference pattern to the scheduling decision of the user. After offline training, it can forecast the ideal scheduling strategy based on statistical CSI without requiring iterative calculation. The novel scheduling network was also resistant to changes in the channel environment and the quantity of transmit antennas owing to the normalization of the input data. The transmission system under consideration was a single-cell massive MIMO system. A uniform linear antenna array (ULA) having M antennas was installed in the BS. The BS could service up to U_t users and was assumed to have L single-antenna users. Two benchmark algorithms were used to compare with the proposed scheduling framework. The first algorithm used was the one in [41] and the second one was the "sum rate-based" scheduling technique, which used exhaustive search to optimize the ergodic sum rate approximation. For simulation, a massive MIMO downlink transmission system having $L = 20$, $M = 64$, $U_t = 6$ was considered. The scheduling network was trained offline using a unique GPU from an NVIDIA RTX2080 Ti card. The loss function was minimized and the network's parameters were settled using the stochastic gradient descent optimizer, which had the maximum forecast precision on the validation set. The regularization coefficient ε was 0.0001, the training period 150, and the batch size 1,024. Every 40 epochs, the original learning rate of 0.1 was multiplied by 0.1. Since the sum rate-based strategy was the ideal scheduling algorithm utilized to provide the training labels for the novel scheduling system, simulation results demonstrated that the algorithm attained the highest mean sum rate. Although the scheduling system's mean sum rate performance could not outperform the best scheduling algorithm, it could be noticed that the scheduling system developed outperformed the method in [41] in the low and middle SNR zones and was almost as effective overall. The novel scheduling network performed somewhat worse than the two benchmark algorithms in the high SNR region. A different user scheduling system having its training data produced under an SNR of 26 dB was developed in order to make up for the performance loss in the high SNR zone. It was found that the novel technique performed better and nearly equal to the optimal scheduling technique in the high SNR region. Additionally, the proposed algorithm significantly decreased the computing time needed to produce the scheduling scheme while maintaining the system's excellent spectral efficiency.

7.5.9 SCHEDULING BASED ON DRL MODEL FOR UAV VIDEO

For the purpose of solving the resource elements (RE) scheduling problem, the authors of this research suggested a deep-Q network (DQN) method constructed

Machine Learning-Based Scheduling in 5G/6G Systems

on the requirements of low time latency and high resource usage rate [42]. In set conditions of time latency and RU rate, the best RE distribution system is developed for four users. In order to prepare the neural network, replay memory – which keeps track of states, rewards, and actions – were used. The Q network was created with three feedforward hidden layers for DQN that are fully connected. A schedule for resource items was founded on the DRL method. As the starting network context for the modeling of bandwidth allocation, 100 MHz was utilized. The three QoS indices of time latency, RB usage rate, and fairness were used to assess reward ratings of actions from one state to another. In essence, the bandwidth resource allocation procedure was a MDP. For simulation, the RB rate was set to 90%, the time latency to 200 ms, and the fairness index to 0.5. The DQN algorithm utilized to estimate the action-value consisted of three fully connected feedforward hidden layers and the quantity of neurons in the layers were 256, 256, and 512, respectively. When the replay memory was fully loaded with 400 transfer samples, the training procedure begins. While the exploration stage was set at 100,000 steps, the observation stage had 300 steps. A random action was chosen from the present set of actions during the observation step. Following that, a choice was made in the exploration stage that had a lower probability than a particular value of ε_k. The following equation was utilized to update the value of ε_k.

$$\varepsilon_k = 0.8 \times \left(1 - \frac{k}{K}\right) \quad (7.21)$$

where K represents the total number of training steps, while k represents the current step.

The DQN structure was created using a Python program based on the aforementioned constraints and modeling parameters, and the Q network was trained using the continuously generated transfer samples. The scheme shown in Figure 7.12 for the ultimate allocation strategy was given to show the impact of RB allocation.

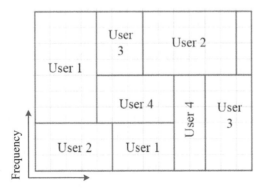

FIGURE 7.12 Resource elements allocation result for four users [42].

7.5.10 Reinforcement Learning Algorithms in Fairness-Oriented OFDMA Schedulers

In order to address the user fairness criteria for scheduling, a reinforcement learning-based system was developed in this work [43]. For the PF scheduler, this framework used feedforward neural systems to translate instantaneous states to the appropriate parameterization choices. To augment the fraction of time (measured in TTIs) during which the next generation of mobile networks (NGMN) fairness criteria was satisfied, a reinforcement learning-based solution was utilized. This solution learns the generalized PF (GPF) parameterization on each instantaneous state. The proposed RL framework communicates with the scheduler environment in an effort to continuously refine its selection of the GPF parameterization scheme founded on a significant quantity of iterations. The developed RL method updates the neural network weights as nonlinear functions up until some error-based convergence criterion is satisfied. In order to identify the GPF parameters reliably and quickly, so that the NGMN fairness provisioning is increased, this study targeted to train nonlinear functions using a selection of RL techniques. This chapter built upon work from [44] and a scheduling framework was proposed that satisfied the NGMN fairness criteria for downlink OFDMA systems.

The RRM-Scheduler simulator [45] was used for simulations using a C++ utility that built upon the long-term evolution simulator (LTE-Sim) by adding new features such as cutting-edge OFDMA schedulers, RL algorithms applied to various scheduling issues, neural network approximation for RL decisions, and CQI compression techniques. Q-Learning, DQ-Learning, SARSA, QV, QV2, QVMAX, QVMAX2, and ACLA were some of the RL algorithms that were employed. Each of these RL methods requires training a total of 11 neural networks at each TTI. For approximately 3,000 seconds, the learning step was carried out on all RL techniques under the same networking settings (i.e., channel conditions, user activity, and data quota in the queues). The entire number of available resource blocks, $B = 100$ and a system bandwidth of 20 MHz were taken into account. In the simulations, a cluster of seven cells and a radius of 1 km was considered. The scheduling performance was assessed in the central cell, while the interference levels were provided by the other cells. During the learning and exploitation phases, the number of users varied between $U_t = [15, 120]$, with $U_{max} = 10$ being the ideal number of maximum users who may be scheduled. At each 1,000 TTIs during the learning phase, a random number between 15 and 120 active users is selected. In both the learning and exploitation phases, each user takes into account a full buffer traffic model and travels at a speed of 120 km/h while employing a random direction mobility model to investigate a wide range of CQI distributions. The parameters used for simulations are illustrated in Table 7.5.

The top five RL algorithms were compared for various windowing factor values. When the average proportion of TTIs with viable zones was quantified, all RL algorithms offered almost the same performance when the windowing factor assumed low values, such as $\rho = 2.0$. The examined RL algorithms offered roughly similar performance by gradually increasing the number of possible TTIs by employing windowing factors in the span of $\rho \in \{2.25, 2.5, 3.0\}$. The best solutions for determining the smallest percentage of TTIs when the scheduler is inequitable were ACLA,

TABLE 7.5
Simulation Parameters

Parameter	Value
Bandwidth/radius of cell	20 MHz (100 RBs)/1,000 m
Speed of user/mobility model	120 kmph/random mobility
Channel	Jakes model
Path loss/penetration loss	Macro cell model/10 dB
Interfered cells/shadowing STD	6/8 dB
Frequency of carrier/DL power	2 GHz/43 dBm
Frame structure	FDD
CQI reporting mode	Full-band, periodic at each TTI
AMC levels	QPSK, 16-QAM, 64-QAM
BLER targeted	10%
Parameter	Value

CACLA-1, and CACLA-2 when the windowing factor was increased to $\rho = 4.0$. When $\rho = 4.25$, ACLA, CACLA-1, and CACLA-2 remained the best choice. However, as opposed to the performance attained for $\rho = 4.25$, CACLA-2 and ACLA delivered the best outcomes when $\rho = 4.5$, with the average percentage of TTIs with over-fair states higher by nearly 10. In comparison to other RL options, CACLA-2 was the best option when the windowing factor was increased to $\rho = 5.0$ by reducing the number of TTIs with unfair states and raising the proportion of possible TTIs with more than 15%. QV2, QVMAX2, ACLA, CACLA-1, and CACLA-2 all performed similarly for low windowing factors (i.e., $\rho \in [2.0, 3.0]$).

7.6 CONCLUSION

Machine learning is a key component in several current and future mobile communication systems. Although machine learning has not yet been fully applied in 5G communication systems, various research done has demonstrated that the use of machine learning techniques such as RL and DL can contribute considerably in improving the performance of 5G systems. The use of machine learning in scheduling for 5G processes can enable the automatic selection of an optimal scheduling algorithms based on several scheduling factors such as radio conditions, KPI, and distance from the user. In this chapter several previous works based on RL, DRL, and multiagent learning have been reviewed. It has been observed that these techniques are very promising when applied to 5G scheduling and several future works can be performed to further improve these schemes by hybridizing these schemes.

REFERENCES

1. Osseiran, A., Parkvall, S., Persson, P., Zaidi, A., and Magnusson, S., April 2020. 5G wireless access: An overview, Ericsson White Paper [Online]. Available at: https://www.ericsson.com/en/reports-and-papers/white-papers/5g-wireless-access-an-overview.

2. Qualcomm Technologies Inc., 2022. Everything you need to know about 5G [Online]. Available at: https://www.qualcomm.com/5g/what-is-5g#:~:text=5G%20can%20be%20significantly%20faster, has%20lower%20latency%20than%204G.
3. Kranz, G. and Christensen, G., 2020–2022. What is 6G? Overview of 6G networks and technology, TechTarget [Online]. Available at: https://www.techtarget.com/searchnetworking/definition/6G#:~:text=6G%20(sixth%2Dgeneration%20wireless), support%20one%20microsecond%20latency%20communications.
4. Jumani, M.A., Mehdi, H., and Hussain, Z., 2022. A detailed overview of 6G and related technologies. *Electrica*, 22(2), pp. 321–328. doi:10.54614/electrica.2022.21069.
5. ETSI, 2018. ETSI TS 138 300 V15.3.1 (2018–10) [Online]. Available at: https://www.etsi.org/deliver/etsi_ts/138300_138399/138300/15.03.01_60/ts_138300v150301p.pdf.
6. Intel, 2018. 5G NR – Driving wireless evolution into new vertical domains [Online]. Available at: https://www.intel.co.uk/content/dam/www/public/us/en/documents/guides/5g-nr-technology-guide.pdf.
7. Sciencedirect, 2018. Physical resource block [Online]. Available at: https://www.sciencedirect.com/topics/computer-science/physical-resource-block.
8. Intel, 2018. 5G NR-driving wireless evolution into new vertical domains [Online]. Available at: https://www.intel.ca/content/www/ca/en/wireless-network/5g-technology/standards-and-spectrum/5g-nr-technology.html.
9. IEEE, 2017. Massive MIMO for 5G [Online]. Available at: https://futurenetworks.ieee.org/tech-focus/march-2017/massive-mimo-for-5g.
10. Sheik Mamode, M.I. and Fowdur, T.P., 2020. Survey of scheduling schemes in 5G mobile communication systems. *Journal of Electrical Engineering, Electronics, Control and Computer Science*, 6(2), pp. 21–30. Corpus ID: 221691434.
11. ETSI TS 138 211 V15.3.0, 2018. 5G; NR; Physical channels and modulation (3GPP TS 38.211, Version 15.3.0, Release 15) [Online]. Available at: https://www.etsi.org/deliver/etsi_ts/138200_138299/138211/15.03.00_60/ts_138211v150300p.pdf.
12. MATLAB, 2022. NR PDSCH throughput example [Online]. Available at: https://www.mathworks.com/help/5g/examples/nr-pdsch-throughput.html.
13. Sheik Mamode, M.I. and Fowdur, T.P., November 2020. Performance analysis of link adaptation with MIMO and varying modulation and coderates for 5G systems. In *2020 3rd International Conference on Emerging Trends in Electrical, Electronic and Communications Engineering (ELECOM)* (pp. 222–228). IEEE. doi:10.1109/ELECOM49001.2020.9297025.
14. Heidari, R., Packet Scheduling Algorithms in LTE Systems, 2017. A thesis submitted to Faculty of Engineering and Information Technology, University of Technology Sydney, New South Wales, Australia [Online]. Available at: https://opus.lib.uts.edu.au/bitstream/10453/123167/1/01front.pdf.
15. Abduljalil, M.A., 2014. Resource scheduling algorithms in long term evolution (LTE). *Journal of Electronics and Communication Engineering (IOSR-JECE)*, 9(6), pp. 50–53. doi:10.9790/2834-09635053.
16. Haidine, A., Salmam, F.Z., Aqqal, A., and Dahbi, A., 2021. Artificial intelligence and machine learning in 5G and beyond: A survey and perspectives. In *Moving Broadband Mobile Communications Forward: Intelligent Technologies for 5G and Beyond*, IntechOpen, p. 47. doi:10.5772/intechopen.98517.
17. Wakefield, K., 2021. A guide to the types of machine learning algorithms and their applications [Online]. Available at: https://www.sas.com/en_gb/insights/articles/analytics/machine-learning-algorithms.
18. Shaw, R., 2017. Top 10 machine learning algorithms for beginners [Online]. Available at: https://www.simplilearn.com/10-algorithms-machine-learning-engineers-need-to-know-article.

19. Chauhan, N.S., 2022. Naïve Bayes algorithm: Everything you need to know [Online]. Available at: https://www.kdnuggets.com/2020/06/naive-bayes-algorithm-everything.html.
20. Subramanian, D., 2019. A simple introduction to k-nearest neighbors algorithm, Towards Data Science [Online]. Available at: https://towardsdatascience.com/a-simple-introduction-to-k-nearest-neighbors-algorithm-b3519ed98e.
21. Radhika, 2020. The mathematics behind support vector machine algorithm (SVM), Data Science Blogathon [Online]. Available at: https://www.analyticsvidhya.com/blog/2020/10/the-mathematics-behind-svm/.
22. Sayad, S., 2022. Decision tree – Regression: An introduction to data science [Online]. Available at: https://www.saedsayad.com/decision_tree_reg.htm.
23. Datarobot, 2020. Semi-supervised learning [Online]. Available at: https://www.datarobot.com/blog/semi-supervised-learning/#:~:text=A%20common%20example%20of%20an,amount%20of%20labeled%20text%20documents.
24. Kanungo, T., Mount, D.M., Netanyahu, N.S., Piatko, C.D., Silverman, R., and Wu, A.Y., 2002. An efficient k-means clustering algorithm: Analysis and implementation. *IEEE Transactions on Pattern Analysis and Machine Intelligence*, 24(7), pp. 881–892. doi:10.1109/TPAMI.2002.1017616.
25. Verma, Y., 2021. A guide to hidden Markov model and its applications in NLP [Online]. Available at: https://analyticsindiamag.com/a-guide-to-hidden-markov-model-and-its-applications-in-nlp/.
26. IBM Cloud Education, 2020. Deep learning [Online]. Available at: https://www.ibm.com/cloud/learn/deep-learning.
27. Patil, A., Iyer, S., and Pandya, R.J., 2022. A survey of machine learning algorithms for 6G wireless networks. *arXiv preprint arXiv:2203.08429*. doi:10.48550/arXiv.2203.08429.
28. Rekkas, V.P., Sotiroudis, S., Sarigiannidis, P., Wan, S., Karagiannidis, G.K., and Goudos, S.K., 2021. Machine learning in beyond 5G/6G networks—State-of-the-art and future trends. *Electronics*, 10(22), p. 2786. doi:10.3390/electronics10222786.
29. Johnson, D., 2022. Reinforcement learning: What is, algorithms, types & examples [Online]. Available at: https://www.guru99.com/reinforcement-learning-tutorial.html.
30. Comşa, I.S., Zhang, S., Aydin, M.E., Kuonen, P., Lu, Y., Trestian, R., and Ghinea, G., 2018. Towards 5G: A reinforcement learning-based scheduling solution for data traffic management. *IEEE Transactions on Network and Service Management*, 15(4), pp. 1661–1675. doi:10.1109/TNSM.2018.2863563.
31. Habiby, A.A.M. and Thoppu, A., 2019. Application of reinforcement learning for 5G scheduling parameter optimization. *arXiv preprint arXiv:1911.07608*. doi:10.48550/arXiv.1911.07608.
32. Gym Documentation, 2022. [Online]. Available at: https://gym.openai.com/.
33. Gu, Z., She, C., Hardjawana, W., Lumb, S., McKechnie, D., Essery, T., and Vucetic, B., 2021. Knowledge-assisted deep reinforcement learning in 5G scheduler design: From theoretical framework to implementation. *IEEE Journal on Selected Areas in Communications*, 39(7), pp. 2014–2028. doi:10.48550/arXiv.2009.08346.
34. Goldsmith, A., 2005. *Wireless Communications*. Cambridge University Press. doi:10.1017/CBO9780511841224.
35. Gu, L., Zeng, D., Li, W., Guo, S., Zomaya, A.Y., and Jin, H., 2019. Intelligent VNF orchestration and flow scheduling via model-assisted deep reinforcement learning. *IEEE Journal on Selected Areas in Communications*, 38(2), pp. 279–291. doi:10.1109/JSAC.2019.2959182.
36. Al-Tam, F., Correia, N., and Rodriguez, J., 2020. Learn to schedule (LEASCH): A deep reinforcement learning approach for radio resource scheduling in the 5G MAC layer. *IEEE Access*, 8, pp. 108088–108101. doi:10.1109/ACCESS.2020.3000893.

37. Yan, M., Feng, G., Zhou, J., Sun, Y., and Liang, Y.C., 2019. Intelligent resource scheduling for 5G radio access network slicing. *IEEE Transactions on Vehicular Technology*, 68(8), pp. 7691–7703. doi:10.1109/TVT.2019.2922668.
38. Zhang, T., Shen, S., Mao, S., and Chang, G.K., December 2020. Delay-aware cellular traffic scheduling with deep reinforcement learning. In *GLOBECOM 2020–2020 IEEE Global Communications Conference* (pp. 1–6). IEEE. doi:10.1109/GLOBECOM42002.2020.9322560.
39. Yuan, M., Cao, Q., Pun, M.O., and Chen, Y., 2020. Fairness-oriented user scheduling for bursty downlink transmission using multi-agent reinforcement learning. *arXiv preprint arXiv:2012.15081*. doi:10.48550/arXiv.2012.15081.
40. Yu, X., Guo, J., Li, X., and Jin, S., 2021. Deep learning based user scheduling for massive MIMO downlink system. *Science China Information Sciences*, 64(8), pp. 1–10. doi:10.1007/s11432-020-2993-6.
41. Li, X., Yu, X., Sun, T., Guo, J., and Zhang, J., 2019. Joint scheduling and deep learning-based beamforming for FD-MIMO systems over correlated Rician fading. *IEEE Access*, 7, pp. 118297–118309. doi:10.1109/ACCESS.2019.2936880.
42. Jiang, J., Qiu, Y., Su, Y., and Zhou, J., March 2021. Low-latency resource elements scheduling based on deep reinforcement learning model for UAV video in 5G network. *Journal of Physics: Conference Series*, 1827(1), p. 012071. IOP Publishing. doi:10.1088/1742-6596/1827/1/012071.
43. Comşa, I.S., Zhang, S., Aydin, M., Kuonen, P., Trestian, R., and Ghinea, G., 2019. A comparison of reinforcement learning algorithms in fairness-oriented OFDMA schedulers. *Information*, 10(10), p. 315. doi:10.3390/info10100315.
44. Comşa, I.S., Zhang, S., Aydin, M., Kuonen, P., Trestian, R., and Ghinea, G., April 2019. Enhancing user fairness in OFDMA radio access networks through machine learning. In *2019 Wireless Days (WD)* (pp. 1–8). IEEE. doi:10.1109/WD.2019.8734262.
45. Comşa, I.S., 2014. Sustainable scheduling policies for radio access networks based on LTE technology (Doctoral dissertation, University of Bedfordshire). doi:10.13140/RG.2.2.15110.52803.

8 Application of Deep Learning Techniques to Modulation and Detection for 5G and Beyond Wireless Systems

Mussawir A. Hosany
University of Mauritius

CONTENTS

8.1 Introduction 309
8.2 Deep Learning with 6G Technology 310
8.3 Deep Learning Classifications 312
    8.3.1 Supervised Learning..312
    8.3.2 Unsupervised Learning 314
    8.3.3 Reinforcement Learning..316
8.4 Deep Learning and Modulation in beyond 5G Networks 316
    8.4.1 CLDNN-Based Modulation...317
        8.4.1.1 Architecture of the CLDNN 318
        8.4.1.2 Principal Component Analysis 318
        8.4.1.3 CLDNN Architecture 320
        8.4.1.4 Error Performance of CLDNN Architecture.............320
    8.4.2 Deep Neural Network Architectures for Modulation
        Classifications 322
8.5 Concluding Remarks 323
References..323

8.1 INTRODUCTION

Wireless communications and networking have undergone exceptional revolutionary progress over the past years. The deployment rate of the fifth generation (5G) mobile technology and its associated standards continue to increase. This has led researchers, standardization bodies, and industry to turn their attention to the realization and implementation of the sixth generation (6G) communication system with target goals such as 1 Tbps peak data rate with 1 ms end-to-end latency and up to 20 years

DOI: 10.1201/9781003205494-10

of battery life. Topics of special interest for driving 5G communications toward the 6G technology include THz communications, big data analytics as well as cell-free networks and AI.

To fully optimize the 6G networks, AI and ML have been chosen as key enablers. AI and ML will be applied to various parts of the 6G network by mapping AI algorithms to the physical, data link, network, and application layers. Thrusts such as novel approaches to dynamic security as well as network slice management are being studied to deliver an intelligent 6G ecosystem. ML algorithms will be applied to various components of the 6G networks that will perform tasks independently as well as precisely determining 6G parameters and supporting interactive decision-making.

8.2 DEEP LEARNING WITH 6G TECHNOLOGY

Models based on DL employ signal processing algorithms to study the behavior or sequences of the 6G communication system. These models originate from explicitly mathematical algorithms such as classifications, regression as well as exchanges of a smart negotiator with the wireless background. It has been shown that when a model recognizes the features of the 6G communication system, also known as a qualified model, then it can learn the features of such a system [1]. Recently, some latest DL models as well as massive data sets and high computing power [2,3] have enabled the application of AI to 6G networks. The academic community is presently focusing on the application of DL algorithms such as supervised, unsupervised, and reinforcement to 6G [4]. The salutary effect of such application will help to process the last amount of metadata with fewer resources and less computational power thus achieving an efficient ecosystem. In the following sections we give a brief explanation of the various deep learning techniques that can be applied to 6G networks. Table 8.1 presents the latest surveys related to DL and mobile networks.

With a view to optimize, monitor as well as manage the broadband spectrum in mobile networks, wireless signal reception (WSR) is employed in mobile networks. Recently, a great deal of research and focus has been made in the area of WSR including modulation and detection. Conventional algorithms of modulation and detection employ the likelihood-based (LB) and feature-based (FB) approaches as

TABLE 8.1
Summary of Existing Surveys in Deep Learning and Mobile Networks

Publication	Summary
LeCun et al. [5]	A milestone overview of deep learning.
Schmidhuber [6]	A comprehensive deep learning survey.
Liu et al. [7]	A survey on deep learning and its applications.
Deng et al. [8]	An overview of deep learning methods and applications.
Deng [9]	A tutorial on deep learning.
Pouyanfar et al. [10]	A recent survey on deep learning.
Arulkumaran et al. [11]	A survey of deep reinforcement learning.
Hussein et al. [12]	A survey of imitation learning.

detailed in [13]. Statistical theory such as hypothesis testing is employed in the LB approach, which results in optimal performance of the wireless system but includes large computational complexity. In practical wireless communication system, it is the FB approach that is suitable due to its low complexity of implementation but the performance achieved is suboptimal.

With the advent of ML algorithms, it has been shown that FB approaches can significantly achieve high performance by employing support vector machine (SVM) [14]. However, there are some degradation rates with ML methods, although advantages such as classification efficiency and performance are obtained. In wireless environment, there are many variations in the conditions of channels due to the obstruction of objects. With DL it has been shown that salutary performance can be achieved in computer vision [15] and natural language processing [16,17]. Moreover, modulation classification has been applied to convolutional neural networks (CNN) model, which validated outstanding performance in terms of efficiency and accuracy [10].

The concept of DL instigates from the study on artificial neural network and the goal is to recognize data by mirroring the contrivance of the human brain [18,19]. There are three major sections of a basic neural network as shown in Figure 8.1. The hidden layer may support more than one sublayer or node, which is a basic operational unit where the input vector is multiplied by a series of weights and the sum value is fed to an activation function f.

There has been significant research progress in deep learning by Hinton [20] and many DL architectures composed of various stacked layers of neural networks have been developed. These layers contribute to meaningful information and high-level representations. Large-scale data is necessary for the training process in a DL model and this data is accessible from various components of the wireless system. In a DL architecture, the feature engineering component is primordial in signal recognition schemes.

It has been reported in [21] that in the FB approach of DL the received signal may be impractical or inaccurate in estimating many parameters but this approach

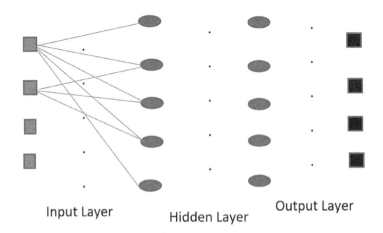

FIGURE 8.1 Layered neural network.

TABLE 8.2
Deep Learning Algorithms and Modulation

Mean/Range	Mean/Range
[23–25]	Deep neural network
[26,27]	Deep belief network
[21]	Convolutional neural network
[28]	Long–short-term memory
[29]	Convolutional restricted Boltzmann machine

performs well for certain solutions that employ self-learning models [22]. Concerning LB approaches, it is well known that optimal performance can be achieved when considering unknown parameters in modulation recognition. This has been possible by using the probability density function (PDF) of the wireless channel. However, there are high computational complexities as well as mismatching of the theoretical system model to the practical one in such approaches. To deal with the high complexity of LB approaches some features of the received signal are extracted and then reasonable classifiers are used to classify different modulation schemes. Table 8.2 illustrates the latest research associated with the application of DL algorithms to modulation and detection in mobile 5G and beyond.

8.3 DEEP LEARNING CLASSIFICATIONS

DL can be classified as supervised, unsupervised, and reinforcement learning.

8.3.1 Supervised Learning

In an artificial neural network (ANN) applied to a 5G wireless communication system the structure of the channel can be used to solve problems such as spectrum and resource allocations [30–33]. It has been shown lately in [33] that ANN can be transformed to deep neural network (DNN) with more applicability and capabilities.

Moreover, in kNN DL algorithms, the distance between various feature values is used to study the system. These algorithms form the basis of classification and regression theories. Consider a situation where most of the neighbors belong to a certain class then the learned sample is assigned to this class. If the DL framework employs Bayes theorem, which consists of a simple probabilistic classification model, then the framework is known to use the Naïve Bayes algorithm [34].

In a random forest supervised learning there are multiple decision trees. The algorithm constructs a tree with branches by randomly selecting a subset of features and each decision tree is assigned a new data set with the data samples to be found classified into a specific class. Figure 8.2 depicts an example of the random forest learning model.

CNN form part of a DL model consisting of neurons that can be optimized by themselves [33]. This model has been extensively applied to image processing and

Modulation and Detection for 5G and Beyond Wireless Systems

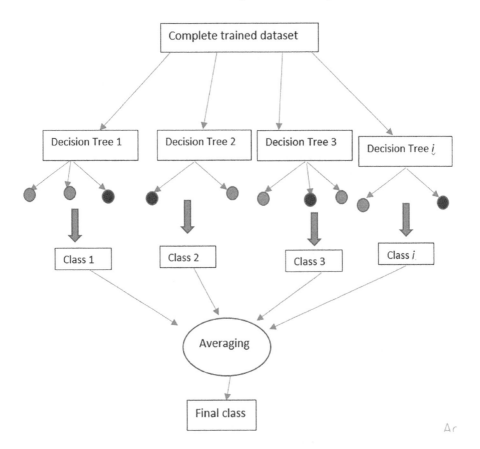

FIGURE 8.2 Random forest learning model.

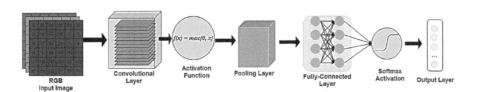

FIGURE 8.3 CNN architecture.

pattern recognition with a complete CNN architecture as shown in Figure 8.3. The architectures of CNN are made up of several distinct layers as can be seen. All the layers have one similarity that the input to them is a 3D volume and undergo a transform through the corresponding layer by employing differential equations and output a 3D volume parameter. The responsibilities of each layer are summarized next.

Input image layer. The raw pixel values of an image that is represented as a 3D matrix with dimensions Width × Height × Depth where Depth refers to the number of color channels in the image.

TABLE 8.3
Major Benefits and Restrictions of Supervised Learning

Supervised Method	Benefits	Limitations
ANN	Strong error acceptance	
Knn	Capable of parallel processing and robust to noise	Depend on the architecture
Naïve Bayes	Employs trial and error	
Decision tree	No training step and nonparametric	
Random forest	Easy to implement	Expensive and subtle to noise
CNN	Needs homogeneous features	
RNN	Has real-time applicability and fast	

Convolutional layer. By using convolutions the output nodes will be computed and dot products are calculated between a set of weights called a filter and the values associated with a local region of the input.

Activation layer. In this layer, the convolved data is fed to an element-wise activation function called the rectified-linear unit (ReLu), which will determine whether an input node will trigger when the input is fed. A maximization function such as $\max(0, x)$ is used in the convolutional process with threshold set to 0.

Pooling and fully connected layers. To decrease the width and height of the output data a downsampling strategy is used in the pooling layer. The convolved and reduced features are then fed to the fully connected layer to every node in the volume of features being fed-forward.

In a recurrent neural network (RNN) sequential or time series data are used to solve ordinal or temporal problems such as natural language processing, speech recognition, and image captioning. A RNN model is one type of long-term short memory (LTSM). RNNs are also used to present issues related to gradient determination, classification, processing, and determining predictions based on time series data. The major benefits and limitations of supervised DL methods discussed earlier are provided in Table 8.3 [35,36].

8.3.2 Unsupervised Learning

In unsupervised learning, a given set of unlabeled data is used to accurately predict the output data. The unsupervised algorithms have been widely used for aggregation as well as clustering and regression problems. Some most commonly used unsupervised algorithms are: K-means, Self-organizing maps (SOMs), hidden Markov model (HMM), auto-encoders (AEs), principal component analysis (PCA), and restricted Boltzmann machine (RBM). In [32] it has been observed that the performance can be enhanced by employing unsupervised learning algorithms. Below is a brief description of some unsupervised learning algorithms.

***K*-means.** It is an algorithm used to generate various different clusters from the unlabeled raw input data [32]. The K-means algorithm allocates each new input data to a cluster with the rule that its distance is from the nearest associated centroid. The allocated data points that have been previously allocated are used to create centroids.

Modulation and Detection for 5G and Beyond Wireless Systems 315

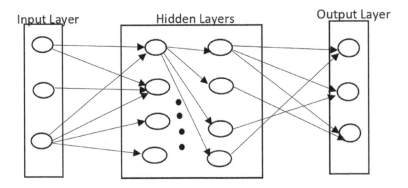

FIGURE 8.4 Self-organizing map model.

Self-organizing maps (SOMs). In issues related to reduction of data clustering and dimensionality the SOM algorithm can be used [32]. The SOM algorithm creates a map during the training process by using a competitive unsupervised learning approach. In [32] it has been shown that during a competition the winning neuron will determine the cluster in which any new input vector is classified. Figure 8.4 shows a conventional SOM model.

Auto-encoders. The goal of the AE unsupervised learning algorithms is to have the minimum possible deviation from some learning circuits that copy inputs into the output data. These algorithms yield excellent performance when applied to problems related to classification and regression. AEs employ stacked approaches and in the training process both supervised and unsupervised learning methods are used. The benefits and limitations of the most commonly used unsupervised DL algorithms are tabulated in Table 8.4.

TABLE 8.4
Benefits and Limitations of Unsupervised Learning

Unsupervised Learning Approach	Benefits	Limitations
K-means	Simple implementation and suitable to large datasets	Parameter k large dependent on performance
SOM	Simple to analyze, design and implement	
Auto-encoders	Capable to work with large complex datasets	Requires big data
PCA	Creates similarities and errors in the sample vectors	
HMM	De-noising training and can reduce dimensionality	Highly complex and prone to overfitting
RBM	Fast and can reduce dimensionality	Unable to work with features that are not linear

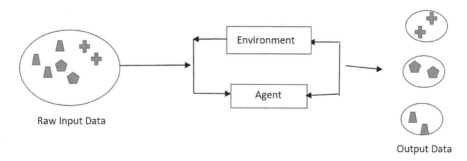

FIGURE 8.5 Reinforcement learning.

TABLE 8.5
Benefits and Limitations of Reinforcement Learning

Reinforcement Learning	Benefits	Limitations
Q-learning	Relatively fast and low cost of implementation	
SARSA	Efficient for offline learning	Not effective for online learning
Policy gradient	High per-sample variance	
Actor–critic	Fast and efficient for online learning datasets	Learning methods that are offline do not offer efficiency

8.3.3 Reinforcement Learning

Inspired by the way a child learns to carry out a new task and based on the principles of behaviorist psychology the reinforcement learning model has been developed [31]. The reinforcement learning algorithm has been extensively studied in [37], which provides excellent performance with algorithms employed such as Q-learning, SARSA, and policy-based algorithms such as policy gradient and actor–critic [33,34,38]. Figure 8.5 illustrates an example of reinforcement learning.

Some advantages and limitations of the most common reinforcement learning algorithms are listed in Table 8.5.

8.4 DEEP LEARNING AND MODULATION IN BEYOND 5G NETWORKS

Modulated waveforms are crucial communication components for dynamic and heterogeneous networks for 5G and beyond wireless technologies. It is imperative to accurately select and identify a specific modulation type, which the transmitter will use at a given time to demodulate and decode the data with high reliability. Hence considerable challenges exist in the DL algorithms for modulation classification. In this section, a DL framework in the form of modified architectures of CNN is applied to the modulation and detection processes of a beyond 5G network [39].

The architectures referred to in this work is called CLDNNs for convolutional LSTM (LTSM) deep neural networks that employ a learning parameter known as the mean cumulative sum (MCS), which significantly improves the modulation classification accuracy. To reduce the dimensions of the DL algorithm a statistical function called the PCA is used with a view to reduce the training time period. It will be shown that the CLDNN model outperforms a conventional CNN one and takes less training time. Moreover, in this novel CLDNN architecture, it has been possible to classify ten modulation schemes, which makes it highly practical for 5G and beyond wireless systems.

Beyond 5G wireless communication systems have multi-path fading channels with many variations in the channel states. For this reason, it is necessary to use adaptive modulation schemes at the base stations to achieve an acceptable error performance in terms of block error rates. Therefore, modulation classification algorithm needs to be designed and implemented in beyond 5G networks. Moreover, with the advent of multiple-input multiple-output (MIMO) systems with very large antenna arrays there are more degradation of the signal waveforms in terms of the power transmitted resulting in the identification of signals at the receiver side [39]. Hence a real-time system to classify the modulation schemes at the receiver is required. Recently, a few algorithms for automatic classifications of modulation and detection have been reported in [40]. These algorithms accurately allocate the modulation scheme but the classification procedure is time-consuming, which results in delays in the communication system.

According to the latency standard defined by the beyond 5G wireless system it is reported in [41] that the desired latency should be of the order of 0.1 ms. To meet this requirement of minimum latency and high accuracy, DL algorithms are employed in the modulation classification process. It has been recently shown in [40] that it is possible to use deep learning-based modulation classifications for MIMO systems employing CNN learning algorithms. In this work, an architecture of DL called CLDNN that combined CNN, LSTM, and DenseNet for modulation classification is described. It will be shown that the memory unit in the LSTM part of the architecture increases the memory unit while the DenseNet section will reduce the latency. To further decrease the latency in terms of training time, the PCA will be used.

8.4.1 CLDNN-Based Modulation

The classification system of modulation schemes is shown in Figure 8.6. Consider the transmitter that sends $s(t; u_i)$ baseband data to a receiver that retrieves signal $r(t)$, which is fed to the analog-to-digital converter (ADC) followed by the PCA

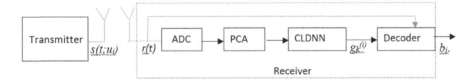

FIGURE 8.6 CLDNN communication system.

component. The complex received signal $r(t)$ can be represented by equation (8.1), where $n(t)$ denotes complex Gaussian noise and $s(t)$ in equation (8.1) can be written as in equation (8.2). The symbol $g_k^{(i)}$ represents k-modulated symbols taken from the ith modulated symbol period and ξ indicates the normalized epoch for the time offset between the transmitter and receiver, respectively. The output of the PCA component is of reduced dimension and it is fed to the CLDNN section whose output is then sent to the decoder that outputs symbol b_i.

$$r(t) = s(t) + n(t) \tag{8.1}$$

$$s(t) = b_i e^{j2\pi\Delta ft} e^{j\varnothing} \sum_{k=1}^{k} e^{j\theta_k} g_k^{(i)} p(t - (k-1)T - \xi T) \tag{8.2}$$

8.4.1.1 Architecture of the CLDNN

The CLDNN architecture proposed in this work is illustrated in Figure 8.7, which combines CNN, LSTM, and dense network models. The CNN with L layers comprising L connections between each layer and its next one thereby providing $L(L+1)/2$ direct connections. Moreover, a LSTM architecture is combined with that of the CNN with the input as time series modulated data. In this architecture, features are extracted and used for the modulation classifications. A typical LSTM architecture introduced in [42] is employed in this work and is shown in Figure 8.8. As depicted in Figure 8.7 a DenseNet architecture is connected to the LSTM one. Also, smaller training set sizes have been observed and the vanishing-gradient problem is mitigated and feature propagation is strengthened.

In this DL framework, there is one additional component used called the MCS. It has been shown in [39] that the combined architecture of CNN, LSTM, and DenseNet along with the MCS pooling provides significantly improved performance and efficiency.

8.4.1.2 Principal Component Analysis

PCA is a multivariate analysis described in [43], also called a factor analysis, and considered as an effective technique to reveal the internal structure of the data in terms of its variance. The eigenvector parameter is principally used in the PCA to represent a picture in a reduced lower dimensional space. It has been demonstrated in [43] that the dimension of the data can be significantly reduced. Figure 8.9 shows the application of PCA to modulation classification scheme employed in this research

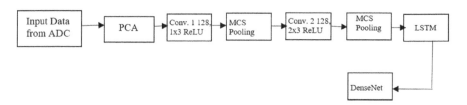

FIGURE 8.7 CLDNN architecture.

Modulation and Detection for 5G and Beyond Wireless Systems 319

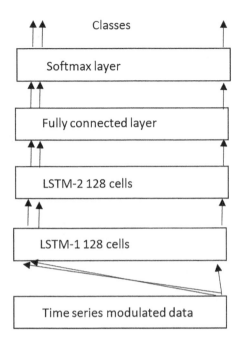

FIGURE 8.8 LSTM architecture.

FIGURE 8.9 Components of PCA to CLDNN.

work. The test data is inputted to the centering block that subtract a mean from it and afterward the covariance matrix is determined from the mean subtracted value. Then, the eigenvector and value are computed and finally the principal component vectors are found from the eigen data and the resulting information is fed as an input to the CLDNN architecture.

Table 8.6 shows the dataset attributes employed in this work. A continuous voice data is used as input to the DL architecture, which consists of acoustic speech with some interludes and off times.

Training and testing phases. The experimental setup, described in [39], was carried out at the Keras library. Almost 840,000 samples have been used for training phase and the remaining 360,000 samples have been reserved for testing phase. An Adam optimizer, which is an optimal-based DL parameter, is used with a learning parameter rate set to 0.0018.

Results and discussions. The simulation results obtained from the experimental setup will now be presented and discussed in detail for each component of the CLDNN architecture.

TABLE 8.6
Dataset Attributes in PCA

Dataset Parameter	Specification
Number of samples	12,000,00
Number of modulation scheme	10
Modulation scheme	CPFSK, BFSK, 16QAM, 64QAM, BPSK, QPSK, 8 PSK, 4PAM, AM-DSB, WB-FM
Temporal resolution	128 μs each
Sampling rate	1 million samples/seconds

The results obtained in [39] clearly demonstrates that less accurate performance at low SNR is observed but at high SNR there is considerable improved performance. It has been shown that at high values of SNR, an accuracy of 72% is achieved in the correct modulation classification scheme.

8.4.1.3 CLDNN Architecture

The simulated data employed by the experiment is tabulated in Table 8.7 for the entire CLDNN architecture.

It has been observed that with low signal power of the modulated signal, that is for a case of SNR at −20 dB, the CLDNN architecture does not perform well and there are few desired features registered. However, when SNR grows above 0 dB there is significant diagonalization indicating that most modulation schemes are being correctly classified. It is worth noting that with this CLDNN architecture an accuracy of 98% is recorded with a low SNR of −1 dB. This clearly indicates that compared to the classical CNN this new architecture performs much better. The accuracy curves shown in [39] were obtained for varying distances and plotted for comparison purposes.

8.4.1.4 Error Performance of CLDNN Architecture

The error rate performance of any communication system is crucial for the evaluation and effectiveness of such system. In this section, the bit error rate probability (P_b), is derived and analyzed for a DL framework that employs the CLDNN architecture. It is assumed that BPSK modulation is used and the channel is modeled as additive

TABLE 8.7
Dataset Attributes in CLDNN

Training and Testing	SNR	Modulation Schemes
70% (840,000) samples, 30% (360,000) samples	−20 to +18 dB	BPSK, AM-DSB, CPFSK, GFSK, PAM

Modulation and Detection for 5G and Beyond Wireless Systems 321

white Gaussian noise (AWGN). Mathematically, the received signal at the demodulator side of the CLDNN network can be found by using equation (8.3), where p is the transmitted signal with BPSK modulation and $\rho \in \{A, -A\}$ assuming amplitudes of the BPSK signal are A and $-A$; n is called the AWGN component defined as $n \sim CN(0, \sigma^2)$; $\sigma^2 = No$ equivalent to the spectral density of noise; and P_ε denotes an error factor defined as $0 < P_\varepsilon < 1$.

$$q = p + n + P_\varepsilon \quad (8.3)$$

Moreover, to introduce the error in a modulation classification scheme the parameter ε is defined as complex and represented as $\varepsilon \{Re(-A, +A) + Im(-A, +A)\}$. Due to a wrong classification scheme then a wrong modulation and constellation will be chosen. Furthermore, P represents a parameter defined as $P = 1-a$, where a is called the accuracy with $P = 1$ indicating that an error always occurs and $P = 0$ then no error is introduced in the modulation and detection processes. The mean and variance of ε is taken to be 0 and $2p^2$, respectively. Hence the bit error probability can be found by using equation (8.6) is given next. Using the Gaussian Q-function it can be shown that (8.6) reduces to (8.7).

$$\eta_{re} = n_{re} + \varepsilon_{re} \sim N\left(0, \frac{\sigma^2}{2} + p^2\right) \quad (8.4)$$

$$\eta_{Im} = n_{Im} + \varepsilon_{Im} \sim N\left(0, \frac{\sigma^2}{2} + p^2\right) \quad (8.5)$$

$$P_b = P(\eta > A) = \int_A^\infty \frac{1}{\sqrt{2\pi\left(\frac{\sigma^2}{2} + p^2\right)}} \exp\left\{\frac{-x^2}{2\left(\frac{\sigma^2}{2} + p^2\right)}\right\} dx \quad (8.6)$$

$$P_b = Q\left(\frac{A}{\sqrt{\frac{\sigma^2}{2} + p^2}}\right) \quad (8.7)$$

Assuming an accuracy of $a = 0.99$ then $P = 1 - a = 0.01$ and if the SNR is varied from 0 to 12 dB then the bit error probability of a BPSK modulation scheme for CLDNN model can be found using (8.5). It can be observed from [39] that at a low error rate of 1×10^{-5} the SNR is at a low value of 10 dB. It has been further shown in [39] that the CLDNN model outperform a simple CNN one since the proposed CLDNN architecture is composed of LSTM and DenseNet models.

8.4.2 DEEP NEURAL NETWORK ARCHITECTURES FOR MODULATION CLASSIFICATIONS

To achieve a high accuracy of correctly recognizing the modulation scheme of signals in the beyond 5G network, it is imperative to fine-tune the CNN architecture and add more convolutional and dense layers [44]. It has been shown that approximately 75% accuracy can be achieved in the wireless signal recognition procedure by employing DL architectures. Improved accuracy of approximately 83.8% has been observed at high SNR in [44]. Furthermore, it is possible to increase accuracy by employing residual networks and densely connected networks.

The architecture of a 7-layer CNN, as illustrated in Figure 8.10, consists of input/output data plane as well as four convolutional layers with two dense layers. The hyper parameters that are optimized are learning rate, drop-out rate, filter size, number of filters per layer, and the network depth. The dense layers have been configured to contain 128 and 11 neurons in order of their depth in the network.

It has been observed in [44] that the DenseNet architecture improves the information flow between layers than ResNet does because each layer obtains additional inputs from all preceding layers and passes on its own feature maps to all subsequent layers. Another efficient DL architecture has been designed and implemented in [44] that significantly improves the accuracy of modulation classifications. The latter consists of eight layers comprising of four convolutional layers, one LSTM layer, and two DenseNet layers. The LSTM layer has 50 computing units, which yield the best accuracy result compared to other layered architectures.

Table 8.8 illustrates the error introduced in the wireless signal modulation classification process using the eight-layered architecture. The number of columns shows

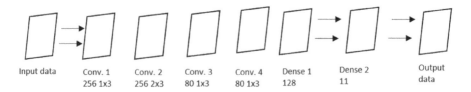

FIGURE 8.10 Seven-layer CNN deep learning architecture.

TABLE 8.8
Accuracy Results for Various Modulation Schemes

Error in Modulation Classification	Percentage
8PSK/QPSK	5.5
16QAM/64QAM	58.48
64QAM/16QAM	20.14
Wideband FM/AM-DSB	59.6
Wideband FM/GFSK	3.3

the % of the left-hand side modulation type that is misclassified as the modulation type on the right-hand side. It can be clearly seen that a low error occurs for 8PSK/QPKS classification and also for the wideband FM/GFSK recognition. However, for 16QAM/64QAM wireless modulation classification as well as wideband AM-DSB there are significant deviations with high percentages in error resulting in low accuracy.

8.5 CONCLUDING REMARKS

DL is presently playing a crucial role in the mobile and wireless networks domain. In this chapter DL models have been discussed, and further detailing how to tailor them to modulation and detection for 5G and beyond 5G networks. We have described the latest findings from several researchers and the basic principle of DL in mobile and wireless communications. The modified architecture of CNN called CLDNN has been explained and applied to modulation and detection in the form of modulation classifications. Moreover, the error rate performance for a DL framework employing the CLDNN is derived and analyzed in the presence of AWGN channel for BPSK modulation. It is shown that accuracy in the modulation classification algorithm is improved and low error rates at low SNR values are achieved. We conclude by hoping that this work will become a comprehensive guide to researchers and practitioners interested in applying DL to solve complex problems in modulation and detection for beyond 5G networks.

REFERENCES

1. J. Wang, C. Jiang, H. Zhang, Y. Ren, K. Chen, and L. Hanzo. Thirty years of machine learning: The road to Pareto-optimal wireless networks. *IEEE Communications Surveys & Tutorials*, 2020, vol. 22, no. 3, pp. 1472–1514.
2. H. Viswanathan and P.E. Mogensen. Communications in the 6G era. *IEEE Access*, 2020, vol. 8, pp. 57063–57074.
3. Y. Liu, S. Bi, Z. Shi, and L. Hanzo. When machine learning meets big data: A wireless communication perspective. *IEEE Vehicular Technology Magazine*, 2020, vol. 15, no. 1, pp. 63–72.
4. B. McMahan, E. Moore, D. Ramage, S. Hampson, and B.A. Arcas. Communication-efficient learning of deep networks from decentralized data. In A. Singh and J. Zhu (Eds.), *Proceedings of the 20th International Conference on Artificial Intelligence and Statistics. In: Proceedings of Machine Learning Research (PMLR)*, USA, 2017, vol. 54, pp. 1273–1282.
5. Y. LeCun, Y. Bengio, and G. Hinton. Deep learning. *Nature*, 2015, vol. 521, no. 7553, pp. 436–444.
6. J. Schmidhuber. Deep learning in neural networks: An overview. *Neural Networks*, 2015, vol. 61, pp. 85–117.
7. W. Liu, Z. Wang, X. Liu, N. Zeng, Y. Liu, and F. E Alsaadi. A survey of deep neural network architectures and their applications. *Neurocomputing*, 2017, vol. 234, pp. 11–26.
8. L. Deng and D. Yu. Deep learning: Methods and applications. *Foundations and Trends in Signal Processing*, 2014, vol. 7, no. 3–4, pp. 197–387.

9. L. Deng. A tutorial survey of architectures, algorithms, and applications for deep learning. *APSIPA Transactions on Signal and Information Processing*, 2014, vol. 3, p. e2.
10. S. Pouyanfar, S. Sadiq, Y. Yan, H. Tian, Y. Tao, M. P. Reyes, M.-L. Shyu, S.-C. Chen, and S. S. Iyengar. A survey on deep learning: Algorithms, techniques, and applications. *ACM Computing Surveys (CSUR)*, 2018, vol. 51, no. 5, pp. 92:1–92:36.
11. K. Arulkumaran, M. P. Deisenroth, M. Brundage, and A. A. Bharath. Deep reinforcement learning: A brief survey. *IEEE Signal Processing Magazine*, 2017, vol. 34, no. 6, pp. 26–38.
12. A. Hussein, M. M. Gaber, E. Elyan, and C. Jayne. Imitation learning: A survey of learning methods. *ACM Computing Surveys (CSUR)*, 2017, vol. 50, no. 2, pp. 21:1–21:35.
13. O. A. Dobre, A. Abdi, Y. Bar-Ness, and W. Su. Survey of automatic modulation classification techniques: Classical approaches and new trends. *IET Communications*, 2007, vol. 1, no. 2, pp. 137–156.
14. L.-X. Wang and Y.-J. Ren. Recognition of digital modulation signals based on high order cumulants and support vector machines. In *Proceedings of the 2009 Second ISECS International Colloquium on Computing, Communication, Control, and Management (CCCM 2009)*, China, August 2009, pp. 271–274.
15. A. Voulodimos, N. Doulamis, A. Doulamis, and E. Protopapadakis. Deep learning for computer vision: A brief review. *Computational Intelligence and Neuroscience*, 2018, vol. 2018, p. 13.
16. S. P. Singh, A. Kumar, H. Darbari, L. Singh, A. Rastogi, and S. Jain. Machine translation using deep learning: An overview. In *Proceedings of the 1st International Conference on Computer, Communications and Electronics (COMPTELIX 2017)*, India, July 2017, pp. 162–167.
17. T. Young, D. Hazarika, S. Poria, and E. Cambria. Recent trends in deep learning based natural language processing. *IEEE Computational Intelligence Magazine*, 2018, vol. 13, no. 3, pp. 55–75.
18. J. Schmidhuber. Deep learning in neural networks: An overview. *Neural Networks*, 2015, vol. 61, pp. 85–117.
19. Y. Lecun, Y. Bengio, and G. Hinton. Deep learning. *Nature*, 2015, vol. 521, no. 7553, pp. 436–444.
20. G. E. Hinton and R. R. Salakhutdinov. Reducing the dimensionality of data with neural networks. *American Association for the Advancement of Science: Science*, 2006, vol. 313, no. 5786, pp. 504–507.
21. K. Yashashwi, A. Sethi, and P. Chaporkar. A learnable distortion correction module for modulation recognition. *IEEE Wireless Communications Letters*, 2018, vol. 8, no. 1, pp. 77–80.
22. R. Li, L. Li, S. Yang, and S. Li. Robust automated VHF modulation recognition based on deep convolutional neural networks. *IEEE Communications Letters*, 2018, vol. 22, no. 5, pp. 946–949.
23. B. Kim, J. Kim, H. Chae, D. Yoon, and J. W. Choi. Deep neural network-based automatic modulation classification technique. In *Proceedings of the 2016 International Conference on Information and Communication Technology Convergence (ICTC 2016)*, Republic of Korea, October 2016, pp. 579–582.
24. F. N. Khan, C. H. Teow, and S. G. Kiu et al. Automatic modulation format/bit-rate classification and signal-to-noise ratio estimation using asynchronous delay-tap sampling. *Computers and Electrical Engineering*, 2015, vol. 47, pp. 126–133.
25. F. N. Khan, C. Lu, and A. P. T. Lau. Joint modulation format/bit-rate classification and signal-to-noise ratio estimation in multipath fading channels using deep machine learning. *IEEE Electronics Letters*, 2016, vol. 52, no. 14, pp. 1272–1274.

26. G. J. Mendis, J. Wei, and A. Madanayake. Deep learning-based automated modulation classification for cognitive radio. In *Proceedings of the 2016 IEEE International Conference on Communication Systems (ICCS 2016)*, China, December 2016, pp. 1–6.
27. F. Wang, Y. Wang, and X. Chen. Graphic constellations and DBN based automatic modulation classification. In *Proceedings of the 2017 IEEE 85th Vehicular Technology Conference (VTC Spring)*, Sydney, NSW, June 2017, pp. 1–5.
28. N. E. West and T. O'Shea. Deep architectures for modulation recognition. In *Proceedings of the 2017 IEEE International Symposium on Dynamic Spectrum Access Networks (DySPAN 2017)*, USA, March 2017, pp. 1–6.
29. L. Zhou, Z. Sun, and W. Wang. Learning to short-time Fourier transform in spectrum sensing. *Physical Communication*, 2017, vol. 25, pp. 420–425.
30. M. Chen, U. Challita, W. Saad, C. Yin, and M. Debbah. Artificial neural networks-based machine learning for wireless networks: A tutorial. *IEEE Communications Surveys & Tutorials*, 2019, vol. 21, pp. 3039–3071.
31. S.J. Nawaz, S.K. Sharma, S. Wyne, M.N. Patwary, and M. Asaduzzaman. Quantum machine learning for 6G communication networks: State-of-the-art and vision for the future. *IEEE Access*, 2019, vol. 7, pp. 46317–46350.
32. S. Zhang and D. Zhu. Towards artificial intelligence enabled 6G: State of the art, challenges, and opportunities. *Computer Networks*, 2020, vol. 183, p. 107556.
33. H. Dahrouj, R. Alghamdi, H. Alwazani, S. Bahanshal, A.A. Ahmad, A. Faisal, R. Shalabi, R. Alhadrami, A. Subasi, and M. Alnory. An overview of machine learning-based techniques for solving optimization problems in communications and signal processing. *IEEE Access*, 2021, vol. 9, pp. 74908–74938.
34. I. Zhou, I. Makhdoom, N. Shariati, M.A. Raza, R. Keshavarz, J. Lipman, M. Abolhasan, and A. Jamalipour. Internet of things 2.0: Concepts, applications, and future directions. *IEEE Access*, 2021, vol. 9, pp. 70961–71012.
35. R.A. Nugrahaeni and K. Mutijarsa. Comparative analysis of machine learning KNN, SVM, and random forests algorithm for facial expression classification. In *Proceedings of the 2016 International Seminar on Application for Technology of Information and Communication (ISemantic)*, Semarang, Indonesia, 5–6 August 2016, pp. 163–168.
36. K.M. Al-Aidaroos, A.A. Bakar, and Z. Othman. Naive Bayes variants in classification learning. In *Proceedings of the 2010 International Conference on Information Retrieval and Knowledge Management (CAMP)*, Shah Alam, Malaysia, 17–18 March 2010, pp. 276–281.
37. M.S. Mollel, A. I. Abubakar, M. Ozturk, S.F. Kaijage, M. Kisangiri, S. Hussain, M.A. Imran, and Q.H. Abbasi. A survey of machine learning applications to handover management in 5G and beyond. *IEEE Access*, 2021, vol. 9, pp. 45770–45802.
38. S. Mohammed, S. Anokye, and S. Guolin. Machine learning based unmanned aerial vehicle enabled fog-radio aerial vehicle enabled fog-radio access network and edge computing. *ZTE Communications*, 2020, vol. 17, pp. 33–45.
39. J.C. Clement, N. Indira, P. Vijayakumar, et al. Deep learning based modulation classification for 5G and beyond wireless systems. *Peer-to-Peer Networking and Applications*, 2021, vol. 14, pp. 319–332.
40. J. Nie, Y. Zhang, Z. He, S. Chen, S. Gong, and W. Zhang. Deep hierarchical network for automatic modulation classification. *IEEE Access*, 2019, vol. 7, no. 94, pp. 604–613.
41. K.M.S. Huq, S.A. Busari, J. Rodriguez, V. Frascolla, W. Bazzi, and D.C. Sicker. Terahertz-enabled wireless system for beyond-5g ultra-fast networks: A brief survey. *IEEE Network*, 2019, vol. 33, no. 4, pp. 89–95.
42. T.N. Sainath, O. Vinyals, A. Senior, and H. Sak. Convolutional, long short-term memory, fully connected deep neural networks. *2015 IEEE International Conference on Acoustics, Speech and Signal Processing (ICASSP)*, 2015, pp. 4580–4584.

43. J.M. Li, Y.H. Hu, and X.H. Tao. Recognition method based on principal component analysis and back-propagation neural network. *Infrared and Laser Engineering*, 2019, vol. 34, no. 6, p. 719.
44. X. Liu, D. Yang, and A. E. Gamal. Deep neural network architectures for modulation classification. In *2017 51st Asilomar Conference on Signals, Systems, and Computers*, 2017, pp. 915–919.

9 AI-Based Channel Coding for 5G/6G

Madhavsingh Indoonundon and
Tulsi Pawan Fowdur
University of Mauritius

Zoran S. Bojkovic and Dragorad A. Milovanovic
University of Belgrade

CONTENTS

9.1 Introduction 328
9.2 Application of ML in 5G NR LDPC Codes 329
    9.2.1 5G NR LDPC Codes 330
    9.2.2 LDPC Codes with Neural Networks 330
        9.2.2.1 Linear Approximation Min-Sum-Based ML for Optimizing LDPC Decoding 330
    9.2.3 Neural MS Decoder for Protograph-Based LDPC Codes 331
    9.2.4 Blind Recognition Method Using Convolutional Neural Networks ... 332
    9.2.5 Deep Learning-Based Unified Polar-LDPC Decoder 333
    9.2.6 Neural Normalized Min-Sum LDPC Decoding Network 334
    9.2.7 Neural 2D NMS (N-2D-NMS) Decoders 335
    9.2.8 Neural Layered MS Decoder for Protograph-Based LDPC Codes ... 336
9.3 LDPC Codes with Reinforcement Learning 336
    9.3.1 Multiarmed Bandit-Based Node-Wise Scheduling (MAB-NS) Scheme 336
    9.3.2 Reinforcement Learning and Monte Carlo Tree Search 337
9.4 Application of ML in 5G NR Polar Codes 338
    9.4.1 5G NR Polar Codes 338
    9.4.2 Polar Codes with Neural Networks 339
        9.4.2.1 Neural Network-Based Frame Error Rate Prediction of Polar Codes 339
        9.4.2.2 Transfer Learning-Based Decoder Training Method 340
        9.4.2.3 A CNN-Based Polar Decoder 340
        9.4.2.4 A Residual NND for Polar Codes 341
        9.4.2.5 A Differentiable Neural Computer-Aided Flip Decoding Algorithm 342
        9.4.2.6 Neural Network-Based Bit Flipping 343
        9.4.2.7 Stacked Denoising Autoencoder for Polar Codes 344
        9.4.2.8 A Machine Learning-Based Multi-Flips SC Decoding Scheme 344

		9.4.2.9	Double Long–Short-Term Memory-Based SC Flipping Decoder 346
	9.4.3	Polar Codes with Reinforcement Learning 346	
		9.4.3.1	Reinforcement Learning for Polar Codes Construction 346
		9.4.3.2	A Reinforcement Learning-Aided CRC-Aided BP Decoder 348
9.5	Gaps in Previous Research 348
9.6	Summary 349
Bibliography 349

9.1 INTRODUCTION

5G New Radio (NR) is the latest generation of mobile communications networks, which uses state-of-the-art technologies to provide high-performance mobile connectivity, capable of supporting a wide variety of vertical industries. The technical specifications of 5G NR are actively being defined by the Third-Generation Partnership Project (3GPP). The 3GPP Release 14 introduced 5G NR with a thorough study of the technology. Release 15 onward consisted of the technical specifications for 5G NR.

5G NR is being designed to suit three main service classes mainly:

1. **Enhanced Mobile Broadband (eMBB).** 5G NR should meet peak data rate requirements of 20 Gbps and 10 Gbps in the downlink and uplink, respectively, and a user plane latency of 4 ms (ITU, 2017). Such specifications allow 5G to support data-intensive services.
2. **Ultra-Reliable Low-Latency Communications (URLLC).** 5G NR should be capable of providing a latency as low as 1 ms with a reliability of 10^{-5} for packets of 32 bytes (ITU, 2017). These specifications should allow 5G to support delay-sensitive services such as autonomous driving and telesurgery.
3. **Massive Machine-Type Communications (mMTC).** 5G NR networks should support a connection density of 1 million connected devices per km^2 (ITU, 2017) with low-energy communications for enabling the Internet of Things (IoT).

Channel codes are essential components in the physical layer of the 5G NR communications system, which have a significant impact on the performance of the latter. Currently, low-density parity check (LDPC) codes and polar codes have been included in 5G NR for the data channel and control channels, respectively, for eMBB services (Zhu et al., 2018). The channel codes used in 5G NR need to meet strict requirements such as a high decoding throughput of at least 20 Gbps, a low decoding latency, good error correction capabilities, flexibility to adapt to various data sizes, and low complexity (Indoonundon and Fowdur, 2021).

Machine learning (ML) is a mechanism widely used for solving high-complexity problems by relying on a machine, which has undergone learning through the automatic identification of meaningful patterns in data. Recently, as high computing

AI-Based 5G/6G Communications

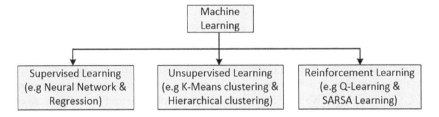

FIGURE 9.1 Types of machine learning algorithms.

power has become more accessible, ML has become of significant help to researchers in several fields by providing effective solutions to hard problems and useful insights into bulks of data. Over the past decade, researchers have also integrated ML with channel codes to provide design and performance enhancements. For example, neural networks are being used to learn and execute tasks in decoders to reduce their computational complexity and ideal decoders' parameters such as scaling factors and normalization factors based on specific channel conditions that are determined by ML algorithms to provide performance gains. An in-depth literature review of such schemes is given in Sections 9.2 and 9.3.

ML algorithms can be categorized as either supervised or unsupervised learning algorithms. Supervised learning algorithms require details about the expected output for given inputs to be fed to the machine for training purposes. This generally implies that additional efforts from a human operator are required. Supervised learning algorithms are mainly used for regression and classification. On the other hand, unsupervised learning algorithms do not require such details. However, to achieve high accuracy, unsupervised learning requires larger datasets to identify similarities and differences between them. Unsupervised learning is mainly performed for data clustering (Fowdur et al., 2021). One ML technique, which has paved its way in numerous fields is reinforcement learning. In reinforcement learning, an agent is placed in a training environment in which it takes random actions to cumulate rewards. The agent learns to maximize its rewards based on the sequences of actions it can take and thus eventually learn the best course of action required to solve a problem. The different types of ML algorithms are summarized in Figure 9.1, along with examples for each of them.

ML algorithms can also be categorized as either shallow learning or deep learning algorithms. While shallow learning consists of three learning steps, namely data collection, manual feature selection, and regression, deep learning requires only the data collection step and combines the feature selection and regression steps in a single step (Fowdur et al., 2021).

9.2 APPLICATION OF ML IN 5G NR LDPC CODES

In this section, an overview of LDPC codes used in 5G NR is first given followed by a review of ML techniques that have been employed to enhance the performance of these codes.

9.2.1 5G NR LDPC Codes

LDPC codes are forward error correction codes, which were introduced by Gallager (1962). They were, however, initially ignored due to their computational complexity and the limited access to high computing power at that time and were rediscovered by MacKay and Neal in 1996. Since then, there has been significant research focus on LDPC codes as they have near Shannon limit performances and were therefore considered to be the most powerful classes of error-correcting codes developed. LDPC codes have been adopted in multiple communications standards such as WiMax, DVBT2, and IEEE 802.11n (Fowdur and Indoonundun, 2017).

More recently, LDPC codes have been chosen as the channel codes for the data channel of 5G NR eMBB services (Session Chairman [Nokia], 2016). 5G NR employs protograph-based quasi-cyclic (QC) LDPC codes based on two rate-compatible base graphs, which can generate parity check matrices using a defined lifting size. The first base graph is optimized for the transmission of large blocks of sizes of up to 8,448 bits at high code rates of up to 8/9. On the other hand, the second base graph is used mainly for the transmission of small blocks of sizes of up to 3,840 bits at codes rates as low as 1/5 (3GPP, 2019).

The LDPC decoding algorithms supported by 5G NR are the belief propagation (BP), layered belief propagation, normalized min-sum, and offset min-sum algorithms. Each of these algorithms has different error performances, computational complexities, and decoding latencies and therefore the choice of the decoding algorithms should be made carefully based on the service to be provided.

Recently, several research studies have proposed LDPC decoding schemes using ML to further enhance the error performance. An overview is given in the following subsections.

9.2.2 LDPC Codes with Neural Networks

9.2.2.1 Linear Approximation Min-Sum-Based ML for Optimizing LDPC Decoding

Wu, Jiang, and Zhao (2018) proposed a new scheme for optimizing simplified LDPC BP decoding algorithms with the use of ML. They introduced a generalized min-sum (MS) algorithm, named the linear approximation min-sum (LAMS) in which the check node outputs are altered by a linear function containing both a normalized and an offset factor. These factors are optimized using a neural network, which runs over a single iteration of the decoding process. After an iteration, the optimized factors determined by the neural network are used to calculate the log-likelihood ratios (LLRs) for the following iteration. This approach of finding the optimal factors for each iteration instead of a globally optimal factor for all iterations allows the neural network to be shallow and thus less complex. The low complexity of the neural network allows the LAMS decoder to be optimized for LDPC codes of diverse code lengths and iteration numbers.

In the input layer of the neural network, the received LLRs are assigned to the next layer similar to the message-passing process from variable nodes to check nodes. In the hidden layer, LLRs are calculated in a similar way as in the check

AI-Based 5G/6G Communications

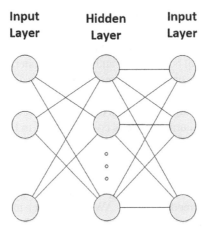

FIGURE 9.2 Structure of proposed neural network (Wu et al., 2018).

node update process. Finally, in the output layer, the LLRs are updated using the linear adjustment method of the magnitude of the channel LLRs. The neural network optimizes the factors in the hidden and output layers (Wu et al., 2018). The structure of the neural network is illustrated in Figure 9.2. The key network requirements for space-to-ground networking capability are flexible addressing and routing, satellite bandwidth capability, admission control of satellites, admission control by satellites, edge computing, and storage.

The authors used TensorFlow to build the neural network, using a learning rate of 0.1 with a total of 500 rounds of training and a batch size of 500. The training dataset consisted of the LLRs obtained from a binary phase-shift keying (BPSK) signal of an all-zero codeword transmitted over an additive white Gaussian noise (AWGN) channel with an Es/N0 of −4 dB. The performance of the trained neural network was then evaluated using simulations. The simulation results show that with 15 iterations, the proposed LAMS decoder outperforms both the conventional normalized min-sum (NMS) and offset min-sum (OMS) algorithms. At SNRs above −2.5 dB, the proposed algorithm is also found to outperform the BP algorithm as the latter is not optimized for low iteration numbers and short codewords (Wu et al., 2018). By providing a better error performance than BP, NMS and OMS which are used in 5G NR for eMBB, the proposed scheme is a good candidate for eMBB in future 5G NR releases.

9.2.3 Neural MS Decoder for Protograph-Based LDPC Codes

A novel neural MS decoder that utilizes the lifting structure of protograph-based LDPC (PB-LDPC) codes has been proposed in Dai et al. (2021). The authors used 5G NR's LDPC codes to demonstrate the performance of their decoder. In PB-LDPC codes, the structural properties of the base graph are identical to that of the lifted graph. This allowed the authors to use a parameter-sharing mechanism whereby groups of edges that are obtained from the same edge from the base graph share the same parameters. This mechanism reduces the decoder's computational complexity by minimizing the parameter array's size and makes it easily applicable to LDPC codes with a variety of codelengths (Dai et al., 2021).

For training their neural network, the authors used an iteration-by-iteration greedy training method in which only the parameters from the decoder's last iteration are learnable. This method resolves the vanishing gradient problems in the MS decoder and it also eliminates message correlation caused by short cycles in the LDPC codes. This method allows the neural decoder's configurations to be determined in a single iteration and therefore reduces the training complexity (Dai et al., 2021).

During the training phase of the neural network, the authors use LDPC codes of different code lengths and code rates, obtained from a single base code to address the overfitting problem, making their network become more generalized. Simulations were performed using an AWGN channel to compare the proposed decoder against the conventional sum-product (SP), MS, NMS, and OMS decoders. Simulation results showed that for short LDPC codes with lifting size, $Z=3$, the proposed neural MS decoder outperforms the conventional decoders whereas, for medium and long LDPC codes, the decoder is slightly outperformed by the SP decoder. The performance gain of the proposed neural MS decoder over the SP decoder for short codes can be attributed to the better approximation that the former provides for the normalization and offset factors. The authors also explained that the performance boost is generally due to the elimination of short cycles, which their algorithm provides using well-learnt parameters. Thus, for longer codewords, the reduced number of short cycles allowed the SP decoder to provide better performance (Dai et al., 2021).

A complexity analysis performed in their research showed that by avoiding hyperbolic tangent functions, the proposed neural MS decoder provides a similar computational complexity to that of NMS and OMS. However, the storage of additional parameters required by the neural MS decoder increases the latter's memory requirements (Dai et al., 2021).

9.2.4 BLIND RECOGNITION METHOD USING CONVOLUTIONAL NEURAL NETWORKS

Blind recognition is a technique used by receivers to estimate channel code parameters such as the code rate and code length. This technique is particularly useful when the parameters are dynamically changed based on the channel conditions. A group of researchers proposed a novel LDPC codes' blind recognition method, which relies on convolutional neural networks (CNNs) (Li et al., 2021). The method involves determining the LDPC matrix based on the received codeword and a candidate set, which is based on five LDPC codes used in 5G NR.

The proposed method consists of two main components. The first one retrieves the demodulator's output to generate a feature matrix, which splits the output into α rows and L columns where α is the smallest possible length of the parity check matrix and L is the number of vectors of size α, which can be obtained from the demodulator's output. The second component is a one-dimensional (1D) CNN, which evaluates the blind recognition outcome. The convolutional filters in the CNN use the feature matrix for feature extraction. The authors used the cross-entropy loss function and the Adam optimizer and trained the CNN using backpropagation (Li et al., 2021).

The authors opted for variable hyperparameters for the 1D CNN. The latter is composed of up to two convolutional layers, 64–256 filters per convolutional layer and a kernel size of 8–16. Rectified linear unit (ReLU) is used as the activation function in the convolutional layers, which are followed by batch normalization and a pooling layer. The convolutional layer's output is pooled by a global average and forwarded to a dense neural network, which performs the classification of the code parameter. The first layer of this dense neural network uses ReLU activation and the second layer employs a softmax function (Li et al., 2021).

To make their method work with several channel models, the author considered three channel models: the AWGN channel, the Rayleigh fading channel, and the Rician fading channel. The two later channels are also cascaded with an AWGN channel to perform SNR simulations. The neural network is trained and then simulations are performed to compare the proposed method with an existing conventional method. In the AWGN channel, the proposed method is found to determine the code parameters with similar accuracy as the conventional method. The authors claim that the accuracy of their method can further be increased by making their neural network deeper. When using the two other channels, the proposed method has a correct parameter identification probability above 70% when the SNR is greater than 1 dB, while the conventional method performs poorly with a correct parameter identification probability below 30% for all tested SNRs. This demonstrates the high performance of the proposed method (Li et al., 2021).

9.2.5 Deep Learning-Based Unified Polar-LDPC Decoder

With a view to save implementation resources, researchers have proposed a deep learning-based unified polar-LDPC decoder (Wang et al., 2018). The input to the neural network is a sequence consisting of the received symbols concatenated with an indication section, which is used to specify the channel coding type. The hidden layers are used to determine the decoding function (polar or LDPC codes) and employ the ReLU activation functions. The output layer employs the sigmoid activation function and outputs an estimation of the information bits. The mean squared error (MSE) loss function between the output estimated bits and the actual information bits is minimized through training

The authors justified their use of the deep neural network (DNN) over CNN and long–short-term memory (LSTM) through simulations where they demonstrated that LDPC and polar codes decoding using DNN provides better BER performance than with CNN and LSTM. The authors attribute this to the better ability of DNN to work with 1D data. Similarly, based on BER simulations, the indicator section is chosen to be of a single bit, which the authors may increase if more channel coding methods are to be included and the layer size of the DNN is opted to be (512, 256, 128), which is found to provide a reasonable BER performance at reasonable complexity. Training is performed using SNR values between 0 and 5 dB in 0.5 dB steps at 50,000 epochs to improve the generalization abilities of the decoder (Wang et al., 2018). The DNN is built to decode polar and LDPC codes of size (16, 8) and the indicator section is configured to have the value of '1' or '−1' for polar codes and LDPC codes, respectively. Testing is performed using 4×10^5 samples making a total of 3.2×10^6 bits (Wang et al., 2018).

When polar coding is used, the proposed DNN-based unified polar-LDPC decoder provides nearly the same performance as the conventional BP polar decoder. At a BER of 10^{-2}, the proposed decoder offers an Eb/N0 loss of only 0.2 dB with respect to the conventional decoder. The authors stated that this gap may be reduced by increasing the size of the neural network at the cost of increased complexity and they also justify the gap with the addition of the indicator section. On the other hand, when LDPC coding is used, the proposed decoder is capable of significantly outperforming the conventional BP LDPC decoder. At a BER of 10^{-2}, the proposed decoder provides an important Eb/N0 gain of 0.8 dB over the conventional decoder (Wang et al., 2018). Compared to the conventional BP decoder, the proposed decoder provides a throughput gain of six orders of magnitude. Furthermore, compared to the traditional isolated DNN-based decoders, the proposed decoder uses 0.3% less overhead at a tradeoff of 0.4% loss in throughput (Wang et al., 2018).

9.2.6 Neural Normalized Min-Sum LDPC Decoding Network

The high complexity of DNNs for long codes decoding has been addressed by a novel neural normalized min-sum (NNMS) LDPC decoding network (Wang et al., 2020), which accelerates the decoding process by unfolding the iterative decoding progress between the check nodes and variable nodes into a feedforward propagation network. The NNMS is then extended to a shared neural normalized min-sum (SNNMS) decoder, which reduces the complexity by reusing the same correction factors in each layer.

The system model used by the author involves executing LDPC encoding on information bits to obtain the LDPC codeword, which is modulated using BPSK before being transmitted through an AWGN channel. The LLRs of the received symbols are calculated and sent to the deep feedforward neural network, which estimates the information bits. In the NNMS network, for each iteration, the messages between the check node and variable node layers are multiplied by different correction factors, which is similar to setting weight values to the edges of the Tanner graph. The check node layer determines the messages sent from the check nodes to variable nodes, whereas the variable node layer determines the messages sent from the variable nodes to the check nodes. The neural network uses three types of neurons. One calculates the check nodes to variable node messages in the check node layer. Another type calculates the variable nodes to check node messages in the variable node layer and the last type calculates the output of the output layer. On the other hand, the proposed SNNMS decoder shares the same correction factors in the same layer, which brings minor changes to the equations used by the three different types of neurones. This technique is also more similar to the one used in conventional LDPC codes (Wang et al., 2020).

A complexity analysis based on the number of mathematical operations used in each decoder shows that the SNNMS decoder has lower complexity than the NNMS and nearly matches the complexity of the conventional normalized MS (CNMS) decoder. Simulations were performed with the SNNMS, NNMS, and CNMS decoders to evaluate their BER performances. The results showed that irrespective of the number of iterations used, the SNNMS outperformed both the two other decoders. At a BER of 10^{-7}, the SNNMS decoder provides a 0.4 dB gain over the NNMS. Hence, the lower complexity and better BER performance make the SNNMS decoder suitable for long LDPC codes (Wang et al., 2020).

9.2.7 Neural 2D NMS (N-2D-NMS) Decoders

Another similar work in which the NNMS decoder is further enhanced and adapted to short codes is the paper in which a family of neural 2D NMS (N-2D-NMS) decoders was introduced (Wang et al., 2021). The authors solve the high memory issue to implement the NNMS decoder by iteratively conducting backpropagation and storing parameters efficiently.

The proposed N-2D-NMS decoders make use of multiplicative weights applied to the messages in the check and variable nodes. The different types of N-2D-NMS they introduced are as follows (Wang et al., 2021):

The type 1 decoder applies similar weights to edges, which have the same variable and check node degrees. The type 2 decoder does not share the weights between the variable and check node degrees. The type 3 and type 4 decoders consider only variable node and check node degrees, respectively. The type 5 decoder assigns the same weight to edges of the same type. The type 6 and type 7 decoders are simplified versions of the type 5 decoder, which assign parameters only based on the horizontal and vertical layers, respectively. The type 8 decoder uses iteration-distinct parameters (Wang et al., 2021).

Simulations are performed to evaluate the proposed enhanced NNMS decoder and family of N-2D-NMS decoders with DVBS-2 LDPC codes and protograph-based raptor-like (PBRL) LDPC codes. BPSK modulation is used and the modulated message is transmitted through an AWGN channel. With DVBS-2 LDPC codes, results showed that the NNMS decoder outperforms the conventional BP decoder with a significant Eb/N0 gain of 1.3 dB. The NNMS decoder was outperformed by Type 1 and 2 decoders. As the variable node weights of the DVBS-2 LDPC codes had a larger dynamic range than the check node weights, the Type 2 and 4 decoders outperformed the Type 3 decoder (Wang et al., 2021).

The simulation results when PBRL LDPC codes were used showed that the NNMS decoder provided a 0.5 dB Eb/N0 gain over the conventional NMS decoder. The NNMS decoder provided the same performance as the Type 1, 2, and 5 decoders but the latter have a lower complexity due to the smaller number of parameters needed. It was also observed that the weight-sharing mechanisms did not result in performance degradation. However, assigning weights based on variable nodes instead of the check nodes resulted in performance degradation. The proposed NNMS decoder slightly outperformed the Type 4,7 decoders by at most 0.2 dB. The complexity reduction of this scheme could make it suitable for 5G mMTC services.

9.2.8 Neural Layered MS Decoder for Protograph-Based LDPC Codes

A group of researchers proposed a neural layered MS decoder for protograph-based LDPC codes, which exploits their lifting structure (Zhang et al., 2021). The decoding process is converted to a neural network in which messages are processed in a feed-forward manner. A conventional layered LDPC decoder processes subsets of check nodes called layers sequentially instead of all of the check nodes at once to improve the convergence speed.

The proposed decoder takes the vector of LLRs obtained from the channel output at the input layer and outputs processing elements at the output layer. The architecture of the conventional layered decoder is imitated by the neural network by splitting the layers into small groups containing a variable node sublayer, a check node sublayer, and a decision sublayer. The neurons in the sublayers forward messages delivered over the edge from the nodes to their adjacent nodes. The decision sublayer notes the most recent update and outputs temporary decoding results, which are used for calculating the loss function. The network's output layer sends the late decision sublayer's output as the final decoding result (Zhang et al., 2021).

The authors provide the flexibility to optimize both weights and biases unlike weighted SP decoder and neural NMS. These weights and biases are also made to vary in each iteration for increasing the level of optimization. Parameters are sharing for some edges within the same iteration, in order to minimize the complexity and the memory requirement of the neural network and to make the network compatible with arbitrary code lengths (Zhang et al., 2021).

MSE was used as the loss function between the actual message and the outputs of the decision sublayers. The loss function was minimized by gradient-based optimization techniques to optimize the decoder. To prevent the vanishing gradient problem, the weights and biases obtained after training one layer are used without any change in subsequent layers and only the weights and biases of the last layer are made learnable (Zhang et al., 2021).

The 5G NR LDPC codes, which are protograph-based, were used to build and test the performance of the proposed decoder. Simulation results showed that the proposed decoder outperforms the conventional layered MS decoder and even the conventional layered SP decoder at high SNRs with both an AWGN channel and Rayleigh fading channel. For LDPC codes with longer code lengths, the proposed decoder outperforms the conventional layered MS decoder but is slightly outperformed by the layered SP decoder. However, the lower complexity of the proposed decoder makes it a good alternative to the SP decoder (Zhang et al., 2021).

9.3 LDPC CODES WITH REINFORCEMENT LEARNING

9.3.1 Multiarmed Bandit-Based Node-Wise Scheduling (MAB-NS) Scheme

In the conventional iterative decoding of LDPC codes, all check nodes and variable nodes are updated simultaneously. This method is called flooding. Another updated method, which has proven to lead to enhanced performances, consists of sending messages from a single check node at a time and consecutive check nodes are scheduled based on the difference between the subsequent messages emerging from the check node (Casado et al., 2010). This update method is called node-wise scheduling (NS) but is known to be more computationally complex than the conventional update method. Habib, Beemer, and Kliewer (2021) proposed a multiarmed bandit-based node-wise scheduling (MAB-NS) scheme, adapted to short LDPC codes, which aims at reducing the computational complexity of NS. Basically, a multiarmed bandit problem is a reinforcement learning problem in which an agent is allocated numerical rewards based on the course of actions it chooses to take. The agent aims to maximize this reward through optimal choices of actions. The proposed scheme is based on Q-learning, which is a model-free reinforcement learning algorithm for learning the value of an action in a given state. As LDPC codes have a large number of check nodes, the authors proposed to group them into clusters to make them more suitable for Q-learning. The clusters are determined such that they have separate state and action spaces and that they have a low dependence on each other.

In the proposed scheme, a single check node sends messages to all of its connected variable nodes, which subsequently send messages to their connected check nodes, in a single message-passing iteration. Check node scheduling is then executed sequentially until a specific stopping condition is met. The decoder relies on an action-value function to determine its scheduling policy to minimize its computational complexity. In this scheme, the Q-learning agent is informed of the state of the decoder and the reward obtained after scheduling a check node. The agent then learns to enhance the reward earned, thus optimizing the decoder (Habib et al., 2021). The authors categorized their Q-learning schemes based on the clustering method used. They denoted the schemes as MQC, MQR, and MQO for contiguous, random, and cycle-optimized clusters, respectively.

Simulations are performed to compare the proposed schemes mainly with the conventional flooding decoder, NS decoder, and an NS decoder based on Thompson sampling (NS-TS). Simulation results show that MQO offers the best BER performance and shares the least amount of messages between check nodes and variable nodes, which implies a reduced message-passing complexity (Habib et al., 2021). The benefits of a good BER performance and a reduced complexity can make the proposed scheme a good option for mMTC services where low-energy processing is required.

9.3.2 Reinforcement Learning and Monte Carlo Tree Search

Deep reinforcement learning has also been tested for the construction of high-performance LDPC codes (Zhang et al., 2018). The authors used reinforcement learning and Monte Carlo tree search (MCTS) to train a neural network to search for edge growth routes in the construction of LDPC codes. The MCTS technique involves constructing and traversing a search tree while taking arbitrary actions at each node until the best solution to the problem is obtained and is commonly used for addressing construction and planning problems. Their method provides a long-term vision, which leads to performance enhancements and flexibility in terms of code length, code rate, and the degree distribution at a lower complexity than the progressive-edge-growth (PEG) method, making it appropriate for 5G NR, which is designed to be flexible.

The DNN is trained using codes constructed by the MCTS. The construction stage is repeated using the trained DNN from the last learning stage to optimize the network. The construction involves updating an all-zero parity check matrix with 1 s at the selected edges sequentially based on the specified degree distribution until the complete LDPC matrix is obtained. The edges are considered as the states whereas the selection of the next edge is considered an action (Zhang et al., 2018). The authors selected the construction of a quasi-cyclic LDPC (QC-LDPC) code to evaluate their proposed method. QC-LDPC codes were chosen because they have a massive structure and hence enhancement in their constructions may be easier to notice. They selected a code length of 64 bits with a code rate of 1/2. The circulant size, variable node degree, and check node degree are chosen to be 8, 3, and 6, respectively. The base matrix of the LDPC code is chosen to be of size 4×8 with a circulant permutation matrix of size 8×8 (Zhang et al., 2018).

During the training, the BER performance of the constructed LDPC code is used to rate the code. The DNN takes a $4\times 8\times 2$ image stack as input and uses three convolutional layers, each with a different number of filters. The output of the last convolutional layer is fed to a policy head and a value head. The policy head consists of one convolutional layer and a fully connected layer, which gives an output using a softmax activation function. The value head consists of one convolutional layer and two fully connected linear layers. All the convolutional layers used in the network are activated using ReLU (Zhang et al., 2018).

The proposed construction method is compared to the PEG method through BER simulations and it was observed that both methods provide the same error performance with the added benefits of flexibility and lower complexity (Zhang et al., 2018).

9.4 APPLICATION OF ML IN 5G NR POLAR CODES

In this section, an overview of polar codes used in 5G NR is first given, followed by a review of ML techniques that have been employed to enhance the performance of these codes.

9.4.1 5G NR Polar Codes

Polar codes are high-performance channel codes introduced by Arikan in 2009. Arikan proved that, unlike other channel codes, polar codes are capacity-achieving. Polar codes make use of the polarization effects of the Kronecker powers of a kernel matrix to create virtual channels having different reliabilities, to transmit one bit each. In the polar encoding process, K message bits are assigned to the most reliable channels while the other remaining N–K channels are filled with frozen bits, which are usually set to 0 (Pillet et al., 2020).

Since its introduction, polar codes have attracted much research attention and have even been deployed in 5G NR's control channel for eMBB services and in 5G NR's broadcast channel (Bae et al., 2019). 5G NR utilizes the CRC-aided (CA) polar codes, which support rate-matching. 5G NR's polar codes are decoded using the CRC-aided successive cancellation list (CA-SCL) algorithm. Unlike the BPA and NMSA LDPC decoding algorithms, the CA-SCL decoding is performed in a SISO manner (Arikan, 2009). This implies that the decoding of bits in the codeword is performed sequentially.

Compared to the simpler successive cancellation (SC) decoding algorithm, the CA-SCL decoding algorithm makes use of more than one decoder working in parallel, providing multiple decoding paths and leading to a better error performance than the SC decoding algorithm. Furthermore, the addition of CRC to the decoder is used for accurate codeword selection. The decoding path having the smallest path metric and passing the CRC is selected as the decoded codeword.

9.4.2 POLAR CODES WITH NEURAL NETWORKS

9.4.2.1 Neural Network-Based Frame Error Rate Prediction of Polar Codes

Leonardon and Gripon (2021) developed a neural network with the ability to predict the frame error rate (FER) of polar codes based on the parameters used for their construction, which can ease the design of high-performance polar codes. The dataset used for training the neural network consisted of frozen bit sequences paired with their corresponding FERs, which are determined by Monte Carlo simulations. The simulations are performed such that no other parameters than the frozen bit sequence are changed across the communication system to obtain the FER. Most of the training dataset consisted of poor performance frozen bit sequences to lead the neural network in predicting high FERs.

The training is performed on both short (256, 128) and large (1,024, 512) polar codes with SCL decoders of list sizes 4 and 32, respectively. The FER data is generated using Eb/N0 values 3.2 and 2.7 dB for short and large polar codes, respectively. 80% of the data set is used for training the neural network whereas 20% is used for validation. After the validation phase, it was observed that the predicted FERs generally do not deviate by more than 6% from the actual FERs. Worst-case scenarios brought this deviation to not more than 30%, which was still considered acceptable by the authors (Leonardon and Gripon, 2021).

The authors experimented with their neural network by changing the training parameters to evaluate the outcomes. The authors found that the best inflation of error (IOE) was obtained by increasing the neurons number, keeping the number of layers to three or nine, or using small gap lengths. The epoch number should not be excessive as this leads to increases in the IOE (Leonardon and Gripon, 2021). For the construction of high-performance polar codes, the authors employed the projected gradient descent (PGD) methodology, which aims at using gradient descent to change the neural network's input such that it minimizes the output, which is the FER. For large codes, the neural network was able to come up with a frozen bit sequence, which provided an FER of 1.01×10^{-5} that is considerably lower than the minimum FER in the training set which was 5.75×10^{-5}. Short codes provided less significant gains mainly because the frozen bit sequence possibilities are very limited due to their size and the training set already contained well-optimized frozen bit sequences (Leonardon and Gripon, 2021).

9.4.2.2 Transfer Learning-Based Decoder Training Method

Lee, Seo, Ju, and Kim used transfer learning to propose an efficient decoder training method (2020). The authors opted for multilayer perceptrons (MLP) and LSTM models for their neural network decoder (NND). To ease the learning process, the authors restricted the codeword size to 16 and 32 (short polar codes). The MLP model used consists of an input layer, an output layer, and three hidden layers of 128-64-32 nodes and is a fully connected feedforward neural network. The hidden layers utilize the ReLU and the output layer utilizes the sigmoid function. The LSTM model uses a single LSTM cell with an output dimension of 256 and the nodes use the sigmoid function. To emulate noisy transmission during training, a noisy layer is added to the input of the neural network. Due to the virtually unlimited amount of data available, the concept of an epoch is ignored and instead, the speed and complexity of the training are evaluated with the number of batches of data used.

The authors set the objective of their experiment to obtain a decoder with performances comparable to the maximum a posteriori (MAP) decoder. They employed transfer learning by transferring the learned decoder state for one codeword to the decoder for the next codeword until the last codeword's decoder's training is executed. MSE and stochastic gradient descent (SGD) are used as the cost function and optimizer, respectively (Lee et al., 2020).

To evaluate their proposed schemes, the authors used TensorFlow for training the neural network. The proposed transfer learning method is compared to a separate learning method. For polar codes of length, $N = 16$, the decoder trained using separate learning had a similar performance as the MAP decoder while the decoder trained using transfer learning provided quicker training. It is observed that the more the code rate is increased, the proposed scheme provides a better bit error rate (BER) performance than that with separate learning. With an information message of length $K = 8$, MLP and LSTM employing the proposed learning scheme were found to achieve coding gains of 0.6 and 0.5 dB, respectively (Lee et al., 2020).

For polar codes of length $N = 32$, the proposed scheme outperformed the standard learning scheme at high code rates whereas, at low code rates, both schemes showed the same performance. At a BER of 10^{-3}, the transfer learning-based MLP decoder provided performance gains of 0.5, 0.7, 0.9, and 1.0 dB for $K = 8, 9, 10,$ and 11, respectively. The transfer learning-based LSTM decoder provided performance gains of 0.2, 0.2, 0.3, 0.5, and 0.5 dB for $K = 8, 9, 10, 11,$ and 12, respectively. The proposed scheme was also found to overcome the underfitting problem of separate learning (Lee et al., 2020).

9.4.2.3 A CNN-Based Polar Decoder

Qin and Liu designed a CNN polar decoder having a structure adapted to decode long polar codes (2019). The authors justified their choice of using CNN based on tests they performed, which showed that a CNN decoder has a closer performance to that of an SC decoder than an MLP decoder. The CNN is adapted to perform polar decoding by changing its input and feature map from a 1D image to a 1D vector and by using a filter size of 1 × 3 instead of a square size.

The CNN consisted of several convolutional, pooling, and fully connected layers. At the convolutional layers, the input codeword was used to generate feature maps. The pooling layers reduced the number of feature maps to half and the fully connected layer output the decoded codeword over which a sigmoid function is applied. Some typical CNN components such as the dropout and batch normalization are excluded from the CNN's structure to minimize the complexity (Qin and Liu, 2019).

The CNN is adapted to long codes by using auxiliary labels to mark the position of frozen bits so that their information can be used in the backpropagation process. Measures were also taken to reduce the size of the CNN for it to work with long polar codes without increasing the complexity too much. By considering that some neural network connections having higher weight can outweigh the contribution of those having lower weight, connections weighing below a set threshold were removed from the convolutional layers, in a layer-by-layer sequence (Qin and Liu, 2019).

The simulation performed consisted of a message, which was encoded using a polar code encoder and then modulated using BPSK. The modulated codeword was sent to the receiver through an AWGN channel and decoding was then performed using the CNN decoder. Training of the CNN decoder was performed by feeding the latter with both the received message from the channel and the actual message that was transmitted. The training and validation data were obtained at SNR values of 0, 1, 2, 3, and 4 dB to generalize the decoder. The authors also concluded from the training phase that using a high filter size did not provide performance gains. Cross-entropy was used to compare the decoder's output with the expected output (Qin and Liu, 2019).

After the training was completed, the performance of the decoder was evaluated. The authors observed that the pruning mechanism provided SNR gains to the system. Results also showed that their proposed decoder was found to perform faster than the conventional SC decoder at the tradeoff of a small BER performance loss (Qin and Liu, 2019). The work can be extended to compare the proposed scheme with an SCL decoder to test its suitability for 5G NR.

9.4.2.4 A Residual NND for Polar Codes

Zhu, Cao, Zhao, and Li (2020) worked on a residual NND (RNND) for polar codes, which uses residual learning to increase the SNR of received symbols. The RNND consisted of a residual learning denoiser placed in front of a NND.

The residual learning denoiser is similar to a commonly used image denoiser but with a 1D input instead. The denoiser consists of stacked weight layers and a shortcut connection. The parameters of the denoiser are optimized such that the difference between its output and the transmitted signal is minimal. The stacked weight layer could be an MLP, a CNN, or an RNN and therefore, the authors used all three of them as different schemes (MLP-RNND, CNN-RNND, and RNN-RNND) to determine the most suitable one. The loss function used for the denoising task is MSE. The NND placed after the denoiser works by categorizing the received symbols as either 0 s or 1 s and also uses MSE as the loss function (Zhu et al., 2020).

The authors also employed a unique multitask learning method so that the entire RNND can perform both denoising and decoding effectively. Hence the denoising and decoding losses are summed to obtain a final loss function with the expectation that the denoiser and decoder can reinforce each other. The learning of the denoiser is enhanced by the gradient signal decoder and the denoiser reduces the noise level making decoding more accurate (Zhu et al., 2020).

The training data were generated by using all the 256 possible codewords that could be obtained from (16, 8) polar codes. The codewords were modulated using BPSK and transmitted over an AWGN channel with an SNR of 0 dB to obtain the training data. Batch-based training was performed by sending the codewords into the model in batches before evaluating the loss function. Backpropagation was used to calculate the gradients and the Adam model was used to update the parameters of the model. After training, testing was performed using symbols obtained from the AWGN with noise levels not used in training (Zhu et al., 2020).

Simulation results showed that the MLP-RNND and the RNN-RNND have the same denoising abilities and they outperform the CNN-RNND. The proposed RNNDs were compared to NNDs and it was observed that the MLP-RNND has the closest BER performance to the SC decoder and it also has a 0.2 dB gain over the MLP-NND at a BER of 10^{-4}. The MLP-RNND was also found to be 100 times faster than the SC decoder. However, all the RNNDs were found to be slightly slower than their NND counterparts. Hence the speed-reliability tradeoff must be taken into good consideration when choosing between the NNDs and the RNNDs for decoding (Zhu et al., 2020).

9.4.2.5 A Differentiable Neural Computer-Aided Flip Decoding Algorithm

Tao and Zhang (2021) proposed a differentiable neural computer (DNC)-aided flip decoding algorithm to provide high error correction performance by bit flipping which SCL decoders perform poorly when bits have long-distance dependencies in the code sequence. DNCs are considered to be more performant than LSTM controllers because the DNCs use an external memory, which allows them to solve problems having long-distance dependencies and many computational steps.

The proposed decoder consisted of two decoding phases. The first phase starts by performing conventional decoding. If the decoding fails, as indicated by the CRC, a flip DNC (F-DNC) ranks the potentially erroneous bit positions based on their probabilities of being erroneous and generates a flip vector, which is then used to perform multi-bit flipping at those bit positions. In case the CRC still fails after the multi-bit flipping, the decoding proceeds to the second phase in which bits are flipped successively and a flip-validate DNC (FV-DNC) is used to verify if each flip is correct. The rationale behind this method is that among the erroneous bit positions determined by the F-DNC, several may be actual erroneous bit positions while some can be error-free bit positions. The sequence of flipping in the second phase is determined by the ranks of the error positions determined in the first phase (Tao and Zhang, 2021).

The training was performed offline so that the complexity of the proposed solution stays low. During the training phase, phase 1's F-DNC was trained to recognize error positions. The state information derived from LLRs and path metrics was paired with the corresponding reference flip vectors to form the training set. To obtain the training set for the FV-DNC in phase 2, supervised flip decoding based on the reference flip vectors was performed. Bits in the first five error positions are flipped one by one and the corresponding state encoding details are recorded and categorized as successful erroneous bit flipping. Then five random non-error bit positions are flipped and the corresponding state encoding details are recorded and categorized as unsuccessful erroneous bit flipping.

The performance of the trained system was evaluated using long polar codes of codelengths 256 and 1,024 bits. Simulation results showed that compared to a conventional SCL decoder and an LSTM-based decoder, the proposed solution has higher bit-flipping accuracy, which leads to improved error performance and requires fewer flip decoding attempts.

9.4.2.6 Neural Network-Based Bit Flipping

Another work which focuses on enhancing the bit-flipping process is the introduction of neural network architecture by Ivanov, Kotov, and Alexey (2022) for generating a bit-flipping critical set, which can be used to enhance the error performance of polar decoders. The authors preferred using CNN instead of LSTM for this work.

The authors defined the bit-flipping problem as a ML problem to be resolved through classification. They used a metric for estimating the probability of error of chosen decoder paths as the input to their neural network. The metric has the ability to use information from both discarded and continued paths, making it very efficient. Before being fed to the network, the metric is standardized by excluding the meaning and using scaling. The neural network outputs the details of bit positions which caused SCL decoding to fail (Ivanov et al., 2022).

Simulations with an SCL decoder were performed at Eb/N0 values between 0 and 2 dB to obtain the training dataset, which consists of samples of values of the metrics paired with the bit positions that require flipping for enhancing the error performance. The authors compared the available types of neural networks and activation functions and chose the one with the highest performance. Their final choice was a convolutional neural network with 1D convolutional layers, a pooling layer, and a normalization layer. The activation function chosen was the Harswish function. The kernel size of the convolutional layers and maximum pooling was set to four and cross-entropy was selected as the loss function. The Adam optimizer was used for training and the learning rate was set to 5×10^{-3}, which was decreased by a factor of five if the loss function did not decrease after five iterations (Ivanov et al., 2022).

The trained neural network was tested in a communication system where BPSK modulation and an AWGN channel were used with polar codes of both codelengths 512 and 1,024. The proposed network was compared with an LSTM neural network and a conventional SCL decoder. The simulation results showed that when the same list size was used in all of the tested decoders, the proposed decoder provides the best error performance with Eb/N0 gains of up to 0.25 dB. The proposed decoder was also able to achieve the efficiency of an SCL decoder with a list size of 32 using only a critical set size of 20. The proposed solution hence provided error performance gains with reduced memory requirements which can make it a good candidate for 5G NR (Ivanov et al., 2022).

9.4.2.7 Stacked Denoising Autoencoder for Polar Codes

Li and Cheng (2020) introduced a stacked denoising autoencoder (SDAE) to improve the FER performance of polar codes over Rayleigh fading channels. The authors' design consisted of two DNNs concatenated with each other and placed before the SC decoder. The DNNs are based on SDAE over which offline training is performed. The inputs of the DNNs are the estimated symbols obtained from the channel. The role of the DNNs is to reduce the impact that the channel had on the symbols. The outputs of the DNNs are sent to the SC decoder for recovering the transmitted information. The authors explained that the ability of the DNNs to extract features from corrupted inputs makes them efficient in performing their tasks.

The first DNN is trained to minimize distortions in the symbols containing information bits and then the second DNN is trained to estimate the complete codeword for more accurate decoding. The use of the second DNN helps by adding a second layer of error correction in the information symbols, hence leading to more successful polar decoding. Training samples for the first DNN consist of sets of actual information bits being transmitted paired with the received symbols corresponding to the information bits whereas the second DNN uses the remaining symbols not used by the first DNN as training samples (Li and Cheng, 2020).

The performance of the proposed solution was evaluated by simulating the transmission of polar codewords of lengths 16 and 32 bits, with code rate 1/2 over a Rayleigh fading channel. The first and second DNNs were configured with two hidden layers. The first DNN had 32 and 16 nodes in the first and second hidden layers, respectively, for the 16 bits polar code and had 64 and 32 nodes in the first and second hidden layers, respectively, for the 32 bits polar code. The second DNN had 64 and 32 nodes in the first and second hidden layers, respectively, for the 16 bits polar code and had 128 and 64 nodes in the first and second hidden layers, respectively, for the 32 bits polar codes. Adaptive moment estimation with a learning rate of 0.001 was used for optimizing the training process. As a reference for comparison, simulations for a conventional polar decoder using the one-dimensional method with SCL decoding were performed. At an FER of 0.001, the proposed scheme provided a significant E_b/N_0 gain of 1 dB over the conventional scheme. The proposed scheme was then compared with a conventional feedforward neural network (FFNN) structure. Results showed that for the same training dataset size used, the proposed scheme outperforms the FFNN scheme by 0.5 dB at an FER of 0.001 (Li and Cheng, 2020).

9.4.2.8 A Machine Learning-Based Multi-Flips SC Decoding Scheme

A team of researchers introduced a machine learning-based multi-flips successive cancellation decoding scheme (ML-MSCF) with better performance and lower latency than the dynamic SC flipping (DSCF) decoding algorithm (He et al., 2020). The scheme consisted of decoding received data using a SC decoder and feeding the LLR sequence and syndromes to an LSTM network if SC decoding fails. The LSTM network generates a flip-bits list. The SC decoder is used again after flipping the first erroneous bit and if the decoded codeword is not error-free, the output of the LSTM

AI-Based 5G/6G Communications

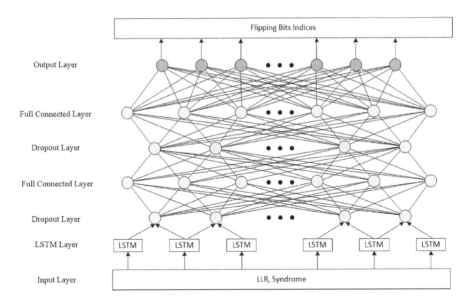

FIGURE 9.3 Proposed LSTM network (He et al., 2020).

network and the syndrome are fed to a Q-table which outputs another flip list. The flip process is continuously performed until the CRC is successful or a maximum flip number is reached.

The LSTM network used consists of an LSTM layer, two fully connected layers, two dropout layers and one output layer, as shown in Figure 9.3. The output layer has the same amount of neurones as the polar codelength and these neurones are activated by a softmax function. Training is performed using data obtained from the Monte Carlo simulation of SC decoding using BPSK modulation and an AWGN channel. The network's input consisted of the LLR sequence, and the syndrome, and the first erroneous bit position for each data sample is specified using one-hot encoding. The output of the LSTM is given in the form of an index list of the flipping bits. The reward function is made dependent on the number of bit errors connected by the network (He et al., 2020).

The training of the LSTM network was performed using 2×10^4 samples, a minibatch of size 1,000, 60 epochs, a dropout layer ratio of 0.05, a hidden layer of size 256, a learning rate of 0.001, a discount factor of 0.7, and exploration factor of 0.2 and 2×10^4 training processes for the Q-table. The Adam optimizer was used in the LSTM network (He et al., 2020).

Simulations using the trained network in the proposed ML-MSCF decoder showed that the latter provides performance gains over the flip-successive cancellation (SCF) and the dynamic SCF (DSCF) decoders when the Eb/N0 is above 2.5 dB. The FER gains over the DSCF are in the range of 0.2–0.3 dB. The performance gain is attributed to the ability of the algorithm to correct more than only the first erroneous bit. Furthermore, the absence of exponential and logarithmic functions in the proposed decoder allows it to have the same processing latency as the DSCF. The proposed scheme, therefore, is a better option than SCF and DSCF for wireless technologies such as 5G NR (He et al., 2020).

9.4.2.9 Double Long–Short-Term Memory-Based SC Flipping Decoder

A double long–short-term memory (DLSTM)-based SC flipping decoder was introduced to address the performance loss encountered by the SC flipping algorithm with short polar codes (Cui et al., 2021). The authors performed an analysis of the error propagation effects of the first, second and third error bits in a polar codeword and observed that the first error bit was the root cause of most subsequent errors in the codeword. Hence the authors made use of their DLSTM network to predict the first error bit.

For (N, K) polar codes, the network is composed of an input layer, a DLSTM layer with $2N$ hidden LSTM units, and an output layer of K fully connected neurones. An LLR sequence which fails the CRC check in the polar decoder is fed at the input of the DLSTM network and the network outputs a vector of probabilities of error for each information bit. Compared to traditional LSTM, the proposed DLSTM network is better at retrieving historical data features and maintaining long-term relevancy (Cui et al., 2021).

The training phase consisted of generating 2×10^6 polar codewords transmitted over an AWGN channel after being modulated using BPSK. Only the LLR sequences of codewords which failed the CRC process are used as training and validation samples, paired with a one-hot encoding of the first error bit. In the output layer, the frozen bits are clipped as flipping them can cause more error propagations. 80% of the samples were used for training and 20% were used for validation. The loss between the expected and predicted values is measured using a cross-entropy function. For reducing the complexity of the decoder, channel reliability ordering details were also used to identify multiple error bits (Cui et al., 2021).

Simulations were performed using an Intel(R) Xeon(R) Gold 6,132 CPU @ 2.60 GHz server with BPSK modulation and an AWGN channel with polar codes of sizes (64, 32) and (16, 8). Performance results showed that the proposed algorithm takes the same amount of computing time as the machine learning-based multi-bit flipping SC (ML-MSCF) algorithm. When compared to DSCF, the proposed algorithm minimizes the use of exponential and logarithmic functions and thus uses up to 27.68% less time for decoding at $Eb/N0 = 1.0$ dB. The proposed decoder used up to 3.64% less time for decoding than CA-SCL with a list size of 2 when $N=32$ and up to 31.26% less time for decoding than CA-SCL with a list size of 4. At a block error rate (BLER) of 10^{-3}, the proposed decoder provided Eb/N0 gains of 0.21 dB, 0.36 dB and 0.55 dB over the ML-MSCF, DSCF and CA-SCL (list size of 2) respectively (Cui et al., 2021).

9.4.3 POLAR CODES WITH REINFORCEMENT LEARNING

9.4.3.1 Reinforcement Learning for Polar Codes Construction

Another approach to constructing high-performance polar codes is by mapping them to a game in which a SARSA (λ) agent is trained to find its way through a maze (Liao et al., 2022). The objective of the ML algorithm was to identify the optimal frozen

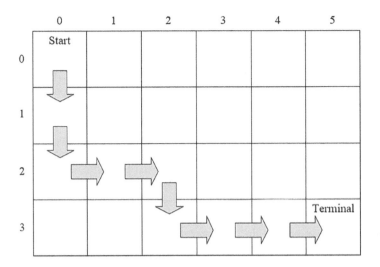

FIGURE 9.4 Reinforcement learning for polar codes construction (Liao et al., 2022).

bit positions in an (N, K) polar code which can enhance the error performance of the polar code.

The learning environment consisted of a maze of width $K+1$ and height $N-K+1$, with the starting position being the upper leftmost cell and the terminal position being the bottom rightmost cell. The rule of the game was that the learning agent had to reach the terminal cell or receive a non-zero reward by making only down and right movements across the environment (Liao et al., 2022). An example of a learning environment and a possible sequence of actions, represented by arrows, is given in Figure 9.4.

The construction of the polar code is based on the actions that the agent takes. The kth bit is set to a frozen bit if the kth step is downwards and if the step is rightwards, the bit is set to an information bit. The allocated reward is linked to the SCL-Genie decoding process and the agent had to maximize the expected return while traversing the maze. The decoding process involves decoding one bit after the other in a sequential fashion and the decoding of a bit is independent of whether the following bits are frozen or not. This allows the reward to be determined instantly. For training, a table of the value function Q is updated by the learning agent using an ε-greedy policy. The exploration rate is decreased after each training episode. The authors highlighted that the devised training process is highly efficient. Only 2,000 training samples are required for constructing (16, 8) polar codes, with each sample being used a single time in the training (Liao et al., 2022).

The polar code constructed by the ML algorithm is evaluated using the conventional SCL and CRC-aided SCL (CA-SCL) with list sizes 2, 4, and 8. For the list size of 2, the generated code performs similarly to standard polar codes whereas for the list sizes of 4 and 8. The generated code outperforms the standard one over the whole tested SNR range. The code generated by the proposed algorithm is found to perform best for long codewords with CRC (Liao et al., 2022). Such a polar code construction method can help design strong polar codes for 5G NR.

9.4.3.2 A Reinforcement Learning-Aided CRC-Aided BP Decoder

Doan, Hashemi, and Gross (2020) defined the factor-graph permutations' selection for polar codes in the CRC-aided BP (CABP) decoder as a multiarmed bandit problem. They designed a reinforcement learning-aided CABP (RL-CABP) which performs online learning, i.e., the agent learns to do the permutations selection task during the decoding process. The authors use the ε-greedy, upper confidence bound (UCB), and Thompson sampling (TS) algorithms to tackle the multiarmed bandit problem. The ε-greedy and UCB algorithms are used to estimate the expected reward value whereas the TS algorithm estimates the distribution of the reward value linked to every possible action. To eliminate any additional latency to suit 5G uRLLC services' requirements which can be added by these algorithms, the latter is performed in parallel with the first CABP decoding attempt.

In case the CRC fails during CABP decoding using the original factor-graph permutations, the proposed decoder performs a random selection of factor-graph permutations. The decoder is awarded either a numerical reward of '1' if the CRC passes with the selected factor-graph permutation or a numerical reward of '0' if the CRC fails. In this way, the learning agent learns to maximize the reward (Doan et al., 2020).

The proposed RL-CABP decoder was evaluated using 5G NR's eMBB polar codes of size 128 bits with 64 information bits and a CRC length of 16. The performance of the proposed decoder was compared to that of the conventional CABP, cyclically shifted permutations CABP (CP-CABP) and random factor-graph permutations CABP (RP-CABP) and CA-SCL decoders. Simulations were performed at Eb/N0 = 3.0 dB to obtain the k value of the k-armed bandit which can result in maximum rewards. This value was obtained as $c = 500$. The ε-greedy algorithm was found to provide the highest average reward whereas the TS algorithm provided the least average reward. The ε-greedy algorithm also has the lowest computational complexity among the three algorithms (Doan et al., 2020).

Experimental results showed that the proposed RL-CABP decoder provides an Eb/N0 gain of 0.62 dB over the conventional CABP decoder and of at least 0.125 dB over CP-CABP and RP-CABP at an FER of 10^{-4}. At that same FER, the proposed decoder significantly outperforms a BP decoding algorithm by 0.92 dB. However, compared to CA-SCL decoders, the proposed decoder can only provide a performance gain of 0.12 dB when the list size is 2 and is even outperformed by the former when the list size is 4 (Doan et al., 2020).

9.5 GAPS IN PREVIOUS RESEARCH

The different discussed works generally show that ML algorithms can provide enhancements in error performance, decoding speed, and computational complexity. The proposals for new solutions are made mainly to address channel coding problems encountered by conventional channel decoders. However, a gap observed in the works discussed in Sections 9.2 and 9.3 is the consideration of the suitability of the proposed solutions for 5G NR based on eMBB, URLLC, or mMTC requirements.

Most of the discussed works focused on demonstrating the effectiveness of ML solutions on short channel codes mainly because schemes, which involve neural networks become significantly more complex as the codelength increases. Research on adapting those schemes which were performant on short codes to longer codes will be very useful for determining their suitability for eMBB services. Furthermore, most of the discussed research focuses on the channel decoder. By designing an encoder jointly with the decoder, with a code construction method which works best with the proposed decoder, better performance gains can be expected.

It was also observed that only a few research included a latency analysis of the proposed schemes as the researchers mostly focused on the error performance and the complexity of their schemes. Thus an extension of the works to assess their suitability for 5G URLLC based on their decoding latency would be useful.

A common practice observed in the discussed works was the use of a single SNR value to train ML algorithms. Some researchers proved that using a wider range of SNRs for training helped generalize their decoder. By adopting this same methodology, better performance gains could have been obtained from the proposed ML schemes. Similarly, instead of limiting the training with BPSK modulation, researchers could have used a wider set of modulation orders to make their proposed schemes more generalized and appropriate to a wider range of requirements.

9.6 SUMMARY

5G NR is being designed to fulfill stringent performance requirements to support the three service classes: eMBB, mMTC, and URLLC. Channel codes are considered to be one of the most essential aspects of the communication system, which can help achieve the stringent requirements and hence, it is crucial to select the right channel codes for the system. The rise in ML and its integration in the channel coding field has till now proven to be promising. ML can provide significant gains in error correction performance, computation complexity and latency in both LDPC and polar codes. While some ML schemes show their suitability for one service class, there is still room for the development of both channel encoders and channel decoders embedded with machine learning algorithms, which can suit the requirements of 5G NR or even of future 6G mobile network technologies.

BIBLIOGRAPHY

3GPP. (2019). TS 38.212: Multiplexing and channel coding. In *3rd Generation Partnership Project; Technical Specification Group Radio Access Network*.

Arikan, E. (2009). Channel polarization: A method for constructing capacity-achieving codes for symmetric binary-input memoryless channels. *IEEE Transactions on Information Theory*, *55*(7), 3051–3073. https://doi.org/10.1109/TIT.2009.2021379.

Bae, J., Abotabl, A., Lin, H., Song, K., & Lee, J. (2019). An overview of channel coding for 5G NR cellular communications. *APSIPA Transactions on Signal and Information Processing*, *8*(1), e17. https://doi.org/10.1017/ATSIP.2019.10.

Casado, A., Griot, M., & Wesel, R. (2010). LDPC decoders with informed dynamic scheduling. *IEEE Transactions on Communications*, *58*(12), 3470–3479. https://doi.org/10.1109/TCOMM.2010.101910.070303.

Cui, J., Kong, W., Zhang, X., Chen, D., & Zeng, Q. (2021). DLSTM-based successive cancellation flipping decoder for short polar codes. *Entropy*, *23*(7), 863. https://doi.org/10.3390/e23070863.

Dai, J., Tan, K., Si, Z., Niu, K., Chen, M., Poor, H., & Cui, S. (2021). Learning to decode protograph LDPC codes. *IEEE Journal on Selected Areas in Communications*, *39*(7), 1983–1999. https://doi.org/10.1109/JSAC.2021.3078488.

Doan, N., Hashemi, S., & Gross, W. (2020). Decoding polar codes with reinforcement learning. In *GLOBECOM 2020–2020 IEEE Global Communications Conference*. https://doi.org/10.1109/GLOBECOM42002.2020.9348007.

Fowdur, T., & Indoonundon, M. (2017). A hybrid statistical and prioritised unequal error protection scheme for IEEE 802.11n LDPC codes. *International Journal of Electrical and Computer Engineering Systems*, *8*(1), 1–9. https://doi.org/10.32985/ijeces.8.1.1.

Fowdur, T., Babooram, L., Rosun, M., & Indoonundon, M. (2021). *Real-Time Cloud Computing and Machine Learning Applications*. New York: Nova Science Publishers, Inc.

Gallager, R. (1962). Low-density parity-check codes. *IRE Transactions on Information Theory*, *8*(1), 21–28. https://doi.org/10.1109/TIT.1962.1057683.

Habib, S., Beemer, A., & Kliewer, J. (2021). Belief propagation decoding of short graph-based channel codes via reinforcement learning. *IEEE Journal on Selected Areas in Information Theory*, *2*(2), 627–640. https://doi.org/10.1109/JSAIT.2021.3073834.

He, B., Wu, S., Deng, Y., Yin, H., Jiao, J., & Zhang, Q. (2020). A machine learning based multi-flips successive cancellation decoding scheme of polar codes. In *IEEE 91st Vehicular Technology Conference (VTC2020-Spring)*. https://doi.org/10.1109/VTC2020-Spring 48590.2020.9128875.

Indoonundon, M., & Fowdur, T. (2021). Overview of the challenges and solutions for 5G channel coding schemes. *Journal of Information and Telecommunication*, *5*(4), 460–483. https://doi.org/10.1080/24751839.2021.1954752.

ITU. (2017). Minimum requirements related to technical performance for IMT-2020 radio interface(s). Report ITU-R M.2410-0.

Ivanov, F., Kotov, F., & Alexey, Z. (2022). Method of critical set construction for successive cancellation list decoder of polar codes based on deep learning of neural networks. *SSRN Electronic Journal*, 1–6. https://doi.org/10.2139/ssrn.4111931.

Lee, H., Seo, E., Ju, H., & Kim, S. (2020). On training neural network decoders of rate compatible polar codes via transfer learning. *Entropy*, *22*(5), 496. https://doi.org/10.3390/e22050496.

Leonardon, M., & Gripon, V. (2021). Using deep neural networks to predict and improve the performance of polar codes. In *11th International Symposium on Topics in Coding (ISTC)*. https://doi.org/10.1109/ISTC49272.2021.9594059.

Li, J., & Cheng, W. (2020). Stacked denoising autoencoder enhanced polar codes over Rayleigh fading channels. *IEEE Wireless Communications Letters*, *9*(3), 354–357. https://doi.org/10.1109/LWC.2019.2954907.

Li, L., Huang, Z., Liu, C., Zhou, J., & Zhang, Y. (2021). Blind recognition of LDPC codes using convolutional neural networks. In *IEEE 4th International Conference on Electronics Technology (ICET)*. https://doi.org/10.1109/ICET51757.2021.9450940.

Liao, Y., Hashemi, S., Cioffi, J., & Goldsmith, A. (2022). Construction of polar codes with reinforcement learning. *IEEE Transactions on Communications*, *70*(1), 185–198. https://doi.org/10.1109/TCOMM.2021.3120274.

MacKay, D. J. C., & Neal, R. M. (1996). Near Shannon limit performance of low density parity check codes. *Electronics Letters*, *32*, 1645–1646.

Pillet, C., Bioglio, V., & Condo, C. (2020). On list decoding of 5G-NR polar codes. In *2020 IEEE Wireless Communications and Networking Conference (WCNC)*. https://doi.org/10.1109/WCNC45663.2020.9120686.

Qin, Y., & Liu, F. (2019). Convolutional neural network-based polar decoding. In *2nd World Symposium on Communication Engineering (WSCE)*. https://doi.org/10.1109/WSCE49000.2019.9040920.

Session Chairman (Nokia). (2016). Chairman's Notes of Agenda item 7.1.5 Channel coding and modulation. In *3GPP TSG RAN WG1 Meeting 87, R1-1613710*.

Tao, Y., & Zhang, Z. (2021). DNC-aided SCL-flip decoding of polar codes. In *IEEE Global Communications Conference (GLOBECOM)*. https://doi.org/10.1109/GLOBECOM 46510.2021.9685277.

Wang, L., Chen, S., Nguyen, J., Dariush, D., & Wesel, R. (2021). Neural-network-optimized degree-specific weights for LDPC minsum decoding. In *11th International Symposium on Topics in Coding (ISTC)*. https://doi.org/10.1109/ISTC49272.2021.9594227.

Wang, Q., Wang, S., Fang, H., Chen, L., Chen, L., & Guo, Y. (2020). A model-driven deep learning method for normalized min-sum LDPC decoding. In *IEEE International Conference on Communications Workshops (ICC Workshops)*. https://doi.org/10.1109/ICCWorkshops49005.2020.9145237.

Wang, Y., Zhang, Z., Zhang, S., Cao, S., & Xu, S. (2018). A unified deep learning based polar-LDPC decoder for 5G communication systems. In *10th International Conference on Wireless Communications and Signal Processing (WCSP)*. https://doi.org/10.1109/WCSP.2018.8555891.

Wu, X., Jiang, M., & Zhao, C. (2018). Decoding optimization for 5G LDPC codes by machine learning. *IEEE Access, 6*, 50179–50186. https://doi.org/10.1109/ACCESS.2018.2869374.

Zhang, D., Dai, J., Tan, K., Niu, K., Chen, M., Poor, H., & Cui, S. (2021). Neural layered min-sum decoding for protograph LDPC codes. In *IEEE International Conference on Acoustics, Speech and Signal Processing (ICASSP)*. https://doi.org/10.1109/ICASSP 39728.2021.9414543.

Zhang, M., Huang, Q., Wang, S., & Wang, Z. (2018). Construction of LDPC codes based on deep reinforcement learning. In *10th International Conference on Wireless Communications and Signal Processing (WCSP)*. https://doi.org/10.1109/WCSP.2018.8555714.

Zhu, H., Cao, Z., Zhao, Y., & Li, D. (2020). Learning to denoise and decode: A novel residual neural network decoder for polar codes. *IEEE Transactions on Vehicular Technology, 69*(8), 8725–8738. https://doi.org/10.1109/ICASSP39728.2021.9414543.

Zhu, H., Pu, L., Xu, H., & Zhang, B. (2018). Construction of quasi-Cyclic LDPC codes based on fundamental theorem of arithmetic. *Wireless Communications and Mobile Computing, 2018*, 1–9. https://doi.org/10.1155/2018/5264724.

Part 3

Artificial Intelligence towards 6G

10 Enabling Technologies and Applications of 5G/6G-Powered Intelligent Connectivity

Tulsi Pawan Fowdur, Lavesh Babooram, Madhavsingh Indoonundon, and Anshu P. Murdan
University of Mauritius

Zoran S. Bojkovic and Dragorad A. Milovanovic
University of Belgrade

CONTENTS

10.1	Introduction	356
10.2	Enabling Technologies of Intelligent Connectivity	359
	10.2.1 Internet of Things	361
	10.2.1.1 Cellular IoT	365
	10.2.1.2 Massive IoT	365
	10.2.1.3 Broadband IoT	367
	10.2.1.4 Critical IoT	367
	10.2.1.5 Industrial Automation	369
	10.2.2 5G Mobile Networks	369
	10.2.2.1 5G-IoT Requirements	371
	10.2.2.2 5G-IoT-Enabling Technologies	372
	10.2.2.3 Wireless Network Function Virtualization	372
	10.2.2.4 Heterogeneous Networks	373
	10.2.2.5 Device-to-Device Communications (D2D)	373
	10.2.2.6 Advanced Spectrum Sharing and Interference Management	374
	10.2.3 Artificial Intelligence	374
	10.2.3.1 Intelligent Networks	375
	10.2.3.2 AI-Enabled Autonomous Systems and Human Interaction	377
	10.2.4 Cloud Computing and Networking	380
	10.2.4.1 Cloud Deployment Models and Service Classes	380
	10.2.4.2 Intelligent CC and Networks	382

DOI: 10.1201/9781003205494-13

    10.2.5 Blockchain 385
        10.2.5.1 Blockchain in 5G 387
        10.2.5.2 Blockchain in IoT 389
        10.2.5.3 Blockchain in AI 390
10.3 Applications of Intelligent Connectivity 390
    10.3.1 Transportation 391
    10.3.2 Industry 4.0 and Manufacturing Operations 392
    10.3.3 Smart Cities 393
    10.3.4 Health Care 394
    10.3.5 Education 395
10.3 Summary 396
References 396

10.1 INTRODUCTION

With the fusion of 5G, the Internet of things (IoT), artificial intelligence (AI), cloud computing, and blockchain, the concept of intelligent connectivity paves the way to expedite technological advancement and gives rise to modern disruptive digital services. The agenda of intelligent connectivity consists of analyzing and interpreting the digital data aggregated by machines, devices, and sensors comprising the IoT, through AI technologies, to generate a more considerable, relevant, and expressive output to users. This translates into a combination of better actions taken as well as customized deliverables to clients, tightening the gap between users and the environment they are interacting with. The continuous progress in computing power, the rise in awareness of data scientists, and the fact that machine learning tools, with intensive documentation, are available at the click of a button have propelled the development of AI practices. These edges the IoTs a step closer to attaining colossal global popularity and attention. 5G acts as the glue that sticks together these technologies such that the intelligent connectivity vision can be achieved. The ultra-high-speed and ultralow-latency characteristics that accentuate 5G networks, coupled with the massive amount of information collected by the IoTs, as well as the refined abilities of AI technologies to make intricate decisions, will lead to modern groundbreaking potentials. This will eventually transcend into every industry sector and seep through our society as well as our daily lives [1]. Figure 10.1 illustrates the main components of intelligent connectivity.

  The evolution from 4G to 5G is the main driving element for intelligent connectivity. The 5G connections are expected to reach 1.2 billion globally by 2025, according to GSMA Intelligence. 5G has the potential to significantly boost network capacity, speed, and responsiveness, allowing operators to customize connections for each application. This sets the stage for a range of forthcoming and prospective applications. The last decade witnessed the dramatic rise of cloud-based applications, which are bound to further boon, with the looming worldwide 5G deployment. Intelligent connectivity is thus referred to as the eventual crossover between the IoT, blockchain, and AI, where the junction corresponds to a collaboration of the involved technologies operating in unison [2].

Applications of 5G/6G Powered Intelligent Connectivity

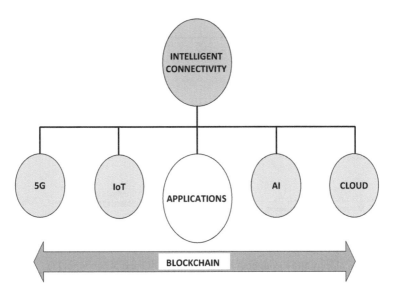

FIGURE 10.1 Components of intelligent connectivity [1].

Equipped with unparalleled responsiveness, 5G networks will allow an immersive and closer experience between the user and the machine, evoking the impression of a more tangible Internet where users can remotely take major decisions at the ease of their fingertips, impacting both businesses and consumers. Operators will be able to practically remove network latency in many circumstances, allowing users to engage virtually with each other and their surroundings through virtual reality (VR) and augmented reality (AR). 5G will bring along novel networking techniques to yield real-time or near-real-time facilities. Split-second decisions required by self-driving vehicles and immersive gaming are unfeasible and unreliable in the era of current 4G networks. Low-latency connectivity will speed up response times, allowing vehicles and devices to promptly intervene, laying the basis for the launch of autonomous cars, drones, and robots. On the same wavelength, a variety of smart city applications will be enabled, such as intelligent traffic management and real-time crime detection, as shown in Figure 10.2 [3].

The jolt caused by COVID-19 has urged digital transformation with regard to almost every industry [4]. Existing digital infrastructure, together with ICT infrastructure expansion policies, can help prevent a K-shaped recovery pattern and fuel a more equal and resilient future development model. The standards refined by COVID-19 in fact point toward the adoption of intelligent connectivity, thus highlighting the following properties:

- Increased demand for high-speed connectivity in teleworking and online education.
- Cloud computing affordability pertaining to corporate operations as well as deployment of scalable frameworks for data storage, processing, and distribution of facilities.

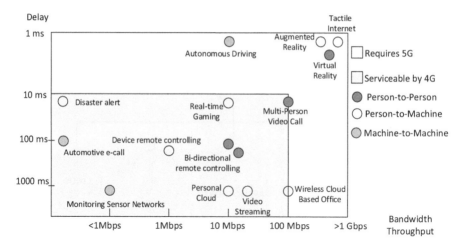

FIGURE 10.2 Delay and throughput requirements of 4G and 5G services [3].

- AI to facilitate accurate decision-making and workflow automation with services such as chatbots.
- IoT gadgets equipped with automation capabilities of processes and deliverables, to refine resistance and recovery abilities.
- Technologies that lower the total cost of processes and operational expenses of telecom networks and data centers. Some examples are AI, IoT, and measures for renewable energy, energy digitization, and cooling practices. The benefits of such technologies include (a) an improvement in deployment practicality and network growth on the supply side and (b) the mitigation of the carbon emissions caused by the ICT industry, into the atmosphere.

Worldwide governments have laid out plans for the integration of the listed technologies, deemed as essential intelligent connectivity facilitators, into their economies. The value generated by intelligent connectivity revolves around collaborative ecosystems. In contrast with the twentieth century where oil was the economy's driving factor, the fuel for the twenty-first century is simply data, acting as a form of viable progress. The amount and properties of generated data dictate the tempo of intelligent connectivity structures, which consist of AI, cloud services, and IoT. A greater availability of unstructured and fitting data means better AI analysis, and thus enhanced intelligent connectivity. This is backed by the collection of large quantities of high-quality input, through which customizable products and services can be developed to increase the number of clients and target real-world issues [5].

In this chapter, a framework as well as an overview of the key enabling technologies for intelligent connectivity will be discussed in Section 10.2. Section 10.3 will elaborate upon the five most important use cases of intelligent technology, and Section 10.4 will conclude the chapter.

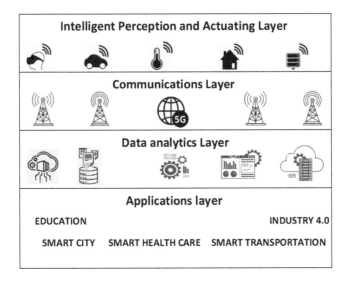

FIGURE 10.3 Intelligent connectivity layers.

10.2 ENABLING TECHNOLOGIES OF INTELLIGENT CONNECTIVITY

A typical reference model for intelligent connectivity with a four-layer representation is shown in Figure 10.3.

The layers of this model are described as follows:

Intelligent perception and actuating layer. The physical sensors present in the IoT, whose purpose is to gather and handle data, make up the first layer, also known as the objects (devices) or perception layer. This layer contains sensors and actuators that trigger different responses, including polling features such as location, weight, temperature, vibration, motion, humidity, and acceleration. Interchangeable plug-and-play techniques need to be implemented by the perception layer for the automatic configuration of a diverse range of devices [6,7]. By means of secure channels, this layer sends digitized data to the network and communications layer. This is also the starting point of the big data created by the IoT.

As the most fundamental operations in 5G/6G networks, sensing and detection are processes where these systems intelligently receive and perceive data from physical surroundings using an array of large devices such as cameras, sensors, vehicles, drones, and smartphones. This may also include gatherings of people. By being substantially closer to the physical environment, AI-enabled sensing and detection provide dynamic and active data collection architectures such as radio-frequency usage identification, environment monitoring, spectrum sensing, infringement detection, and more [8]. The collection of large amounts of scalable data is thus possible.

Due to the ultrahigh-reliability and ultralow-communication latency required by 5G/6G networks, high-accuracy sensing, real-time sensing, and robust sensing are all of paramount importance. Furthermore, dynamic 6G networks introduce variability in spectrum characteristics, making robust and accurate sensing challenging. AI approaches can provide precise, real-time, and resilient spectrum sensing, where fuzzy support vector machine (SVM) and nonparallel hyperplane SVM are resistant to instabilities in the environment. Likewise, sensing precision and correctness can be improved by CNN-based cooperative detection with low intricacy. By combining K-means clustering and SVM, real-time detection can be accomplished by using low-dimensional input samples for the training stage. Similarly, by handling heterogeneous data fusion, Bayesian training may solve considerably large and diverse detection challenges [8].

Communications layer. The connection to other smart things, network appliances, and servers relies on the network layer. It contains support and implementation to transmit and process sensor data, which can be transmitted through multiple technologies including RFID, GSM, 3G, UMTS, Wi-Fi, infrared, ZigBee, Bluetooth Low Energy, etc. However, in the case of intelligent connectivity, 5G networks are used for the data transfer as it provides ultrahigh reliability and low latency. It also has support for massive IoT [9].

Data analytics layer. This layer is in charge of interpreting the data collected by the perception layer and transmitted by the communications layer. It performs a number of machine learning tasks such as prediction, classification, and clustering, among others, on different types of data including big data and real-time data. The analytics can be done on cloud-hosted servers and the cloud databases may be used for the storage of gathered data. Based on the analytics, this layer can send results to end users as well as instructions to the actuators found at the perception layer to trigger certain actions. These could range from controlling a watering plant to a robot's arm in a production environment [10].

Applications layer. The application layer handles the queries made by the clients. For example, in an interaction where a user asks for temperature and air humidity measurements, the application layer is responsible for handling that particular provision. This means that this layer has the ability to manage high-quality smart services, to meet the requirements of the customers, which sums up its importance. A wide range of vertical sectors is governed by the application layer, including smart homes, smart buildings, factory equipment, and intelligent healthcare [10].

The driving force of AI can achieve smart programming and strategic planning to sustain more diverse high-level intelligent and advanced applications, such as automated processes, smart industry, smart transportation, smart city, smart grid, and smart health while also catering for the worldwide effective management of all sorts of smart systems. The smart application layer manages all of the operations of smart devices, terminals, and facilities in 6G networks using AI approaches to achieve self-organization capabilities in networking [8].

This layer also aims to assess service performance, by gauging a variety of criteria and factors such as QoS, QoE, and the quality of both gathered data and acquired knowledge. Simultaneously, cost dimension metrics should also be considered, in terms of resource efficiency, including spectral utilization ability, computational effectiveness, energy efficiency, and storage efficiency [8].

The main enabling technologies of intelligent connectivity are 5G, IoT, AI, cloud, and blockchain, which are covered in the subsections that follow.

10.2.1 Internet of Things

The IoT consists of interconnections between three possibilities: (i) people to people, (ii) people to machine/things, and (iii) things/machine to things/machine. The primary purpose of the IoTs is to ensure a constant connection between things, at anytime, anywhere, with anything, and anyone, preferably via any route/network, and any service. Objects become identifiable and gain intelligence by making or facilitating context-related judgments due to their ability to convey information about themselves [11]. They can acquire intelligence obtained by other devices or act as bits of sophisticated services. An overview of IoT capabilities is given in Figure 10.4.

IoT-enabling technologies and components. IoT is not a standalone technology, but rather a combination of several hardware and software technologies. It holds the key to the integration of information technology, which corresponds to the whole architecture involving both software and hardware used for storage, retrieval, and processing of data. This framework also consists of communication paradigms, i.e., electronic systems for transmission of information between individual users, or groups. An exhaustive list of enabling technologies for IoT is given in Figure 10.5 [11].

From these enabling technologies, the main elements of IoT are formed as depicted in Figure 10.6 [12].

These elements are defined as follows.

Identification. In order for the IoT to link the services required to their demand, identification is essential. Several such systems are accessible, such as electronic product codes (EPC) and ubiquitous codes (uCode) [13]. Moreover, it is imperative that the IoT objects undergo addressing, such that the Object ID and its network address can be differentiated. The Object ID corresponds to a device's name, e.g., a temperature sensor, while its address pertains to its identifier in a communications system. Methods for addressing include IPv6 and IPv4. A compression method is

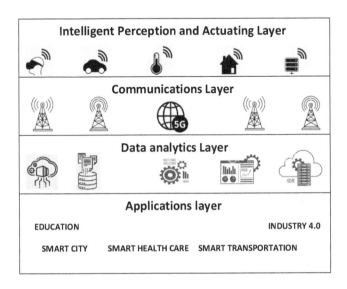

FIGURE 10.4 IoT capabilities [11].

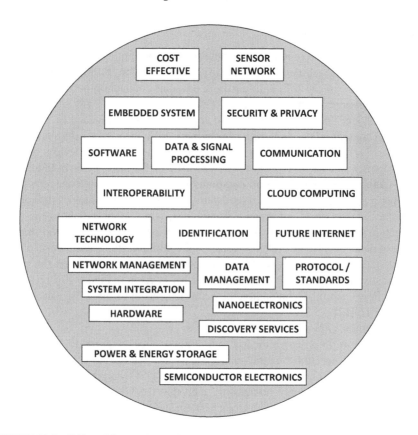

FIGURE 10.5 IoT-enabling technologies [11].

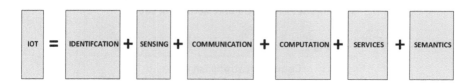

FIGURE 10.6 IoT elements [12].

provided by 6LoWPAN [14,15], through IPv6 headers where the latter's addressing is made suitable for low-power wireless systems. Allocating a unique address to an object is essential since identification patterns around the world have not been standardized, giving rise to the crucial importance to differentiate between the object and its address. Furthermore, networks can contain objects that use public IP addresses instead of private ones. Thus, identification hands out crystal-clear identities to objects such that they are easily identifiable.

Sensing. Generally, IoT sensing encompasses the processes involved in collecting data from objects in a system and transmitting it to a data warehouse, database, or cloud framework. The required services then dictate the course of action to be taken. Smart sensors, actuators, and wearable data collection devices are examples of IoT sensors. Wemo, revolv, and SmartThings are example of companies that supply intelligent hubs and mobile applications to remotely track and evaluate smart gadgets and devices indoors [16–18]. Single board computers (SBCs) with built-in sensors and TCP/IP architectures along with security frameworks are mostly utilized to produce IoT products. Some examples are Raspberry PI, Arduino Yun and BeagleBone Black [13]. Such devices are often linked to a single administration site to supply the necessary services queried by the customers.

Communication. Intricate and customized smart services are made possible by connecting together a diversified range of objects that make up the IoT communication model. The IoT nodes usually function with low power within communication systems filled with noise and losses. Common communication systems that govern IoT are Wi-Fi, Bluetooth, IEEE 802.15.4, Z-wave, LTE-Advanced, and 5G [11], while more distinct ones include RFID, near-field communication (NFC), and ultra-wide bandwidth (UWB). With relation to intelligent connectivity, the main focus is on cellular IoT with emphasis on 5G. An overview of cellular IoT is given in Section 10.2.

Computation. Computation represents the processing capabilities of the IoT, i.e., the "brain," consisting of units such as microcontrollers, microprocessors, and FPGAs, as well as software aspects. To execute IoT applications, many hardware platforms were designed, e.g., Arduino, FriendlyARM, Raspberry PI, Arduino, UDOO, Gadgeteer, Intel Galileo, Cubieboard, BeagleBone, Mulle, Z1, WiSense, and T-Moke Sky. These are topped up with software platforms that render IoT functionalities possible. Since the operating system runs throughout the device's lifetime, it is of utmost importance during the designing stage. Various real-time operating systems (RTOS) are suitable for the development of RTOS-based IoT applications. For example, the Contiki RTOS is often employed in IoT environments [19]. For settings requiring lightweight OS implementations, TinyOS [20], LiteOS [21], and Riot OS [22] are fitting. Additionally, the realm of Internet of vehicles (IoV) is headed for improvement with the established Open Auto Alliance (OAA) between auto manufacturing leaders and Google, to integrate new functionalities on Android [23]. An IoT microcontroller is essentially an embedded system comprising computer hardware equipment paired up with appropriate software. It can be either standalone or part of a larger architecture. An embedded system is often implemented in massive automatic electrical or mechanical models and is therefore automated by a controller, thus making up the RTOS. It is generally employed as a fully functioning device that contains both hardware and mechanical components [24]. On the same spectrum, cloud platforms are of equal importance in the computational sphere [12]. Cloud systems are responsible for the storage of data sensed by smart objects, after which big data is analyzed in real-time before returning the requested response to the exact query demanded by the end user.

Services. Overall, IoT services can be summarized by four classes [25,26]:

- Identity-related services – They are the most basic, yet possibly one of the most significant services supplied to an IoT application. Some form of identity-related service is integrated and present in every IoT application. This is

due to the necessity of the IoT to map the physical realm to the digital one, requiring the application to identify all connected devices [27]. This type of service is handled by most IoT applications through the use of RFID technology. The RFID tag holds a device-specific identifying code. The latter is scanned by the RFID reader, which searches the RFID server for the device, before delivering the detailed information required by the application [28].
- Information aggregation services – These gather and consolidate raw sensory readings for processing and reporting to the IoT application. By combining identity-related services with other systems such as wireless sensor networks (WSNs) and access points, they gather and transmit data to the application for analysis. This means that the information aggregation service simply routes the collected data to the application. Between the collection terminals such as sensors and RFID tags to the application, the sensed data may also be processed.
- Collaborative-aware services – They sit just above the information aggregation services, using the collected information to respond appropriately. The primary distinction between these two types of adjacent services is the application of acquired data to make choices and alter the course of action in environments such as a smart house. Speed, network security, and terminal computing capability are all required for developing a collaborative-aware service. It is no longer convenient for terminals to be solely sensors that have no capabilities of affecting the surrounding based on analysis. There is a growing requirement for embedded devices within the same network that can take decisions based on collected data [27].
- Ubiquitous services – These services up the ante by aiming to ensure that collaborative-aware services are allocated to anyone, anytime, and anywhere [27]. They are thus the epitome of the IoTs. Ubiquitous services intend to enable total access and control over everything around us, irrespective of the device in question. A smart city figures on the list of applications requiring ubiquitous services.

Semantics. In the context of IoT, semantics is the concept of intelligently extracting meaningful information from different devices. Deriving important knowledge from information consists of using resources and generating models. It also revolves around identifying and evaluating facts in order to make the proper decision to deliver the best service [29]. Consequently, semantic serves as the IoT's brain, routing requests to the appropriate resource. The resource description framework (RDF) and the web ontology language (OWL) form part of semantic web technologies, which supports this requirement. The World Wide Web Consortium (W3C) recommended the Efficient XML Interchange (EXI) standard in 2011 [30]. The EXI format is deemed crucial for IoT architectures with its ability to dynamically improve XML applications where resources are scarce. This occurs by minimizing bandwidth requirements without compromising collateral resources such as battery life, code size, processing power, and memory allocation. To accommodate for this optimization, XML messages are transformed to binary by EXI, to reduce required bandwidth and storage space.

10.2.1.1 Cellular IoT

The adoption of cellular IoT was witnessed internationally when early IoT applications were deployed on 2G and 3G networks. The 4G era brought about increased bandwidth, lower response times, and enhanced assistance for highly populated cells. With the presence of 5G on the horizon, these features will be further strengthened, starting originally with the 5G New Radio (NR) standard. Support for progressively important and responsive applications is expected with the advent of 5G networks that will allow ultrareliable low-latency communications (URLLC).

Hitting two birds with one stone, cellular IoT is equipped with the potential of solving both the comparatively simpler needs of the colossal IoT market, in addition to the more tailored and delicate requirements of tricky systems and applications. Thus, cellular IoT connectivity supported by narrowband IoT (NB-IoT) and long-term evolution for machines (LTE-M) is proliferating. By 2024, the number of machines linked by massive IoT and other upcoming cellular technologies is expected to reach 4.1 billion. Cellular IoT is a fast-expanding sector centered around 3GPP international norms, with support from a growing number of mobile telecom companies, gadgets, microprocessors, components, and network infrastructure suppliers [31].

In terms of unrivaled worldwide coverage, cellular IoT outperforms other low-power wide area (LPWA) network technologies [31]. It also offers improved QoS, flexibility, reliability, and the versatility to address the various needs for a wide scope of uses. This expansion in IoT networking is likely to intensify due to two primary factors:

- An attempt is being made to computerize and modernize businesses such as industrial, automobile, and utility companies.
- Telecommunication companies are increasingly interested in expanding their present business beyond mobile Internet.

According to Ericsson, cellular IoT can be categorized as massive IoT, broadband IoT, critical IoT, and industrial automation IoT. This is summarized in Figure 10.7 [31].

An overview of these four categories of cellular IoT is discussed in the following sections.

10.2.1.2 Massive IoT

Massive IoT connects low-sophisticated IoT devices to cellular networks using narrow band-IoT and Category M (CAT-M) technology. It is aimed at large numbers of devices with limited complexity that communicate data rarely. The traffic generated is frequently forgiving of delays, and common use cases include low-cost detectors, monitors, gadgets, and trackers. Devices of such nature are often employed in tough radio and wireless environments such as a facility's basement, which translates into the imperative requirement of extended coverage. They may also depend entirely on stored battery energy, putting high demands on the device's battery life.

In Release 13, three giant new innovations were established by 3GPP. They include EC-GSM-IoT, LTE-M, and NB-IoT. LTE-M enhances LTE by adding additional

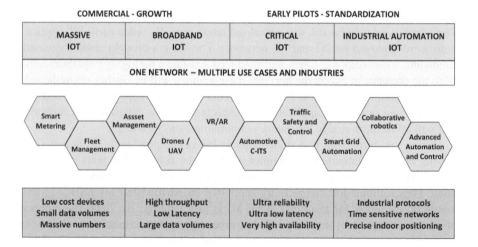

FIGURE 10.7 Cellular IoT categories.

capabilities such as enhanced battery life, expanded reception distance, and support for devices with solely simple features, called CAT-M. NB-IoT is an independent radio access technology based on LTE foundations that offer exceptional coverage and prolonged battery life for ultralow-complexity gadgets. The radio range per cell tower is increased by repeating messages and taking advantage of flexible data rate and delay constraints.

To improve battery life, techniques such as extended discontinuous reception (eDRX) and power saving mode (PSM) are used by putting the device in sleep mode for prolonged periods of time. The use of restricted bandwidths, half-duplex processing, and the integration of a singular transceiver antenna on the unit keep the intricacy of CAT-M and NB-IoT devices to a minimum. NB-IoT devices are relatively less performing than CAT-M ones and therefore, less complex. NB-IoT has a bandwidth of 200 kHz, while that supported by CAT-M is 1.4 MHz with CAT-M1, and 5 MHz with CAT-M2.

Despite existing functionalities for operating CAT-M devices in full-duplex mode, the current CAT-M infrastructure only allows half-duplex mode in order to restrict device intricacy and battery consumption. The use cases for CAT-M and NB-IoT should be complementary. CAT-M is ideal for applications seeking greater bandwidth, lower response times, linked mode mobility, improved geolocation positioning, and voice communications. Wearables, detectors, monitors, alarm panels, and helpdesk consoles are examples of typical CAT-M use cases, all of which offer voice and data interactions. NB-IoT is better adapted for extremely low throughput applications that can tolerate latency but necessitate substantial reach, including basic utility devices and sensors stationed in difficult radio circumstances.

Another benefit for telecom operators is that NB-IoT may be placed in an LTE carrier's guard-band, utilizing otherwise idle frequencies. CAT-M and NB-IoT are regarded as long-term plans with respect to 5G systems [32,33]. They can seamlessly cooperate with 5G NR in the same band and already meet all 5G large machine-type communications (MTC) standards. As summarized by the IMT-2020 and 3GPP standards, the same specificities can be used with regard to features such as distance reach, latency, transmission rate, battery life, and number of connections per cell [34,35].

10.2.1.3 Broadband IoT

Broadband IoT leverages mobile broadband access for IoT, delivering faster data rates and lower delay than massive IoT. It employs features tailored to MTC to enhance coverage and improve the unit's battery life. Built around 4G and 5G NR radio access platforms, this category addresses a wide range of applications in transportation, unmanned aerial vehicles (UAVs), augmented reality/VR (AR/VR), services, industrial, and wearable gadgets. LTE includes a variety of device types that are well-suited for such applications. For instance, it already provides mobile connectivity to millions of modern vehicles. The extent achievable by broadband IoT has been enlarged with the advent of NR. NR-based broadband IoT functions in both previous and modern spectrums through substantially larger bandwidths. It also contains additional features to handle notably higher throughputs, achieving the threshold of tens of Gbps and lowering delay to around 5 ms [31].

10.2.1.4 Critical IoT

Critical IoT paves the way to incredibly low radio interface delay, as low as 1 ms, or high reliability of up to 99.9999%, with stringent delay boundaries at a range of transmission speeds. Reliability pertains to the probability of successfully delivering information within a limited delay. Certain applications in smart power systems, intelligent industry, smart vehicles, intelligent healthcare, and immersive AR/VR necessitate ultralow latency in the range of 5–20 ms, along with reliability achieving heights of 99.9999%. Such situations require continuous real-time connectivity and communications between devices [36]. These include:

- Automation of power transmission.
- Fault diagnosis and repair in smart grid environments.
- Real-time management of industrial robots.
- Synchronization of self-driving cars and transport systems in real time.

Furthermore, some applications also require human involvement, notably remotely controlled vehicles and teleoperated surgeries, demanding even more reliability. However, due to the significant difference in awareness, response, and reaction times between humans and machines, latency is not as crucial as reliability.

5G NR is unquestionably a remarkable technology for supporting critical IoT. Even in its initial 3GPP deployment, Rel-15 in 2018, NR offers more capabilities than LTE for supporting URLLC. NR uses a wider spectrum of bands and significantly bigger bandwidths than LTE to deliver considerably higher throughputs to a greater number of machines while maintaining relatively low latencies and ultrahigh consistency. The evolution route laid down for NR is crystal clear. As per 3GPP Rel-16, the establishment for NR-based, improved URLLC is already in the works. As shown in

	WIDE AREA USE CASES	LOCAL AREA USE CASES	
High bands (24 GHz – 40 GHz)			-Extremely low latency -Ultra-high reliability -High capacity -Limited coverage
Mid bands (1 GHz – 6 GHz)			-Extremely low latency (with FDD / latency favorable TDD) -Ultra-high reliability -Decent coverage and capacity
Low bands (sub 1 GHz)			-Extremely low latency -Ultra-high reliability -Wide area coverage -Limited capacity

FIGURE 10.8 Critical IoT support in different bands by NR.

Figure 10.8, NR can accommodate critical IoT across all of its frequencies, enabling broad area and small area use scenarios.

In low radio frequencies with corresponding bandwidth allocations, due to optimal radio wave propagation, NR frequency division duplex (FDD) yields exceptionally low response times and ultrahigh reliabilities with large geographical areas per cell tower. However, link transmissions in the lower frequency regions are constrained. These frequencies should thus specifically serve vast regions.

NR revolves around a wide variety of enablers for URLLC. The ultralow latency characteristics are:

- Communications with extremely short intervals.
- Approaches for instantaneous delivery to reduce data communication latencies.
- Strategies for rapid retransmission that mitigate feedback latencies from a receiver to an emitter.
- Mechanisms for immediate cancellation and prioritizing.
- Handovers between base towers to occur without interruption.
- Devices and base stations equipped with high processing performances.

A wide variety of communication designs that still thrive in difficult radio environments is imperative to cater for the ultrahigh reliability requirements. This high link connection consistency is assured by redundancy methods such as multi-site connectivity and carrier aggregation. There is a gradual growing prospect to enhance the link budget with sophisticated antenna designs. At the core of NR lies the vendor-specific scheduling techniques and link adaptation algorithms, such that the proper resource maintenance and usage are guaranteed.

There are inherent compromises when it comes to factors such as delay, reliability, distance covered, and spectral efficiency, with regard to a particular rollout. Ultra-short broadcasts, e.g., that aim for extremely low response times, might diminish the network strength that accommodates for each base station. As a result, strong signal emissions necessitate additional unused spectrum, lowering throughput per base station. NR

Applications of 5G/6G Powered Intelligent Connectivity 369

allows for a significant deal of flexibility in optimizing these trade-offs. As per 3GPP Rel-15, studies and assessments have previously verified that the NR radio interface is capable of sending a small payload in a maximum delivery timeframe of 1 ms, with 99.999% consistency with regard to both uplink and downlink [37,38]. The 99.9% to 99.9999% consistencies are being further analyzed by 3GPP Rel-16 where latency is found in the range of 1–7 ms, with transmission speeds ranging between Kbps and Mbps [39].

10.2.1.5 Industrial Automation

The concept of industrial automation IoT revolves around implementations for production frameworks as well as control regimes for metros together with the generation of electricity and its consequent supply. These processes require sensors placed in the surroundings, along with monitoring gadgets for stock and delivery supervision. There is thus a growing need for automated guided vehicles (AGV) and state-of-the-art real-time monitors, along with robot expert systems in the production process. These pieces of equipment are usually wired.

Therefore, an industrial network is made up of a blend between massive, broadband, and critical IoT, along with industrial automation IoT, centered around 5G NR, to fill the shortcomings in performance. However, this void in capabilities is envisioned to handle a deterministic system where both ultralow response times and consistent secured delivery are needed. In short, 5G is anticipated to embed functionalities that allow minimal latency fluctuations and remarkable low loss, both ensured at once.

10.2.2 5G MOBILE NETWORKS

The gap between mobile broadband and massive IoT is being bridged by the advent of fifth-generation wireless networks (5G). The point at which 5G demarcates itself from today's 4G and 4.5 (LTE-advanced) apart from upturns in transmission speeds is the basis laid out for welcoming new IoT and life-depending communication scenarios. For context, small response times are the pillar that enable real-time exchanges with services that contact the cloud. This can be observed in the case of autonomous vehicles. Similarly, the necessity for recurrent human intervention is removed with low power consumption capabilities, thus allowing gadgets to extend their uptimes to months or years [37].

Contrary to current norms where performance trade-offs are often employed as solutions for wireless technologies including 3G, 4G, and Wi-Fi, 5G includes built-in and seamless functionalities to enable massive IoT. Figure 10.9 outlines the advancements with regard to transmission speeds, starting with 3.5G.

Figure 10.10 shows the eight main targets, which 5G aims to achieve namely with regard to data rate, availability, coverage, energy usage, battery life number of connected devices, bandwidth, and latency.

Figure 10.11 depicts the ecosystem of forthcoming 5G networks consisting of the trident: (i) enhanced mobile broadband (eMBB), (ii) ultrareliable and low-latency communications (URLLC), and (iii) massive MTC (mMTC) [38,40].

The coming of every new generation wireless network marked the arrival of new possibilities, as illustrated in Figure 10.12.

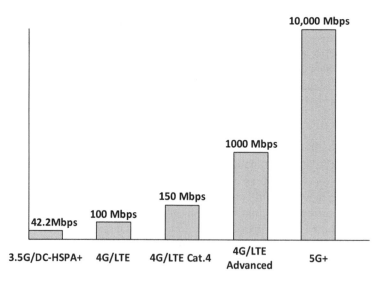

FIGURE 10.9 Speed-wise evolution of mobile technologies [37].

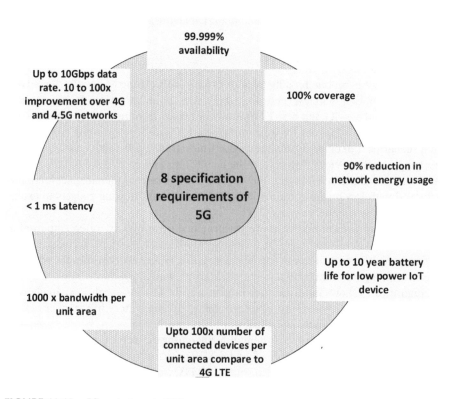

FIGURE 10.10 5G main targets [37].

Applications of 5G/6G Powered Intelligent Connectivity

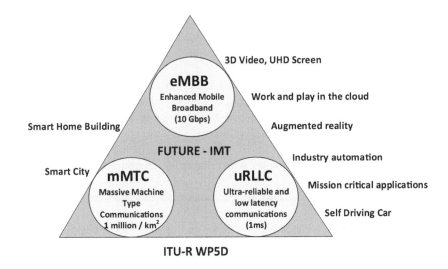

FIGURE 10.11 5G ecosystem [38,40].

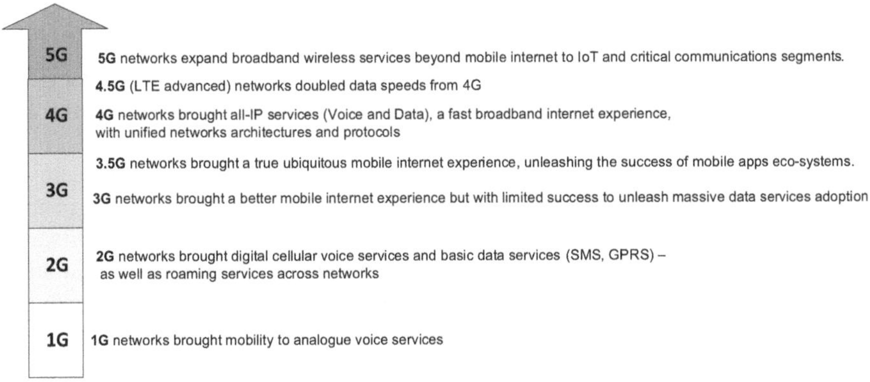

FIGURE 10.12 Mobile standards capabilities evolution [38,40].

10.2.2.1 5G-IoT Requirements

The revolutionary experience surrounding the surge of 5G can be classified as real-time, programmable, exclusively online, and loaded with social-oriented innovations [38].

These definitions call for an armature of end-to-end coordination capabilities, with rapid, autonomous, and smart adaptive tasks set up at every step [41]. The following list summarizes the expectations of 5G:

- Establish conceptually separate networks depending on application needs.
- Rebuild radio access infrastructures (RAN) using cloud-based radio access networks (Cloud-RAN).

- Deliver extensive links bearing different protocols while enforcing on-demand installation of RAN purposes needed by 5G.
- Enable on-demand programming of network functions to simplify core network ecosystem [41].
- Provide support for high transmission speeds in an era where high-quality video streaming is established as a social norm and VR presses for increasingly high data rates of the order of 25 Mbps for up-to-par performances [41].
- Exhibit ultralow latencies in 5G-IoT use cases including AR, tactile Internet, and VR video games where latencies around 1 ms are considered acceptable.
- Extend coverage distance and improve handover smoothness for IoT devices and customers.
- Implement cutting-edge security policies and concepts for IoT applications such as mobile banking and electronic wallets.
- Effectively implement energy-saving methodologies to feed low-power and low-cost IoT devices over sweeping time periods.
- Support high connection density and mobility for interconnected communication exchanges.

10.2.2.2 5G-IoT-Enabling Technologies

Starting from the physical paradigm and ascending all the way up to the complete IoT application, the 5G-enabled IoT sphere is held uptight by various key factors that fall into the following five components [37]:

5G-IoT architecture. Since the 5G-IoT rests on 5G wireless infrastructures, the framework typically consists of two planes, notably the data plane and control plane. The former concentrates on monitoring and recording data through software-defined front-haul networks while the latter is geared toward network orchestration tools, customizable services, and applications providers [42]. The following fields are dependable on the 5G-IoT architecture and their demands should be met:

- Network function virtualization (NFV), expandability, and cloud migrations.
- Refined and complex network management including mobility control, intrusion detection and prevention, and effective network virtualization.
- Inclusion of big data analysis for intelligent services providers.

10.2.2.3 Wireless Network Function Virtualization

Compatible with 5G networks, wireless network function virtualization (WNFV) will shoulder the virtualization of complete network functions, in turn simplifying the implementation of 5G-IoT. NFV will fractionate scalable and malleable hardware such that 5G-IoT is enabled on broadscale cloud servers [43]. Likewise, NFV will allow tailored network slicing on top of distributed cloud to generate customizable networks for 5G-IoT scenarios [44]. It allows a physical medium to be concatenated into several virtual networks, as depicted in Figure 10.13. Thus, the devices can simply be reprogrammed to exhibit the characteristics required by the IoT application.

Applications of 5G/6G Powered Intelligent Connectivity

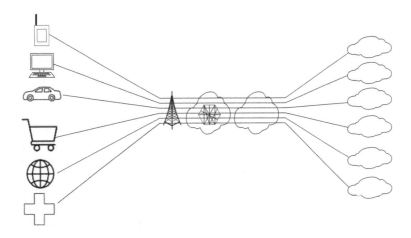

FIGURE 10.13 5G NFV technology [44].

NFV technology renders the applications faster, more intensive, and equipped with optimized coverage in the logically sliced networks such that requirements pertaining to a singular network slice can be realized.

10.2.2.4 Heterogeneous Networks
Heterogenous Networks (HetNet) is a pioneering networking architecture suggested to attain the needs of service-driven 5G IoT. HetNets operate with regard to the high-speed rates of 5G-IoT applications. Several 5G HetNet solutions have been developed recently. With the expected deployment of a colossal number of resource-capped devices, various HetNet solutions have been suggested [45,46].

10.2.2.5 Device-to-Device Communications (D2D)
The typical macrocell base station (MBS) is synchronized in HetNet to deliver low-energy BSs. For short-range transmissions between two gadgets, device-to-device communication (D2D) is a novel suggested data communication method. This will bring about improvements in 5G-IoT with regard to reduced energy usage, load balancing, and improved QoS for end users. D2D facilitates exchange of information between user equipment, bypassing the use of BSs, thus functioning as a "cell tier" in 5G-IoT [37].

Over 60% of the IoT systems demand minimal power, higher battery capacity, and extensive network coverage. Current technological solutions, among the likes of Bluetooth Low Energy (BLE), Zigbee, Wi-Fi, and 2G/3G/4G, do not go hand in hand with these requirements. To therefore meet the aforementioned demands, emerging innovations include LPWA, NB-IoT, LoRa, Sigfox, LTE-M, etc. D2D communication is quickly ramping up the technology ladder when it comes to short-range wireless communications of less than 200 m. It satisfies the necessity of low power and high QoS surrounding massive device communication [37].

As an add-on to NB-IoT uplinks, the D2D protocol is used to set up mobile links through NB-IoT [47–49]. D2D is implemented in IoT in conjunction with cellular

NB-IoT end devices. In short, D2D is equipped with the potential of improving energy and spectrum efficiency in 5G-IoT.

10.2.2.6 Advanced Spectrum Sharing and Interference Management

To address the discrepancies orbiting around network coverage and traffic congestion, 5G-IoT will be thoroughly doped with a massive number of interconnected devices. Consequently, a proper structure and accommodation is required for the management of spectrum and interference with densely populated and congested areas. HetNet seems to assertively address handle interference control in 5G-IoT.

Massive MIMO is critical to attain higher spectrum efficiency. A variety of sophisticated MIMO approaches, including multiuser MIMO (MU-MIMO), very large MIMO (VLM), and others, have been introduced in recent years. By capitalizing on an abundance of antennas at the BS, the 3GPP LTE-A involved MU-MIMO to considerably boost network coverage and power [50]. The fate of 5G-IoT frameworks resides on the amalgamation of technologies such as millimeter wave (mmWave), software-defined networking (SDN), MTC, multiaccess edge computing (MEC), narrowband IoT (NB-IoT), and NFV.

10.2.3 Artificial Intelligence

As global networks flourish in terms of novelty, magnitude, and intricacy, network traffic is growing at a sky-rocketing pace with the mushrooming number of interlinked devices. The near future is expected to bring about autonomous sensing, computing, learning, analytics, and even decision-making with regard to business without human intervention. The coupling of AI with networking is pivotal to the inception of automation, complexity and scalability management, and exploitation of live information obtained from distributed architectures.

The AI sphere is surrounded by multifaceted approaches involving machine learning, deep learning, game theory, optimization theory, and meta-heuristics [51]. Machine learning can be further categorized into supervised learning, unsupervised learning, and reinforcement learning. With a constant need to monitor and manage wireless networks in terms of resources, security, and performance, the ever-increasing widespread involvement of machine learning and deep learning does not go unnoticed. An overview of the main subfields of AI is as follows [52]:

Supervised learning. Supervised learning makes use of pre-categorized training instances to put together the learning model through the training process. The two main offshoots are (i) regression and (ii) supervised learning. Supervised learning makes use of pre-categorized training instances to put together the learning model through the training process. The two main offshoots are regression and classification. Classification deals with the prediction of outcomes that are typically classes or categories by first analyzing a pre-labeled dataset, before predicting the class for an unlabeled set of inputs. Some prevalent algorithms are decision trees (DT), SVM, and K-nearest neighbors (KNN). On the other hand, regression deals with the forecasting of variables that change with time by analyzing a particular number of inputs. Some common mathematical concepts are support vector regression (SVR) and Gaussian process regression (DPR).

Unsupervised learning. The motive behind unsupervised learning is to uncover trends that happen between the lines and are otherwise unnoticed to superficial judgment. It is broadly classified into clustering and dimension reduction. The basic premise of dimension reduction is to reduce the number of variables, characteristics, or features making up a dataset, in a way that the new data remains useful. Some techniques for dimensionality reduction are principal component analysis (PCA) and isometric mapping (ISOMAP). Likewise, with the aim of grouping together similar entities, clustering primarily comprises K-means clustering and hierarchical clustering.

Reinforcement learning (RL). Generally, in RL, an agent dynamically analyses and exploits its environment to learn and therefore act heuristically, via trial and error in an attempt to optimize long-term payoff. Q-learning, policy learning, Markov decision process (MDP), actor–critic (AC), deep reinforcement learning (DRL), and multi-armed bandit (MRB) are conventional instances of RL methodologies.

Deep learning. Deep learning seeks to tighten the gap between computers and the human brain by mimicking the latter's analytical approaches to create artificial neural networks. Equipped with several layers of neurons, the training model falls into the supervised, semi-supervised, and unsupervised categories. Deep neural network (DNN), convolutional neural network (CNN), recurrent neural network (RNN), and long-short-term memory (LSTM) are all examples of typical deep learning concepts.

10.2.3.1 Intelligent Networks

Telecommunication providers in every corner of the world are either preparing for, or already deploying 5G. With aforementioned qualities such as autonomous sensing, computing, training, and decision-taking on the horizon, the emergence of zero-touch cognitive networks is anticipated [53]. Fresh from the oven, the endless possibilities of modern networks will yield opportunities for exploitation as well as open tons of doors for the management of the rapid surge of data flows. Despite the inevitable complexity of network topologies associated with the plethora of innovative use cases of network edge and cloud, future networks are expected to comfortably stay afloat. However, this breakthrough comes packaged with new challenges with respect to intricacy, upgradability, toughness, and reliability as well as installation factors such as the provision of up-to-par network services itself, design, and deployment [54]. Conventionally, telecom specialists develop and maintain mobile infrastructures, relying significantly on their vast expertise of network architecture, customer mobility, traffic consumption trends, and wave transmission models to create and implement the protocols and rules that fundamentally ensure the continuous operation of the immense network. The complexity of network topologies has significantly increased, given the arrival of smaller cells and state-of-the-art radio technology. Not only have traffic trends grown harder to predict but radio transmission models have also become more difficult to evaluate considering the new spectrum frequencies and densely populated networks. AI therefore sits at the throne when it comes to providing support to service providers, to ease the deployment and maintenance process of 5G networks. Moreover, AI is considered indispensable and primordial to ensure zero-touch provisioning of these sophisticated networks, in addition to be entrusted with autonomous analysis and decision-making functionalities.

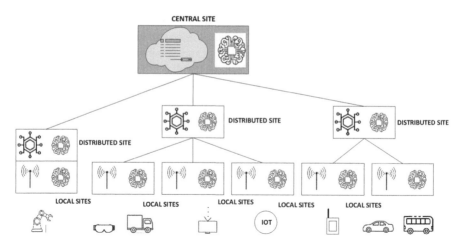

FIGURE 10.14 Learning at local and global levels with autonomous decision-making in widely distributed networks [53,55].

A mobile network is embedded with distributed and decentralized properties. Figure 10.14 depicts the addition of AI into the networking framework where data processing is envisioned. In this case, this can happen both locally and centrally. Therefore, AI can be integrated with both central and local edge sites. Similarly, local learning and outcome generation occur at distributed sites, whereas data and understanding are merged throughout locations to provide a full global overview of links between networks, facilities, and network operations.

Every local site provides quantitative information on the state of the various components, the timeframes of events, and related data. This data may be harnessed to create representations and patterns of local behavior. However, computational logic is necessary to interpret the information acquired throughout sites and extrapolate meaningful understandings of the whole framework. Ultimately, the information and findings obtained at a single site may be steered into valuable insights to forecast those of other sites. Network characteristics are constantly changing and there is an increasing need to embed real-time support via modular and customizable analytics solutions to address the amount, speed, and diversity of real-time information. Algorithms are gradually evolving by being provisioned with capabilities for real-time decisions.

The integration of sophisticated intelligence to networks, applications, and business systems will result in a gradual shift toward a statistics-geared strategy, which will promote a superior level of automation, reliability, and productivity. Equipped with increased control, telecommunication providers will be able to act upon its networking systems with greater efficiency, based on analytics on captured data. By tailoring and eventually mapping specific functions to network behavioral patterns, each service can be rendered more secure, reliable, as well as robust, further propelling the provision of mobile services to higher gear with regard to both industrial and societal advancements [55].

Within a massive distributed architecture, decision-making processes exist at several sites and stages. Some actions are determined by strict-control systems with low delay and are dependent on local data, while some require intensive and well-planned decisions that alter the behavior of the whole network and are therefore fueled by the collection of data from multiple sites. Such decisions commonly fall into the critical category, thereby needing immediate, real-time attention. Examples include power-grid outages and collateral transient faults. To continuously and exquisitely feed this global and monumental architecture, the governing intelligence should be built around the distributed characteristic of the topology.

In many cases, data created at the edge, i.e., within the device's perimeter will require immediate local computation. Transferring this data to a centralized cloud may not always be the best option, with potential laws stating the location at which data can only be stored. This infers that data may not, in certain cases, undergo transfer at all. The decisions revolving around certain applications are usually lightweight and the algorithms can be housed at the local site itself. The trade-off, however, is a total misjudgment during analytics, leading to a setback in performance and in turn, delivery of services. This necessitates an upscale in distribution and the building of adequate local models capable of accurately meeting established requirements, while at the same time transferring the forecasts to other sites for increased model growth.

A possible workaround of this complication is to learn about global data trends directly from multiple networking nodes without having to transfer it to a centralized system. This gives rise to federated learning, a way of training the algorithms across multiple decentralized edge devices, bypassing the need to exchange data. Distributed training patterns include vertical federated learning and split learning. Machine learning models are thus more prone to understanding the purpose for which they were designed for, by closely adhering to transfer and computing requirements, along with memory and resource allocation, while guaranteeing exceptional performance. Given the variable nature of networks, intensive research is essential to bring about stronger and more reliable conclusions with regard to model reliability and security.

The conditions that would satisfy the giant necessities of both local and global data frameworks are planned to be distributed and decentralized. Finding proper ways to disseminate acquired knowledge while ensuring timely delivery across devices is also imminent. In the wireless oasis, the partnership between machine learning, AI algorithms, and such frameworks can pave the way for technologies such as self-management, self-optimization, and self-evolution.

10.2.3.2 AI-Enabled Autonomous Systems and Human Interaction

With increasingly complex decisions needed to be made by machines, either on their own, or with human assistance, these devices require a fundamental understanding of the problem such that they can act upon the solution effectively. A clear distinction between observation by sensors, and the action required, needs to be made. Since humans set the onus on the machine, it needs to be fitted with more autonomous features that will allow a fine line to be drawn between what to do and what to achieve. They thus need to be inculcated with relevant domain knowledge. These declarative

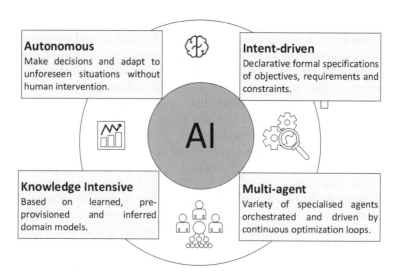

FIGURE 10.15 Properties of future systems that take automation to the next level [56].

objective specifications, also known as intents, are characterized as being functional, nonfunctional, and operational objectives; challenges; and requisites [56].

Formal designs are used in knowledge-driven structures to understand apparent conditions and implement actions based on intelligent and smart decisions. The different methods to generate such models are by either using machine learning algorithms created by experts or logically inferring through a search.

Using a completely conventional data science technique, in which a model is developed using all accessible data, has various drawbacks. To begin with, a typical technique may not develop, thus becoming unrealistic. The variable matrix might include an outsized range of attributes, yielding limitless permutations of scenarios for training during analytics. Furthermore, since there is no input on which the training model can be based, such a strategy cannot handle unforeseen circumstances. The problem should instead be partitioned into smaller, more achievable blocks through an agent-based approach, as shown in Figure 10.15. These agents are in the form of conventional machine learning (ML) elements, mathematical methods, expert standards, etc. A continuous optimization pattern can shoulder the orchestration of agents by correlating present and required states (defined by intent) based on the system's perception of the network state.

It is crystal clear that hybrid techniques will bring about better effectiveness in next-generation smart systems. Intricate models trained through robust learning with mathematical notation will offer knowledge bases, logic, and interpretation capabilities. For instance, the expertise could include fundamental laws of the universe or the most well-known procedures in a certain sector. Automated architectures must be bestowed with the power to make independent judgments to achieve specific goals, the resilience to handle a challenge in multiple perspectives, and the versatility in reasoning by leveraging diverse elements of both pre-stored and acquired intelligence.

Applications of 5G/6G Powered Intelligent Connectivity 379

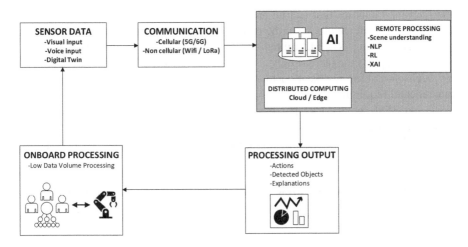

FIGURE 10.16 Properties needed for human–machine collaboration [57].

Interconnected, smart sensors of many forms, spanning from virtual personal assistants to industrial robotics or collaborative robots, are seemingly seeping through our lives [57]. It is helpful for better interaction that these technologies properly interpret human needs and intentions. Moreover, all machine-related data should be ready for provision for situation-based insights. AI efficiently bridges the gap between humans and machines to improve abilities and enable collaboration.

Natural language processing and computer vision advancements have enabled robots to read human input more accurately. Nonverbal human traits such as body language and voice tonality are factors that have made this possible. Precise emotion recognition is currently emerging and can help identify more complicated feelings such as fatigue and focus. Furthermore, evolution in areas such as better environment understanding and the extraction of semantic information is critical to establishing a thorough map of the ecosystem. Figure 10.16 shows the properties needed for human–machine interaction. The devices uphold the responsibility of extrapolating the maximum amount of possible information from the surroundings for the most favorable collaborative. Bearing such properties is RL where a system is entrusted to critically analyze current states before taking measures to affect the future states. The adoption of RL is increasingly gaining popularity [57]. To prevent dangerous scenarios, solutions such as safe AI are being researched to make security and safety arrangements throughout the RL model life cycle.

With the incorporation of digital twins, AI has also facilitated a comprehensive overview of machine operation. The use of extended reality (XR) equipment and facilities is increasingly becoming important to generate mixed-reality setups that contain several technologies including AI, thus enabling interaction with digital twins. This elevates machine breakdown, analysis, and comprehension to a whole new level, allowing the anticipation of a chain of events. Explainable AI (XAI) can be added as a building block to further understand why a particular decision was taken by a device. XAI accounts for transparency in the AI sphere to produce logs and relevant databases that can be produced for stakeholders, explaining the decisions taken. This technology can be paired up with several AI frameworks such as supervised learning, RL, ML mixed with reasoning, etc. [58]. XAI thus acts as the pillar that supports the increasingly crucial use of AI models in networks by providing decision-making justification to AI customers, and is therefore regarded as a fundamental requirement in cases where worldwide standardization parties such as IEEE and ETSI demand transparency and highlight the credibility and integrity of smart communication systems.

For collaboration between machines and humans to be effective, rapid response criteria must be met. A distributed intelligence approach is needed to deliver real-time interactions since the AI techniques used in the shared environment may have significant processing intricacy while machines may have hardware-limited resources. This translates into the networking system acting as the pivot in the process by accommodating for high-reliability and ultralow-latency communication paradigms.

10.2.4 Cloud Computing and Networking

According to the ITU, cloud computing (CC) can be referred to as [59] "A Paradigm for enabling network access to a scalable and elastic pool of shareable physical or virtual resources with self-service provisioning and administration on-demand." These resources can be operating systems, servers, software, networks, storage equipment, and applications.

Cloud computing is a methodology for providing technology solutions through the dynamic utilization of virtual machines, with upgradability and management features. The properties that make up the cloud as we know it are summarized in Table 10.1 [60].

10.2.4.1 Cloud Deployment Models and Service Classes

The four main types of cloud deployment architectures are defined as follows by the NIST [61]:

Private cloud. A single enterprise with several users is granted entirely private access to the cloud architecture (e.g., business units). The infrastructure may be local or off-shore, while still being governed, controlled, and run by the company.

Community cloud. A group of customers from firms with similar challenges is given exclusive access to the cloud infrastructure. The shared concerns may include objectives, security needs, protocols, and compliance requirements. This means that it may be governed, controlled, and run by one or more community groups, a foreign member, or a mixture of both. The infrastructure can be local or off-shore.

Public cloud. The public is authorized to use the cloud platform without restriction. A commercial, educational, governmental body, or a combined party may own, administer, and run it. It is present on the cloud provider's property.

Hybrid cloud. This type of cloud platform is made up of two or more different cloud environments ranging from private to public to community, each of which is independent and continues to exist as a separate entity, but which are connected by regularized or exclusive technologies that allow data and application migration and mobility. An example enabling load balancing is cloud bursting between clouds.

TABLE 10.1
Features of Cloud Computing

Feature	Description
Massively scalable infrastructure	From the subscriber's standpoint, massive scalability indicates that the end user manages the complete provision of computing or storage resources as necessary. By offering quick virtualized resource deployment, streamlined hardware, and continuous data storage features, cloud technology significantly simplifies this procedure.
Universal access	Universal accessibility is another feature that accentuates cloud computing. Despite private email services once needing local connectivity or virtual private network (VPN) accessibility, we now have global access to email from several companies over the Internet. A similar tactic may be used to provide universal access to cloud computing services.
Fine-grained usage controls and pricing	Given the rapidly evolving nature of online tasks, processing resources can be bought and subscribed to, through cloud computing platforms. Likewise, storage is also purchased depending on what is truly required in the moment. Purchasing choices are no longer bound to factors specific to a single component such as maximum capacity requirements for a server. When demand is sky-high, more cloud resources can be dynamically and flexibly allocated and eventually unsubscribed to, after achieving the said demands. In turn, clients are only charged for what is actually used.
Standardized resources	Cloud computing offers regularized hardware, virtualization, and software services. However, regularization does not infer homogeneity. Cloud computing provides a great deal of flexibility when it comes to providing a variety of tailored solutions. For instance, cloud technologies provide services ranging from configured server, to different operating systems, to an array of application stacks among the likes of Linux, Microsoft Oss, and LAMP (Linux, Apache HTTP Server, MySQL database and Perl/Python programming languages).
Management support services	The lack of proper support and administration, strips cloud computing of its full potential of delivery for all its services. Management support services assist with both the managerial and operational elements of cloud computing. They offer the insights required to fine-tune cloud services. An example is the generation of reports detecting low usage in certain areas where more servers, than required, have been deployed. Network traffic statistics and storage reports may be produced to, e.g., better choose between either moving data to and from the cloud constantly, or employing a service that solely focuses on permanent uplink storage.

The most common CC services privately provided or by external vendors are: software-as-a-service (SaaS), platform as a service (PaaS), and infrastructure as a service (IaaS), which are defined as follows by the NIST [61]:

Software-as-a-service (SaaS). The client is given access to the cloud provider's applications that ultimately run on the platform. A variety of thin client interfaces such as web browsers or graphical user interfaces allow clients to access such applications and services. Apart from the likely exclusion of a small number of particular software configuration choices to the discretion of the client, the latter does not handle the actual building blocks of the cloud framework, which includes networks, servers, operating systems, storage, or even tailored application capabilities.

Platform as a service (PaaS). The consumer is handed the ability of deploying custom-made or procured applications, onto the cloud platform, through application development software, libraries, servers, and in-house tools. Although the hosted programs and perhaps the configuration options for the framework are within the control of the customer, the latter has no further management rights or influence over the actual cloud infrastructure.

Infrastructure as a service (IaaS). The consumer is given access to processing, storage, networks, and other essential computing services, on which the client can install and execute required software such as operating systems and custom-made applications. Likewise, the consumer has no control over the cloud infrastructure that upholds the services with the exception of a small number of network elements such as host firewalls.

However, software service architectures have evolved and several variants have been developed as outlined in Table 10.2.

10.2.4.2 Intelligent CC and Networks

The trend followed by IT companies for the management of their departments is fueled by the rise of AI and ML paired up with CC aspects, also deemed as revolutionary for the foreseeable future. The coupling of AI and ML with cloud networking indicates smarter networks that are mostly geared toward reducing workload and optimizing performance. Network administrators will then only have to perform the most elementary setups, from which AI and ML can pick up with self-optimization features. The power of AI and ML will be considerably felt, even in intricate cases such as BYOD or IoT where devices can be identified and mapped to their respective users, through fingerprints or traffic patterns, after which they are allocated appropriate user profiles and undergo access control mechanisms. Another feature on the list is self-healing, allowing issues to be automatically resolved before spreading [67].

Similar to how the power grid provided households with electricity, thus leading to the second industrial revolution, the partnership between cloud and networking to yield smart cloud networks will drive forward the digital economy by dynamically and globally distributing computing resources and smart components to every sector. This result will be a surge in production capability for corporates, as illustrated in Figure 10.17 [68].

The three unique features making up the intelligent cloud network are [69]:

Network digitalization. Through the use of digital technologies, the state of the whole network may be determined and digitally duplicated, allowing for uniform and standard cloud storage and network design for any architecture. This paves the way for standardized and regularized network management on the cloud, with real-time breakdowns of network behavior.

Network intelligence. In order to enhance the cloud–network partnership, technological advances such as AI and big data may be applied when network digitalization is established. The outcome is packed with novel features such as intelligent

TABLE 10.2
Cloud Service Models [62]

Feature	Description
Data-as-a-service (DaaS)	Real-time industry and client data are provided via innovated DaaS solutions. DaaS separates data from its connected services without taking into account the ecosystem or site in order to provide consumers with useful information. Enterprise resource planning (ERP) systems, data warehouses, transactional databases, and customer relationship management (CRM) tools are just a few examples of the diverse datasets that may be extracted using the techniques that DaaS combines together.
Data-analytics-as-a-service (DAaaS)	In order to supply analytical techniques in scalable format, data-analytics-as-a-service (DAaaS) or analyticsas-a-service (DAaaS) employs a cloud-based delivery approach [63]. This service is packed with a multitude of analytical tools for homogeneous data analysis. Housed with machine learning concepts as its backend, corporate data collected at the client's side is uploaded to the cloud where analysis occurs to yield useful information.
DataBase-as-a-service (DBaaS)	By seemingly bypassing the configuration and setup of physical hardware, DBaaS provides a cloud computing service model where users have the ability to access and use actual databases. As one of the fundamental rules of cloud computing, in DBaaS, consumers only pay for what they really use, i.e., the capacities and tools associated with the subscribed database. Through an application programming interface (API), the DBaaS database management module handles the actual backend database instances.
Hadoop-as-a-service (HaaS)	Numerous big data initiatives and companies use Hadoop as their foundational framework. This mechanism for processing data storage makes it possible to store information, share documents, analyze statistics, and more. Hadoop is widely employed by businesses including Facebook and Yahoo as worldwide social media and Internet usage sky-rocket. Local, on-shore Hadoop has been superseded by HaaS.
Big-data-as-a-service (BDaaS)	With BDaaS, businesses may implement big data solutions from beginning to end. It is ultimately a hybridization between HDaaS, DaaS and DAaaS. One of the main forces at play in this market is the enormous expansion of data.
Information-as-a-service (INaaS)	INaaS is a service that enables the systematic and safe creation, management, interchange, and extraction of useful knowledge from all existing data at the appropriate time and in the suitable manner. This relates to the accessibility to APIs as well as the usage of dedicated hosted content.
Business-process-as-a-service (BPaaS)	Enterprises are using streamlining processes to boost productivity and specify precise corporate goals. The use of web interfaces via web browsers on devices including PCs, smartphones, and tablets to engage in corporate processes such as payroll systems, printing, and e-shopping is referred to as a business-process-as-a-service (BPaaS).
Security-as-a-service (SECaaS)	SeCaas is elevated by software-as-a-service (SaaS) but is, however, only applicable to certain information security services. SeCaas is an information security leasing model where services such as antivirus software are distributed across the Internet. Likewise, it may also apply to cybersecurity services that are offered internally by a third-party company.

(Continued)

TABLE 10.2 (*Continued*)
Cloud Service Models [62]

Feature	Description
Testing-as-a-service (TaaS)	TaaS is an outsourced strategy where testing tasks related to company operations are delegated to a foreign entity with expertise in mimicking real-world testing conditions in accordance to customer needs. TaaS may employ specialists to assist and counsel staff members or only outsource a fraction of assessments to the third-party. This unravels new room for growth in the business sector, as well as difficulties and expectations for creative service models, validation strategies, and QoS regulations [64].
Communication-as-a-service (CaaS)	CaaS points to the offloading of a corporate communications service. The onus of providing and managing the required software and hardware lies on the CaaS vendor, which is the provider of this kind of cloud solution. The backbone enables services such as voice over IP (VoIP), instant messaging (IM), and video conferencing. The entirety of hardware and software deployed is managed by the CaaS supplier, who is in turn fully responsible for reliability and performance. This calls for an accord for a particular service-level where CaaS providers are expected to provide assured QoS [65].
Network-as-a-service (NaaS)	NaaS is a cloud deployment architecture that makes network management straightforward for companies. It allows for substantial adaptability and versatility together with the option to shift from CapEx to OpEx. This makes it possible for enterprises to get desired results without having to own, construct, or manage infrastructure. Without the requirement of a proper network management structure, NaaS is described by Cisco as a cloud architecture that allows participants to effortlessly administer the network and obtain the results that they anticipate from it [66].

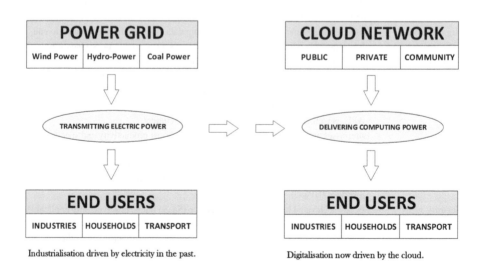

FIGURE 10.17 Cloud networking and digitalization compared with power grid and industrialization [68].

network O&M, drastically improved diagnostic effectiveness, smart security defense strategies, and a better equilibrium between cloud and network resources.

Network-as-a-service (NaaS). Network services can be purchased at the ease of our fingertips, immediately uplifting network rapidity, responsiveness, and agility by adopting cloud capabilities. The latter matches the high demands of corporates by favoring open network programming and more versatile cloud compatibilities. The union between the cloud, networking, and security leads to the provision of better all-rounded cloud–network services.

The integration of both CC and AI makes up an intelligent cloud with several interesting capabilities that would incorporate:

Cloud agents. By replacing certain human interactions such as making healthcare appointments or searching for the best deals on shopping items, cloud agents will be performing at unprecedented heights, with limitless computational resources, logic and intelligent data.

Sensors everywhere. The presence of sensors and monitoring devices in every nook and cranny of the world calls for the necessity of being able to communicate among themselves, as well as activate one another in a network. Similarly, they are closely bound by the intelligent cloud, which allows fluent communication.

Robot assistants. Machines will need to be extremely precise, trustworthy, and be able to adapt to changing circumstances with the widespread demands for robotic companions, autonomous cars, drones, industrial bots, and humanoid employees. Connectivity, intelligent analytics, sensory engagement, contextual understanding, and deep learning will all be available through the intelligent cloud.

AR heads-up displays. The near future will possibly unfold highly envisioned capabilities such as the implantation of an AR layer in our eyewear, corrective lenses, or even retinas to improve our contextual and surrounding awareness. With our eyes being one of the main biological sensory "equipment" that takes a massive inflow of information within a small timeframe, the intelligent cloud will provide the basis for such accommodation.

10.2.5 Blockchain

Blockchain is an important component of intelligent connectivity as it can enhance processes in 5G, IoT, and AI.

Blockchain is widely recognized as the underlying technology that powers Bitcoin [70,71]. It is built upon a decentralized foundation, which translates into its databases being distributed and dispersed over a number of users, rather than being housed in a single place. For records maintained on blockchains, this decentralized model offers exceptional resilience and confidentiality with the absence of a single point of failure. Furthermore, all participants in the network witness every transaction, making the blockchain visible. This is made possible via a process developed as consensus, which consists of a set of guidelines to ensure that all parties are in accord over the state of the blockchain ledger. The blockchain concept is illustrated in Figure 10.18.

Blockchains can typically be divided into two categories: public (permission-less) and private (requires privileges) ones [72]. A public blockchain is open to all users, who may join, execute business, and partake in the agreement procedure. Bitcoin

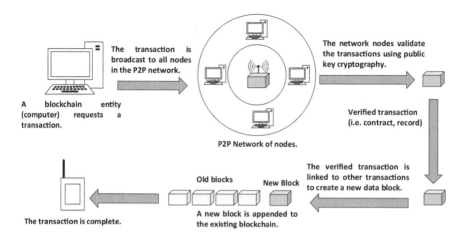

FIGURE 10.18 The concept of blockchain operation [70].

and Ethereum are two of the most renowned public blockchain implementations. The other side of the coin pertains to private blockchains where access is only possible via invitations, and it is generally run by a single organization. A validation method must grant authorization to a member. A list of major elements found in blockchain is discussed next.

Data block. A blockchain is a collection of interconnected blocks that begins with a parent node and forms a link with each subsequent update. A hash label connects each block to the preceding one, establishing a secure relationship between them and eliminating the danger of alteration. Each data block is made up of two fundamental components, which involve a blockchain header and the transaction records [28]. The sequence of events is organized in a Merkle tree format comprising all previously made transactions. Additionally, the block header consists of the hash value, Merkle root, nonce value, and timestamp, making a total of four small elements. Figure 10.19 depicts a characteristic blockchain configuration [70].

Distributed ledger (database). By first duplicating the database, it is dispersed among all network users. It keeps track of and maintains user-generated events, and the mining process ensures functional agreement, which in its entirety, corresponds to the "proof-of-work" (PoW) concept. The distributed ledger enforces a distinct cryptographic signature upon each entry, not linked with a timestamp, making the ledger immune to modification. Furthermore, consensus algorithms prevent security loopholes such as double-spending invasions by dedicating the responsibility of moving a block over a chain to all users, instead of a single entity managing it [29]. This thus eradicates vulnerabilities with regard to security. From a blockchain perspective, this consensus acts as an agreement basis, guarding safe transactions among parties. For example, the fundamental consensus technique for managing transactions in Bitcoin is the PoW methodology [7]. Nodes that have high processing features can participate in mining while simultaneously competing with one another to win the race of validating a transaction. The winner is then given a set quantity of coins as a prize for their mining work. New consensus techniques such as proof-of-stake (PoS)

Applications of 5G/6G Powered Intelligent Connectivity 387

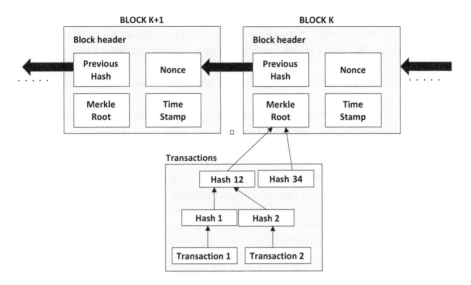

FIGURE 10.19 The data block structure [70].

and Byzantine fault tolerant (BFT), among others have emerged with the development of blockchain. Further details about these methods can be obtained in [72,73].

Smart contracts. A smart contract is an autonomous program embedded on a blockchain, which is triggered when predetermined conditions are met. Following the first Ethereum-based smart contract in 2015 [73], it gained massive popularity over the past years. Smart contracts are typically used to automate the execution of a consensus where the involved parties are immediately exposed to the outcome, without any loss of time. They are triggered by the receipt of a transaction or message from users. In addition, smart contracts are fed by coins, which in Ethereum terms, is known as "gas." The clauses governing a smart contract cannot be changed or affected by any party, nor is its functioning run and managed by any foreign entity. These traits are what renders smart contracts highly resilient to outsiders and foreign attacks [74]. With such properties, smart contracts are favorable in a multitude of scenarios, such as banking transactions, exchange of medical records, and business logistics management [75].

10.2.5.1 Blockchain in 5G

There are several key characteristics of blockchain, which makes it suitable for 5G networks as described subsequently [70]:

Immutability. It refers to the irreversible nature of transaction data once made on the blockchain. This feature is embedded in the blockchain ledger. As aforementioned, modifying the contents of a particular block escalates into the successive blocks being invalid, which requires their hashes to also be updated. This renders falsifications nearly impossible, since invalid data will quickly be detected. As a result, a particular transaction cannot be repudiated since the latter is governed by digital signatures, a pair of cryptographic keys as well as consensus methods observed by miners. With all nodes witnessing any changes made by a particular individual, non-repudiation is thus ensured. This feature of blockchain enables safe data exchange and retention in 5G applications, such as private network virtualization, confidential spectrum access, and D2D connectivity. Immutability goes hand in hand with 5G's needs to perform accounting operations such as unchangeable database sharing, global, and crystal-clear visibility and robust security implementations for the user. Irreversible transaction ledgers thus lead to networks inhibiting strong and secure paradigms that eventually allow extensive heterogenous networking and computing, as well as IoT partnerships or edge computing across non-transparent and shady IoT ecosystems.

Decentralization. It refers to the absence of a central or singular party for database management. In a decentralized P2P architecture such as blockchain, information flows through all nodes in the network and can thus be monitored and processed easily without any intervention. This eradicates the need for third-party management terminals, with the use of the PoW concept where the security of both the secure chain and database is maintained. Such characteristics permit the development of a storage platform with excellent non-repudiation, resilience, and fast data access properties. Further, foreign third parties are not required for maintaining regulations and policies. This therefore bypasses frequency licensing, spectrum administrators, and database supervisors in resource management. Smartphone communications do not require central/edge service managers. 5G UAV networks can function without a UAV management center, while also eliminating complex cryptographic measures in 5G IoT architectures. The absence of a central point of failure, combined with hashed blocks and transaction ledgers keeps data safe, secure, and reliable over the network.

Transparency. All transactional data on blockchains concerning permission-less ones is observable and viewable by all parties, thereby contributing to a blockchain's visibility. To put it differently, a clone of transactional records is replicated throughout the whole community of nodes for public verification. With data gathered at local nodes, a user can always access it from anywhere and in turn share their ability to validate events based on their roles in the chain. For example, miners will cater for verifying the mining process while meta data checking is performed by slight nodes. Transparency is thus achieved and maintained with the involvement of each and every node, giving rise to increased data integrity. Such features are accentuated by 5G application requirements where availability and equality are important prerequisites. For example, blockchain can be paired up with cooperative cognitive network slicing in IoT systems by providing the same transparent ledger databases to allow service suppliers to monitor and trace back transactions to their origin. Furthermore, features such as smart contracts can be used to trigger tasks such as mobile resource trading in 5G IoT. This means more visibility and reliability when different service suppliers come together with IoT customers.

Security and privacy. Public and private keys dictate the security of blockchain. The implementation of asymmetric cryptography, which uses random sequences to create keys so that hackers cannot intercept events, is the pillar underpinning secure blockchain practices [32]. Additionally, the non-repudiation and consensus features guarantee the confidentiality of any information held. Users have a bird'-eye view of the network, enabling transaction tracing and monitoring. The bottleneck caused by hashing and mining lead to data being preserved inside the block. By offering distributed secure architectures along with increased access validation, blockchain uplifts the security of 5G networks, which allows 5G facilities to defend themselves and guarantee confidentiality. Cybercriminals face a huge brick wall when trying to get past authentication systems since 5G systems store IoT metadata over several devices in the network. Additionally, smart contracts might enable 5G solutions such as data validation, user identification, and protection of resources from threats [70].

10.2.5.2 Blockchain in IoT

Blockchain plays several crucial roles in boosting the domain of IoT applications. These are summarized as follows [76]:

Providing high scalability. With the massive influx of devices onto IoT networks for sensing and monitoring 5G environments, networks are filled with lag and eventually become money pits. This bottleneck is due to the microtransactions requiring authentication and validation. The consensus concept around blockchain instigates high scalability, allowing latencies during verification processes to be mitigated [77,78]. On the same wavelength, in totalitarian areas where some organizations dominate the control, processing and possession of a massive repository of people's details, the shift from a centralized to a P2P distributed paradigm eradicates monopolistic behaviors, central breakdown points, and expanding capabilities [79]. Improved architecture flexibility, uptime, and high availability are further benefits of decentralizing the design. The vision of improving IoT scalability would also assist in reducing IoT silos.

Preserving full data privacy. With different levels of access rights determined by public or private access for reading and writing transactions according to the consensus, IoT users benefit from the use of fictitious names, also known as pseudonyms, which translates into the prevention of connecting transactions to specific individuals [80]. Blockchain promotes transaction anonymity among its regulations, with enhanced privacy-preserving methods. Secure multiparty computing, zero information checks, commitment plans, ring signatures, and homomorphic concealing are some of the most significant cryptographic methods and strategies for identity protection in blockchain [81].

Orchestration of connected IoT devices. The blockchain coordinates the administration of smart contracts and IoT technology in a particular setting. The choices that are pertinent to IoT applications are defined by smart contracts. The amalgam of devices and their communications, access management, data production and processing, flexible location adaptation, and container changes including software services are all aided by blockchain [82]. The scheduling of specific IoT operations is linked to a variety of integrity and perceived risks, including documenting the identities of all IoT nodes in the network, tracking the source of sensor data as well as that of other parties joining the platform, in addition to using smart contracts to track the stages [83].

Ensuring interoperability. Interoperability refers to the possibility of connecting and sharing data between two or more wholly separate IoT structures. With blockchain in the picture, cross-chain communication allows blockchain-based IoT systems to merge and connect with one another [84–86]. Integration with current systems is guaranteed, together with the start of activities on other chains, making transactions with other links, and thereby connecting the application on the same chain. The two major methods used for archiving data in cross-chain communication are atomic swaps and stateless simplified payment verifications (SPVs). The former involves the exchange of the same cryptocurrency in straightforward P2P proceedings. Atomic swaps therefore bind two parties through a transaction event, through proper coordination, despite it not being regarded as a real type of cross-chain communication. Smart contracts may validate a portion of the transaction history logs using stateless SPVs. Block headers and other events are validated by relays on another chain. By using a relay chain, combined consensus makes it possible for chains to communicate with each other in both directions. A chosen set of trustworthy members is permitted to authenticate the occurrences of one chain by another in the case of alliances.

10.2.5.3 Blockchain in AI

The combination of blockchain and AI in fact leads to a synergy whereby these two techniques can bring added values to each other. The combined benefits of blockchain and AI are as follows [87]:

Authenticity. Explainable AI is addressed by blockchain's digital ledger, which provides a backstory on the theories related to AI as well as the origin of the material it uses. This boosts trust in the accuracy of the data and, consequently, in the suggestions and conclusions that AI generates. Combining blockchain with the deployment of AI facilities leads to the availability of an audit trail as well as improved security measures.

Augmentation. Blockchain-based corporate systems are envisioned to be revolutionized with the thorough comprehensiveness and dynamic data correlation abilities of AI, which is lightning-paced. Blockchain enables AI to expand by managing data consumption and model exchange, facilitating access to vast amounts of data from both inside and outside enterprises, and producing a reliable and open data market.

Automation. Bringing together AI, blockchain and automation may provide new value to business operations involving numerous stakeholders by reducing friction, and boosting speed and productivity. For instance, AI models built into smart contracts that run on a blockchain can highlight recalled goods that are past their expiry dates, carry out events such as stock investments or ordering based on predetermined baselines and circumstances, settle arguments, and choose the most environmentally friendly delivery option.

10.3 APPLICATIONS OF INTELLIGENT CONNECTIVITY

Intelligent connectivity has several application areas and is even playing a major role in driving the UN SDGs as reported in the works of [88,89].

The five following sectors are predicted to witness the significant impact of intelligent connectivity [1]:

- transportation and logistics
- industrial and manufacturing operations
- smart city

- health care
- education

An overview of the role of intelligent connectivity in these application areas is given in the following subsections.

10.3.1 Transportation

Road transportation will be significantly better and more efficient if every vehicle and individual in the environment can seamlessly communicate among themselves. Real-time positioning information from automobiles, bicycles, and pedestrians may be shared with other people in the area, allowing AI-integrated expert systems to prevent accidents. In fact, according to recent study by Bosch, connectivity among cars will prevent 11,000 deaths by 2025, thus minimizing annual road accident casualties by 350,000 [90]. Additionally, the integration of traffic management systems in the 5G realm can elevate driving itself to a much safer level where cars have real-time updated information about when to reduce their speeds or when they are free to hit the gas. This even gives rise to the talk of the possible removal of speed cameras and traffic lights. The inclusion of radar and computer vision in 5G will thus further edify and smoothen traffic systems.

When it comes to the driving dynamics, 5G will bring about significant improvement by disseminating sensitive details from adjacent cars and road-related entities along the route. In situations where surrounding vehicles suddenly hit their brakes, data is relayed instantaneously to neighboring ones, where the integrated processor is also instructed to appropriately reduce speed. With embedded eCall services, cars can automatically make the required help and hotline calls upon particular impacts. The interconnected mesh network operates with a continuous feed of live information being sensed by real-time road interactions in particular areas, by drivers, cyclists, and pedestrians [91]. This is where AI can then be employed to analyze weather- and road-related constraints to warn users in real-time about possible trajectory and location information. It is paramount that all information about road users inhibit anonymity, privacy, and security features [3].

With intelligent connectivity, the combination of D2D communication, smart traffic systems, and public transportation, traffic bottlenecks will be lessened, while uplifting the longevity, reliability, and efficiency of traditional transport methods. The combination of these technologies can be referred to as smart-transportation systems (STSs). Also known as the IoV, the architecture involves an interconnection between each passenger vehicle in a given area via D2D and vehicle-to-vehicle (V2V) data transmissions. With the aid of IoT technology, every car may be monitored during emergencies. Previously collected data can also be used to forecast and determine optimum paths at a given moment. The near future expects mobile applications such as Google Maps to integrate real-time data about vehicles [92].

5G will ultimately lead to trustworthy autonomous vehicles. The on-board computer chip will be AI-based, with a range of interconnected local sensors as well as those ready to receive information from other road-related events, which all form this colossal 5G network. This allows the vehicle to be aware of its surroundings and therefore thrive and overcome any circumstance. As for cloud functionalities, autonomous vehicles will give rise to a novel mobility-as-a-service model, comparable to the current Uber services that already exist, but adapted for driverless transportation. The simplicity of requesting for a self-driving ride instead of buying your personal vehicle will sky-rocket, given the array of associated benefits [93]. A few simple taps on your mobile phone summons a driverless taxi, waiting at your doorsteps for pickup. The near-future networked automobile will be equipped with the necessary sensitive on-board sensors and intelligent chips to transport humans securely and effectively thereby providing trusted and ultralow delay communication [1,3].

10.3.2 Industry 4.0 and Manufacturing Operations

In this age of machines, intelligent connectivity has laid the basis for a fourth industrial era where robots and computers drive manufacturing operations, with continuous production upgrades and massively scalable industries. Some of the essential propellers of Industry 4.0 are AI, CC, and IoT. By exploiting licensed frequency bands, edge computing, and network slicing, which are products of the mentioned combination, 5G will contribute to the delivery of tremendously reliable and ultralow delay connections, in turn allowing the dynamic prioritization of certain applications. Mobile carriers, partnered with edge computing may deliver incredibly consistent services, allowing industries to ease their reliance on cables, thus increasing adaptability. By being remotely controlled without requiring physical human presence, production plants can upscale their output, in turn generating larger sales and meeting client demands more easily.

The hallmark of Industry 4.0 will be the self-sustaining manufacturing plants that analyze constant information intakes, to dynamically adjust to situations, such as a supply deficit, indicators of a probable equipment malfunction, or a new client request. Specialized robots will be designed to create 3D structures on command, allowing manufacturers to replace damaged parts. To turn this idea into a reality, the facility's main controller unit requires a comprehensive and elaborate digital twin of every component that lies on the production lines such that real-time reporting and maintenance are possible. With industries embarking on the automation train, employees can access and control equipment remotely. Eventually, firms will have far more freedom to choose where to establish manufacturing facilities since fewer personnel will be required on-premise, enabling manufacturers to shift their focus toward different priorities rather than qualified employees. Consequently, this will require uninterrupted connections, smart clouds, and top-notch connectivity among IoT devices.

The digitization of manufacturing applications and the remote management of devices such as robotic systems will benefit from 5G's ultrahigh speeds, ultralow latencies, and exceptional stability. For instance, ML algorithms may gather information from sensors and video feeds along a production line to instantly notify a worker about any discrepancies in the system, after which AI systems may promptly and autonomously repair the error or take actions. Using interconnected equipment such as touch-enabled gloves and virtual or augmented reality (VR/AR) headgear, employees will be able to virtually watch and modify Industry 4.0 operations from a distance while interacting with machinery in a live fashion, giving this feeling of a more tangible Internet.

Performing operations from anywhere around the world, including audits, servicing, and equipment adjustments would also be possible thanks to haptic Internet applications powered by intelligent connectivity. As a consequence, remotely controlling machines in risky, difficult-to-reach, or hostile places, such as nuclear facilities, oil rigs, or mining areas, leads to minimal expenses and prevents associated threats to human life. The same gear may be used to carry out or assist staff training as well as mimic complicated scenarios in a secure virtual setting.

10.3.3 SMART CITIES

The conquest of 5G has expanded and reached the doorstep of companies, enabling massive commercial prospects and digital business. This erected the vision for novel smart city architectures, in light of bettering life for city residents [94]. The foundation of smart cities is the omnipresence of IoT devices, diverse networking systems, scalable data storage capabilities, and the deployment of increasingly specific services. Buildings and cities are already witnessing the proliferation of mobile connectivity, being rendered smarter, safer, and more financially viable. This pattern will be expedited with the deployment of 5G, i.e., the sky-rocketing presence of interconnected sensors and actuators onto our networks. As a result, local governments, businesses, and private entities will have a continuous eagle's view of their properties and manage access.

Intelligent connectivity may also be leveraged to remotely manage physical entry to a facility without dependence on physical keys or access cards that can often be lost. Instead of requiring employees to return their keycards when they leave a firm, it is now feasible to communicate updated biometric credentials or a passcode to the locking device. Smart locks have the potential to revolutionize access control by eliminating necessity for an abundance of keys, which are time-consuming to handle since keys can be misplaced while locks need to be maintained and updated. On the same wavelength, customers will have the ability to remotely operate their houses with their smartphones, through smart home platforms with built-in 5G support. For instance, Orange envisions the connection of smart home gadgets to the wireless router such that they can be remotely controlled. In collaboration with Groupama, Orange also plans to provide a security service called Protected Home that will interconnect homes to a video-monitoring hub that can ping enforcement agencies if necessary [3].

An exorbitantly high volume of monitoring information will be continually produced by pervasive live video feeds in 5G smart cities. On a wide scale, it is exceedingly difficult to instantly recognize unusual items and happenings, or discern harmful behavior from millions of video frames. To accomplish large-scale data processing in this situation, the construction of distributed edge computing systems is extremely effective [95,96]. From a security standpoint, blockchain would be a logical choice for establishing decentralized security mechanisms by graciously linking edge devices, IoT gadgets, and city inhabitants, whereby data exchange, processing, and commercial transactions can be handled on the blockchain ledger platform. In terms of reduced delay, power usage, improved customer service, quicker user responsiveness, and privacy and security commitments, it is also shown that the adoption of decentralized blockchain delivers more advantages than centralized designs with a single cloud platform [97].

10.3.4 HEALTH CARE

Intelligent connectivity wrapped around the healthcare sector will contribute to the delivery of more efficient preventive care at a lower cost, since healthcare facilities can be more easily managed and operated. Intelligent connectivity could also bring about remote consultation and treatment, possibly redefining healthcare access, which is currently constrained by the physical locations of medical specialists [98].

In light of the COVID-19 pandemic, substantial emphasis has been given to 5G's ability to enhance telemedicine solutions, with examples ranging from virtual health check-ups to doctor appointments, which offer tremendous advantages in terms of removing the need for patients to physically visit medical centers. The current Internet backbone, along with 4G connectivity easily accommodate simple, one-on-one, low-touch interactions. That being said, 5G technology is expected to significantly consolidate these exchanges. For instance, by incorporating sensing devices and VR technology into videoconferencing, medical professionals will be able to remotely monitor patients' health status during live sessions. Additionally, with 5G's ability of conveying staggering data flows, it is also feasible to employ cloud-connected scanning devices to continuously monitor patients with illnesses, for abnormalities in their pulse, glucose levels, and heart rate. These developments would then open up new perspectives on the general treatment of mankind.

Another major element that 5G-integrated equipment could track is the mobility of healthcare specialists, personnel, and patients around the hospital, along with core health parameters. The facility's electronic medical records (EMR) system might then be upgraded to incorporate these findings, allowing for the visualization and refined management of healthcare proceedings at a pace and level never achieved before. The increased transparency would serve as the foundation for very successful managerial improvement. From a futuristic standpoint, 5G may impact the way healthcare professionals give medical treatment and affect how patients and physicians interact. The rise of telemedicine is only starting to be felt and will eventually become a norm, with the increasingly intensifying adoption of wearable technology such as advanced fitness trackers. This collaboration between 5G and the healthcare industry, facilitated by wearables, has been coined "Internet of the body."

The most highlighted prospect of 5G, however, is the possibility for remote operations. Despite the current possibilities of performing visual presentations of surgeries and broadcasting the procedures in real time, all eyes remain locked on the introduction of the "tactile Internet." The latter would revolutionize healthcare by allowing doctors to operate on patients physically present in another geographical location. The most awaited breakthrough is the possibility of physicians conducting operations in one geographical location through mere movements, resulting into the registered actions to be immediately replicated by complex computerized equipment, physically connected to the patient at another location. This heavily addresses the scarcity of doctors that work on intricate surgeries [98].

10.3.5 EDUCATION

The education sector is envisioned to be also hit by the massive 5G wave, where improved education access and opportunities are expected, thereby raising the overall standard of teaching and learning. Having gained immense popularity among gamers, VR and AR gear can potentially be applied in the education realm. For instance, trainees, engineers, technicians, and medical students could be exposed to thorough tasks, tailored to match real-world scenarios with VR and AR simulators. Likewise, the teaching of certain subjects such as biology and geography is simplified by adapting information collected in the real world, in the form of imagery, to simulate tailored VR environments. This is efficient in the case of highly specific concepts that can be aided through visualization.

5G has enabled the ongoing development of ideas that decades ago would have sounded absurd, leading toward the possibility of intriguing and fascinating advancements in education and on-the-job training. Current research on the interconnection of digital technologies with the teaching sphere has been yielding promising insights. For example, education is being steered toward a more hands-on approach with the lurking likes of AR, VR, and extended reality (XR). Equipped with super low delay and high-speed attributes, 5G technology provides students and trainees additional possibilities to further grasp their field of study in a more immersive and dynamic setting, which bears similar characteristics as issues they expect to face in the workplace. This encourages educators to favor mixed-reality applications in all nooks and crannies of the education spectrum. The benefit of mixed reality in educational spaces is that it fosters a richer intuitive comprehension of the topic. Educators may communicate difficult and hypothetical situations to pupils in a stimulating and virtual context through headsets, visors, and sensors to add a realistic dimension to the learning experience [99].

The partnership between British Telecom and North Lanarkshire Council in 2021 resulted in the provision of the first 5G-equipped classroom to Scotland [100]. With BT's EE 5G network as the foundation, the classroom is brought to life by being rendered into a 360-degree room where the student can explore a range of environments and land masses, resulting into a more involved and "real" experience. Moreover, instructors may essentially live broadcast from anywhere with little to no interruption with the possibility of connecting way more devices onto networks where 5G is bestowed with increased bandwidth. Students can participate in a comprehensive learning opportunity that ranges from closely witnessing the Northern Lights to being on top of Mount Everest the next second and transitioning to a safari tour the next instant. BT simply exhibits how cutting-edge technology and ultra-fast data flow can elevate communication and experiences in all sorts of ways.

5G will also help workers in different fields to master their profession and add skills to their repertoire. Mixed-reality applications can increase employee involvement, cut training expenses, and make learning faster and easier, with other benefits such as constant skill development and more opportunities for career progression. A smoother transition from theoretical concepts to practical implementations is made possible in some industries such as manufacturing and logistics, with AR and VR technology. Moreover, XR headsets may deliver real-time, step-by-step guidance and enable teachers to immediately critique practical sessions. This is especially useful for practical work that requires hands-on training since it enables quicker, more frequent training while providing a safe, regulated setting. With the idea of tactile Internet marking its territory in every sector, training around manual tasks is envisioned to be rendered easier. 5G will also enable real-time haptic feedback for applications where mobility dictates actions, with achievable requirements for latencies in the order of 20 ms.

10.3 SUMMARY

In this chapter a review of revolutionary technologies such as 5G, AI, IoT, CC, and blockchain has been made. These technologies have the potential of revolutionizing the daily life of people by making connectivity omnipresent, extremely reactive and adaptable. The quality of the daily human life will be elevated, with intelligent connectivity affecting the way people study, work, and engage as a community. In a realm where the ability to access any information and service relies on a few taps, companies and governments are on their way to increased productivity and efficiency. With intelligent connectivity, the underlying of 5G architecture will catapult the IoT toward greatness, where it will significantly expand by accommodating real-world data that can be fed to AI models. Sophisticated AI technology will be geared toward providing a broad spectrum of innovative, novel solutions, that in the long run, is able to forecast the desires and requirements of individuals, thereby even assisting them in overcoming obstacles in life. In parallel scalable storage and processing power from the cloud along with enhanced security provided by blockchain will further facilitate the deployment and reach of several intelligent connectivity applications. 5G also acts as one of the fundamental components for a brighter development. The transportation industry for example can provide higher safety and efficiency levels with the inclusion of 5G in autonomous vehicles, by enabling several links at once, allowing individuals to have a global overview of the system. In fact several other use cases such as health care, education, Industry 4.0, and smart cities are expected to thrive with the advent of intelligent connectivity. Ultimately, data acquired by 5G networks will assist mankind in addressing a handful of its urgent issues, such as global warming, aging populations, and the rise of both persistent and contagious illnesses.

REFERENCES

1. Eugenio Pasqua. "How 5G, AI and IoT enable intelligent connectivity". IOT ANALYTICS, 2019. Available Online: https://iot-analytics.com/how-5g-ai-and-iot-enable-intelligent-connectivity/.
2. Sterlite Technologies Limited. "Don't just change transform with intelligent connectivity". 2019. Available Online: https://www.stl.tech/mwc19/pdf/01_Intelligent_Connectivity_Whitepaper_16_01_19_web.pdf.
3. GSMA. "Intelligent connectivity – How the combination of 5G, AI, Big Data and IoT is set to change everything". GSMA, 2022. Available Online: https://www.gsma.com/ic/report/.
4. Huawei. "Shaping the new normal with intelligent connectivity". *GCI*, 2020. Available Online: https://www.huawei.com/minisite/gci/en/.

5. Huawei. "Powering intelligent connectivity with global collaboration". *GCI*, 2019. Available Online: https://www.huawei.com/minisite/gci/assets/files/gci_2019_whitepaper_en.pdf.
6. Z. Yang, Y. Peng, Y. Yue, X. Wang, Y. Yang, and W. Liu. "Study and application on the architecture and key technologies for IoT". In *Multimedia Technology (ICMT), 2011 International Conference*, pp. 747–751, 2011.
7. R. Khan, S.U. Khan, R. Zaheer, and S. Khan. "Future internet: The internet of things architecture, possible applications and key challenges". In *Frontiers of Information Technology (FIT), 2012 10th International Conference On*, pp. 257–260, 2012.
8. H. Yang, A. Alphones, Z. Xiong, D. Niyato, J. Zhao, and K. Wu. "Artificial intelligence-enabled intelligent 6G networks". *IEEE Network*, vol. 34, no. 6, pp. 272–280, Nov./Dec. 2020. doi:10.1109/MNET.011.2000195.
9. A.A. Bahashwan, M. Anbar, N. Abdullah, T. Al-Hadhrami, and S.M. Hanshi. "Review on common IoT communication technologies for both long-range network (LPWAN) and short-range network". In F. Saeed, T. Al-Hadhrami, F. Mohammed, E. Mohammed (Eds.), *Advances on Smart and Soft Computing. Advances in Intelligent Systems and Computing*, vol. 1188. Springer, Singapore, 2021. doi:10.1007/978-981-15-6048-4_30.
10. P. Sethi and S.R. Sarangi. "Internet of things: Architectures, protocols, and applications". *Journal of Electrical and Computer Engineering*, vol. 2017, Article ID: 9324035, p. 25, 2017. doi:10.1155/2017/9324035.
11. M. Zennaro. "Intro to Internet of Things: ITU ASP COE training on 'Developing the ICT ecosystem to harness IoTs'". 13–15 Dec. 2016, Bangkok, Thailand. Available Online: https://www.itu.int/en/ITU-D/Regional-Presence/AsiaPacific/SiteAssets/Pages/Events/2016/Dec-2016-IoT/IoTtraining/IoT%20Intro-Zennaro.pdf.
12. A. Al-Fuqaha, M. Guizani, M. Mohammadi, M. Aledhari, and M. Ayyash. "Internet of things: A survey on enabling technologies, protocols, and applications". *IEEE Communications Surveys & Tutorials*, vol. 17, no. 4, pp. 2347–2376, Fourthquarter 2015. doi:10.1109/COMST.2015.2444095.
13. N. Koshizuka and K. Sakamura. "Ubiquitous ID: Standards for ubiquitous computing and the internet of things". *IEEE Pervasive Computing*, vol. 9, no. 4, pp. 98–101, Oct.–Dec., 2010. doi:10.1109/MPRV.2010.87.
14. N. Kushalnagar, G. Montenegro, and C. Schumacher. "IPv6 over low-power wireless personal area networks (6LoWPANs): Overview, assumptions, problem statement, and goals". *RFC4919*, vol. 10, Aug. 2007.
15. G. Montenegro, N. Kushalnagar, J. Hui, and D. Culler. "Transmission of IPv6 packets over IEEE 802.15. 4 networks". *Internet Proposed Standard RFC 4944*, 2007.
16. K. Pilkington. "Revolv teams up with home depot to keep your house connected". *CNET – News*, 2014. Available Online: http://ces.cnet.com/8301-35306_1-57616921/revolv-teams-up-with-home-depot-to-keep-your-house-connected/.
17. SmartThings. "Home automation, home security, and peace of mind", Sept. 2014. Available Online: http://www.smartthings.com.
18. BusinessWire. "Belkin brings your home to your fingertips with WeMo home automation system". *2012 International CES*. Available Online: https://www.businesswire.com/news/home/20120109005995/en/belkin-brings-your-home-to-your-fingertips-with-wemo-home-automation-system.
19. A. Dunkels, B. Gronvall, and T. Voigt. "Contiki – A lightweight and flexible operating system for tiny networked sensors". In *Local Computer Networks, 2004. 29th Annual IEEE International Conference On*, pp. 455–462, 2004.
20. P. Levis, S. Madden, J. Polastre, R. Szewczyk, K. Whitehouse, A. Woo, D. Gay, J. Hill, M. Welsh, E. Brewer, and D. Culler. "TinyOS: An operating system for sensor networks". In *Ambient Intelligence Anonymous Springer*, pp. 115–148, 2005.

21. Q. Cao, T. Abdelzaher, J. Stankovic, and T. He. "The LiteOS operating system: Towards Unix-like abstractions for wireless sensor networks". In *Information Processing in Sensor Networks, 2008. IPSN'08. International Conference On*, pp. 233–244, 2008.
22. E. Baccelli, O. Hahm, M. Gnes, M. Whlisch, and T.C. Schmidt. "RIOT OS: Towards an OS for the internet of things". In *Computer Communications Workshops (INFOCOM WKSHPS), 2013 IEEE Conference On*, pp. 79–80, 2013.
23. Open Auto Alliance. 20 Sept. 2014. Available Online: http://www.openautoalliance.net/.
24. https://iotbyhvm.ooo/iot-enabling-technologies/.
25. X. Xiaojiang, W. Jianli, and L. Mingdong. "Services and key technologies of the internet of things". *ZTE Communications*, vol. 2, p. 011, 2010.
26. M. Gigli and S. Koo. "Internet of things: Services and applications categorization". *Advances in Internet of Things*, vol. 1, pp. 27–31, 2011.
27. https://pdfs.semanticscholar.org/425e/bae20e5a277a97719d86b26f083c6e605536.pdf.
28. J. Gao, F. Liu, H. Ning, and B. Wang. *RFID Coding, Name and Information Service for Internet of Things*, 2007.
29. P. Barnaghi, W. Wang, C. Henson, and K. Taylor. "Semantics for the internet of things: Early progress and back to the future". *International Journal on Semantic Web and Information Systems (IJSWIS)*, vol. 8, pp. 1–21, 2012.
30. T. Kamiya and J. Schneider. "Efficient XML interchange (EXI) format 1.0". *World Wide Web Consortium Recommendation REC-Exi-20110310*, 2011.
31. A. Zaidi, Y. Hussain, M. Hogan, and C. Kuhlins. "Cellular IoT evolution for industry digitalization". *Ericsson White Paper, GFMC-19:000017 UEN*, Jan. 2019.
32. "Evaluation of LTE-M towards 5G IoT requirements". *GSMA, 2018*. Available Online: https://www.gsma.com/iot/evaluation-of-lte-m-towards-5g-iot-requirements/.
33. "Mobile IoT in the 5G future – NB-IoT and LTE-M in the context of 5G". *GSMA, 2018*. Available Online: https://www.gsma.com/iot/mobile-iot-5g-future/.
34. "IMT-2020 self-evaluation: mMTC coverage, data rate, latency & battery life". *Ericsson and Sierra Wireless*. 3GPP R1-1814144, Nov. 2018. Available Online: http://www.3gpp.org/ftp/tsg_ran/WG1_RL1/TSGR1_95/Docs/R1-1814144.zip.
35. "LTE-M and NB-IoT meet the 5G performance requirements". *Ericsson Blog Post*, Dec. 2018. Available Online: https://www.ericsson.com/en/blog/2018/12/lte-m-and-nb-iot-meet-the-5g-performance-requirements.
36. 3GPP TS22.104. "Service requirements for cyber-physical control applications in vertical domains". V16.0.0, Jan. 2019. Available Online: https://portal.3gpp.org/desktopmodules/Specifications/SpecificationDetails.aspx?specificationId=3528.
37. Gemalto. "Introducing 5G networks – Characteristics and usages". Gemalto, 2016. Available Online: http://repositorioiri5g.iri.usp.br/jspui/bitstream/123456789/106/1/tel-5G-networks-QandA.pdf.
38. S. Li, L.D. Xu, and S. Zhao. "5G internet of things: A survey". *Journal of Industrial Information Integration*, vol. 10, pp. 1–9, 2018. doi:10.1016/j.jii.2018.01.005.
39. 3GPP TR 38.824. "Study on physical layer enhancements for NR ultra-reliable and low latency case (URLLC)". V1.0.0, Oct. 2018. Available Online: https://portal.3gpp.org/desktopmodules/Specifications/SpecificationDetails.aspx?specificationId=3498.
40. N. Jaiswal and A. Mason. "5G: Continuous evolution leads to quantum shift". Telecom Asia, 2014. Available Online: https://www.telecomasia.net/content/5g-continuous-evolution-leads-quantum-shift/.
41. I.F. Akyildiz, S. Nie, S.-C. Lin, and M. Chandrasekaran. "5G roadmap: 10 key enabling technologies". *Computer Networks*, vol. 106, pp. 17–48, 2016. ISSN 1389-1286. doi:10.1016/j.comnet.2016.06.010.
42. I.F. Akyildiz, P. Wang, and S.C. Lin. "SoftAir: A software defined networking architecture for 5G wireless systems". *Computer Networks*, vol. 85, no. C, p. 118, 2015.

43. I.F. Akyildiz, A. Lee, P. Wang, M. Luo, and W. Chou. "A roadmap for traffic engineering in SDN-OpenFlow networks". *Journal of Computer Networks*, vol. 71, p. 130, 2014.
44. SDX Central. "How 5G NFV will enable the 5G future". SDxCentral, 2017. Available Online: https://www.sdxcentral.com/5g/definitions/5g-nfv/.
45. M. Hasan and E. Hossain. "Random access for machine-to-machine communication in LTE-advanced networks: Issues and approaches". *IEEE Communications Magazine*, vol. 51, pp. 86–93, 2013.
46. X. Ge, H. Cheng, M. Guizani, and T. Han. "5G wireless backhaul networks: Challenges and research advances". *IEEE Network*, vol. 28, no. 6, p. 611, Nov. 2014.
47. J. Liu, N. Kato, J. Ma, and N. Kadowaki. "Device-to-device communication in LTE-advanced networks: A survey". *IEEE Communications Surveys & Tutorials*, vol. 17, no. 4, p. 19231940, 2015.
48. P. Mach, Z. Becvar, and T. Vanek. "In-band device-to-device communication in OFDMA cellular networks: A survey and challenges". *IEEE Communications Surveys & Tutorials*, vol. 17, no. 4, p. 18851922, 2015.
49. A. Pyattaev, J. Hosek, K. Johnsson, et al. "3GPP LTE-assisted Wi-Fi direct: Trial implementation of live D2D technology". *ETRI Journal*, vol. 37, no. 5, p. 114, Nov. 2015.
50. S. Talwar, D. Choudhu, K. Dimou, E. Aryafar, and B. Bangerter. *Enabling Technologies and Architectures for 5G Wireless.* IEEE Xplore, Jun. 01, 2014. Kenneth Stewart Intel Corporation, Santa Clara, CA.
51. S.J. Russell and P. Norvig. *Artificial Intelligence – A Modern Approach.* Pearson Education, 2010.
52. T.P. Fowdur, L. Babooram, M.N.I. Nazir Rosun, and M. Indoonundon. *Real-Time Cloud Computing and Machine Learning Applications.* Nova Science Publishers, Jul. 2021, Computer Science, Technology and Applications Book Series, ISBN: 978-1-53619-813-3. Available Online: https://novapublishers.com/shop/real-time-cloud-computing-and-machine-learning-applications/.
53. V. Berggren, R. Inam, L. Mokrushin, A. Hata, J. Jeong, S.K. Mohalik, J. Forgeat, and S. Sorrentino. "Artificial intelligence in next-generation connected systems". *Ericsson White Paper, 2021*. Available Online: https://www.ericsson.com/en/reports-and-papers/white-papers/artificial-intelligence-in-next-generation-connected-systems.
54. E. Ekudden. "To deliver cognitive networks, we build human trust in AI". *Ericsson Blog Post, 2021*. Available Online: https://www.ericsson.com/en/blog/2021/5/cognitive-networks.
55. G. Wikström, et al. "6G – Connecting a cyber-physical world". *Ericsson White Paper, 2022*. Available Online: https://www.ericsson.com/en/reports-and-papers/white-papers/a-research-outlook-towards-6g.
56. "Intent in autonomous networks v1.0.0, TM forum document TIG1253", May 2021. Available Online: https://www.tmforum.org/resources/how-to-guide/ig1253-intent-in-autonomous-networks-v1-0-0/.
57. A. Hata, R. Inam, K. Raizer, S. Wang, and E. Cao. "AI-based safety analysis for collaborative mobile robots". In *2019 24th IEEE International Conference on Emerging Technologies and Factory Automation (ETFA)*, pp. 1722–1729, 2019. doi:10.1109/ETFA.2019.8869263.
58. R. Inam, et al. "Explainable AI – How humans can trust AI". *Ericsson White Paper, 2021*. Available Online: https://www.ericsson.com/en/reports-and-papers/white-papers/explainable-ai--how-humans-can-trust-ai.
59. B. Jamoussi. "Cloud computing & Big Data ITU-T standardization". *ITU-T, 2018*. Available Online: https://www.itu.int/en/ITU-D/Regional-Presence/CIS/Documents/Events/2018/06_Tashkent/Presentations/ITU%20Workshop%2019%20June%202018%20-%20Bilel%20Jamoussi.pdf.

60. D. Sullivan. *The Definitive Guide to Cloud Computing*. Real Time Publishers, Jul. 2009. Available Online: http://eddiejackson.net/web_documents/The_Definitive_Guide_to_Cloud_Computing.pdf.
61. P. Mell and T. Grance. "The NIST definition of cloud computing". *NIST Special Publication 800-145, 2011*. Available Online: https://nvlpubs.nist.gov/nistpubs/Legacy/SP/nistspecialpublication800-145.pdf.
62. V. Raghavendran, G. Naga Satish, P. Suresh Varma, and G. Jose Moses. "A study on cloud computing services". *International Journal of Engineering and Technical Research*, vol. 4, p. 67, 2016. ISSN: 2278-0181, ICACC-2016 Conference Proceeding. Available Online: https://www.ijert.org/research/a-study-on-cloud-computing-services-IJERTCONV4IS34014.pdf.
63. "Data analytics as a service: Unleashing the power of cloud and Big Data". *Ascent White Paper*, Mar. 2013. Available Online: https://atos.net/wp-content/uploads/2017/10/01032013-AscentWhitePaper-DataAnalyticsAsAService.pdf.
64. J. Gao, X. Bai, W. Tsai, and T. Uehara. "Testing as a service (TaaS) on clouds". In *2013 IEEE 7th International Symposium on Service-Oriented System Engineering*, pp. 212–223, 2013. doi:10.1109/SOSE.2013.66.
65. G. Kulkarni, M. Jadhav, S. Bhuse, and H. Bankar. "Communication as service cloud". *International Journal of Computer Networking, Wireless and Mobile Communications (IJCNWMC)*, vol. 3, pp. 149–156, 2013.
66. Cisco. "Network as a service". 2021. Available Online: https://www.cisco.com/c/dam/en/us/solutions/enterprise-networks/network-as-a-service/pdf/nb-06-plus-naas.pdf.
67. Extreme. "Cloud networking". 2020. Available Online: https://www.veracomp.ro/wp-content/uploads/2020/06/2020CloudNetworkingGuideBG.pdf.
68. Huawei. "Why do we need intelligent cloud-networks?" Available Online: https://e.huawei.com/en/solutions/enterprise-networks/intelligent-ip-networks.
69. P.U. Reesha. "Intelligent cloud". *International Journal of Engineering Research & Technology (IJERT) – NSRCL*, vol. 3, no. 28, 2015. Available Online: https://www.ijert.org/intelligent-cloudJournal>.
70. D.C. Nguyen, P. Pathirana, M. Ding, and A. Seneviratne. "Blockchain for 5G and beyond networks: A state of the art survey". *Journal of Network and Computer Applications*, vol. 166, p. 102693, 2020. doi:10.1016/j.jnca.2020.102693.
71. S. Nakamoto, et al. "Bitcoin: A peer-to-peer electronic cash system", 2008. Available Online: https://bitcoin.org/bitcoin.pdf.
72. W. Wang, D.T. Hoang, P. Hu, Z. Xiong, D. Niyato, P. Wang, Y. Wen, and D.I. Kim. "A survey on consensus mechanisms and mining strategy management in blockchain networks". *IEEE Access*, vol. 7, pp. 328–370, 2019.
73. K. Christidis and M. DevetsikIoTis. "Blockchains and smart contracts for the internet of things". *IEEE Access*, vol. 4, pp. 2292–2303, 2016.
74. J. Liu and Z. Liu. "A survey on security verification of blockchain smart contracts". *IEEE Access*, vol. 7, pp. 77894–77904, 2019. doi:10.1109/ACCESS.2019.2921624.
75. S. Rouhani and R. Deters. "Security, performance, and applications of smart contracts: A systematic survey". *IEEE Access*, vol. 7, pp. 759–779, 2019.
76. A. Abdelmaboud, A.I.A. Ahmed, M. Abaker, T.A.E. Eisa, H. Albasheer, S.A. Ghorashi, and F.K. Karim. "Blockchain for IoT applications: Taxonomy, platforms, recent advances, challenges and future research directions". *Electronics*, vol. 11, p. 630, 2022. doi:10.3390/electronics11040630.
77. F. Wen, L. Yang, W. Cai, and P. Zhou. "DP-hybrid: A two-layer consensus protocol for high scalability in permissioned blockchain". In *Proceedings of the International Conference on Blockchain and Trustworthy Systems*, 6–7 Aug., pp. 57–71, 2020.

78. D. Perez, J. Xu, and B. Livshits. "Revisiting transactional statistics of high-scalability blockchains". In *Proceedings of the ACM Internet Measurement Conference*, Pittsburgh, PA, USA, 27–29 Oct., pp. 535–550, 2020.
79. S.P. Veena Pureswaran, S. Nair, and P. Brody. "Empowering the edge-practical insights on a decentralized internet of things". IBM Corporation, 2015. Available Online: https://www.ibm.com/downloads/cas/2NZLY7XJ (accessed on 8 Jan. 2022).
80. P. Shah, D. Forester, D. Polk, and M. Berberich. "Blockchain technology: Data privacy issues and potential mitigation strategies". 2019. Available Online: https://www.davispolk.com/files/blockchain_technology_data_privacy_issues_and_potential_mitigation_strategies_w-021-8235.pdf (accessed on 23 Nov. 2020).
81. J.B. Bernabe, J.L. Canovas, J.L. Hernandez-Ramos, R.T. Moreno, and A. Skarmeta. "Privacy-preserving solutions for blockchain: Review and challenges". *IEEE Access*, vol. 7, pp. 164908–164940, 2019.
82. W. Shbair, M. Steichen, and J. François. "Blockchain orchestration and experimentation framework: A case study of KYC". In *IEEE/IFIP Man2Block 2018-IEEE/IFIP Network Operations and Management Symposium*, IEEE, Taipei, Taiwan, 2018.
83. C. Pahl, N. El Ioini, S. Helmer, and B. Lee. "An architecture pattern for trusted orchestration in IoT edge clouds". In *Proceedings of the 2018 3rd International Conference on Fog and Mobile Edge Computing (FMEC)*, Barcelona, Spain, 23–26 Apr., pp. 63–70, 2018.
84. B. Pillai, K. Biswas, and V. Muthukkumarasamy. "Cross-chain interoperability among blockchain-based systems using transactions". *Knowledge Engineering Review*, vol. 35, p. E23, 2020.
85. M. Madine, K. Salah, R. Jayaraman, Y. Al-Hammadi, J. Arshad, and I. Yaqoob. "Application-level interoperability for blockchain networks". *TechRxiv*, 2021. [Preprint].
86. H. Wang, Y. Cen, and X. Li. "Blockchain router: A cross-chain communication protocol". In *Proceedings of the 6th International Conference on Informatics, Environment, Energy and Applications*, Jeju, Korea, 29–31 Mar., pp. 94–97, 2017.
87. IBM. "Defining blockchain and AI". Available Online: https://www.ibm.com/topics/blockchain-ai.
88. M. Matinmikko-Blue, S. Yrjölä, P. Ahokangas, et al. "6G and the UN SDGs: Where is the connection?" *Wireless Personal Communications*, vol. 121, pp. 1339–1360, 2021. doi:10.1007/s11277-021-09058-y.
89. T.P. Fowdur, M. Indoonundon, M.A. Hosany, D. Milovanovic, and Z. Bojkovic. "Achieving sustainable development goals through digital infrastructure for intelligent connectivity". In Z. Boulouard, M. Ouaissa, M. Ouaissa, S. El Himer (Eds.), *AI and IoT for Sustainable Development in Emerging Countries. Lecture Notes on Data Engineering and Communications Technologies*, vol. 105. Springer, Cham, 2022. doi:10.1007/978-3-030-90618-4_1.
90. Bosch. "Connected car effect 2025". Bosch Presse, 2017. Available Online: https://www.bosch-presse.de/pressportal/de/en/bosch-study-shows-more-safety-more-efficiency-more-free-time-with-connected-mobility-82818.html.
91. D.E. Kouicem, A. Bouabdallah, and H. Lakhlef. "Internet of things security: A top-down survey". *Computer Networks*, vol. 141, pp. 199–221, 2018.
92. A. Khare, G. Merlino, F. Longo, A. Puliafito, and O.P. Vyas. "Design of a trustless smart city system: The# SmartME experiment". *Internet Things*, vol. 10, p. 100126, 2020.
93. GSMA. "Intelligent connectivity how the combination of 5G, AI and IoT is set to change the Americas". GSMA, 2018. Available Online: http://www.innovation4.cn/library/r31490.

94. K.E. Skouby and P. Lynggaard. "Smart home and smart city solutions enabled by 5G, IoT, AAI and CoT services". In *2014 Internationa lConference on Contemporary Computing and Informatics (IC3I)*, pp. 874–878, 2014.
95. S.Y. Nikouei, R. Xu, D. Nagothu, Y. Chen, A. Aved, and E. Blasch. "Real-time index authentication for event-oriented surveillance video query using blockchain". In *2018 IEEE International Smart CitiesConference (ISC2)*, pp. 1–8, 2018.
96. R. Wang, W.-T. Tsai, J. He, C. Liu, Q. Li, and E. Deng. "A video surveillance system based on permissioned blockchains and edge computing". In *2019 IEEE International Conference on Big Data and Smart Computing (BigComp)*, pp. 1–6, 2019.
97. A. Damianou, C.M. Angelopoulos, and V. Katos. "An architecture for blockchain over edge-enabled IoT for smart circular cities". In *2019 15th International Conference on Distributed Computing in Sensor Systems (DCOSS)*, pp. 465–472, 2019.
98. PWC. "5G in healthcare". PwC, 2020. Available Online: https://www.pwc.com/gx/en/industries/tmt/5g/pwc-5g-in-healthcare.pdf.
99. Telecoms.com. "How 5G will transform learning from early years to in-work training". Telecoms, 2021. Available Online: https://telecoms.com/opinion/how-5g-will-transform-the-learning-experience-from-early-years-to-in-work-training/.
100. BT.com. "New immersive classroom brings limitless potential to North Lanarkshire". BT Group, 2021. Available Online: https://newsroom.bt.com/new-immersive-classroom-brings-limitless-potential-to-north-lanarkshire/.

11 AI-Assisted Extended Reality Toward the 6G Era
Challenges and Prospective Solutions

Girish Bekaroo
Middlesex University Mauritius

Viraj Dawarka
Staffordshire University London

CONTENTS

11.1	Introduction	404
11.2	Background: The Convergence of XR, AI, and Cellular Data Networks	405
	11.2.1 Augmented Reality	407
	11.2.2 Virtual Reality	407
	11.2.3 Mixed Reality	407
	11.2.4 Artificial Intelligence	408
	11.2.5 Machine Learning	408
	11.2.6 Deep Learning	408
11.3	AI-Assisted XR: Opportunities in the 5G and 6G Era	408
	11.3.1 AI-Assisted AR	409
	11.3.1.1 Camera Calibration and Pose Estimation	409
	11.3.1.2 Detection and Tracking of Real Objects	410
	11.3.1.3 Creation of Virtual Objects	411
	11.3.1.4 Displaying Virtual Objects	411
	11.3.2 AI-Assisted Virtual Reality	412
	11.3.2.1 Content Creation	413
	11.3.2.2 Optimization and Rendering	413
	11.3.2.3 Interaction in the Virtual World	414
	11.3.3 AI-Assisted Mixed Reality	414
11.4	Challenges and Prospective Solutions	415
	11.4.1 Portability and Compatibility Issues	415
	11.4.2 Skepticism in Adoption of XR Solutions	416
	11.4.3 Issues with Visual Interfaces	416
	11.4.4 Processing Requirements and Resource Constraints	417
	11.4.5 Technological Limitations	417

11.4.6 Power and Thermal Efficiency.. 418
11.4.7 Challenges Related to the Creation of Contents 418
11.4.8 Lack of Skills and Competencies 418
11.4.9 Challenges Related to Application of AI 419
11.4.10 Cellular Data Network Challenges and Deployment Issues 420
11.4.11 Privacy and Security Issues 420
11.5 Conclusions.. 420
References.. 421

11.1 INTRODUCTION

Advances in display and novel interaction technologies have led to computational devices being able to overlay computer-generated elements (e.g., text, images, and video) onto the real environment or embed aspects of the real world into computer-generated virtual environments. The umbrella term referring to immersive technologies associated with both real and virtual environments is referred to extended reality (XR), and encompasses technologies including augmented reality (AR), virtual reality (VR), and mixed reality (MR). During recent years, XR technologies have undergone rapid development and are increasingly being used in various domains including training, education, and manufacturing, among others. Among the XR technologies, the AR market is expected to grow from US$14.7 billion in 2020 to US$88.4 billion by 2026, particularly due to the increasing demands for AR devices, growth in investments in AR market, and quickly rising number of applications in sectors including healthcare and commerce (Markets and Markets, 2021). Similarly, an increase in the number of global VR device shipments is expected to grow from 13.48 million units in 2020 to 112.62 million units by 2026 particularly due to increasing VR applications in training and adoption of VR gaming at home (LLP, 2021).

As the 5G era has united high bit rate, low latency, high-energy efficiency, increased availability, and intelligent networks, various opportunities for XR technologies have emerged, which are also facilitated by advances in artificial intelligence (AI; Yang et al., 2020). The convergence of these distinct technologies has the potential to create unique applications and experiences for end users within different areas including medical training, travel and tourism, education, and even gaming. Furthermore, due to the promising future of the combination of these technologies, Mark Zuckerberg, co-founder and CEO of Meta Platforms presented the company's vision of the metaverse, which is an integrated network of 3D virtual worlds, as the successor to the mobile Internet. Although the metaverse is still in the early stages, the concept has created a major boost in the advancement of AI-assisted XR, which are key technologies involved in the metaverse along with cellular data networks. Nevertheless, the absolute XR experience is still challenging to be achieved in the 5G era because certain features require high-quality video and 3D animations to be generated and displayed in real-time thereby requiring a minimum of 10 Gbps data rate (Mumtaz et al., 2017). The enhanced capabilities of the sixth-generation cellular data networks can potentially enable such features to be realized and bring further advancements to XR technologies as well as boosting their proliferation. Furthermore, over the previous two decades, AI and machine learning have progressed dramatically, and their applications are widespread, notably within intelligent agents, robot control systems, and chatbots, among others (Jordan & Mitchell, 2015). Such advances were also driven by the rapid growth in the networked and mobile computing systems having the ability to collect and transmit huge amounts of data within different contexts. These data have been successfully used as training data within machine learning by scientists and engineers in order to derive useful insights, predictions, and decisions for different kinds of problems. With the improved capabilities in the 5G and 6G eras, more significant progress in AI and machine learning are anticipated and can significantly extend capabilities of XR technologies. As such, during the 5G and 6G eras, the convergence of XR technologies with AI and the new generations of cellular data networks are expected to progressively strengthen, to the benefit of XR technologies and applications whereby creating new opportunities at various levels. However, the development of AI-assisted XR systems and solutions in these eras, their diffusion, and adoption are not going to be without challenges. These issues are important to be investigated and addressed in a timely manner in order to contribute to the achievement of the absolute XR experience. This chapter discusses the opportunities created by the integration of AI and advancements in cellular data networks within XR technologies, toward advancing AI-assisted XR in the 5G/6G eras. In addition, the key challenges that are expected to hamper the development of AI-assisted XR toward 6G are discussed, along with prospective solutions to these challenges.

11.2 BACKGROUND: THE CONVERGENCE OF XR, AI, AND CELLULAR DATA NETWORKS

The fifth generation of cellular networks (5G) has brought a huge step ahead in terms of connectivity, network speed, capacity, and scalability in addition to reduction in latency and energy consumption as compared to 4G. These include high data rates of up to 100 times faster than 4G, ranging between 1 and 10 Gbps in real networks, in addition to reduced latency of 1 ms round trip time as compared to the 10 ms round trip time within 4G (reduction by ten times). Moreover, 5G brought enhanced connectivity, availability, and geographical coverage whereby enabling support of billions of connected devices within the network. However, the rapid proliferation of data-centric and automated systems is expected to exceed the capabilities of 5G wireless systems (Chowdhury et al., 2020). For example, certain technologies such as VR require features beyond 5G for high-quality video and 3D animations to be generated and displayed in real-time thereby requiring a minimum of 10 Gbps data rate (Mumtaz et al., 2017). In order to address these limitations of 5G, a sixth-generation (6G) wireless system with enhanced capabilities is required. This new generation of wireless systems will encompass enhancements in features such as higher data rates, higher throughput, enhanced energy efficiency, higher reliability, massive connectivity, and network densification. Such enhanced capabilities are expected to bring advancements and help proliferation of various technologies such as XR, wearables, AI, and intelligent environments, among others, which generate and process massive volume of data and require high data rate connectivity per device (Chowdhury et al., 2020). A comparative summary of the key capabilities of 5G and 6G is provided in Table 11.1.

TABLE 11.1
A Comparison of 5G and 6G

Feature	5G	6G
Peak data rate	10 Gbps	1 Tbps
Latency	1 ms	1 μs
Frequency bands	Sub 6 GHz mmWave for fixed access	Sub 6 GHz mmWave for mobile access exploration of THz bands (above 140 GHz) non-RF bands
Maximum spectral efficiency	30 bps/Hz	100 bps/Hz
Mobility support	Up to 500 km/h	Up to 1,000 km/h
Reliability	99.999%	99.99999%
Connection density	10^6 devices/km^2	10^7 devices/km^2

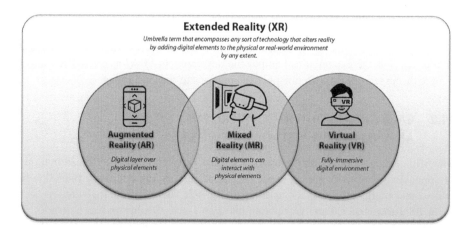

FIGURE 11.1 Extended reality.

The vision of extended reality is expected to accelerate due to advances in the cellular networks as illustrated in Table 11.1 and this will produce improved XR experiences to users (Patil & Kumar, 2019). Founded on the reality–virtuality continuum, XR is an umbrella term that encompasses any sort of technology that alters reality by adding digital elements to the physical or real-world environment by any extent, as illustrated in Figure 11.1. Within the acronym XR, X is a variable englobing current and future spatial computing technologies implemented under this umbrella term and include AR, VR, and MR. In the last few years, XR has received renewed attention within academia and the industry due to recent technological developments that led to cheaper and lighter devices that have more powerful processing capabilities than previous generations (Çöltekin et al., 2020). This expanded the spread of the associated group of technologies from the lab and specialized domains to a wider uptake within businesses and the society for applications in training, emergency preparedness, gaming, and aviation, among others. The key technologies under the umbrella term XR are as follows.

11.2.1 AUGMENTED REALITY

AR provides an interactive experience of the real-world environment where computer-generated elements (e.g., images, 3D models, animations) and meaningful digital information are superimposed on the user's view of the real world with the aim to enhance the perception of reality. A common example of AR is Pokémon Go, where players need to physically explore the physical world to find locations of digital Pokémon that are superimposed on the screen of their mobile devices. In addition, AR has shown prospects in medical training and surgical procedures whereby enabling computer-generated images or 3D models to be overlaid in real time on the body parts of patients, models, or even images of body parts in order to enable visualization of unapparent anatomical details and obtain key information (Pessaux et al., 2015).

11.2.2 VIRTUAL REALITY

VR is an artificial and interactive computer-generated virtual environment, which provides a completely immersive experience to the end user. Within the environment, the user has the ability to look around the computer-generated world, move and interact with virtual features or items. Such experience is complemented principally with visual and auditory feedback, in addition to other sensor feedback such as haptic. VR has various applications in practice including training and education, gaming and entertainment, among others. As an example, in the film industry, cinematic VR (CVR) enables a viewer to watch omnidirectional movies whereby freely choosing viewing direction using head-mounted displays (HMD) or other VR devices.

11.2.3 MIXED REALITY

MR is a blend of AR and VR, where the real and virtual worlds are combined to produce new environments and experiences to the end user. Within the environment, the end user can interact with both physical and digital objects that coexist in real time. MR has prospects in various industries and its applications are proliferating. An example of MR involves its use in medical training where students at Case Western can study human anatomy using Microsoft HoloLens. In this application, students are exposed to a life-sized, standing 3D image and they can then walk around, manipulate, dissect with hand gestures, listen to sounds, and examine different parts from unobstructed points of view, which is challenging to achieve when learning with cadavers since one organ must be removed in order to view another one.

In addition, XR technologies are expected to benefit from the huge recent developments in AI toward furthering intelligence in XR solutions. Although the application of AI within the context of AR, VR, and MR is not new, the capabilities brought by the developments in cellular networking technologies are expected to resolve earlier challenges such as enabling the processing of computationally intensive tasks on the edge, fog, or cloud servers, thereby facilitating the integration of intelligence. These capabilities are further discussed in later sections of this chapter, in which different AI-related terms are involved, which are defined as follows:

11.2.4 Artificial Intelligence

AI relates to intelligence demonstrated by a machine that is able to make decisions in a manner similar to human intelligence. In other words, AI refers to systems or machines that mimic human intelligence to perform tasks and that can iteratively improve themselves based on the information collected. An example of AI includes recommender systems integrated into e-commerce websites that recommend items for purchase based on products viewed or recently bought by a customer. Similarly, chatbots utilize AI in order to understand customer problems quickly and respond in the most effective manner.

11.2.5 Machine Learning

It is a sub-branch of AI where machines process data and learn on their own, without constant human supervision. Machine learning is behind applications such as chatbots, language translation applications, and recommender systems, among others. There are three subcategories of machine learning, notably, supervised, unsupervised, and reinforcement learning. Supervised machine learning is the most common type among the three and involves training models with labeled data sets in order to allow the models to learn and grow more accurately over time. On the other hand, within unsupervised machine learning, a program looks for patterns in unlabeled data. Finally, reinforcement machine learning trains machines through trial and error to take the best action by establishing a reward system.

11.2.6 Deep Learning

Deep learning is a subset of machine learning and distinguishes itself from machine learning by the type of data that it works with and the methods in which it learns. Deep learning involves the application of artificial neural networks, algorithms inspired by the human brain, to learn from large amounts of data. Deep learning drives many AI solutions that enable automation, perform analytical and physical tasks without human intervention, such as automatic credit card fraud detection or self-driving cars.

11.3 AI-ASSISTED XR: OPPORTUNITIES IN THE 5G AND 6G ERA

In the 5G and 6G eras, the integration of AI and cellular data networks within XR technologies is expected to progressively strengthen, thereby enabling further advancements in XR technologies. As such, various opportunities are anticipated during different stages involved within the creation of AI-assisted XR solutions in both eras and these are discussed in the following sections.

11.3.1 AI-Assisted AR

Even though AR, AI, and cellular networks are distinct technologies, their combination can enable application developers to create unique experiences for end users. In the past, reliable connectivity, bandwidth, and latency have been major hurdles that inhibit high-fidelity telepresence and collaborative applications that implement innovative technologies such as AR (Orlosky et al., 2017). Since AR systems involve camera image processing where rendered images with augmentations are projected onto the display technology used, integration of AI and machine learning to AR systems offer various benefits including adaptability and versatility of such systems (Sahu et al., 2021). These benefits are expected to be boosted with the support of 5G and 6G, where enhanced connectivity can bring improved cloud support for processing or storage, and these can be leveraged in order to extend features of AI-assisted AR applications.

In the 5G and 6G eras, developments in AI-assisted AR are expected to generate voluminous information through camera-imagery files (images, videos, audio, etc.) and connectivity enhancements provided by 5G and 6G can potentially resolve such issues to enable voluminous and high-speed camera-imagery data to be sent to and retrieved from remote servers for processing and real-time display. In addition, AI techniques such as machine learning and deep learning can benefit from continuous data from camera and multiple sensors integrated with the device including GPS or accelerometer to further train algorithms being used and improve their accuracy and reliability. In the 5G and 6G eras, such voluminous data needs to be rapidly communicated to the cloud for timely processing and feedback to the user (Wang et al., 2018) and the high data rates of such networks will be key requirements. Furthermore, continuous localization of the end user could proliferate further in the 5G and 6G eras. The next sections further discuss how AR is expected to be enhanced with advancements in AI within the 5G and 6G eras, in relation to different stages of development of AR applications.

11.3.1.1 Camera Calibration and Pose Estimation

In the initial stages involved in the development of AR-based systems, camera calibration is essential. This process involves preparing the virtual camera by correctly configuring the internal parameters in order to effectively detect and track real-world objects. This task can be conducted in both online and offline modes and can be supported by both AI and cellular networking technologies. For instance, AI techniques such as convolutional neural network (CNN) and neural network (NN) can be utilized in order to automatically determine key parameters from a camera. A previous study proposed a CNN-based approach that estimates intrinsic parameters of a camera (Bogdan et al., 2018), where the CNN is trained using omnidirectional images retrieved from the Internet and applied to estimate parameters of the camera. Similarly, another study attempted to estimate parameters of a camera such as depth, camera intrinsics, and object motion, following input of multiple consecutive image frames from videos to neural networks (Gordon et al., 2019). Fetching images from the Internet and cloud processing can be greatly benefited with improved data rates and reduced latency in the 5G and 6G eras to enable timely configurations and thus enhancing user experience. Furthermore, a key objective of proper camera calibration is to effectively detect and track objects so that the pose can be correctly estimated. Pose estimation involves getting the position and orientation of the camera from the scene in relation to an object. In this endeavor, AI provides opportunities for directly estimating camera-pose in AR and work has been undertaken in this area. For instance, a pretrained GoogLeNet has been used to directly estimate the 6-DoF pose consisting of depth and out-of-plane rotation of a monocular camera in real-time (Kendall et al., 2015). Again, implementation and use of such models involve cloud processing, which are facilitated with enhanced capabilities of cellular networks and are expected to develop further in the 5G and 6G eras.

11.3.1.2 Detection and Tracking of Real Objects

A key purpose of AR systems involves augmenting the real-world with superimposed meaningful information. As such, a core task within AR systems involves effective real-time detection, identification, and tracking of targets to be augmented in the real world. The two common methods for object detection are marker-based and markerless detection. The marker-based approach involves placing fiducials on objects identifiable by the AR system. These fiducials have characteristics such as distinct shapes or patterns that are easily detectable by the camera. Although the fiducials are easily trackable and provide robustness, particularly in poor lighting conditions, placing fiducials on trackable objects is a manual process which is not desirable in various contexts (e.g., industrial settings). On the other hand, markerless AR involves detection and tracking of specific points or natural features within objects. To achieve this, AR development platforms predominantly used traditional computer vision techniques such as simultaneous localization and mapping (SLAM) where visual features between camera frames are compared in real time in order to track changes in the environment.

In the object detection process, high accurate trackers are needed because even minor tracker errors can lead to significant issues in placement and alignment of virtual objects in relation to the real objects. For accurate detection and tracking of targets within AR-based applications, proper illumination is necessary. This is a challenging task in scenarios where adequate light is not available around targeted objects or within scenes. Various works have been undertaken that estimate distribution of light around objects or within scenes (Gruber et al., 2012) to eventually apply AI techniques and algorithms to attempt correction of illumination and more effectively detect objects in real-time. As such, while the application of AI necessitates high volumes to train models, 5G and 6G can enable delivery of large training sets quickly such that illumination can be corrected within the least possible duration during the use of AR systems. In addition to proper illumination, other challenges that hinder effective detection of targets in the real world include camera quality and even characteristics of objects, where it is challenging to identify large and shiny objects, as well as differentiating between objects that look alike (Bekaroo et al., 2018). As a promising solution to address this issue and to facilitate real-world object recognition, CNN within deep learning can be used. These techniques can enhance accuracy and robustness of calibration, object identification, recognition, and tracking within different lighting conditions. While deep learning has good learning capabilities in feature extraction to even find complex structures in data, the application of CNN even showed excellent performance in image feature extraction, which is important, especially in real-time image recognition and rendering (Cheng et al., 2020). The bandwidth, low latency, and reliable connections provided within 5G and 6G are key to resolve networking-related challenges involved in the transmission of image files and other data required in the process.

11.3.1.3 Creation of Virtual Objects

Virtual objects are essential components within AR applications and are typically used to display meaningful information to end users. These virtual objects can be in the form of text, images, voice, animations, and 3D models, among others. The virtual objects can be created manually during the development phase and displayed accordingly when real objects are identified by camera during runtime of the AR application. A more innovative approach involves the application of AI-based methods to create virtual objects during runtime. For instance, 2D images can be generated using AI-based techniques where images from the camera can be fed to AI-generated models in order to produce appropriate augmentations. For example, different attempts have been made to produce sketches of real objects identified by the camera using AI-based techniques, including CNN, RNN, and even reinforcement learning (Muhammad et al., 2018). In terms of practical applications, ClipDrop enables users to scan real-world objects using their mobile phone camera and convert them into digital objects, where AI is used to accurately remove any background from the scanned object. Similarly, SketchAr is a drawing application that scans pen movements during drawing to create digital sketches, which can be converted into other forms (e.g., cartoons) through the application of AI. In addition to images, AI has also been utilized in sound synthesis and audio recognition within the context of AR applications where for example if a user says a particular word such as 'pet', a virtual object of a dog can be created and brought on the visual display. Moreover, AI-based 3D reconstruction methods are available to produce 2D models, point clouds or meshes when the application is running, based on image fed by the camera. Machine learning and deep learning can be used in the development of adaptable augmentations within virtual object creation and to also assist in the selection of data to be displayed during rendering. In order to reduce the size of the AR application, processing can be performed over the edge or cloud where images of the real object are sent so that appropriate algorithm can be applied in order to generate the sketch and sent back to the display interface. In such features necessitating networking capabilities, 5G and 6G provide various benefits discussed earlier.

11.3.1.4 Displaying Virtual Objects

After generating the virtual objects, their placement and alignment with regard to the user's environment are essential. This process is called registration and involves the various parameters recorded during camera calibration so that the virtual objects are correctly placed and displayed on the visual interface (e.g., screen of mobile phone or AR glasses). Within AR applications, misregistration can also occur where the virtual object is not correctly positioned and aligned with regard to the corresponding object in the real world. For example, mismatch in virtual and physical distances can take place as the distance to real objects in relation to the display interface can vary when the user moves around. Consequently, incorrectly positioned and aligned overlays provide misleading information to the user as the context assumed by the user is inaccurate and as such, user experience deteriorates quickly when accuracy is lost. Misregistration can occur thereby resulting in incorrect focus, while also impacting display parameters of the virtual object (e.g., contrast and brightness), which is recognizable by human eyes (Olshannikova et al., 2015). Various parameters are involved in order to prevent such misregistration problems such as hardware requirements in the form of long-range accurate sensors, camera quality, and even external factors such as illumination (Wang, 2009), as discussed earlier. Moreover, AI has its potential to resolve registration errors (e.g., improper spacing between the real and virtual objects, depth, or angular errors), where different techniques can be applied to correctly determine the placement and alignment of virtual objects based on parameters such as camera calibration settings. In the 5G and 6G eras, there are further opportunities pertaining to the application of AI to effectively resolve registration errors.

After registration, the virtual objects are displayed onto the main interface of the AR application and this process is called rendering. In this process, two important decisions need to be taken, notably what data to display and how to represent the data to be displayed. A key challenge that often arises during rendering is limited display capacity and as such, it is essential to determine what information to display on the AR interface. In order to overcome this challenge, filtering techniques and visualization layouts are particularly useful, and these can be assisted through AI techniques that can be better executed over the cloud with advances in cellular data networks. While AI-assisted filtering techniques can help to reduce the amount of information to be presented into more meaningful argumentations, AI-driven visualization layouts can help to better arrange and represent the information to be displayed to the end user. An example of the application of such techniques in practice involves 3D rendering of MRIs using AI techniques in order to accurately overlay such images upon the actual heart of patients to assist surgeons prior and during surgery (Teich, 2020). In this endeavor, the surgeon needs to have minimal delay in response time since much of the computation takes place in the cloud or even in the edge, and transmissions can be enhanced by the new generations of cellular networks beyond 5G.

11.3.2 AI-Assisted Virtual Reality

The major aim of fully immersive VR is to create a digital real-time experience, which imitates human perception to its fullest, whereby recreating every image seen by human eyes, every little sound heard by human ears, along with various other cognitive aspects, including touch and smell. In other words, the more the VR model resembles the human brain model, the more the user feels immersed and being present in the simulated virtual world. This technology is characterized by high immersion, dynamic interactive response, and multidimensional information digitization. The convergence of technologies including AI, VR, and advanced cellular networks is expected to create innovative VR applications that are seemingly endless and to the benefit of various industries including medical training and gaming. The application of AI in the different stages of development of VR applications is discussed as follows.

11.3.2.1 Content Creation

In this phase, the visual elements of the VR application, notably, 2D, 3D, and animation are created. In this process, tools such as Unity 3D and Unreal Engine are involved to translate the design into a prototype. In other words, technical development is conducted by the programming team who produce the codes for determining what happens when events are triggered and translate the actions that occur when the user interacts with virtual objects, makes choices, or changes environment. In this endeavor, machine learning provides powerful features to integrate intelligence to the contents that developers attempt to create. For instance, machine learning can automate slow and difficult processes while also removing technical and creative hurdles. Examples include photoconversion, where images can be digitized for integration in virtual environments, enhancing resolution of images, smartly adding filters and corrections, among others. In addition, AI also automates various processes, which are labor-intensive and time-consuming where some include skinning and rigging. While rigging attempts to create the skeleton of a person or living thing, skinning attempts to draw the exterior around the skeleton up to the skin, all to be utilized in game engines or animations. In this endeavor, deep learning is helpful in automating involved processes. Other examples of 3D content creation using AI include the NVIDIA tool called Audio2Face, which creates expressive facial animation from an audio source. Such tool typically simplifies the generation of animated 3D character to match the voice-over track. Similarly, DeepStream from NVIDIA enables the application of AI to create human pose for animated characters.

11.3.2.2 Optimization and Rendering

During recent years, advancements in computer graphic techniques have significantly enhanced VR applications in general and the integration of intelligence along with enhanced connectivity has the potential to significantly improve experience and immersion. In order to achieve such experience and immersion, VR systems should be able to display images at a realistic rate. In other words, real-time rendering is essential as it involves creating sensory images fast enough so as to be perceived as continuous to the eyes of the end user for smooth representation of the virtual world. This process should be an optimized one, whereby adjusting and finding the balance between different parameters pertaining to the quality of display, performance, and development time to also ensure that visuals are synchronized with other forms of feedback such as sound. Recently, AI-driven cloud-based rendering was found to provide various benefits as it reduces overheads on the client side (Li et al., 2020). Such approach allows the user to choose between various platforms with high processing capabilities in order to provide the optimum experience for the user. In the process, the application of AI can help to provide the balance between the parameters such as display quality and performance. In addition, with 5G and 6G, the high bandwidth and low latency capabilities can make real-time rendering smoother than ever to deliver realistic and high-quality experience within VR applications.

11.3.2.3 Interaction in the Virtual World

A key element of any VR experience involves interaction with a virtual world. At the most fundamental level, the display of the virtual world should correspond with the user's physical movement or corresponding actions triggered by the end user. There are different ways in which end users can interact within a VR experience, notably direct user control, physical control, virtual control, and agent-based control (Mine, 1995). AI can be of assistance in the different forms of interaction in order to more effectively capture the intention of the user in order to prepare the VR system for potential events or actions. In addition to the use of specialized hardware for interaction with the virtual world, research and development has actively been conducted involving the use of body movements, voice, eye tracking, and gestures to send signals in human–computer interaction. For all these interaction mechanisms, AI and cellular networks can provide enhanced capabilities. For instance, previous works have been undertaken involving the application of AI, such as KNN rapid recognition method, in order to obtain feature values during human gestures and movements (Zhang, 2021) toward enhancing human interaction within the virtual world. Likewise, the application of machine learning to improve eye tracking in the context of VR has been studied (Pettersson & Falkman, 2020). In the 5G and 6G eras, further developments are also anticipated in the area of AI-assisted interaction, which is key to achieving the metaverse vision.

11.3.3 AI-Assisted Mixed Reality

Mixed reality, which is also referred to as hybrid reality, is the most recent development within the reality technologies, notably, after AR and VR. In MR, new environments and visualizations are produced, where physical and digital objects coexist and interact in real-time, somewhere between AR and VR. In other words, MR takes place not only in the physical or the virtual world but is a mix of actual reality and VR. MR takes place in two main forms, where the first one involves the user starting with the real world, where virtual objects are overlaid and that the user can interact with both real and virtual objects, to create a MR environment. In the second form of MR, the user starts in a virtual environment where the real world is replaced by a digital environment. Within the same virtual environment, some virtual objects are also physically represented by corresponding real-world objects, that the user can interact with, thus creating a MR environment. As such, MR development inspires a lot from AR and VR, and thus the convergence of AI and 5G/6G technologies discussed in earlier sections are applicable to AI-assisted MR in the 5G and 6G eras. In other words, AI is applicable to different tasks involved in the creation of MR solutions and these include camera calibration, modeling of the space and simulation, object detection, recognition and tracking, registration and rendering, as discussed earlier, and all these benefit from enhancements in the new generations of cellular data networks.

As such, in the 5G and 6G eras, MR services are expected to use enormous cloud computing capacity due to their high computational complexity where HMD providing such services will be wirelessly connected and major processing taking place on

AI-Assisted Extended Reality Toward the 6G Era

the edge, fog, or the cloud (Doppler et al., 2017). These services will be enhanced in the 5G and 6G eras with higher data rates and lower latencies, to even bring major enhancements in free viewpoint videos, thereby giving the user complete freedom of movement when experiencing MR solutions through HMDs. AI is expected to enhance such capabilities to effectively adapt perspective and display digital contents when the user is walking around and trying to view contents at different angles.

11.4 CHALLENGES AND PROSPECTIVE SOLUTIONS

An absolute and truly exciting visualization experience involving the XR technologies necessitates a combination of aspects including computing, hardware, storage, and transmission, in addition to sensory aspects notably, human senses (vision, hearing, and touch). While the convergence of 5G networks and AI are expected to further advance XR technologies, the absolute XR experience is expected to be only achievable after the sixth-generation mobile network (Minopoulos & Psannis, 2022). Nevertheless, even in the 5G era, XR technologies have prospects to advance, and a plethora of applications are expected to emerge. The development of such systems, their diffusion and adoption are however not going to be without challenges. As such, it is important to investigate and address key challenges in a timely manner in order to prepare toward the absolute XR experience. These challenges as well as prospective solutions are discussed as follows.

11.4.1 Portability and Compatibility Issues

Various XR systems have cross-compatibility issues such that applications implemented on one platform cannot be easily converted to another platform. For instance, various companies offer their own platforms to create VR environments, which have their hardware or software specificities, and compatibility issues may arise in case of changes in such components (Velev & Zlateva, 2017). For instance, a VR solution for one IoT application cannot be easily translated or even deployed and utilized in another application directly due to lack of cross-platform support (Hu et al., 2021). Portability is further adversely impacted because of strict hardware requirements and configuration settings that prevent deployment on other hardware.

Furthermore, deployment is limited in case hard-coding is thoroughly used during the creation of XR systems. For instance, in order to create AR systems, authoring tools are involved that enable the integration of different functions to control the relationships between real and virtual objects. Authoring tools to build AR-based systems can be broadly categorized into tools meant for programmers and authoring tools for non-programmers. The former are code libraries that necessitate knowledge in computer programming, while the latter include drag-and-drop interfaces for creating applications without writing any lines of codes. A key limitation in the authoring process is that determining what information to display to the user is usually hard-coded and this also particularly due to the absence of suitable software libraries for generating complex virtual objects (e.g., animations) in real-time, thereby further limiting the portability of applications. Moreover, backward compatibility issues may also arise when new versions of authoring tools are released on the market, thus introducing new challenges pertaining to the upgrade of XR solutions.

In order to address cross-platform issues, it has been envisioned that web-based solutions with natural cross-platform benefits can potentially solve this issue and has become active areas of research pertaining to system design of XR solutions (Qiao et al., 2018). For instance, research and development of XR cross-platform tools has already started where the GhostBustXR cross-platform tool was created at the MIT Reality Hack that can potentially work across all XR device platforms and similar platforms are anticipated to emerge as XR technologies are advancing. On the other hand, to address the challenge of hard-coding virtual objects, AI techniques mentioned in the previous section can be utilized in order to better incorporate context awareness within solutions. Also, with the boost in remote processing capabilities of XR solutions in the 5G and 6G eras, the need for hard-coding is anticipated to diminish.

11.4.2 Skepticism in Adoption of XR Solutions

Even though a growing number of XR-based solutions are being released on the market, acceptance by the public and enterprises is still slow due to various hurdles (Avasant, 2022). Also, even for many users who adopt the solution, the duration of use is considered to be shorter than expected and as such, the technology is not regarded to be useful for the long term due to poor quality of contents as well as low standards designs, typically for AR mobile applications (Jha et al., 2019). In order to get people to adopt XR-based solutions, acceptance-based issues need to be investigated and addressed, where these range from unobtrusive fashionable appearance and costs of equipment used (e.g., gloves, helmets, etc.) to enlightening the benefits of the technology and addressing privacy concerns (Hughes et al., 2005). In the 5G and 6G eras, better quality of contents and designs of XR solutions are anticipated as the technologies are advancing. More up-to-date features and contents through crowd-sourcing and application of AI can also potentially boost the adoption of XR solutions in the same eras.

11.4.3 Issues with Visual Interfaces

A major issue with the implementation of XR-based systems is the technological gap involving the use of XR display interfaces. For instance, the experience of using an AR system on an AR gear is completely different as compared to using the same application on a smartphone. Some AR gear systems have also been reported to be bulky, while at the same time expensive, thus adversely impacting adoption by the public as discussed earlier. Similarly, VR predominantly displays visualized information on screens and the displays on the VR helmets can lead to discomfort due to close-to-the-eye proximity principally driven by low display resolution and high graininess (Olshannikova et al., 2015). In addition to being bulky and expensive, some visual interfaces could be of small screen size thus limiting the amount of information to be displayed to the user. A key challenge remains the ability to display concise information to the user to also prevent information overload (Pascoal &

AI-Assisted Extended Reality Toward the 6G Era   417

Guerreiro, 2017). Having too much information on screen, especially in the context of AR applications, could also possibly harm the user due to physical security risks, as attention is also shifted away from the real world as a consequence of lack of details. Manufacturers are aware of issues pertaining to visual interfaces and actions are being taken in order to enhance underlying hardware so that visual experiences when using XR solutions are improved. Moreover, prior to building AR and VR systems, it is important to follow design guidelines and best practices in order to prevent issues such as information overload or visuals that overly distract end users when utilizing such XR applications (Kourouthanassis et al., 2015).

11.4.4 Processing Requirements and Resource Constraints

For realizing smooth experience, immersion, and interaction within XR applications, intensive graphic capabilities are required, which may be challenging to be implemented via standard computer equipment (Velev & Zlateva, 2017). Moreover, such graphics are also challenging to be rendered on resource constrained display interfaces. The challenges due to intensive graphic and processing requirements of XR solutions as well as resource constraints of equipment introduce various issues in such systems, including latency and errors, which adversely impact user experience.

As a solution to address the intensive graphic and processing requirements, computationally intensive tasks can be migrated to more resourceful cloud, fog, or edge servers. This solution does not only increase computational capacity of low-cost equipment used in XR systems but can also save battery energy, which is essential to portable devices involved (Bastug et al., 2017). In order to address system delays or latency, AI techniques discussed in the previous sections could be applied, in addition to techniques such as pre-calculation, temporal stream matching, prediction of future viewpoints, and image warping (Van Krevelen & Poelman, 2010).

11.4.5 Technological Limitations

Technological limitations remain a key obstacle to the development, widespread adoption, and ultimate experience of XR systems. For instance, AR applications need highly accurate trackers to prevent improper placement and alignment (registration) between real and virtual objects. Similarly, VR systems require specialized hardware for end users to experience such technology and to interact with the virtual world. Moreover, wearable device-based VR equipment has integrated processor and battery thus making such hardware heavy to wear, inconvenient to carry, and are costly at the same time (Hu et al., 2021). Furthermore, as discussed earlier, the absolute XR experience is only expected to be achievable only after the sixth generation of cellular data networks given the high-quality video and graphics playback requirements. The advancement of these technologies depends heavily on both industrial and academic efforts in various related domains such as hardware optimization and algorithm developments to improve underlying technological-related issues. The projected growth in the XR markets as discussed earlier has already led to a major boost research and development (R&D) by key stakeholders including leading enterprises such as Microsoft and Meta platforms, along with the leading universities and research communities, among others. As such, technological limitations are anticipated to be progressively addressed that can potentially lead to enhanced development, improved adoption, and the ultimate experience of XR systems.

11.4.6 Power and Thermal Efficiency

Since XR applications and gadgets are expected to process high-quality images from camera in real-time and require intensive graphic capabilities to display contents such as generated images and animations for lengthy durations, huge battery life is a key requirement, with the need for rapid revival equivalent to more than 100 times than typical charging in some contexts (Patil & Kumar, 2019). The power requirements also imply the need for effective cooling. As key opportunities and solutions to address this issue, manufacturers should work to improve battery limit and cooling strategies. In addition to content optimization and rendering, AI techniques can be helpful to effectively analyze battery utilization in order to better apply power-saving techniques and cooling strategies.

11.4.7 Challenges Related to the Creation of Contents

As highlighted earlier, the duration of use of XR systems is considered to be shorter than expected due to poor quality of new and updated contents (Jha et al., 2019). Existing 3D content creation software is considered to be slow and complex, whereby further adding to the challenges related to content creation (Mindy Support, 2021). Similarly, in the current VR market, there are limited UHD video sources owing to the lack of video acquisition equipment and the lengthy video shooting process (Hu et al., 2021). Another challenge is that computer-generated contents tend to look fake due to invariant coloring that do not completely look like objects in real-life (Patil & Kumar, 2019). In order to address such challenges, AI and computer vision can potentially be used to compare and apply illumination algorithms in order to dynamically generate, render, and overlay more realistic virtual contents.

Furthermore, a potential solution to address the challenges related to contents creation in the 5G and 6G eras involve crowdsourced content generation. Earlier, such approach was limited by bandwidth, storage, and processing capabilities, which are expected to be facilitated with the key benefits brought by the new generations of cellular network. More content-sharing platforms are expected to emerge where on one end, developers can post details on required content for different kinds of XR systems, and on the other end, content creators globally can attempt to fulfill demands. Such platforms also provide opportunities for new business models to emerge related to content creation and sharing.

11.4.8 Lack of Skills and Competencies

Since this group of technologies is yet to completely mature and because of a lack of suitable authoring tools, development of commercial XR applications is often regarded as a long and non-intuitive task that involves experts in the domain (Jee et al., 2014). The creation of XR solutions as well as their contents necessitates different kinds of skills and proficiency with different kinds of hardware and software. Since these technologies are still developing and new kinds of hardware (including visual interfaces and equipment) and software (such as authoring tools, libraries, content creation software, etc.) regularly emerge on the market, developers of such solutions need to effectively update themselves and build key competencies. As such, it also means that developers competent in one platform may not be familiar to various other platforms, and such lack of skills often pose challenges to companies requiring these skills for the development of new products. In order to address such demand in competencies, a growing number of courses are being provided on online learning platforms and are also integrated within curriculum of higher education programs at both undergraduate and postgraduate levels. Course contents also include the development of AI-assisted XR systems using cellular networking technologies whereby focusing on teaching the use of 3D modeling tools and 3D tools for creation tool of photoreal visuals and immersive experiences, among others.

11.4.9 CHALLENGES RELATED TO APPLICATION OF AI

Training data is regarded as the core of machine learning systems since without sufficient data, it is challenging to accurately build the model in order to effectively solve problems. Even though various communities have made important contributions on open datasets, it is still considered as not enough for effectively mining key features and for accurately training models in relation to different contexts pertaining to XR systems (Hu et al., 2021). In addition, data sets potentially have various limitations such as missing data within datasets, lack of coverage, domain specificity, and labeling issues, among others (Davahli et al., 2021). These limitations have been actively researched during previous years and techniques such as data augmentation and synthetic data have commonly been used, as these techniques are expected to further transform AI toward extending applications in various domains (Toews, 2022). Moreover, with advances in various technologies discussed in this chapter, there are opportunities for more datasets to be produced and made available publicly, to the benefit of the wider AI community.

Due to the challenges to train AI using real-world data, learning environments are often used as substitute for training models using experimental data (Reiners et al., 2021). For instance, training AI for urban self-driving vehicles in the urban area necessitates significant logistical arrangements and resources, among others. A safer and cost-efficient solution is to conduct the training in a learning environment. However, following the implementation of the solution, a key challenge involves transferring results from the learning environment to the real world. This is because various aspects can be difficult to capture precisely and respond to in the real world (Amini et al., 2020). A combination of XR, AI, and cellular networks is expected to provide opportunities for further developments in domain adaptation so that results transfer from the learning environments to the real world are more effective.

11.4.10 CELLULAR DATA NETWORK CHALLENGES AND DEPLOYMENT ISSUES

The cellular data networks (5G and 6G) have their own limitations, which can potentially adversely impact the development and deployment of XR solutions. For instance, key 5G issues that need further attention include deployment in dense heterogeneous networks, multiple access techniques as well as full-duplex transmission (Li et al., 2018). Deployment of 5G has been done progressively as significant investments is required in the process to install the required infrastructure (e.g., deploying 1,000 of new cell sites) as well as to upgrade required software, thus slowing down the process in some countries. Also, existing hardware used in XR has to be compliant with 5G to be able to use the network technology. Similar challenges are expected in the early stages of the 6G era that can potentially delay the integration of the networking technology within XR solutions. Nevertheless, upgrading from one generation of cellular network technology is a familiar process for telecommunication companies and Internet service providers often came up with effective deployment strategies driven by cost-benefit analyses to the advantage of consumers of such networks.

11.4.11 PRIVACY AND SECURITY ISSUES

With enhanced connectivity, XR systems can potentially create, collect, and analyze voluminous data for the purposes of deriving intelligence, and this could adversely lead to privacy and security concerns (Hu et al., 2021). The risks are even higher with complex XR applications that necessitate always-on sensing and camera, which are also connected to the cloud. As an example, a malicious AR software could illegitimately leak user's data including location, images, and video feed to its servers. This has been implemented in the visual malware named PlaceRaider that produces 3D models of indoor environments of users by utilizing data from smartphone sensors collected in a stealth manner (Templeman et al., 2012). Moreover, AR applications that require always-on cameras and sensory data such as location can create privacy issues for bystanders since such risks have been considered to hamper widespread adoption of AR (Van Krevelen & Poelman, 2010).

In order to address privacy and security risks, designers of XR systems must consider security and privacy issues right from the design phase of such systems whereby considering the granularity of permissions and security of data being transmitted over networks. This can be facilitated by the adoption of privacy and security preserving frameworks where different risks are individually treated, with appropriate mechanisms implemented toward securing XR systems. For instance, such frameworks can include mechanisms to anonymize or blur bystanders in real-time within applications that require always-on camera through the use of AI techniques.

11.5 CONCLUSIONS

AI-assisted XR technologies are expected to advance massively in the 5G and 6G eras, thereby creating multitude innovations and opportunities across areas such as medial training, gaming, industrial settings, and even the social media through the metaverse. However, the absolute XR experience is challenging to be achieved in the 5G era because certain features require high-quality video and 3D animations to be generated and displayed in real-time. These features require more than 10 Gbps data rate and are only attainable after the 6G era. As such, the integration of AI and cellular data networks to XR technologies is intended to be progressive toward the ultimate XR experience. AI and the new generations of cellular data networks have the potential to massively advance XR technologies, notably, AR, VR, and MR in order to create opportunities during different stages involved within the creation of AI-assisted XR solutions including camera calibration, content creation, and rendering. These various opportunities have been discussed in this chapter and can be used as reference for advancing R&D in relation to AI-assisted XR technologies in the 5G/6G eras. Nevertheless, advancing toward the truly exciting XR visualization experience is not going to be without challenges, which are also expected to grow as the technologies are advancing. Some of the key challenges discussed in this chapter include issues with visual interfaces, adoption, application of AI, processing and resources, privacy, and security, among others. It is important to investigate and address key challenges in a timely manner in order to prepare toward the absolute XR experience envisioned by the world.

REFERENCES

Amini, A. et al., 2020. Learning robust control policies for end-to-end autonomous driving from data-driven simulation. *IEEE Robotics and Automation Letters*, 5(2), pp. 1143–1150.

Avasant, 2022. *Virtual and augmented reality show great promise despite slow mainstream adoption* [Online]. Available at: https://avasant.com/report/virtual-and-augmented-reality-show-great-promise-despite-slow-mainstream-adoption/ [Accessed 20 Jun. 2022].

Bastug, E., Bennis, M., Medard, M., & Debbah, M., 2017. Toward interconnected virtual reality: Opportunities, challenges, and enablers. *IEEE Communications Magazine*, 55(6), pp. 110–117.

Bekaroo, G. et al., 2018. Enhancing awareness on green consumption of electronic devices: The application of augmented reality. *Sustainable Energy Technologies and Assessments*, 30, pp. 279–291.

Bogdan, O., Eckstein, V., Rameau, F., & Bazin, J., 2018. DeepCalib: A deep learning approach for automatic intrinsic calibration of wide field-of-view cameras. In *Proceedings of the 15th ACM SIGGRAPH European Conference on Visual Media Production*, ACM, pp. 1–10.

Cheng, Q. et al., 2020. Augmented reality dynamic image recognition technology based on deep learning algorithm. *IEEE Access*, 8, pp. 137370–137384.

Chowdhury, M., Shahjalal, M., Ahmed, S., & Jang, Y., 2020. 6G wireless communication systems: Applications, requirements, technologies, challenges, and research directions. *IEEE Open Journal of the Communications Society*, 1, pp. 957–975.

Çöltekin, A. et al., 2020. Extended reality in spatial sciences: A review of research challenges and future directions. *ISPRS International Journal of Geo-Information*, 9(7), p. 439.

Davahli, M. et al., 2021. Controlling safety of artificial intelligence-based systems in healthcare. *Symmetry*, 13(1), p. 102.

Doppler, K., Torkildson, E., & Bouwen, J., 2017. On wireless networks for the era of mixed reality. In *Proceedings of the 2017 European Conference on Networks and Communications (EuCNC)*, Oulu, Finland, 12–15 Jun., IEEE, pp. 1–5.

Gordon, A., Li, H., Jonschkowski, R., & Angelova, A., 2019. Depth from videos in the wild: Unsupervised monocular depth learning from unknown cameras. In *Proceedings of the IEEE International Conference on Computer Vision*, Seoul, Korea, 27 Oct.–2 Nov., pp. 8977–8986.

Gruber, L., Richter-Trummer, T., & Schmalstieg, D., 2012. Real-time photometric registration from arbitrary geometry. In *2012 IEEE International Symposium on Mixed and Augmented Reality (ISMAR)*, IEEE, pp. 119–128.

Hu, M. et al., 2021. Virtual reality: A survey of enabling technologies and its applications in IoT. *Journal of Network and Computer Applications*, 178, p. 102970.

Hughes, C., Stapleton, C., Hughes, D., & Smith, E., 2005. Mixed reality in education, entertainment, and training. *IEEE Computer Graphics and Applications*, 25(6), pp. 24–30.

Jee, H., Lim, S., Youn, J., & Lee, J., 2014. An augmented reality-based authoring tool for E-learning applications. *Multimedia Tools and Applications*, 68(2), pp. 225–235.

Jha, G., Singh, P., & Sharma, L., 2019. Recent advancements of augmented reality in real time applications. *International Journal of Recent Technology and Engineering*, 8(2S7), pp. 538–542.

Jordan, M. & Mitchell, T., 2015. Machine learning: Trends, perspectives, and prospects. *Science*, 349(6245), pp. 255–260.

Kendall, A., Grimes, M., & Cipolla, R., 2015. Posenet: A convolutional network for real-time 6-DOF camera relocalization. In *Proceedings of the IEEE International Conference on Computer Vision (ICCV)*, Santiago, Chile, pp. 2938–2946.

Kourouthanassis, P., Boletsis, C., & Lekakos, G., 2015. Demystifying the design of mobile augmented reality applications. *Multimedia Tools and Applications*, 74(3), pp. 1045–1066.

Li, M. et al., 2020. *A Virtual Reality Simulation System for Coal Safety Based on Cloud Rendering and AI Technology*. Cham, Springer, pp. 497–508.

Li, S., Da Xu, L., & Zhao, S., 2018. 5G Internet of Things: A survey. *Journal of Industrial Information Integration*, 10, pp. 1–9.

LLP, M.I., 2021. *Virtual reality (VR) market – Growth, trends, COVID-19 impact, and forecasts (2021–2026)*. ReportLinker.

Markets and Markets, 2021. *Augmented reality market with COVID-19 impact analysis, by device type, offering, application, technology and geography – Global forecast to 2026*. ReportLinker.

Mindy Support, 2021. [Online] Available at: https://mindy-support.com/news-post/using-virtual-reality-and-machine-learning-to-create-3d-content/ [Accessed 10 Dec. 2021].

Mine, M., 1995. *Virtual Environment Interaction Techniques*. University of North Carolina.

Minopoulos, G. & Psannis, K., 2022. Opportunities and challenges of tangible XR applications for 5G networks and beyond. *IEEE Consumer Electronics Magazine*. (Early access).

Muhammad, U. et al., 2018. Learning deep sketch abstraction. In *2018 IEEE/CVF Conference on Computer Vision and Pattern Recognition (CVPR)*, pp. 8014–8023.

Mumtaz, S. et al., 2017. Terahertz communication for vehicular networks. *IEEE Transactions on Vehicular Technology*, 66(7), pp. 5617–5625.

Olshannikova, E., Ometov, A., Koucheryavy, Y., & Olsson, T., 2015. Visualizing Big Data with augmented and virtual reality: Challenges and research agenda. *Journal of Big Data*, 2(1), pp. 1–27.

Orlosky, J., Kiyokawa, K., & Takemura, H., 2017. Virtual and augmented reality on the 5G highway. *Journal of Information Processing*, 25, pp. 133–141.

Pascoal, R. & Guerreiro, S., 2017. Information overload in augmented reality: The outdoor sports environments. In *Information and Communication Overload in the Digital Age*, IGI Global, pp. 271–301.

Patil, S. & Kumar, R., 2019. Accelerating extended reality vision with 5G networks. In *2019 3rd International Conference on Electronics, Communication and Aerospace Technology (ICECA)*, pp. 157–161.

Pessaux, P. et al., 2015. Towards cybernetic surgery: Robotic and augmented reality-assisted liver segmentectomy. *Langenbeck's Archives of Surgery*, 400(3), pp. 381–385.

Pettersson, J. & Falkman, P., 2020. Human movement direction classification using virtual reality and eye tracking. *Procedia Manufacturing*, 51, pp. 95–102.

Qiao, X., Ren, P., Dustdar, S., & Chen, J., 2018. A new era for web AR with mobile edge computing. *IEEE Internet Computing*, 22(4), pp. 46–55.

Reiners, D., Davahli, M., Karwowski, W., & Cruz-Neira, C., 2021. The combination of artificial intelligence and extended reality: A systematic review. *Frontiers in Virtual Reality*, 2, p. 114.

Sahu, C., Young, C., & Rai, R., 2021. Artificial intelligence (AI) in augmented reality (AR)-assisted manufacturing applications: A review. *International Journal of Production Research*, 59(16), pp. 4903–4959.

Teich, D., 2020. *Technology comes together: Artificial intelligence, augmented reality, and 5G combine to aid surgeons* [Online]. Available at: https://www.forbes.com/sites/davidteich/2020/08/20/technology-comes-together-artificial-intelligence-augmented-reality-and-5g-combine-to-aid-surgeons/?sh=71cb3c4c67ae [Accessed 7 Jan. 2022].

Templeman, R., Rahman, Z., Crandall, D., & Kapadia, A., 2012. PlaceRaider: Virtual theft in physical spaces with smartphones. *arXiv:1209.5982*. [Preprint].

Toews, R., 2022. *Synthetic data is about to transform artificial intelligence* [Online]. Available at: https://www.forbes.com/sites/robtoews/2022/06/12/synthetic-data-is-about-to-transform-artificial-intelligence/?sh=ac6095575238 [Accessed 17 Jun. 2022].

Van Krevelen, D. & Poelman, R., 2010. A survey of augmented reality technologies, applications and limitations. *International Journal of Virtual Reality*, 9(2), pp. 1–20.

Velev, D. & Zlateva, P., 2017. Virtual reality challenges in education and training. *International Journal of Learning and Teaching*, 3(1), pp. 33–37.

Wang, X., 2009. Augmented reality in architecture and design: Potentials and challenges for application. *International Journal of Architectural Computing*, 7(2), pp. 309–326.

Wang, Y., Kung, L., & Byrd, T., 2018. Big Data analytics: Understanding its capabilities and potential benefits for healthcare organizations. *Technological Forecasting and Social Change*, 126, pp. 3–13.

Yang, H. et al., 2020. Artificial-intelligence-enabled intelligent 6G networks. *IEEE Network*, 34(6), pp. 272–280.

Zhang, T., 2021. Application of AI-based real-time gesture recognition and embedded system in the design of English major teaching. *Wireless Networks*, 6, pp. 2693–2699.

12 An Integrated 5G-IoT Architecture in Smart Grid Wide-Area Monitoring, Protection, and Control
Requirements, Opportunities, and Challenges

Vladimir Terzija
University of Manchester

CONTENTS

12.1 Introduction 425
12.2 Requirements 426
12.3 5G-IoT Standardization and Interoperability 428
12.4 Use Cases 432
    12.4.1 SmartZone 433
    12.4.2 VISOR 434
    12.4.3 EFCC and MIGRATE 436
12.5 Integration of Advanced IoT Architectures 437
12.6 Concluding Remarks 441
References 442

12.1 INTRODUCTION

In recent years, many novel smart grid (SG) system architectures have been proposed and published on integration of high-speed bidirectional data communications network in the power grid. On the other hand, the research on Internet of things (IoT) systems is in its infancy stage. It was found that SG places real and diverse challenges on new 5G network technologies. On the one hand, power grids represent one of key critical

DOI: 10.1201/9781003205494-15

infrastructures of each country, and its planning and operation must be done very carefully, considering a number of aspects, e.g., costs, reliability, efficacy, and quality of the service. To enable optimal operation of power grids, particularly optimal integration of renewable energy sources, implementation of novel sensor and communication infrastructure is a must. They are enablers of novel applications supporting reduction of CO_2 emission, but also optimal utilization of all available energy resources, also those going beyond pure power grids, e.g., gas, or heat networks. Integration of different energy networks is leading to integrated energy networks, possessing higher flexibility, as well as resilience. New technologies must have interoperability and compatibility with existing technologies. The existing infrastructure of the power network should be utilized optimally and modifications should be compatible with the 5G standards.

The recent generation of 5G mobile networks supports the integration of highly reliable and low-latency wide-area monitoring, protection, and control (WAMPAC) laying a solid foundation for the realization of integrated systems [1–3]. The SG requirements of real-time interconnection and two-way interaction determine the development of electric power communications network in the direction of massive connections, high reliability, and low latency [4–6]. The key infrastructure supporting the development of SGs is the information and communications technology (ICT) [7]. The 5G-IoT technology is the main development direction of the next generation of machine-type (MTC) wireless communications systems. The first year of 5G deployment was 2019, while the second phase, standardization of technical specifications and system architecture, was finalized in 2020.

We are very keen to provide a detailed discussion on needs, opportunities, and challenges related to the application of 5G and IoT technologies in modern power systems, power systems with high penetration of renewable energy resources, and systems with significantly changed dynamics and expectations. Here, e.g., aspects related to changed system inertia and fault level will be addressed. These two are directly related to power system stability in general terms, more specifically frequency and voltage stability. At the same time, the importance of the application of novel sensor and communication technologies will be discussed, from the perspective of opportunities and challenges. Examples of the following high-impact practical projects, in which the first author of the chapter acted as PI on behalf of academia, will be presented: VISOR, EFCC, Supergen Multi-Energy Networks Hub, and MIGRATE projects.

This chapter is organized as follows. In the first part, we will investigate requirements and symbiotic relationship between two of the modes in 5G, namely massive machine-type communications (mMTC) and ultrareliable low-latency communications (uRLLC) and the SG communication, which are essential to apply IoT. In the second part, we emphasize unification of standardization efforts to avoid deployment issues such as interoperability. In the third part, we present high-impact practical projects and 5G-IoT integration issues in future SG application scenarios.

12.2 REQUIREMENTS

A smart grid SG is a new generation of electric power systems that integrate advanced sensing, information, control, and energy storage technology, enabling the reliable, economical, and efficient operation of electrical grids. SG includes five major stages:

electricity generation, transmission, transformation, distribution, and consumption. The trend of SGs research and development is clean electricity generation, safe and efficient electricity transmission and transformation, flexible and reliable electricity distribution, and diverse and interactive electricity consumption. In the electricity generation stage, renewable energy has gradually replaced traditional fossil energy, through transformation from centralized energy supply to cooperation with distribution, and interactivity of supply and demand. The traditional single energy network is transforming into a multi-energy complementary system, where electrical grids and ICT are deeply integrated into various types of equipment in multiple nodes and wide coverage. Therefore, massive terminals need to be monitored and controlled in real-time two-way communication.

The SG usage of scenarios of wireless communications is collection and control. The collection-type scenarios mainly include advanced metering, quality assurance, and the application of videos in smart grids, whereas the control-type scenarios include the automation of distributed intelligent electricity distribution, demand-side response to electrical load, and distributed energy regulation. Recent developments in SG applications were enabled by advanced smart metering and measurement devices. The phasor measurement unit (PMU) plays an essential role in monitoring long transmission lines, supporting different applications contributing to enhanced system stability, or prevention of cascading events, or potential blackouts. PMUs are slowly becoming a part of modern control systems in distribution networks. Intelligent electronic devices (IED) are used to detect faults, protection relaying, event recording, measurement, control, and automation aims in power network. A smart meter (SM) records near real time and report regularly information such as consumption of electric energy, voltage levels, current, and power factor.

The core of a smart grid is real-time situational awareness, traditionally considered as real-time monitoring. The inspection of important corridors at all levels of the grid introduces new visualized and real-time operation methods. State estimation (SE) aims to provide an estimate of the system state variables (voltage magnitude and angles) at all the buses of the grid from a set of remotely acquired measurements. The centralized (classical and SCADA-supported) SE schemes may prove inapplicable to emerging decentralized and dynamic power grids, due to large communication delays and high computational complexity that compromise their ability for real-time operation. Just-in-time monitoring the state of power grid is supported by integration of widely deployed PMUs with communication and advanced computations. Accuracy of PMU measurements for phase angle, frequency, and voltage is approximately 0.05° and 0.5% for voltage angle and magnitude for most measurements. A PMU should be able to reliably measure frequency up to 0.001 Hz most of the time. PMUs may require calibration adjustments from time to time to achieve these measurements. Dynamically changing conditions of a power system or nearing the limits of a PMU's measurement range will likely degrade measurements to the limits specified in the standard. Experience of authors on SE will include recent publications [8–11].

At present, the PMU services are mainly applied to the backbone network and supported by optical fiber communications, but the deployment of optical fibers in the distribution network is costly. Through 5G networks, the current and voltage phases in the electrical distribution can be fed back to the control center in time. Network needs of distribution PMU scenario are 2 Mbps bandwidth and 3–20 ms delay, reliability 99.999%, and connection density (10–2,000)/km^2. A terminal needs timing accuracy 10 μs, available through 5G network.

New 5G requirements of achieving low latency and high reliability for many IoT use cases are very important, besides the enhanced mobile broadband (eMBB). In the context of new use cases, IoT applications have been categorized into two classes: massive machine-type communications (mMTC) and ultrareliable low-latency communications (uRLLC). The former consists of a large number of low-cost devices with high requirements on scalability and increased battery lifetime. In contrast, uRLLC requirements relate to mission-critical applications, where uninterrupted and robust exchange of data is important. Both 3GPP (Third Generation Partnership Project) and the ITU (International Telecommunication Union) have defined requirements for mMTC. Technical specifications LTE-M and NB-IoT are candidates for fulfilling these requirements and thus to be considered as 5G technologies. NB-IoT are oriented to cover the MTC scenario with extended coverage, support of massive number of low throughput devices, ultralow device cost, low device power consumption, and optimized network architecture.

12.3 5G-IoT STANDARDIZATION AND INTEROPERABILITY

Since its conception, 5G is anticipated to become a major IoT enabler. The robust 5G infrastructure will act as the backbone to which different types of 5G devices will be connected, forming a massive 5G IoT network. The implementation of such a network is complex as the latter must be compliant with the standards and specifications developed by 5G standardization bodies. One such major 5G standardization body is the 3GPP, which is also known for the standardization of previous generations of mobile networks. An appropriate solution to implement an integrated IoT system is to utilize 5G machine-type connectivity and to incorporate an interoperable data model [12–16].

The 5G requirements defined by ITU-R broadly cover three main service classes:

- uRLLC, which includes critical IoT communications for mission-critical applications and industrial IoT (IIoT).
- mMTC, which can support a high density of connections, provide a wide network coverage, and enable long device operational lifetimes.
- eMBB, which improves significantly access to high-quality multimedia content, enabling services such as remote video inspection and video conferencing.

All of the above service classes are covered by 3GPP in the final 5G specifications.

The requirements for IoT usage scenarios are extremely diverse. To address these requirements, the following three connectivity segments, which are allowed to coexist in a single 5G network are defined:

- Critical IoT connectivity. This segment is defined for time-critical communications, which have strict latency requirements. The segment must be supported by the 5G uRLLC technology and shall flexibly adapt to various data rate requirements. Reliability in this context can be further defined as the rate of successful and timely data transmission. Moreover, in this segment data exchange must not be congested even in heavily loaded networks.
- Massive IoT connectivity. This segment is utilized to allow a large number of low-cost, narrow-bandwidth devices to communicate infrequently using small volumes of data. Since 2017, NB-IoT and LTE-M have been co-existing with LTE in 4G networks and they already fulfill all of the mMTC requirement defined for 5G by ITU and 3GPP. The NB-IoT standalone radio access technology can support high system capacity and has high spectrum efficiency while LTE-M can provide extended coverage that may be accessed by low-complexity devices. There is also already a panoply of devices that are supported by the NB-IoT and LTE-M network on the market.
- Broadband IoT connectivity. This segment aims at providing high data throughputs and low latencies to enable IoT capabilities such as extended coverage, high uplink throughputs, and high-precision device positioning. Typical service scenarios that are supported by this segment are remote inspection using drones and voice communications (manual maintenance and inspection).

The critical IoT use cases require latency and reliability enhancements of networks, devices, and applications for their proper implementation. The end-to-end latency in a network is the sum of the latencies contributed by the radio, transport and core networks, and the overall reliability of the network is largely dependent on the reliability of the weakest component in the network. However, the reliability and the latency of a network can be enhanced by the allocation of more spectrum resources.

The 5G uRLLC is the most appropriate wireless communications technology for supporting critical IoT. The 5G New Radio (NR) can provide low communication latencies below 1 ms and reliability of up to 99.9999%. The latency contributed by 5G's core network is insignificant compared to that contributed by 5G's transport network. The transport network latency is dependent on transmission distances and the channel and therefore varies widely between regions. Latency in the transport layer is usually optimized by switching copper connections to fiber ones and by decreasing the number of router hops. For example, the round-trip time between two cities 1.3 km apart in a European country has been halved by incorporating fiber connections. The latency is today just 16 ms, which is near the 13 ms theoretical optical fiber latency. Other technologies that may reduce the transport network latency to implement critical IoT use cases are edge computing, distributed anchor points, local break-out, and on-premise full-core deployments. On-premise full-core deployment dedicates network resources that allow local data to be managed autonomously locally, hence satisfying ultralow latency and ultrareliability requirements. Furthermore, roaming for mobile IoT devices can be enabled by connecting the local area network to a public network (Figure 12.1).

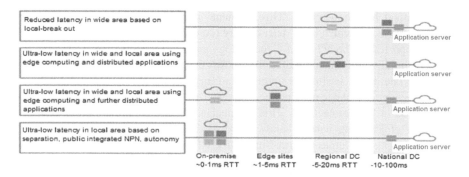

FIGURE 12.1 Core network deployment examples to support critical IoT use cases.

The Service and System Aspects Technical Specification Group (TSG-SA) is actively developing the overall architecture and service capabilities of 5G-IoT systems that are based on 3GPP specifications. The 3GPP technical standards are published in the form of releases. URLLC specifications for 5G NR and 5G-Core (5GC) were introduced in Release 15 and evolved further in Release 16 and Release 17.

To promote and accelerate the standardization process of 5G, ITU released a global schedule for 5G standardization works in 2015, which the 3GPP uses as a reference for delivering the 5G standards. Furthermore, the 3GPP discussed 5G applications, requirements, and key technologies, and formulated a work plan for 5G standardization in September 2015 during the 5G workshop held in Phoenix, United States. Subsequently, in the 3GPP Release 14 issued in February 2016, 3GPP published research works on the targets, requirements, and technical solutions for 5G. The 3GPP also issued a research report on 5G in December 2016. In December 2017, the Radio Access Network (RAN) working group released standards for 5G new air interface, which will operate in non-standalone (NSA) network mode and the working group on business and SA (Stand-Alone) released 5G's core network architecture and 5G's process standard. At the 80th plenary meeting of 3GPP held in June 2018, 5G's independent networking standard was officially frozen and released by the RAN working group. During that same meeting, the Core Network and Terminals (CT) working group released Release 15's new core network design standards for 5G independent networking. These events marked the completion of the standardization for 5G SA networks, which are expected to satisfy the requirements from the vertical industries that will bring unprecedented opportunities to carriers and collaborating industries. However, there were still some challenges, which 5G from Release 15 needs to face to completely satisfy the throughput, latency, and reliability requirements of certain vertical industries. Hence, 3GPP organized more than 70 research projects, which focused on the standardization of the uRLLC and mMTC in the final phase of Release 16, which was completed in June 2020.

The 3GPP technologies are considered to be suitable for IoT applications due to their almost worldwide availability and their use of licensed spectrum.

5G-IoT Architecture in Next-Generation Smart Grid 431

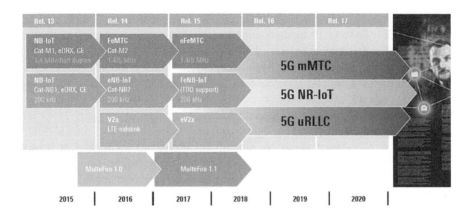

FIGURE 12.2 Timeline of 3GPP standardization process.

In 3GPP Release 13, enhanced MTC (eMTC) and narrowband IoT (NB-IoT) were introduced and these two distinct standards were further enhanced in subsequent releases. The NB-IoT and LTE-M 3GPP standards are known to satisfy the long-term 5G low power wide-area networks (LPWAN) requirements. The 3GPP technology is also investigating methods to use LTE-M and NB-IoT RANs with the 5G core network (Figure 12.2). The development of 6G technology is expected to enable new use cases and applications, and to address the rapid growth of traffic, for which contiguous and wider channel bandwidth is desirable than currently available for mobile systems. As the amount of spectrum required for mobile services increases, it becomes increasingly desirable to harmonize existing and newly allocated and identified spectrum. A continuous and wider spectrum from a range of frequency bands, aligned with future technology development, would support achieving the objectives of the 6G systems, reducing device complexity, avoiding interference, and developing ecosystems.

- NB-IoT is a fast-growing 3GPP mMTC standard that was introduced in Release 13. The latter addresses the LPWAN requirements optimized for the low-end massive IoT. It was standardized and classified as a 5G technology by 3GPP in 2016, and it will continue evolving with the 5G specification.
- LTE-M LPWAN technology provides coverage of up to 11 km in remote regions with a bandwidth of up to 20 MHz. The technology's objectives were introduced in Release 12 and are as follows: low device price, long battery life, wide radio coverage, and variable throughputs ranging from 10 kbps to 1 Mbps depending on coverage requirements. In-band deployment of LTE-M is supported and this allows it to coexist with other services and existing base stations may be upgraded to cater for its support.

To meet the requirements for 5G mMTC and the emergence of new MTC applications, the 3GPP Release 15 focused on the optimization of even further enhanced MTC (eFeMTC) and further enhanced NB-IoT (FeNB-IoT).

The success of 5G-IoT depends fundamentally on having robust interoperability characteristics. Challenges encountered in IoT are constantly being addressed by several academic and industry proposals, and the development of a panoply of devices and IoT platforms has substantially been made in the last few years. However, interoperability issues crop up by the infrastructure, devices, and data format differences between the different solutions. Some IoT device suppliers implement vendor lock-in to restrict customers from choosing products from different suppliers. IoT device suppliers also limit the use of their devices to their proprietary IoT platforms, which lead to more interoperability issues. Such issues are caused by the lack of standards and have become a burden in large-scale IoT deployments. The 5G-based IoT solutions will be capable of addressing interoperability issues by offering a scalable architecture through the use of software-defined networking (SDN) and network functions virtualization (NFV), which will be flexible enough to cater for the needs of different use cases.

The increasing amount of attention that IoT has received is causing significant research to be conducted to improve IoT by addressing interoperability issues. IoT standardization, frameworks, and novel applications are reviewed in [17]. Various aspects including devices, semantics, networks, platforms, and interoperability have been surveyed [18]. The following EU-funded Horizon programs addressing IoT interoperability issues were organized among others.

- SymbIoTe (Symbiosis of smart objects across IoT environments) – For designing an interoperability framework for different IoT platforms.
- BIG IoT (Bridging the Interoperability Gap of the IoT) – For addressing interoperability issues using a unified web application programming interface (API).
- VICINITY – For providing a platform for supporting interoperability as a service across IoT ecosystems used around the world.
- INTER-IoT – For supporting voluntary interoperability across heterogeneous IoT platforms.

However, more research covering the energy and data flow aspects need to be performed [19].

12.4 USE CASES

Energy networks exist primarily to exploit and facilitate temporal and spatial diversity in energy production and use. As technology changes the planning, design, and operation of energy networks need to be revisited and optimized. Current energy networks research does not fully embrace a whole systems approach. The energy networks community would strongly benefit from a more diverse, open, supportive community such as computing science, statistics, and applied mathematics to help implement a whole systems approach. Application of data analytics and artificial intelligence are on its way to support monitoring, protection, and control of future electricity networks.

12.4.1 SMARTZONE

SmartZone, a geographic entity located in the area of the city of York, is a typical example of a wide-area monitoring in a segment of the Great Britain (GB) network, which includes nonsynchronous renewable energy resources. A number of operational problems related to, e.g., frequency control, are introduced. The developed SmartZone app monitors the capacity transfer (of active and reactive powers) over the existing transmission network. For this purpose, PMUs were installed in key power system substations at the 400 kV voltage level. These PMUs provided information about voltage and current phasors to the central data concentrator, at which information obtained was processed. First, the system state was estimated, and after that, the capacity margin, taking into account the weather conditions and information received from the Meteorological Office, has been calculated. The entire process is visualized and provided available as Web app (Figure 12.3) [1]. Adequate situational awareness was used for controlling a so-called operational tripping scheme (OTS), responsible for intentional disconnection of generation resources in cases in which the capacity margin is not maintained at the acceptable level.

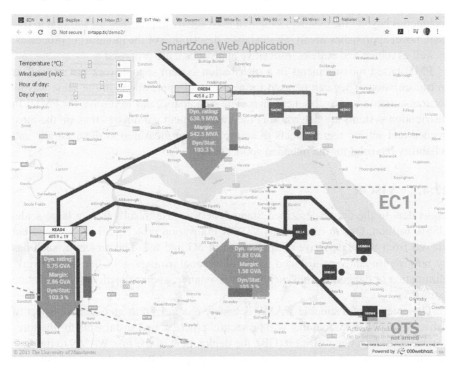

FIGURE 12.3 SmartZone web app https://svtapp.000webhostapp.com/demo2/.

Recent studies have explored algorithms for optimal placement of PMUs to minimize the number of devices required to collect sufficient information. PMU-based wide-area monitoring systems (WAMSs) use the global position system (GPS) to synchronize PMU measurements. Such synchronized measurements allow two quantities to be compared in the real-time analysis of grid conditions. Through wide-area monitoring and synchronization, PMUs have made great strides in power system stability, which was often hindered by SCADA's slow state updates. The implementation of synchrophasors has also allowed voltage and current data from diverse locations to be accurately time-stamped in order to assess system conditions in real time. Synchrophasors are also available in protection devices, but since requirements for protection devices are fairly restrictive, the full integration of synchrophasors into line protection is still debated. The increasing application of synchrophasors in wide-area monitoring, protection and control systems, post-disturbance analyses, and system model validation has made these measurement tools invaluable.

The need for situational awareness also motivated the development of sensor networks. Phasor measurements provide a dynamic perspective of the grid's operations because their faster sampling rates help capture dynamic system behavior. PMUs measure voltage and current phasors and can calculate watts, vars, frequency, and phase angles 120 times per power line cycle. PMU data immediately enhance topology error correction, SE for robustness and accuracy, faster solution convergence, and enhanced observability. Hence, one interest of the community is shifting from centralized to distributed SE algorithms based on more sophisticated optimization techniques beyond the classical weighted least square approaches. In addition to this goal, novel approaches involving stochastic description of the monitoring processes, as well as robust and outliers insensitive approaches, are proposed.

PMU-based measurements in a distribution system should work just as well as in transmission. They can be used for a variety of applications in the same way as for transmission or possibly new ones. The principal issues to resolve for PMUs in distribution systems are determining the measurement parameters that are the most needed and setting requirements to assure they are measured at the accuracy and reliability required for the intended applications.

12.4.2 VISOR

Equivalent to the above-described SmartZone use case, results from a large-scale, US$10 million, Ofgem-funded [15] VISOR innovation project [16] is presented but viewed from the perspective of application of 5G and IoT. The project as such has by now significantly contributed to the situational awareness of the GB network, elimination of dangerous subsynchronous resonance, and application of hybrid state estimator.

The creation of a suitable WAMS is widely recognized as an essential aspect of delivering a power system that will be secure, efficient, and sustainable for the foreseeable future. In Great Britain (GB), the deployment of the first WAMS to monitor the entire power system in real time was the responsibility of the visualization of real-time system dynamics using enhanced monitoring (VISOR) project. The core scope of the VISOR project is to deploy this WAMS and demonstrate how WAMS

5G-IoT Architecture in Next-Generation Smart Grid 435

applications can in the near term provide system operators and planners with clear, actionable information. In Figure 12.4, the architecture of the VISOR-WAMPAC system is presented [17].

Measurements of voltage and current phasors are recorded by PMUs installed across a wide-area power system and time tagged at the point of measurement using

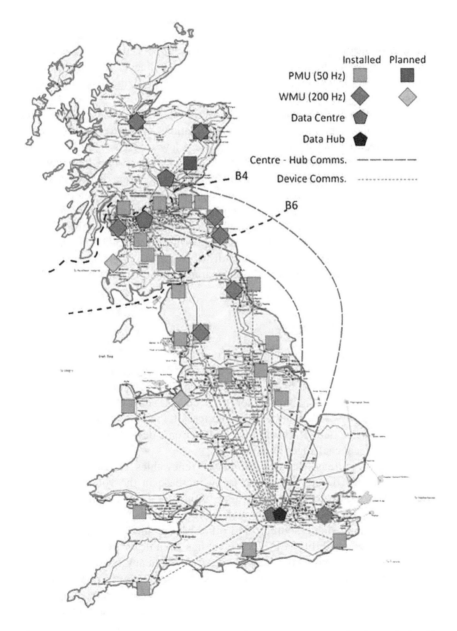

FIGURE 12.4 VISOR-WAMPAC architecture for power system monitoring and visualization of real-time dynamics.

a common time reference. Synchronizing the angle measurements to a common time reference allows them to be combined into a single data record in real time that represents a snapshot of the system at that time, which with existing technology is updated at a rate of up to once per cycle. With suitable supporting communication and computing resources, a sequence of these snapshots can be used to visualize system dynamics in real time. However, these synchronized snapshots can be used for far more than visualization of dynamics. Advances in the computing resources available to power system engineers have enabled the development of a wide range of new algorithms that process WAMS data online to support the operation of power systems. Examples of these online WAMS applications include the real-time estimation of oscillation parameters (e.g., interarea oscillations), the dynamic rating of transmission lines, and hybrid and linear SE.

Many WAMS applications have been deployed in power systems and many more proposed. The application for monitoring of oscillations and deliver real-time monitoring, visualization, and alarming in range: very low frequency (VLF), low frequency (LF), and subsynchronous. The application for model-based simulation of dynamic behavior is critical to the proper planning and operation of a power system. It is used for both steady-state and post-fault contingency analyses to determine if the system is operating within security margins and quality of supply standards. Model inadequacies can thus have real and significant consequences for the power system. The focus of the hybrid SE application is on improving the reliability of convergence in post-processing, integration, fusion, and distributed HSE. Application correlation based for improving the accuracy of the line parameters of power system could have real benefits in stability assessment.

12.4.3 EFCC AND MIGRATE

The authors are ambitious to demonstrate opportunities of implementing 5G and IoT technologies to results achieved through massive, large-scale GB/European Horizon 2020 projects, respectively, US$13 million EFCC [18] and US$20+ million MIGRATE [19] projects. The first author of the chapter acted as the PI on behalf of academia (University of Manchester, Manchester, UK) in both projects. The aim of MIGRATE (Massive InteGRATion of power Electronic devices) is to find solutions for the technological challenges in the grid.

A significant increase in the volume of renewables providing electricity reduces system inertia and gives rise to an increase in the volume and speed of frequency response to maintain system frequency. It will require the development of new, significantly faster and coordinated response solutions using renewables, demand-side resources, and other new technologies. The enhanced frequency control capability (EFCC) project has been designed to find a resolution to electricity system challenge in wide-area frequency control. The aim of the EFCC project was to develop and demonstrate an innovative new monitoring and control system (MCS), which obtains accurate frequency data at a regional level, calculates the required rate and volume of fast response, and then enables the initiation of this required response within 0.5 seconds of a detected system frequency event (Figure 12.5).

5G-IoT Architecture in Next-Generation Smart Grid

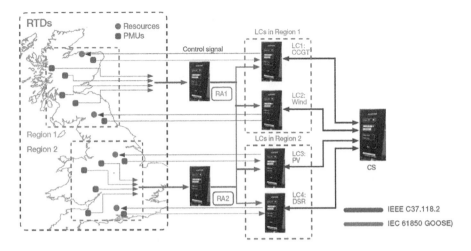

FIGURE 12.5 EFCC block diagram (note that RTDS hardware in the loop testing facilities were used for validation of a new frequency control paradigm).

12.5 INTEGRATION OF ADVANCED IoT ARCHITECTURES

In this section, a comprehensive overview of the interoperability of advanced IoT architectures is presented. With the major advancements in the development of IoT technologies and the global access to renewable energy resources, energy generation and distribution techniques are evolving. The integration of the IoT architecture, ICT-based end-to-end digital energy chains, energy-aware real-time platforms, and intelligent monitoring and control algorithms to SGs allows energy supply and demand to be balanced in real-time [20]. Some IoT-based SG architectures that are already available are listed below:

- Smart Grid Architecture Model (SGAM),
- Three-layered architecture (perception, network, and application layers),
- Four-layered architecture (perception, network, platform, and application layers),
- Cloud-based architecture,
- Web-enabled architecture.

The layered architectures are very generic and do not meet all the requirements of SG, whereas the cloud- and web-enabled SG architectures can support IoT along with additional useful services.

Advanced IoT architectures are based on huge dynamic global network infrastructure with the connectivity, data formatting, and network analytics. Intelligent applications are classified by network type, scalability, coverage, flexibility, and heterogeneity. Applications include SG, smart monitoring, and integrated multi-energy systems. An SG is composed of a data communications network integrated with a power grid to perform collection and analysis of data retrieved from transmission lines, distribution substations, and end users.

The use of uRLLC and network slicing (NS) in SG communication for energy monitoring, control, and distributed automation represents a large business potential for 5G. The 5G infrastructure ensures that 5G communication links are extremely robust and provide guaranteed low latency and high reliability. Furthermore, uRLLC can be a major step toward SG automation. The 3GPP specifies that uRLLC may be used in wide-area monitoring and control systems for SGs in the Technical Report (TR) 22.862 titled "Feasibility study on new services and markets technology enablers for critical communications" [21]. SG systems are appropriate examples of mission-critical use case in wide-area monitoring and control systems. For services requiring ultrareliable communications, acceptable levels of reliability and latency should be committed to the service at all times. Support for mission-critical services requires significant improvements in end-to-end latency, security, robustness, availability, and reliability compared to normal telecommunication services.

Network slicing (NS) is a key enabler of 5G-IoT. It allows the creation of several logical networks over mutually shared radio, core, and transport network infrastructures, which helps in improving cost-efficiency, scalability, and flexibility. The two methods that network operators can opt for to provide IoT connectivity are as follows:

- The operator can allocate one network slice for each cellular IoT segment offered to multiple enterprises.
- The operator can allocate one network slice to support multiple IoT connectivity segments at once, serving different enterprises.

To provide optimal performances, network slices should be configured with appropriate network resources and slicing should be performed end-to-end. The radio, transport, and core network must all be sliced accordingly. Network resources can be physical or virtual and can be reserved to a specific slice or shared between different slices. The creation of slices on demand shall also be supported [22].

Proper isolation of different use cases from each other can ease their management. For each slice, the underlying network needs to be reconfigured and the appropriate network functions must be enabled on an on-demand basis. Such requirements in 5G are satisfied by the use of virtualization and softwarization. Software-defined networking (SDN) is an example of softwarization. It is a paradigm where network functions are implemented using software and they are managed in a logical central authority known as an SDN controller, which dynamically manages network elements and links by defining routes and policies that adapt the network capabilities to the requirements of the network operator. Therefore, SDN significantly simplifies network management, monitoring, and reconfiguration. Consequently, SDN improves the flexibility of slicing by providing an effective way to install links requested by virtual network functions (VNFs) for service function chains (SFCs). An SFC is in fact a set of VNFs, which shall be executed in a specific order. Additionally, path separation or traffic isolation can be provided by SDN to avoid traffic from one slice to affect traffic in another slice.

A virtual phasor measurement unit PMU (vPMU) network function enables precise state measurements to be made across an entire gird. In principle, fast sampling will be enriched with software-based synchronization algorithms.

Network data analytics of data extracted from network monitoring tools, applications, and devices may provide valuable insights on the network and thus help optimizing it. The 5G network data analytics functions were introduced in the 3GPP Release 15, and they were further enhanced in Release 16. They provide the following functions:

- Detection of anomalies originating from devices by searching for abnormal traffic patterns.
- Generation of a suitable policy for background data transfer based on the analysis of the network's key performance indicators such as traffic volume, congestion level, and load status details in the specific network area.
- Determination of the optimal traffic routes dynamically by analyzing the network status, the source of the packets, and their corresponding destination.
- Optimization of the performance of applications having predictable network performances by analyzing the users' throughputs, the devices' locations, and the network's health.
- Enhancement of network automation capabilities by providing the network slice orchestrator with analyzed data about the network status so that resource allocation for the slices can be optimized.

In SGs, 5G NS network slicing allows different slices to be customized with different network functions and different service-level agreement (SLA) assurances based on the different service requirements across the SGs. We point out two typical SG application scenarios that may require 5G wireless communications and NS [23].

DA service scenario. Distribution automation makes use of an integrated information management system to regulate energy consumption and to enhance the reliability of power supplies. In the centralized distribution automation scenario, the network mainly transmits data such as telemetry and tele-indication information uploaded from terminals to primary sites and routine or remote control commands sent from primary sites to terminals for the isolation and restoration of faulty lines. With evolutions in communication systems, additional automatic control features are being added in DA systems. Such functions enable the DA systems to integrate a supervisory control and data acquisition (SCADA) system, a power distribution geographic information system, DSM, a dispatcher scheduling simulator, a fault call service system, and a work management system. Additionally, a distribution management system (DMS) may be included. This system supports more than 140 functions such as capacitor bank regulation control, substation automation, remote meter reading, feeder section switch control and user load control. The prerequisites for DA networks are as follows: ultralow latency of a few milliseconds, high isolation (distribution automation is a service in the I/II production area of the power grid, and it must be completely isolated from services in III/IV management areas), and high reliability of 99.999%.

LD service scenario. Millisecond-level precise load control allows interruptible loads of low importance to be disconnected upon power grid failures, whereas in traditional power distribution networks, there is no flexibility to choose which load to be disconnected and instead, the entire power distribution line needs to be disconnected. The stability control system can promptly disconnect the load to minimize damage to the power grid. In scenarios where the precise load control system based on the stability control technology is used, emergency handling requirements may be met as uninterruptible load within enterprises that are not disconnected, thus minimizing the financial loss and bad customer experience. The prerequisites for millisecond-level load control networks are as follows: ultralow latency of a few milliseconds, high isolation (precise load control is a service in the I/II production area of the power grid, and it must be completely isolated from services in III/IV management areas), and a high reliability of 99.999% (Table 12.1).

Based on the application scenarios of smart grids and the architecture of 5G NS, the overall architecture of 5G SG design and management is shown in Figure 12.6. The technical specification requirements of different service scenarios are met using the slices of information acquired from intelligent distributed feeder automation, and millisecond-level precise load control. Domain-specific slice management and integrated end-to-end (E2E) slice management are used to meet service requirements in these scenarios [23–25].

TABLE 12.1
5G Network Slices Requirements in DA (Distribution Automation) and LD (Millisecond-Level Precise Load Control) Smart Grid Scenarios

Service Scenario	Communication Latency	Reliability	Bandwidth	Terminal Quantity	Service Isolation	Service Priority	NS Type
DA	+++	+++	+	++	+++	+++	uRLLC
LD	+++	+++	+/++	++	+++	++/+++	uRLLC

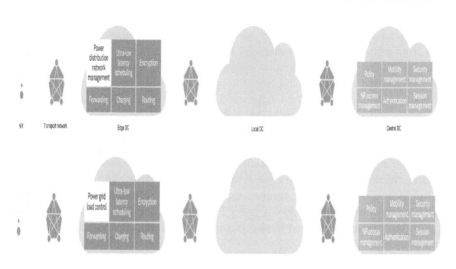

FIGURE 12.6 5G network slicing architecture of smart grid scenarios: (a) distribution automation (DA) and (b) millisecond-level precise load control (LD).

Cost-efficiency may be improved by managing and maintaining the various SG network slices in a unified manner. The analysis of SG application scenarios shows that their service requirements and technical specifications are not uniform. Operation enterprises and network equipment vendors should come up with their network architecture designs based on their security, service isolation, end-to-end latency, and technical specification requirements and they should perform proper technical verification and demonstration of the solutions [26–32].

12.6 CONCLUDING REMARKS

In recent years there is a continued demand for information and communication technologies that enable the Smart Grid with highly reliable and low-latency wide area monitoring, protection and control. Wireless telecommunication technology offers ease deployment, low-latency communication, standard-based solutions, huge data-carrying capacity, and excellent network coverage capability. IoTs provide an interactive real-time network connection and the cooperation required to realize two-way and high-speed data sharing across various applications, enhancing the overall efficiency of a SG. The application of the IoT in SGs can be classified based on the three-layered IoT architecture: IoT smart devices for the monitoring of equipment states, IoT for information collection from equipment with the help of its connected IoT smart devices through communication networks, and IoT for controlling the SG through application interfaces. Significant challenges come from application requirements as well as the infrastructure complexity and heterogeneity. In the SG generation level, the IoT helps to increase the controllability and observability from an operator point of view. In distribution layer, the implementation of IoT improves the observability of lines, which results in better monitoring of the transmission grid. In smart cities, smart buildings, and smart homes as well as industrial IoT, different types of sensors and ICT infrastructure help efficient management of resources and assets.

The new 5G mobile networks are emerging and are rapidly being deployed worldwide enabling huge capacity, zero delay, faster service development, elasticity and optimal deployment, less energy consumption, and enhanced security. Perhaps the highest novelty is radio access technology (RAT) design, which includes applications requirements of SG. The 5G manila supports distributed monitoring and control functionality applications. However, the analysis of communication delay characteristics is necessary for designing a reliable network. Also, 5G cellular networks have a great potential for support massive IoT machine interconnections and transmit data very quickly with ultrareliability and low-latency. We pointed out a symbiotic relationship between two 5G mMTC and uRLLC modes in SG communication.

However, the lack of standards and interoperability could slows the 5G-IoT deployment. IoT-based SG is a complex system that needs heterogeneous communication technologies to fulfill diverse requirements. Hence, the primary objective of communication standardization is to achieve interoperability among different components of SG system. 5G is focused on supporting various vertical industries, such as energy, manufacturing, and transport. The energy sector is one of the most demanding 5G use case, which poses significant new requirements. In this chapter, we focus on the exploration of requirements and identification of 5G innovative concepts toward the realization of a SG use case that will stress 5G-IoT current results. We point out the main types of 5G network slices: massive Machine Type Communication (mMTC), and ultra-Reliable and Low-Latency (uRLLC), and enhanced Mobile Broadband (eMBB).

To investigate the opportunities and challenges of integrated SG architectures, we first analyze the two-way communication delays in a wide-area monitor system (WAMS, which may cause power system fluctuations and expand the fault area. The highest novelty in 5G is uRLLC, which aims to provide extremely robust links with guaranteed latency and reliability. The uRLLC service transmits data within 1 ms and contributes to improve real-time state awareness capabilities, SE state estimation, and load forecasting. Therefore, fast real-time algorithms of state awareness, prediction algorithms of cascading failures, and decision-making algorithms of controlling massive action terminals should be further studied. Furthermore, collected big data requires analytics and visualization in an intuitive way for operators and users to manage their devices.

Next, we pointed out that the interoperable 5G-IoT solutions are fully compliant and aligned with standards developed by 3GPP consortia. Technical specifications of LTE-M and NB-IoT are fulfilling ITU requirements and thus considered as 5G MTC technology for IoT devices. The 5G-based IoTs can doubtless provide better infrastructure for advanced SG use cases. However, there are still many issues, which need to be addressed by the standardization community to deal with interoperability issues of IoT-based SG systems. Moreover, our analysis shows that the service requirements based on technical specifications vary greatly according to scenarios. Researchers should further quantify network technical specifications and integrated architecture design, including further quantifying 5G network slice requirements, and end-to-end uRLLC latency requirements as well as performing technical verification of the solutions.

REFERENCES

1. V. Terzija, D. Cetenovic, D. Milovanovic, and Z. Bojkovic, "Integration of multienergy systems into smart cities: Opportunities and challenges for implementation on 5G-based infrastructure", Chapter 11 in *5G Multimedia Communication: Technology, Multiservices, and Deployment*, pp. 211–233, CRC Press, 2020.
2. V. Terzija, G. Valverde, D. Cai, P. Regulski, V. Madani, J. Fitch, S. Skok, M. Begovic, and A. Phadke, "Wide area monitoring, protection and control of future electric power networks", *Proceedings of IEEE*, vol. 99, no. 1, pp. 80–93, 2011. doi:10.1109/JPROC.2010.2060450.
3. S. Chakrabarti, E. Kyriakides, B. Tianshu, C. Deyu, and V. Terzija, "Measurements get together", *IEEE Power and Energy Magazine*, vol. 7, no. 1, pp. 41–49, Jan.–Feb. 2009. doi:10.1109/MPE.2008.930657. (This article was one of six articles from 2009 reprinted in a 2010 special issue of the magazine.).
4. D. Cai, P. Wall, M. Osborne, and V. Terzija, "A roadmap for the deployment of WAMPAC in the future GB power system", *IET Generation, Transmission & Distribution*, vol. 10, no. 7, pp. 1553–1562, 5 May 2016. doi:10.1049/iet-gtd.2015.0582.

5. V. Terzija, G. Valverde, D. Cai, P. Regulski, P. Crossley, J. Fitch, C. McTaggart, and R. Adams, "Wide area monitoring, protection and control practices in the United Kingdom", in *CIGRE Study Committee B5 Colloquium*, Jeju Island, Korea, pp. 318.1–318.9, 19–24 Oct. 2009.
6. V. Terzija et al., "FlexNet wide area monitoring system", in *2011 IEEE Power and Energy Society General Meeting*, San Diego, CA, pp. 1–7, 2011, doi:10.1109/PES.2011.6039929.
7. Available online: https://www.ncl.ac.uk/supergenenhub/.
8. J. Zhao, M. Netto, Z. Huang, S. Yu, A. Gomez-Exposito, S. Wang, I. Kamwa, S. Akhlaghi, L. Mili, V. Terzija, and A.P. Meliopoulos, "Roles of dynamic state estimation in power system modeling, monitoring and operation", *arXiv preprint:2005.05380*, IEEE TPWRS, accepted on 27 Sept. 2020.
9. A.S. Dobakhshari, S. Azizi, M. Paolone, and V. Terzija, "Ultra fast linear state estimation utilizing SCADA measurements", *IEEE Transactions on Power Systems*, vol. 34, no. 4, pp. 2622–2631, Jul. 2019. doi:10.1109/TPWRS.2019.2894518.
10. J. Zhao, A. Gómez-Expósito, M. Netto, L. Mili; A. Abur, V. Terzija, I. Kamwa, B. Pal, A. Kumar Singh, J. Qi, Z. Huang, and A.P. Sakis Meliopoulos, "Power system dynamic state estimation: Motivations, definitions, methodologies, and future work", *IEEE Transactions on Power Systems*, vol. 34, no. 4, pp. 3188–3198, Jul. 2019. doi:10.1109/TPWRS.2019.2894769.
11. G. Valverde, S. Chakrabarti, E. Kyriakides, and V. Terzija, "A constrained formulation for hybrid state estimation", *IEEE Transactions on Power Systems*, vol. 26, no. 3, pp. 1102–1109, Aug. 2011. doi:10.1109/TPWRS.2010.2079960.
12. D. Milovanović, V. Pantović, and G. Gardašević, "Converging technologies for the IoT: Standardization activities and frameworks", Chapter 3 in *Emerging Trends and Applications of the Internet of Things* (Eds. P. Kocovic, M. Ramachandran, R. Behringer, R. Mihajlovic), IGI Global Publishing, pp. 71–103, 2017.
13. M. Noura, M. Atiquzzaman, and M. Gaedke, *Interoperability in Internet of Things: Taxonomies and open challenges, Mobile Networks and Applications*, Springer, pp. 1–14, 2018.
14. Y. Wu, Y. Wu, J.M. Guerrero, J.C. Vasquez, E.J. Palacios-Garcia, and J. Li, "Convergence and interoperability for the energy Internet: From ubiquitous connection to distributed automation", *IEEE Industrial Electronics Magazine*, vol. 14, no. 4, pp. 91–105, Dec. 2020.
15. Available online: https://www.ofgem.gov.uk/.
16. Available online: https://www.spenergynetworks.co.uk/pages/visor.aspx.
17. P. Wall, P. Dattaray, V. Terzija, P. Mohapatra, J. Yu, M. Osborne, P. Ashton, D. Wilson, and S. Clark, "VISOR project visualisation of real time system dynamics using enhanced monitoring", in *IET Conference Preparing for the Grid of the Future*, pp. 1–7, Birmingham, Apr. 2015.
18. Available online: https://www.nationalgrideso.com/future-energy/innovation/projects/enhanced-frequency-control-capability-efcc.
19. Available online: https://www.h2020-migrate.eu/.
20. A. Ghasempour, "Internet of Things in smart grid: Architecture, applications, services, key technologies, and challenges", *Inventions*, vol. 4, no, 1, p. 22, MDPI, Mar. 2019.
21. 3GPP TR 22862 v14.1.0, *Feasibility study on new services and markets technology enablers for critical communications*, Release 14, Sept. 2016.
22. M. Garau, M. Anedda, C. Desogus, E. Ghiani, M. Murroni, and G. Celli, "A 5G cellular technology for distributed monitoring and control in smart grid", in *Proceedings of IEEE International Symposium on Broadband Multimedia Systems and Broadcast (BMSB)*, pp. 1–6, Jun. 2017.

23. A.M. Escolar, J.M. Alcaraz-Calero, P. Salva-garcia, J.B. Bernabe, and Q. Wang, "Adaptive network slicing in multi-tenant 5G IoT networks", *IEEE Access*, vol. 9, pp. 14048–14069, Jan. 2021.
24. S. Wijethilaka and M. Liyanage, "Survey on network slicing for Internet of Things realization in 5G networks", *IEEE Communications Surveys & Tutorials*, vol. 23, no, 2, pp. 957–994, 2021.
25. Available online: https://www.gsma.com/futurenetworks/wp-content/uploads/2020/02/5_Smart-Grid-Powered-by-5G-SA-based-Network-Slicing_GSMA.pdf.
26. H. Huia, Y. Dinga, Q. Shib, F. Lib, Y. Song, and J. Yane, "5G network-based Internet of Things for demand response in smart grid: A survey on application potential", *Applied Energy*, vol. 257, pp. 1–15, Elsevier, Jan. 2020.
27. H.C. Leligou, T. Zahariadis, L. Sarakis, and E. Tsampasis, "Smart grid: A demanding use case for 5G technologies", in *Proc. IEEE International Conference on Pervasive Computing and Communications Workshops (PerCom Workshops)*, pp. 1–6, Mar. 2018.
28. H.T. Mouftah, M. Erol-Kantarci, and M.H. Rehmani (Eds.), *Transportation and Power Grid in Smart Cities*, Wiley, 2018.
29. Y. Saleem, N. Crespi, M.H. Rehmani, and R. Copeland, "Internet of Things-aided smart grid: Technologies, architectures, applications, prototypes, and future research directions", *IEEE Access*, vol. 7, pp. 62962–63003, Apr. 2019.
30. H. Hao, Y. Wang, Y. Shi, Z. Li, Y. Wu, and C. Li, "IoT-G: A low-latency and high-reliability private power wireless communication architecture for smart grid", in *Proceedings of IEEE International Conference on Communications, Control, and Computing Technologies for Smart Grids (SmartGridComm)*, pp. 1–6, Oct. 2019.
31. J. Tao, M. Umair, M. Ali, and J. Zhou, "The impact of Internet of Things supported by emerging 5G in power systems: A review", *CSEE Journal of Power and Energy Systems*, vol. 6, no. 2, pp. 344–352, Jun. 2020.
32. E. Esenogho, K. Djouani, and A. Kurien, "Integrating artificial intelligence Internet of Things and 5G for next-generation smart grid: A survey of trends challenges and prospect", *IEEE Access*, vol. 10, pp. 4794–4831, Jan. 2022.

13 Privacy Requirements in a Hyper-Connected World

Data Innovation vs. Data Protection

Myriah Abela
Betsson Group

CONTENTS

13.1 Introduction 445
13.2 Evolution of Security and Privacy Issues in Wireless Systems 446
    13.2.1 What Is Classified as Personal Data? 449
    13.2.2 Principles for Data Processing 451
13.3 Data Privacy by Design 455
13.4 AI Regulation and Accountability 458
References 460

13.1 INTRODUCTION

As new use cases for processing of personal data continue to emerge with the evolution of mobile telephony, this chapter aims to explain some of the functionalities of mobile telephony technologies to explore the implications and risks to data privacy and review current and proposed legislation setting out rights and obligations for the actors involved (all relevant industries, authorities and users). Mobile telephony is today a pervasive tool used in almost every aspect of a user's daily life. This evolution of digital application, artificial intelligence (AI) and the unlimited possibilities offered to both entities and consumer groups has introduced the need for specialised privacy legal protection. However, in order not to limit AI innovation, while ensuring such technology does not pose a threat to the rights and freedoms of users, a normative response is necessary to maintain the right balance.

    The General Data Protection Regulation (GDPR) EU 2016/679, which came into effect on 25 May 2018, is a technology-neutral law whose application adapts with society's technological progress, including through AI. AI cannot serve the public good without strong rules in place. GDPR enforces user rights relating to the processing of their personal data as well as imposes obligations on the various actors (controllers and processors), thereby shaping the way AI is developed and applied. The principles of privacy and data protection by design and default, restriction to automated decision-making without human intervention and rights to meaningful information on the logic involved are just some of concepts, which shall be further analysed in this chapter.

In 2018, the European Commission announced a programme pledging to make the EU fit for the digital age with the Commission publishing a White Paper on AI in 2020 [1]. The strategy includes a proposed AI Regulation laying down harmonised rules with the aim of developing an ethical approach towards human-centric, secure and trustworthy AI. AI adoption continues to increase at a rapid rate across every sector of the economy with the COVID-19 pandemic accelerating the automation of processes. At the same time, we can expect the challenges posed by big data and digital technologies, including the threat of cybercriminals abusing these technologies (e.g., spreading disinformation through deepfakes, AI-supported password cracking and hacking, social engineering, human impersonation, facilitation of terrorism through AI-empowered crypto-trading by terrorist groups, morphed passports), to continue emerging over the coming years. A regulatory framework safeguards the fundamental rights and freedoms of natural persons, sets ethical standards, enhances growth, embeds trust in businesses and allows users to fully reap the benefits of such new technologies.

13.2 EVOLUTION OF SECURITY AND PRIVACY ISSUES IN WIRELESS SYSTEMS

Each decade since 1980 has undergone a significant advancement in wireless communication technology, with each generation introducing innovative functionalities supporting the demand for enhanced performance, efficiency and novelty. However, this progress has also brought with it new threats to user privacy and data security. Through the generations, security became an increasingly primary concern due to the potentially severe consequences. When assessing security in mobile systems, the main objective is in preventing the loss of confidentiality and integrity of a user's or operator's data, as well as ensuring the availability of data.

During the 1980s the 1G network was established to provide mobile device users with the functionality for transmission of data over analog signals, supporting only voice calls with a maximum speed of 2.4 Kbps. 1G prioritised function and disregarded data privacy as the network was not encrypted. This caused the network to be prone to several threats including eavesdropping and impersonation attacks [2]. In 1991, 2G was introduced based on digital modulation techniques moving away from the analog signals used for 1G. 2G follows the Global Systems for Mobile Communications (GSM) standard which further defined how communication from one mobile station (MS) to another is performed. This system is still used to this day as the baseline for modern communication. The maximum speed increased to 14.4 Kbps allowing for the possibility to have both voice calls and short message services (SMS). The 2G network evolved into 2.5G with the introduction of the General Packet Radio Service (GPRS) standard which supported the Internet Protocol (IP) allowing mobile devices to connect to the Internet through the Wireless Application Protocol (WAP). The maximum speed allowed by 2.5G went up to 53.6 Kbps allowing the

Privacy Requirements in a Hyper-Connected World   447

user to use the Multimedia Message Service (MMS) whereby media such as pictures could be sent. 2G networks provided for anonymity through the usage of temporary identifiers and authentication via encryption performed through the Subscriber Identity Module (SIM) card. Nevertheless, this was a one-way authentication with the network identifying the user, but there was no way for the user to authenticate the network. This allowed for rogue base stations to disguise themselves as legitimate networks to access user data, rendering the network susceptible to eavesdropping. On top of this, recent discoveries have shown that the 2G network has backdoors implemented, revolving around a supposedly 64-bit encryption key used by GEA-1 which in reality is as effective as a 40-bit length key, thus making the key much less secure than proclaimed [3].

In less than 10 years, the 3G network was introduced in 1998 providing a high-speed data transmission. The speed depended on the device mobility reaching a maximum speed of 384 Kps for moving devices and 2 Mps for non-moving devices. This allowed for the introduction of certain services including video calling, access to mobile Internet and TV streaming. While the security framework of 3G is based on 2G technology, some of the weaknesses noted in the 2G system were remediated through the introduction of a new technology called the Universal Mobile Telecommunication System-Authentication and Key Agreement (UMTS-AKA) protocol [4]. As already noted, the 2G network only had a one-way authentication. This was remediated via AKA which introduced two-way authentication between the mobile phone and the 3G network. Like 2G, the 3G network evolved into 3.5G and 3.75G as more features were introduced before the emergence of the 4G network. Nevertheless, the 3G network, which is still in use primarily as a fall-back to newer networks, remains vulnerable to attacks which can be performed on the IP, including threats to the confidentiality, integrity and availability of data:

- Eavesdropping – messages intercepted without detection.
- Masquerading – use of impersonation to gain unauthorised access.
- Traffic analysis – observation of metadata to identify a user's location.
- Message Forgery – also known as man-in-the-middle attacks, involving the interception of messages and modification of content without the user's knowledge.
- Misuse of privileges – exploitation of privileges to gain unauthorised access to services/information.
- Distributed Denial of Service (DDOS) – use of several hosts to launch an attack by sending excessive data to a network, beyond its intended capacity, with the intention of bringing down a network, thereby disrupting availability.
- No cryptographic separation of keys – session keys generated for a network are valid for another network due to no cryptographic separation of security keys between roaming networks.

In 2009, the 4G technology, also known as Long-Term Evolution (LTE) network, was introduced. LTE is a series of upgrades to the UMTS technology, offering high speed and low latency to allow for more advanced services including IP Telephony VoIP,

High-Definition (HD) mobile television, video conferencing with multiple active participants, fast mobile web access, gaming and cloud computing. 4G introduced a new set of cryptographic algorithms such as EPS Encryption Algorithms (EEA) and EPS Integrity Algorithms (EIA), and the Evolved Packet System-Authentication and Key Agreement (EPS-AKA). However, despite the efforts to enhance the network's security, the security architecture still suffered from various weaknesses. Since the 4G network is heavily based on IP technology, the threats noted in LAN and WAN networks were transferred to the 4G network. Furthermore, the large-scale use of the network resulted in an increase in the scale and frequency of these threats. The 4G SIM communicates its identity to the network in plain text. This can pose problems as attackers are able to intercept the data being transferred. This feature has been remediated in the 5G network by adding a layer of encryption embedded within the 5G SIM called a SUCI (Subscription Concealed Identifier). As a result, unencrypted data such as the IMSI is not sent across 5G networks, allowing for the approval of lawful interceptions in specific cases. For example, where the court issues subpoena for investigation, operators may intercept communications on behalf of authorised law enforcement agents. The 5G network was globally implemented by 2020 and is the first technology to use unified authentication, supporting Wi-Fi, cable and 3GPP networks, offering speeds 100 times faster than 4G. Three disruptive technologies for 5G technology: virtualisation, edge computing and geolocation allow for new applications in various sectors such as autonomous driving cars, telehealth solutions, cloud investments for financial services and augmented reality. Through Multi-Access Edge Computing (MEC), execution resources for applications with networking are moved closer to the end user, typically within the mobile telephony network of a telecommunications operator, or may be placed in premises such as factories, homes and in vehicles (such as planes, trains and cars). Instead of sending all data to the cloud for processing, the network edge analyses, processes and stores the data, resulting in low latency, high bandwidth, as well as trusted computing and storage. This allows for capabilities close to real time, which are essential for certain use cases such as autonomous vehicles, smart homes, smart energy and remotely assisted surgery. Furthermore, with 5G location data is more accurate due to the smaller cells (areas covered by one antenna), providing very detailed data to network operators.

The technological leap is now occurring from 5G to the 6G wireless network technology which is expected to be implemented around the year 2030 enabling hyper-connectivity between people and things. It is also expected to extend mobile communication to a new generation of services using AI as its main driver, machine learning (ML), enhanced edge computing and distributed ledger (DL) technologies such as blockchain. Data processing, threat detection, traffic analysis and data encryption are considered the most critical issues in 6G networks. With the considerable increase of coverage and network heterogeneity, there are concerns that 6G security and privacy can be more impacted than the previous generations. For example, the involvement of connected devices in everyday life poses serious risks of security attacks which are not limited to pecuniary or reputational damage but could also be fatal (e.g., fatal car crash due to attacks on autonomous driving). Further, the achievements of AI can be misused for large-scale online surveillance.

Risks to users' data privacy present in previous technologies continue to increase with the pervasiveness of new technologies, including risks of precise user geolocation; profiling including automated decision-making; existence of multiple actors in the chain of processing potentially leading to ambiguity and unaccountability with respect to responsibility for the data processing; lack of a homogeneous security framework (as 5G allows for the existence of multiple parties in the chain of communication); and cross-border data processing implications [5]. To ensure this technological revolution is ethical and considerate of the primary societal values of privacy, security and transparency, entities shall take into account the following obligations:

- Analysis of the functional roles of the parties in the data processing chain to assign responsibilities accordingly. The concepts of controller, joint controller and processor play an essential role in the application of the GDPR, since they establish who shall be responsible for compliance with different obligations, and how users can exercise their rights in practice.
- Information provided to users, including the purpose of processing, the legal basis, the reliance on automated decisions, including profiling, as well as the users' data subject rights, must be laid out in a clear and understandable manner.
- Carrying out a data protection impact assessment (DPIA) where the data processing is likely to result in a high risk to the rights and freedoms of data subjects, and at least in the cases of: (i) systematic and extensive evaluation of data subjects, including profiling, with decisions producing legal or similarly significant effects on the user; (ii) processing of special category data on a large scale (e.g., health data); and (iii) systematic monitoring of public areas on a large scale.
- Application of privacy by design and default from the very early stages of the design process of 5G products and services.

13.2.1 What Is Classified as Personal Data?

With 5G technology permeating most areas of our lives to facilitate sectors including financial, energy, transport, health and the entertainment industry, this amounts to new purposes and means for collecting and processing of a huge volume of data, including personally identifiable information. With the prospects of 6G technology, the potential for new business models, novel services and data processing could be limitless. Existing technology-neutral privacy regimes address such data processing and lay down the requirements to ensure that any such data processing is done fairly, lawfully and in a transparent manner.

The GDPR has come into effect during a period of digital change where users (also known as data subjects) are faced with considerable risks to their privacy rights and freedoms. GDPR specifically regulates the processing of personal data, which is defined as any information relating to an identified or identifiable natural person who can be identified, directly or indirectly. The definition for personal data is purposely wide to ensure GDPR's horizontal rules to protect users consistently regardless of the technology or industry. As per Recital 26 of the GDPR, personal data which has only undergone pseudonymisation, and which can therefore be attributed to a natural person by use of additional information, is still be considered as 'personal data'. Such additional information must be separately retained and secured through adequate organisational or technical measures, such as encryption.

Throughout its text, the GDPR also uses the term 'data', which should be understood as data other than personal data as defined above. This non-personal data consists of either data which originally does not relate to any identified or identifiable natural person (e.g., information on weather conditions), or data which was originally personal data and was rendered anonymous. Adequately anonymised data must be distinguished from pseudonymous data as it cannot be attributed to a specific person, even if combined with additional data. Such 'data' is out of the Regulation's scope. In the case of IoT, AI environments and technologies enabling big data analytics, a dataset is more likely to be composed of both personal and non-personal data, referred to as a 'mixed dataset'. Where the non-personal data and the personal data parts are 'inextricably linked', the rights and obligations enforced by GDPR shall apply to the entire mixed dataset, even where the personal data comprise only a small percentage of the whole dataset. The concept of 'inextricably linked' refers to a scenario where the separation of personal data from non-personal data is economically inefficient, not technically feasible or would otherwise decrease the dataset's value. For example, in the case of mobile health applications, the separation between personal and non-personal data is not so clear-cut. Nevertheless, while the value of such data sources to improve on healthcare is undisputed, the strict requirements relating to special category data under GDPR must be adhered to [6].

As a result of the higher capacities in 5G technology and high density of small cells, the knowledge of the cell, which is related to a user, can divulge more detailed personal data. High-efficiency device positioning has led to the processing of more precise geolocation information of the device's user which can reveal further data by cross-checking information about a location. As a result, possible identification of personal data could be used for user profiling and tracking. While location-tracking data is highly valuable for legitimate purposes such as advertising and surveillance, without having adequate controls in place users may suffer from significant harm (e.g., such data may be used by unauthorised parties for purposes ranging from unsolicited advertising communication to targeting vulnerable data subjects for political, religious persecution or other criminal purposes).

Big data is at the heart of 5G technology which can support massive connectivity across billions of devices generating huge amounts of data. Big data is often described in terms of five main components: (i) the volume of data generated; (ii) the variety of data sources and formats of data; (iii) the velocity with which the data is produced and the speed at which the data flows; (iv) the veracity/quality of data; and (v) the value of data [7]. This involves the creation and analysis of vast datasets which may also include personal information, e.g., personal data derived from monitoring devices in clinical trials, geolocation information and biometric data from body-worn devices. Big data analytics also can generate new personal data, e.g., car sensors produce information about the vehicle itself but can also be used to analyse patterns in people's driving behaviour which can be used in determining insurance premiums. In certain cases, personal data is irrevocably removed prior to the analysis, and the now anonymised data is aggregated to obtain insight about the target population as whole. This is the case with data from clinical trials which undergoes anonymisation before it is used for data analysis. Anonymisation should not only be regarded to avoid regulatory burdens since it is out of scope of GDPR, but is ultimately a means towards risk mitigation, allowing big data analytics to give users the assurance that the data will not identify them.

13.2.2 Principles for Data Processing

With its extraterritorial reach the GDPR does not only apply to organisations established within the EEA but extends to non-EEA-established organisations which offer goods or services and/or monitors the behaviour of individuals in the European Union (EU). The GDPR enshrines seven foundational principles regarding personal data: (i) lawfulness, fairness and transparency; (ii) purpose limitation; (iii) data minimisation; (iv) accuracy; (v) storage limitation; (vi) integrity and confidentiality; and (vii) accountability. Although not specifically mentioned in the GDPR, its provisions apply to the new ways of personal data processing enabled by AI. However, the traditional foundational principles present a challenge to their application to AI processing.

Firstly, AI systems entail the processing of personal data for multiple purposes that may not be known at the time of collection. For example, when big data analytics is engaged, unexpected correlations of data are discovered which are used for a new purpose. Under the purpose limitation principle, the GDPR requires controllers to distinguish between these purposes and identify an adequate legal basis for each one, as without proper control such multiple purposes may result in the unforeseen use of personal data. However, the purpose limitation principle is not compatible with big data operations [8]. This is where the GDPR throws big data processes a lifeline by providing for the scenario of 'further processing'. Via Article 6(4) of the GDPR, entities are permitted to further process the personal data for compatible purposes by carrying out a compatibility assessment to guarantee a balance between their own interests and the rights and freedoms of affected users, unless the lawful bases of consent or legal obligation can be relied upon. Where the further processing is compatible no additional legal basis from that which allowed the collection of the personal data in the first place is required. Through the compatibility assessment entities must consider five factors to establish whether the reuse of personal data is compatible with the original purpose, including (i) any link between the purposes of the initial and further processing; (ii) the reasonable expectation of the user; (iii) the nature of the personal data; (iv) the potential consequences of the further processing; and (v) whether adequate safeguards will be implemented such as pseudonymisation and encryption.

The reuse of data for scientific or historical research or statistical purposes (e.g., big data applications aimed at market research) is presumed by GDPR to be compatible with the initial purpose and would therefore be admissible unless it involves risks for the user which cannot be mitigated. The GDPR specifically clarifies that the use of personal data for scientific purposes should be broadly interpreted to include technological development and demonstration, fundamental research, applied research, privately funded research and studies conducted in the public interest in public health. Nevertheless, this broad interpretation must not be stretched beyond its common understanding that 'scientific research is a research project set up in accordance with relevant sector-related methodological and ethical standards, in conformity with good practice' [9]. As noted, the GDPR specifies that this exception subsists provided that adequate safeguards are applied. These technical and organisational measures include controls to ensure data minimisation (such as anonymisation or pseudonymisation of data), aggregate results and that these results are not used in support of measures or decisions regarding a specific user. Yet, this marks another challenge for big data analytics as results are sometimes used in a unique interaction with a user. For example, while a company may benefit from this exception to measure the customer engagement rate, it shall not be able to use the results to target specific customers with direct marketing without identifying an adequate legal basis.

In certain cases, consent would be required, otherwise 'further processing' would not be deemed compatible, e.g., tracking and profiling for purposes of direct marketing, behavioural advertisement, data-brokering, location-based advertising or tracking-based digital market research [10]. The use of consent for 'further processing' necessitates the same requirements as in the case of consent for initial collection. Entities must be aware of the high bar set for consent to be deemed valid. Not only can users withdraw their consent at any time, but to be able to rely on consent, organisations must ensure that the consent is freely given, specific, informed, unambiguous, involves a clear affirmative act on the part of the user and is documented. This means that users must have all required information to make a free determination on how their data can be used without any detriment. This may pose a challenge to operations involving AI as their dynamic and complex nature severely contradicts the notion of transparency and explainability. Furthermore, intellectual property rights and competition issues may restrict the publication of such information. In certain cases, transparency may also hinder both public and private entities from carrying out their duties (e.g., predictive policing systems or anti-money laundering controls of subject persons) or may prejudice the data controller's information security controls by allowing users to bypass such controls. Against this context, the GDPR Article 13(1)(f) stipulates the requirement for the data controller to disclose 'meaningful information about the logic involved, as well as the significance and the envisaged consequences' for the user. Data controllers may reveal the type of input data and the intended output, explaining also the variables, rather than providing specific detail on the correlation each variable has in regard to the output. For users to make a free determination, the relationship must not be characterised by a power imbalance. This is likely within the public sector and the workplace, where the data subject would not have any realistic alternative to accepting the terms set by the controller. In such cases, the controller must assess whether there are other lawful bases which are more appropriate for the processing operation in question. Where the user has given consent for specific 'further processing', the controller may further process the personal data irrespective of the compatibility of the purposes.

The principle of data minimisation requires entities to only process data which is adequate, relevant and limited to what is strictly necessary in relation to the purposes for which they are collected and/or further processed, as well as for the data to be kept for no longer than is necessary. This again poses a challenge for big data which as stated earlier involves the processing of a large amount of data from a variety of data sources which may be used to discover unexpected correlations. Due to the volume and variety of datasets available, the issue is not only whether the data processed is excessive but also whether it is relevant. There are various techniques that can be implemented to ensure organisations only process the minimum personal data necessary for their purpose. During the training phase, entities must analyse whether all input variables included in a dataset are relevant for the purpose through feature selection techniques. Non-informative or redundant features should be removed to not only cater for this principle but improve the model's statistical accuracy. Training data can be biased if it represents previous discriminatory decisions, due to intentional discrimination, feature selection or overrepresentation of a particular group. Imbalanced training data may be balanced out with the addition or removal of data on the underrepresented segments. For example, the use of facial recognition technology has been criticised for violation of privacy rights, perpetuating gender norms and racial bias. These systems were found to generally work best on middle-aged White men, but performed poorly on people of colour, women, children or the elderly [11]. The curse of dimensionality, as coined by Richard E. Bellman in 1961, observes that when a model is supplied with the optimal number of features, this is said to increase the model's statistical accuracy. By adding a large number of features, the model is more likely to overfit the training data, thereby producing a model that performs particularly well during testing scenarios but fails to classify correctly once deployed. This means that data minimisation not only safeguards users' data privacy but also ensures that classification results that can discriminate against minority groups are eliminated from the model's features during the training phase. Moreover, AI models which are trained with biased data will reproduce that bias which may be discriminatory on the basis of racial or ethnic origin, political opinion, religious beliefs, trade union membership, genetic or health status or sexual orientation [12]. The case of 20-year-old Dylan Fugett and 21-year-old Bernard Parker illustrates this classification discrimination. In 2013 Fugett and Parker were arrested in Florida, United States for drug possession. They both had previously committed crimes of the same nature. The model determined the probability of recidivism by giving a score between 1 and 10 (10 being highly likely to re-offend). While Parker was scored as 10, Fugett was scored as 3, with the only discernible difference being that Parker was Black while Fugett was White. Fugett was later arrested for drug possession, while Parker did not commit any subsequent offences [13].

On the other hand, removing certain features might have the unintended adverse effect of reducing the model's statistical accuracy. While the principle of accuracy under GDPR requires entities to take all reasonable measures to ensure personal data is kept up to date, in the context of AI statistical accuracy refers to the proportion of answers which the AI model guesses correctly. This means that having a 100% statistically accurate model is not necessary to comply with the principle of accuracy, but it is rather necessary for compliance with the fairness principle, i.e., the personal data is not used in a manner which could have unjustified adverse effects on the user. The GDPR does refer to statistical accuracy in recital 71, requiring entities to use appropriate mathematical or statistical procedures for the profiling of users to mitigate the risk of inaccuracies, errors and discrimination. Entities should therefore test statistical accuracy at the design stage but also review the measures applied throughout the model's lifecycle. Privacy-preserving techniques such as anonymisation or pseudonymisation may be adopted to ensure compliance without compromising statistical accuracy.

AI may produce automated decisions which have a legal or similarly significant effect on users, and as a result, specific lawful bases need to be applied. Solely automated decision-making can be only carried out when the decision is (i) necessary for entering into or performance of a contract; (ii) based on the individual's explicit consent; or (iii) authorised by law. Even when the first and second exceptions apply, organisations are required to implement adequate measures when the automated decision is made by giving users clear information about the logic behind the AI model, explain the significance and envisaged consequences of the processing, introduce simple means for users to request human intervention or challenge a decision and carry out regular checks to ensure the model is working as intended. Furthermore, entities cannot base automate decisions on special category data (data on race, health and sexual orientation, amongst others) unless the user gave explicit consent or the processing is necessary for substantial public interest on the basis of law. For 'human intervention' to be deemed adequate entities must ensure that this oversight is carried out by someone who has the authority and competence to overturn the decision. Without any actual influence on the decision, this would still be a decision based solely on automated processing. The forthcoming proposed AI Regulation also includes the requirement for 'human oversight' for high-risk AI systems which need to be designed in a way that can be effectively overseen by natural persons before placing them on the market. While the AI Regulation does not specify which measures must be taken to implement this human oversight, the European Commission's Whitepaper on AI provides potential mechanisms which may be applied to comply with this requirement: (i) the AI system output is not effective unless it is reviewed and validated by a human; (ii) the AI system output is immediately effective but human intervention is ensured afterwards; (iii) monitoring the AI system while in operation with the ability to intervene in real time and deactivate; and (iv) imposition of operational constraints on the AI model at the design stage [14]. As for the requirement to implement adequate measures to safeguard users, generally the output of an AI model is not to be considered as factual information on the user but rather represents a statistically informed prediction. Errors or bias in the data or in the automated decision-making process can result in incorrect predictions and may impact the user negatively. In order not to misinterpret such outputs as factual information, entities should explain that the model makes statistically informed predictions. Moreover, users should be given the opportunity to request the inclusion of additional explanatory information in their record to counter any incorrect inferences.

13.3 DATA PRIVACY BY DESIGN

As 5G technology becomes increasingly widespread, entities are required to ensure that it is used in a way that respects users' data privacy rights. GDPR enshrines the principle of data protection by design and by default as well as the obligation of carrying out DPIA to ensure appropriate technical and organisational measures are in place to mitigate possible risks to users' data privacy rights. In line with the risk-based approach, carrying out a DPIA is not mandatory for every processing operation, but only for high-risk processing activities in relation to personal data. Where AI technology involves the processing of personal data, a DPIA needs to be carried out to address the inherent risks in AI systems to ensure non-discrimination, fairness, equity and security. AI actors are accountable for the design and implementation of AI systems in such a way as to ensure that personal data is protected throughout the life cycle of the AI system.

First, entities involved in the development and deployment of AI systems must analyse what role they play in the data processing chain. The GDPR concepts of 'controller', 'joint controller' and 'processor' are functional concepts which aim to assign responsibilities according to the actual roles of the parties. The controller generally determines the purposes and means of the processing, i.e., the why and how of the processing, while some more practical aspects of implementation are left to the processor. It is also not necessary for an entity to have access to the data that is being processed to be qualified as a controller. On the other hand, for an entity to qualify as a processor it must be a separate entity in relation to the controller and it processes personal data on the controller's behalf and on its instructions. The controller's instructions may leave a certain degree of discretion on how to best reach the controller's ultimate purpose for processing, allowing the processor to choose the most adequate technical and organisational means. In the context of AI systems, an entity could be designated as controller where it is ultimately taking decisions with respect to the broad kind of ML algorithm that will be used to create the model (e.g., regression models, decision trees, random forests, neural networks); feature selection; the model's parameters; evaluation metrics and loss functions (e.g., the trade-off between false positives and false negatives); decisions on the source and nature of personal data used to train the model; the purpose(s) of processing; the target variables of the model; and for how long the data is to be retained. A processor may still be afforded some discretion to make decisions to support the AI model, including the programming language and code libraries the tools are written in; the configuration of storage solutions; how to optimise learning algorithms to minimise their consumption of computing resources; the details on how the model will be deployed (e.g., choice of virtual machines, microservices and APIs); and the graphical user interface [15]. The roles are not as clear-cut in those cases where an AI model is continuously being upgraded and enhanced while in use by the provider's clients. If an AI service provider isolates client-specific models, this enables its clients to act as the ultimate decision-maker for the model, rendering them as sole controllers, while the provider as the processor.

AI also raises notable risks for the rights and freedoms of users, as well as compliance challenges for entities. The data protection implications of AI depend on the specific use case, the characteristics of the user population and the varying likelihood and severity of the rights and freedoms of natural persons. A DPIA is an important tool for demonstrating accountability as it allows entities to manage these risks resulting from the processing of personal data by adopting risk-based controls. The requirement for a DPIA is also included within Article 29(6) of the EU's proposed AI Regulation where users of high-risk AI systems are required to perform a DPIA. In the context of AI, it is unrealistic to adopt a 'zero tolerance' approach to risks to users' rights and freedoms. Entities are rather required to ensure that these risks are identified and mitigated. Generally, the use of AI will involve data processing which is likely to result in a high risk to users (e.g., use of new technologies or novel application of existing technologies, data matching, geo-tracking or behavioural tracking) and therefore shall require a DPIA. Entities which determine that their use of AI does not subject them to this obligation must still document their justification.

A sandbox-based approach would be appropriate to assess the AI system in a fully controllable environment, prior to its deployment and widespread use. By means of the DPIA entities are to document: (i) a description of the data flows and stages when AI processes and automated decisions may produce effects on users; (ii) the statistical accuracy of the model which may affect the fairness of the personal data processing; (iii) the degree of human involvement in the decision-making process and at which stage this takes place; (iv) the compatibility assessment for any further processing; (v) the assessment of any existing or potential trade-offs (e.g., data minimisation and explainability versus statistical accuracy); (vi) how individual rights are ensured; (vii) the assessment of risks to users such as discrimination as a result of bias and inaccuracy of datasets used, misuse of data analytics to target users based on their race, political opinion, gender (e.g., to manipulate swing voters), as well as risk scenarios related to insufficient security measures; and (viii) the current and proposed controls to mitigate the identified risks (e.g., data minimisation, storage limitation and purpose limitation or compatibility assessment for further processing, and providing users the option to opt-out of the data processing). The risk assessment must identify possible attack paths that a threat actor can use to exploit the vulnerabilities [16], including the following:

- Misconfiguration of infrastructure allowing for the penetration of threat actors via external interfaces resulting in the compromise of data.
- Lack of access controls allowing unauthorised actors to perform adverse actions leading to confidentiality/integrity/or availability breach.
- Espionage by actors using malware abusing the vulnerability of low-quality server components.
- Actors may target end users, e.g., through scam messages as part of a large-scale phishing attack.
- Exploitation of low security devices such as IoT (e.g., sensors, home appliances, etc.) to attack the network by overwhelming its signalling plane.
- Network failure due to interruption of electricity supply or natural disasters.

On a risk-based approach, entities are to identify which contractual, technical and organisational controls need to be implemented to mitigate the identified risks, including the implementation of strict access controls on a need-to-know basis; physical security; software updates and patch management; business continuity plans; incident response management procedures; performing pre-contractual supplier due diligence, and ongoing audits on third-party providers where AI systems are purchased, licensed or leased; entering into adequate agreements to ensure the personal data is safeguarded when processed by third parties on behalf of the controller; ensuring appropriate data transfer mechanisms are applied and transfer impact assessments are carried out in the case of transfers of personal data outside of the EU/EEA. Recital 78 of the GDPR is relevant in this respect as it encourages entities developing AI systems to take into account data privacy at the design stage. The controller is ultimately responsible for the fulfilment of the privacy by design obligation for the processing carried out by its processors and sub-processors. At procurement stage, controllers should set out the system requirements and conduct an evaluation of the trade-offs as part of the supplier due diligence process. If the applied controls do not mitigate the inherent high risk to an acceptable level, entities must first consult with their competent supervisory authority prior to the commencement of the processing. The necessary efforts must likewise be made for the identification and the mitigation of new risks through risk management procedures throughout the AI system lifecycle.

Major 5G network and IoT device manufacturers are in third countries outside the EU/EEA, where personal data may be subject to a greater risk as the data protection standards are not 'essentially equivalent' to those of the EU under GDPR. In most cases, the 5G function would require the cooperation of numerous network providers located in different jurisdictions worldwide. In the Schrems II judgement (C-311/18), the Court of Justice of the European Union (CJEU) clarified that the protection granted to personal data in the EU/EEA must travel with the data wherever it goes. The CJEU clarified that the level of protection in third countries does not need to be identical to that guaranteed under GDPR but must be 'essentially equivalent'. The transfer mechanisms under Article 46 of the GDPR are the first step in ensuring the transfer is lawful, but these do not operate in a vacuum and must be accompanied by a transfer impact assessment. The CJEU noted that the data exporters in collaboration with the importer in the third country are responsible for verifying if the law or practice of the third country impinges on the effectiveness of these transfer mechanisms. In those cases, exporters must implement supplementary measures to elevate the protection afforded to data up to the level required by EU law [17]. In the context of AI entities must consider measures for the de-identification of data. While data which has been irreversibly anonymised would be out of scope of GDPR, since it is no longer personal data, with large datasets processed for big data analytics there are uncertainties as to whether anonymisation can be truly achieved. A Massachusetts Institute of Technology (MIT) study of anonymised credit card data established that four vague pieces of information were enough to identify 90% of users in a dataset recording 3 months of credit card transaction by 1.1 million users in 10,000 shops in a single country. The bank had removed their customer names, credit card numbers, shop addresses and the exact time of the transactions. The remaining information included the amounts spent, the name and location of the shops at which the purchases took place and a random identification number representing each customer. Since each customer's spending pattern is unique, the data had a very high 'unicity' rendering it vulnerable to 'correlation attacks', i.e., through correlation of the available data with information about the users from outside sources, user identity could be revealed [18]. This means knowing whether anonymisation has been truly achieved is never clear-cut, and furthermore anonymisation decreases the data utility. To preserve levels of utility while safeguarding the data, entities should consider the following supplementary measures:

- the personal data transferred is pseudonymised, i.e., processed in such a manner that the personal data can no longer be attributed to a specific user, nor be used to single out the user in a larger group without combining the data with other data sets.
- this additional information is held exclusively by the data exporter and kept separately within the EEA or in a third country offering an essentially equivalent level of protection to that guaranteed within the EEA.
- the data exporter must retain sole control of the algorithm or repository that enables re-identification of the pseudonymised data using the additional information.
- transport encryption is used with state-of-the-art encryption protocols providing effective protection against attacks from state actors within the third country, including testing for software vulnerabilities and potential backdoors.
- where transport encryption is not effective by itself (due to vulnerabilities of the infrastructure or software used) personal data must also be encrypted end-to-end on the application layer.

13.4 AI REGULATION AND ACCOUNTABILITY

As we continue evolving beyond 5G, the landscape is becoming more complex. The massive amounts of data generated by IoT connections and devices will allow for new opportunities for data analytics and AI services. Nevertheless, with such rapid technological advancement, maintaining public trust is key to sustainable development. The European Commission has been working towards the concept of 'Trustworthy AI' by seeking to maximise the benefits of AI through human-centric systems while preventing and minimising the risks to the user. Trustworthy AI is comprised of three main components, which must be met throughout the AI system's life cycle. The system must be (i) lawful, complying with all applicable laws; (ii) ethical, ensuring adherence to ethical principles and values including fairness and respect to human autonomy; and (iii) robust, both from a technical and social perspective by taking into account the context and environment in which it operates [19]. To date, AI systems are already subject to a number of horizontal legally binding rules at European, national and international level including EU primary law (the Treaties of the European Union and Its Charter of Fundamental Rights), EU secondary law

Privacy Requirements in a Hyper-Connected World   459

(such as the GDPR, the Product Liability Directive, the Regulation on the Free Flow of Non-Personal Data, anti-discrimination Directives and consumer law), the UN Human Rights treaties and the European Convention on Human Rights and laws adopted by EU member states. Furthermore, there are various industry-specific rules which apply to particular AI applications such as the Medical Device Regulation for the healthcare sector. In April 2021, the European Commission submitted its proposal for the EU regulatory framework on AI. The Artificial Intelligence Act (AIA) [20,21] represents the first attempt to horizontally regulate AI. The AIA follows a risk-based approach by distinguishing between four levels of risk in AI: prohibited AI systems which cause an unacceptable risk, high-risk AI systems which are subject to stricter requirements, limited and minimal risk systems. High-risk systems include AI technologies implemented in the following use cases:

- critical infrastructures (e.g., road traffic, supply of water, gas, heating and electricity) that could put the life and health of users at risk.
- educational or vocational training (e.g., scoring of exams) that may determine the access to education and professional course of users.
- safety components of products (e.g., AI application in remote-assisted surgery).
- recruitment (e.g., CV-sorting software) and performance assessment of employees.
- essential private and public services (e.g., credit scoring denying users opportunity to obtain a loan).
- law enforcement practices (e.g., assessing the risk of offending/re-offending).
- migration, asylum and border control management (e.g., verification of the authenticity of travel documents).
- administration of justice and democratic processes (e.g., applying the law to a concrete set of facts).

The AIA requires providers of high-risk AI systems to conduct a conformity assessment before placing it on the market, to adopt technical documentation providing all information necessary on the system with clear information provided to the user, to implement appropriate human oversight measures, to use high-quality datasets to minimise issues with statistical accuracy and to ensure an overall high level of robustness and security. After a high-risk system is put into use, the providers must adopt a post-launch monitoring system to ensure continuous compliance and implement any necessary controls throughout the high-risk AI system's lifecycle. Limited-risk AI systems which include those intended to interact with humans (e.g., chatbots), systems used for emotion recognition and biometric categorisation systems or those that generate or manipulate content (e.g., deep fakes) are only subject to transparency obligations. The main requirement is to design the system in a way that people are informed they are interacting with, or using, an AI system, unless this is obvious from the circumstances and the context of use. This overlaps with the GDPR transparency requirements which already require entities processing personal data to be transparent about the use of profiling and automated decision-making. As for minimal risk systems (e.g., AI-enabled video games), the

AIA allows for their free use and suggests that these are mainly regulated through voluntary codes of conduct.

Apart from adopting a similar risk-based approach as to the GDPR, there are other commonalities between the GPDR and the AIA. The extraterritorial reach of the AIA echoes the approach taken with respect to the protection of personal data under the GDPR. The AIA applies to not only users of AI systems located within the EU but also providers that place AI systems on the EU market irrespective of whether they are established in the EU. Furthermore, third-country providers and users of AI systems fall within the scope of the AIA if the output produced by such systems is used in the EU. The AIA extends the principle of accountability on all operators involved in the supply chain by placing horizontal obligations on providers of high-risk AI systems but also provides for proportionate obligations for users and other players within the AI chain, such as importers, distributors and authorised representatives.

Whether the AIA will trigger a ripple effect on a global level much like the GDPR, which has been dubbed as the golden standard for data protection, is yet to be seen. It is evident that the use of novel, often-opaque, AI practices requires higher transparency, ethical consideration as well as accountability on the part of different actors. Much will depend upon the uniform application of the requirements and adequate enforcement. Furthermore, given the complexity of interpreting specific data protection requirements in the context of AI, ML and big data, data protection authorities need to engage in dialogues with stakeholders to develop appropriate guidelines. Consistent application of data protection principles, when combined with the ability to use AI technology to its full potential, can contribute to the success of AI by generating trust and mitigating risks.

REFERENCES

1. EU Commission, *White paper on artificial intelligence – A European approach to excellence and trust*, 2020.
2. M. Wang, T. Zhu, T. Zhang, J. Zhang, S. Yu, and W. Zhou, "Security and privacy in 6G networks: New areas and new challenges", *Digital Communications and Networks*, vol. 6, no. 3, pp. 281–291, 2020.
3. C. Beierle, P. Derbez, G. Leander, G. Leurent, H. Raddum, Y. Rotella, D. Rupprecht, and L. Stennes, "Cryptanalysis of the GPRS encryption", in *Annual International Conference on the Theory and Applications of Cryptographic Techniques*, pp. 155–183, 2021.
4. R. Borgaonkar, L. Hirschi, A. Park, and A. Shaik, "New privacy threat on 3G, 4G, and upcoming 5G AKA protocols", *Proceedings on Privacy Enhancing Technologies*, vol. 2019, no. 3, pp. 108–127, 2019.
5. Agencia Española de Protección de Datos, *Introduction to 5G technologies and their risks in terms of privacy*, 2020.
6. EU Commission, *Guidance on the regulation on a framework for the free flow of non-personal data in the European Union*, Brussels, 2019. [Online]. Available at: https://eur-lex.europa.eu/legal-content/EN/TXT/?uri=COM:2019:250:FIN.
7. Information Commissioner's Office, *Big Data, artificial intelligence, machine learning and data protection*, ICO Version 2.2, 2017.

8. T.Z. Zarsky, "Incompatible: The GDPR in the age of Big Data", *Seton Hall Law Review*, vol. 47, no. 4, pp. 995–1020, 2017.
9. *Guidelines 05/202 on consent under regulation 2016/679*, European Data Protection Board, 2020.
10. Data Protection Working Party, *Opinion 03/2013 on purpose limitation*, Article 29, 2013.
11. J. Buolamwini and G. Timnit, "Gender shades: Intersectional accuracy disparities in commercial gender classification", *Proceedings of Machine Learning Research*, vol. 81, pp. 77–91, 2018.
12. F. Zuiderveen Borgesius, *Discrimination, artificial intelligence and algorithmic*, Council of Europe, 2018.
13. J. Alwin, J. Larson, L. Kirchner, and S. Mattu, *ProPublica: What algorithmic injustice looks like in real life*, 25 May 2016. [Online]. Available at: https://www.propublica.org/article/what-algorithmic-injustice-looks-like-in-real-life.
14. European Commission, *On artificial intelligence – A European approach to excellence and trust*, Brussels, 2020.
15. Information Commissioner's Office, *What are the accountability and governance implications of AI?* [Online]. Available at: https://ico.org.uk/for-organisations/guide-to-data-protection/key-dp-themes/guidance-on-ai-and-data-protection/what-are-the-accountability-and-governance-implications-of-ai/#howshouldweunderstand.
16. European Union Agency for Cybersecurity, *5G supplement to the guideline on security measures under the EECC*, 2021.
17. *Recommendations 01/2020 on measures that supplement transfer tools to ensure compliance with the EU level of protection of personal data*, European Data Protection Board, 2021.
18. Y.-A. De Montjoye, L. Radaelli, V.K. Singh, and A. Pentland, "Unique in the shopping mall: On the reidentifiability of credit card metadata", *Science*, vol. 347, no. 6221, pp. 536–539, 2015.
19. *Ethics guidelines for trustworthy AI*, High Level Expert Group on AI, European Commission, 2019.
20. *Proposal for a regulation of the European parliament and of the council laying down harmonized rules on AI*, 2021.
21. European Data Protection Board, *Recommendations 01/2020 on measures that supplement transfer tools to ensure compliance with the EU level of protection of personal data*, 2021.

14 Evaluation of Representative 6G Use Cases

Identification of Functional Requirements and Technology Trends

Zoran S. Bojkovic and Dragorad A. Milovanovic
University of Belgrade

CONTENTS

14.1 Introduction	463
14.2 Emerging Use Cases and Applications	464
14.2.1 Immersive Multimedia and Holographic Communication	465
14.2.2 Tactile/Haptic-Based Communication	468
14.2.3 Space–Terrestrial Integrated Networks	468
14.3 Evolution of Usage Scenarios and Technology Trends	469
14.3.1 Requirements on Network Operation	473
14.3.2 Definition of System Capabilities	474
14.3.3 Focus Areas for Further Study	475
14.4 Concluding Remarks	477
Bibliography	478

14.1 INTRODUCTION

The latest generation of mobile network technology (5G) is being implemented in many regions of the world from 2020. The second phase 5G-Advanced continues to evolve in 3GPP R17 and R18 versions from 2022 to 2024. In parallel, 6G is starting to create a new mobile platform for 2030. If we want to prepare for all these new challenges, technological research that enables improved and new usage scenarios is necessary. The key performance indicators (KPI) of the 5G system can also be used for the development of new 6G technologies. However, it is necessary to critically analyze KPIs, and to evaluate new indicators.

The development of 5G is a response to the anticipated demand for increased capacity from users, as well as productivity requirements from industrial sectors. Technical success relies on delivering wider range of data rates to much wider set of devices and users. The 6G technology adopts much more holistic approach, involving all community to profile representational requirements. This approach identifies the trends, demands, and challenges fronting society until 2030, thus avoiding the definition of the system capabilities based only on commercial criteria in the global market. Despite the fact that 5G development is based on the requirements of a number of vertical industry sectors, the dominant implementation is driven by mobile network operators (MNOs). The new 6G ecosystem will be based on efficient short-range connectivity solutions beyond traditional MNOs. Smartphones will be probably replaced by immersive virtual experiences through lightweight wearable glasses that enable excellent resolution, frame rate, and dynamic range. Telepresence is enabled by high-resolution imaging and sensing, wearable displays, mobile robots and drones, and next-generation wireless networks. Autonomous vehicles for environmentally sustainable transport and logistics are enabled by integrated space–ground networks and distributed artificial intelligence (AI) and sensors. By applying AI and machine learning (ML), users are supported to connect the physical and digital world in real time, and to capture, retrieve, and access large amount of information. The 6G AI delivers intelligent services to every person in every moment.

Industry and academia have dedicated enormous funds and resources for 6G research. The ultimate goal is to improve network performance and efficiency. In the present period, the results published in large number of journal publications and conferences, keynote talks, and panel discussions at conferences, workshops, seminars, as well as working groups of standardization bodies are important. ITU-R WP5D recommendation IMT 2030 Vision defines the framework and overall objectives of the development of international mobile telecommunications (IMT) until 2030. An analysis of the overall trends of requirements from the user's perspective and the definition of the key capabilities of the 6G system are presented. At the macro level, the driving forces toward 2030 usage scenarios are based on market forecast, network requirements, and 5G gap analysis. At the micro level, service challenges, choice of typical use cases, service models, and users define a set of key system capabilities associated with representative usage scenarios. Key areas of future 6G development are also presented.

In this chapter, we evaluate representative 6G use cases, immersive multimedia and holographic communication, tactile and haptic communication, as well as space–ground integrated networks. In the second part, we analyze usage scenarios and technology trends, identify network requirements, and define system capabilities, as well as key areas for further study.

14.2 EMERGING USE CASES AND APPLICATIONS

Until the advent of 5G system, the focus in the development of mobile communications has been on communication aspects, while other services have had low priority, they are introduced in the only final stages of system design. The consequence is that performance is not optimal or system capabilities are not fully utilized. Co-design of the necessary drivers is not only desirable, it is critical for the high performance of future network services. Mixed reality (MR) is very demanding with numerous components for positioning, 3D mapping, fusion of digital content with a physical model, and extremely fast low-latency communication. Such integrated services are inherently supported by dense, wireless networks, with high-frequency antenna arrays and lots of computing power at the edge. An additional challenge is to achieve this in an energy-efficient manner.

The 5G systems are inevitably based on trade-offs in energy, cost, hardware complexity, throughput, latency, and end-to-end reliability. For example, the requirements of mobile broadband access and ultrareliable low-latency communications are achieved by different configurations of 5G networks. On the contrary, 6G development makes it possible to jointly meet strict network requirements (ultrahigh reliability, capacity, efficiency, and low latency) in a holistic approach. The applications and services enabled by future wireless communication technology altogether will connect not only people but also machines and various things.

14.2.1 Immersive Multimedia and Holographic Communication

The human-centered multimedia and communication will deliver an immersive physical experience through interaction with the digital world. Mixed reality (MR) and holographic telepresence will become the norm for both work and social interaction, enabling an immersive 3D experience with a tight integration of virtual and reality, as well as a quality of interaction very close to reality. Digital twins (DT) enable virtual experiences for humans and computer control for machines. It has the potential to provide ubiquitous tools and platforms for modeling assets, resources, environments, and situations, and to enable monitoring, design, management, analysis, diagnostics, simulation, navigation, and interactive mapping. It uses advanced technology such as the integration of communication with AI, sensing, and computing. DT will also synchronize the digital world with the physical world and provide connections between digital replica components. It also supports applications such as human digital twin, construction planning, real estate management, and smart city (SC). DT technology is still in the early stages of development, so reaching its full potential requires addressing significant limitations and challenges such as cost, complexity and information maintenance, lack of standards and regulations, and issues of cybersecurity and communications.

Enhanced human-centric communication enables highly immersive experience and multisensory interactions as well as remote telepresence with seamless user experience. New human–machine interfaces are being developed that extend the user experience across multiple physical and virtual platforms, where interactive haptic and multisensory communication will support teleoperation. Telepresence enables interaction with both physical and digital objects, enabled by wearable devices with intuitive interfaces that are fully context-aware. Immersive human–machine interaction will be possible in applications, for example, in telesurgery and medicine, remote control of machines in industry, and transportation. Applications facilitate remote work from home, out-of-office collaboration, teacher–student interaction, and improved diagnostics in teleconsultations enabling full human interaction.

The ultimate immersive entertainment experience is made possible by immersive multimedia and remote live performances. All of this can be associated together through the metaverse and cyber-physical systems (CPS), which are developed on the Internet platform to provide users with truly immersive experiences via appropriate interactive activities between physical space, cyberspace, and users.

The 6G platform supports the emergence of exciting immersive applications and technology trends.

- Virtual reality (VR) refers to computer-generated 3D environment in which users can explore and interact virtually, using specialized equipment. VR can be used to simulate realistic or artificial environments in a full 360° viewing panorama, and therefore provides great flexibility to content creators and allows users to engage in new interactive experiences. There are a variety of devices and sensors designed to fully immerse users in a virtual environment.
- Augmented reality (AR) and mixed reality (MR) refer to overlaying computer-generated virtual content over the user's view of the real environment to create an augmented version of reality. Virtual content consists of images or videos and can even be expanded to multiple sensory modalities (visual, auditory, haptic). One of the main advantages of AR is that it does not change the user's perception of the real environment, and the integration of immersive sensory stimuli that are perceived as an actual part of the environment.

5G has driven the early adoption of AR/VR. However, scaling the use of AR/VR applications in 6G requires system capacity above 1 Tb/seconds, in contrast to the peak data rate of 20 Gb/seconds specified for 5G. In addition, to meet the latency requirements that enable real-time user interaction in immersive environment, AR/VR cannot be compressed (encoding and decoding is a time-consuming process). Therefore, the data transfer rate per user is up to 1 Gbps, as opposed to the relaxed target of 100 Mbps for 5G systems. The latency requirement reaches sub-millisecond, and 1,000s of synchronized viewing angles are necessary, as opposed to the few required for AR/VR. Moreover, to fully realize the immersive experience at a distance, all five human senses are intended to be digitized and transmitted through future networks, increasing the overall target data rate. The data rate requirement for a 3D holographic display (raw hologram, without any compression, with color, full parallax, and 30 fps) is 4 Tb/seconds.

A hologram is a 3D picture, created with imaging projection. Holographic and MR communications have entered the market with the development of 5G, giving consumers first-hand experience in MR applications. Today's applications for holograms are limited to static images, then moving to dynamic images later. An ideal holographic display should be based on the naked eye. When necessary, it can be realized with the help of technologies such as AR/VR. Holographic technology on 6G networks can be assumed to be integrated into many application scenarios such as communication, telemedicine, office design, and entertainment games. The characteristics and advantages of the technology are the ability to integrate real-world and virtual information that contains real-time interactivity and at the same time brings brightness to the users. AR holographic technology continues to develop and many companies have focused their research on AR and 3D technology. Holographic images can be used not only in entertainment, art, and education, but also in media science, technology design, and AR.

Holographic displays enable a realistic 3D experience for the end user delivering 3D images from one or more sources to multiple destinations. As consequence of interactive holographic service in the network, combination of very high data rates and ultralow latency is necessary. Depending on how the hologram is constructed, the data transfer rate is determined. Also, the type of screen and the required number of images that need to be synchronized are important. The data rate required for hologram transmission can be reduced by using 3D data compression techniques. On the other hand, holograms require a huge bandwidth from tens of Mbps to 4 Tbps for a human-sized hologram based on image processing to generate the hologram.

Holographic telepresence is an emerging technology that projects 3D volumetric human beings and objects in full motion, in real-time, with free-viewing position and orientation, and high-resolution at a remote location. By placing objects/people in remote locations, holographic telepresence removes physical boundaries and revolutionizes the way people communicate with each other and interact with the physical world. Despite its great potential, the realization of holographic telepresence faces the great challenge of delivering extremely large amounts of 3D point cloud data in real time under inherent bandwidth limitations. Holographic networking is becoming a new frontier in emerging 3D point cloud compression research, 3D human reconstruction, and 3D streaming video communication in real time.

- Holographic usage scenarios require ultralow latency. If sub-millisecond latency is required, haptic capabilities are added. There are numerous cases where synchronization needs to be associated with holographic communications. For example, when streams include video, audio, and tactile data, precise/strict interstream synchronization ensures timely packet delivery.
- Security requirements depend on the application (remote surgery, security coordination of multiple co-flows).
- Resilience refers to minimizing packet loss, jitter, and delay. Relevant service metrics are availability and reliability for holographic communication. Resilience is the main parameter for maintaining QoS requirements.
- Computing implies significant computational challenges in real time at processing steps of hologram generation and reception. There is a trade-off between a higher level of compression, computational bandwidth, and latency, which needs to be optimized. Note that high data rates cannot take advantage of existing compression techniques since holographic signals are fundamentally different from moving image video.

14.2.2 TACTILE/HAPTIC-BASED COMMUNICATION

The tactile internet (TI) use case enables human–machine haptic interaction services with high reliability and sub-millisecond latency as limiting factor for many developing services. Internet-based applications empower real-time remote physical interaction with physical or virtual objects at long distance from the operator. Sensors, actuators, robotics, computer components, and dedicated hardware are the basic components of human–machine interaction, which can be organized into three domains: main domain, network domain, and controlled domain.

In industrial applications, tactile communication allows flexibility and lower maintenance costs despite the fact that they require low latency and high reliability. Cloudification in cases of industrial use represents a significant step toward reducing pollution, unplanned stops of industrial activities, and zero accidents. An advanced real-time robotics scenario in production requires maximum latency in the communication link of about 100 ms, while the round-trip reaction time will be 1 ms.

In autonomous driving, constrained latency of several milliseconds is necessary for collision avoidance and remote driving. Advanced driver assistance and fully automated driving are the key 6G application areas with the first components implemented in the 3GPP R16 standard specification.

Examples of 6G applications in healthcare are tele-diagnosis, remote surgery, and tele-rehabilitation. Using advanced tele-diagnosis tools, medical consultations would be available anywhere and anytime, regardless of the location of the patient and the doctor. Remote and robotic surgery is an application in which the surgeon monitors real-time audio-visual information about the patient undergoing operation.

The data transfer rate depends on the application requirements. The corresponding reaction times of the human brain at the input of the sensor are from 1 to 100 ms. It takes up to 100 ms to decode audio, 10 ms to understand visual information, and 1 ms to receive tactile signals. Robotics and industrial machines need sub-millisecond latency. It is necessary to synchronize real-time inputs from different locations. Machine control has fast reaction time so the inputs need to be also synchronized.

14.2.3 SPACE–TERRESTRIAL INTEGRATED NETWORKS

This use case scenario is about on Internet access via the seamless integration of terrestrial and space networks. The idea of providing the Internet from space using large constellations of low Earth orbit (LEO) satellites has regained popularity in recent years (previous attempts, such as the Iridium project in the late 1990s, were unsuccessful). The key benefits of using Telesat, OneWeb, and SpaceX satellite systems are ubiquitous Internet access on a global scale, including mobile platforms (planes, ships, etc.), extending the Internet path based on domain-boundary traversal protocols relative to the terrestrial Internet, and ubiquitous caching on the edges and computer science. Mobile devices for these integrated systems enable satellite access without relying on ground-based infrastructure that is limited by geographic layout.

In this example, the goal is to use interconnected low earth orbit (LEO) satellites and other non-terrestrial network nodes and platforms to build a parallel Internet network that can be compared to its terrestrial counterpart. The integrated framework has numerous advantages: ubiquitous access to the Internet on a global scale, including rural areas such as oceans, deserts, as well as mobile platforms such as ships and airplanes; enriched Internet paths that could lead to better data delivery performance compared to those over the terrestrial Internet determined by cross-domain border gateway protocol (BGP) configurations; ubiquitous edge caching, and computing services that provide lightweight, embedded computing and storage resources on LEO satellites.

The key network requirements for space-to-ground networking capability are flexible addressing and routing, satellite bandwidth capability, admission control of satellites, admission control by satellites, edge computing, and storage.

14.3 EVOLUTION OF USAGE SCENARIOS AND TECHNOLOGY TRENDS

The 6G usage scenario is expected to expand beyond communications and include significant improvements, as well as new innovations and paradigms, such as scaling sensing, intelligent interaction, immersive media experiences, and multisensory communications. Six usage scenarios are predicted, with the first three scenarios being extension of enhanced mobile broadband (eMBB), ultrareliable low-latency communications (uRLLC), and massive machine-type communications (mMTC) as shown in Figure 14.1. The last three extend 6G into new domains of global broadband, spatiotemporal services, and compute-AI services. Each usage scenario is associated with a set of key system capabilities.

- An immersive multimedia and multisensory communication usage scenario extends eMBB and covers the future of intense human–machine interaction. Typical applications are immersive XR and holographic communications, remote multisensory telepresence, tactile feedback, and industrial robot tactile feedback and control. Standalone support of voice services is an integral part of immersive communication. New communication devices, such as smart glasses, will gradually change the traditional

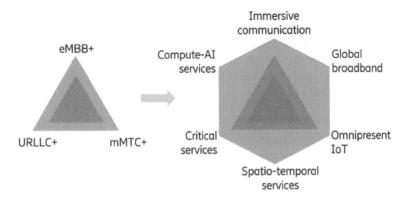

FIGURE 14.1 Extension and expansion of 6G usage scenarios.

eMBB ways of acquiring information and interacting. With very high sampling density of environment data, reliable online computing will be used to offload computationally complex processing and rendering or to remotely access rendered images in real time. The scenario requires extremely high data transfer rates, as well as lower latency and larger system capacity. It covers all types of deployments, from dense urban hotspot to rural.
- Critical use scenario applies to use cases with very strict transmission reliability and availability requirements, such as extreme uRLLC (time sensitive, trust). It is also applicable to nonconnectivity services, for example, dependable compute, accurate positioning or characterization of connected or disconnected objects, data distribution, AI native RAN design, and other network platform functionalities. Data transfer rate, latency, jitter sensitivity, power constraints, device connection density, etc. are also requirements which vary depending on the use case being considered. Limitation or at least predictability of performance variation for individual applications is vital for this usage scenario. Typical use cases include smart industry within all work domains, as well as new applications such as automation and control, autonomous and teleoperation, interaction and collaboration, digital representation, collaborative robotics, drone swarm, different human–machine real-time interaction, autonomous driving in intelligent transportation systems, smart energy, smart household, personalized digital human for precision medicine, remote medical surgery, in-body distributed sensor, and communication networks. This set of use cases is characterized by situations where the failure of a communication service can cause catastrophic or even life-threatening consequences for (safety) critical services.
- The usage scenario of omnipresent Internet of things (IoT) includes large number of sensors, where not only the number but also the geographical distribution is high, for example in manufacturing, smart cities/digital twins, transport, material and goods tracking, or environment monitoring. All this places high requirements on coverage. Also, different types of devices (sensors or actuators) will have significantly different requirements, where some harvest energy and depend on low-energy protocols, while others will require higher data rates. In remote areas, it is necessary to support devices that really do not require maintenance. Key capabilities for omnipresent IoT usage scenario are coverage and support for a wide range of devices (device diversity) (Table 14.1).

6G is based on diversified and advanced requirements. The technological requirements for implementation depend on the improvement of existing technologies, as well as on new technologies that were not taken into account during the previous 5G development (Table 14.2).

The development of 6G is expected to enable new use cases and applications, and to address the rapid growth of traffic, for which contiguous and wider channel bandwidth is desirable than currently available for mobile systems. As the amount of spectrum required for mobile services increases, it becomes increasingly desirable to harmonize existing and newly allocated and identified spectrum. A continuous and

TABLE 14.1
Immersive, Intelligent, and Ubiquitous Usage Scenarios with Candidate Use Cases Trends

Scenarios	Use Cases	Trends
Immersive	Holographic communication (extremely immersive experience)	• User experienced data rate (Gbps/Tbps) • Latency (<10 ms) and synchronization (strictly) • Computing and intelligence (large computing power) • Security (sensitive information)
	Immersive cloud XR (broad virtual space)	• User experienced data rate (Gbps) • Air latency <2.5 ms • Multidimensional sensing (accuracy of location/touch) • Computing and intelligence (distributed cloud computing) • Energy efficiency (low-power terminals)
	Intelligent interaction (interactions of feelings and thoughts)	• User experienced data rate 100 Mbps • Reliability 99.99999% • Computing and intelligence (ubiquitous AI services) • Security (privacy of user data)
Intelligent	Proliferation of intelligence (ubiquitous smart core)	• Transmission (massive training data), area traffic capacity • Distributed connection (collaborative learning) • Connection density 10–1,000/km^2 • Computing (real-time efficient computing power) • AI service accuracy/efficiency >90%
	Digital twins (digital mirror of physical world)	• Data rate (peak 100 Gbps, user 1 Gbps) • Air traffic capacity 0.1–10 Gbps/m^2 • Latency (sub-ms) • Connection density (10^7–10^8 device connections per km^2)
	Machine control and communication (fusion of DT, edge computing, AI)	• User experienced data rate 100 Mbps, reliability 99.99999% • Latency <0.1 ms jitter microsec-level • Computing (distributed computing supports sensing & control) • Sensing (multidimensional, accuracy, & transmission latency)

(*Continued*)

TABLE 14.1 (*Continued*)
Immersive, Intelligent, and Ubiquitous Usage Scenarios with Candidate Use Cases Trends

Scenarios	Use Cases	Trends
Ubiquitous	Sensory interaction (fusion all senses)	• User experienced data rate <100 Mbps • Latency <1–10 ms and synchronization <10–20 ms • Computing and intelligence (large computing power) • Reliability (99.999%)
	Communication for sensing (extending functions of converged communication)	• Sensing accuracy (cm-level) and resolution (cm-level) • Computing and intelligence (ubiquitous AI services) • Detection rate and false alarm rate
	Global seamless coverage (3D connections)	• User experienced data rate (100 Mbps/Gbps) • Latency 100 ms • Coverage (extension to global coverage) • Mobility >100 kmph • Link availability >90%

TABLE 14.2
The Candidate 6G Technologies

Technology	Trends
New network with native AI	• AI native–air interface • AI-driven RAN architecture and optimization • Radio network for AI
Enhanced wireless air interface technologies	• Basic physical layer technologies • Ultra-massive MIMO • In-band full duplex • Efficient spectrum utilization • Energy-efficient RAN
Wireless transmission technologies on new physical dimensions	• Natively support real-time communications • Reconfigurable intelligent surface (RIS) • Orbital angular momentum (OAM) • Intelligent holographic radio (IHR)
Terahertz and visible light technologies	• Terahertz communications (THz) • Visible light communications (VLC)
Integrated communications and sensing	• Sensing-based ultrahigh accuracy positioning and localization
Distributed autonomous network architecture	• Self-synthesizing networks
Convergence of communications and computing architecture	• Split computing • Pervasive compute • Ubiquitous computing and data services
Native network security based on multilateral trust model	• AI is inherent part of the infrastructure, and of the network management and operations

Evaluation of Representative 6G Use Cases 473

wider spectrum from a range of frequency bands, aligned with future technology development, would support achievement of the objectives of the 6G systems, reducing device complexity, avoiding interference, and developing ecosystems.

14.3.1 Requirements on Network Operation

Future wireless networks are expected to support a wide range of sometimes conflicting requirements. 6G is expected to become the first wireless standard requiring hyperfast links with peak throughput per link exceeding the terabits per second (Tbps). 6G use cases, such as wireless factory automation, require highly sophisticated operations such as ultrahigh-reliability, ultralow-latency communication, high-resolution localization (at the centimeter level), and high-accuracy interdevice synchronicity (within 1 ms). Different reliability requirements specific to individual use cases are expected. One of the most extreme is industrial control where only one erroneous bit is allowed in a billion transmitted bits with a delay of 0.1 ms (Table 14.3).

TABLE 14.3
Representative Use Cases and Key Network Requirements

Use Cases	Requirements
Holographic-type communications (HTC)	High bandwidth, low latency, multistream synchronization, edge computation, security, and reliability
Digital twin (DT)	Connect many more sensors and devices, the high-speed ubiquitous connectivity, the improved reliability and redundancy, and ultralow power consumption
Tactile Internet for remote operations	Ultralow latency, ultralow loss, ultrahigh bandwidth, strict synchronization, differentiated prioritization levels, reliable transmission, security
Industrial IoT (IIoT)	High reliability, low latency, flexibility, and security (E2E latency of 1–2 ms and reliability of 99.999%)
Space–terrestrial integrated network	New addressing and routing mechanisms, bandwidth capacity at the satellite side, admission control by satellites, edge computing, and storage
Network and computing convergence	New addressing and routing mechanisms, bandwidth capacity at the satellite side, admission control by satellites, edge computing, and storage
Intelligent operation network	Intelligent closed-loop control, instantaneous high-volume data collection for network status, programmability and softwarization, low-latency event-driven response with data prioritization

TABLE 14.4
Relative Scores of Network Requirement in Selected Use Cases

Use Cases	Bandwidth	Time	Security	AI
Holographic-type communications	+++	++	+	+
Digital twin	++	++	+++	+
Tactile Internet for remote operations	+	+++	++	+
Industrial IoT	++	+++	++	+
Space–terrestrial integrated network	+	++	++	+
Network and computing convergence	+	+++	+++	+
Intelligent operation network	+	++	++	+++

We have compared the requirements of the selected use cases by relative scoring the proposed dimensions (Table 14.4).

Based on the analysis, the most prominent requirements of each case can be easily extracted, and their accumulated statistics further provide network designers with a high-level perspective of the various dimensions of future use cases.

14.3.2 Definition of System Capabilities

6G systems are expected to expand and support the various usage scenarios and applications that represent the 5G evolution. Moreover, the wide range of possibilities is closely related to the selected 6G usage scenarios and applications. The capabilities of systems necessary to support usage scenarios can be seen as an extension of those for 5G. There will be classic capabilities that were used to define 5G, but in addition new ones to support extended and expanded usage scenarios.

Different aspects of the wide range of capabilities are differently relevant and applicable in reference use scenarios. Key design principles include supremacy, efficiency, flexibility, trust, native intelligence and automation, and sustainability to serve diverse use cases in the future of connected intelligence.

The overall capabilities of the 6G system include those with quantitative indicators and those with only functional indicators. Under each of the categories, one or more key capabilities are defined, which will lead to performance and feature requirements: performance, localization and sensing, connectivity, services, and terminal/device.

We anticipate significant increase in data traffic and the number of connected things with 6G. Device density grows to hundreds of devices per cubic meter, which impose stringent requirements on area or spatial spectral efficiency and required

TABLE 14.5
6G vs. 5G KPI Key Capability Indicators

	KPI	Target Range
Performance	User experience data rates	Gbps (10–100×)
	Area traffic capacity	01.–10 Gbps (10–1,000×)
	Peak data rate	100 Gbps–Tbps (5–100×)
	Connection density	10^7–10^8/km² (10–100×)
	Air latency	0.1–1 ms (1/10×)
	Jitter	Micro-second-level
	Mobility	1,000 kmph (2×)
	Reliability	99.99999% (100×)
	Coverage	Extension to global coverage
	Sensing/positioning accuracy	Centimeter level
	AI services accuracy/efficiency	>90%
Efficiency	Spectrum efficiency	1.5–3×
	Energy efficiency	20×
	Cost-efficiency	Low cost in network construction and O&M
	Trustworthiness	Balanced security, durable privacy protection, and advanced system resilience

frequency bands for connectivity. Security, privacy, and reliability are important emerging KPIs. 6G is hyper-secure with demanding requirements for industrial and high-end users, while at the same time it will be low cost and low complexity for IoT applications (Table 14.5).

Improved 6G system capabilities support new use cases, including applications that require very high data rate communications, large numbers of connected devices, and ultralow-latency, high-reliability applications (Figure 14.2).

Significant improvements are needed in the quality of user experience and peak data rates, energy and spectral efficiency, air latency, connection density, and reliability. 6G systems are expected to reuse existing 5G capabilities whenever possible, with improved requirements and KPIs. New capabilities are introduced as needed when existing capabilities are insufficient to support 6G usage scenarios. It is essential that the overall capabilities provide design flexibility and system optimization. Realization of the technical possibilities implies significant challenges that should be overcome.

14.3.3 Focus Areas for Further Study

Development planning of 6G takes into account the deadlines associated with their implementation, which depend on a number of factors such as user trends, user requirements and demand, technical capabilities and technology development, standards specification and their enhancement, spectrum issues, regulatory considerations, and system deployment. All of these factors are interrelated. The first five have been and

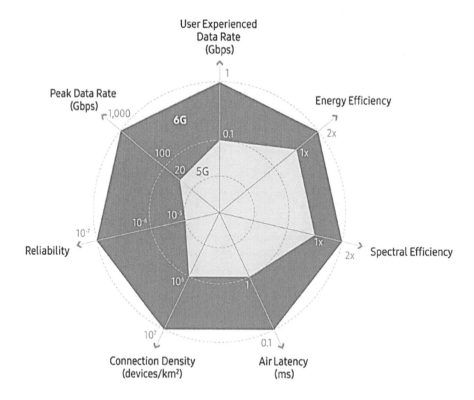

FIGURE 14.2 Improvements of 6G vs. 5G key capabilities (KPI).

will continue to be addressed within ITU. The development and implementation of the system relates to the practical aspects of deploying new networks, taking into account the need to minimize additional investments in infrastructure and to allow time for users to adopt the services of the new system.

- In the mid-term period (up to the year 2030), it is predicted that the future development of 5G systems will progress with the continuous improvement of initial implementation capabilities, as demanded by the global market in addressing the needs of users and as permitted by the status of technical development.
- The long-term period (starting around the year 2030) is associated with the potential introduction of 6G systems that could be deployed around 2030 in some countries. The research is encouraged to focus on the key areas such as radio interface(s) and their interoperability, access network issues, spectrum issues, and traffic characteristics.

6G services are suitable for offering new spectrum bands. It should be noted that the existing bands for 5G will continue and are reframed for 6G. For 6G systems, the spectrum from 100 GHz to 1 THz is considered. Using all windows will not be suitable for all use cases. For example, the first window of interest from 140 to 350 GHz (high mmWaves) has advantages because there are many tens of GHz of bandwidth that are currently unused while developing ultra-massive multiple input–multiple output (MIMO) antenna arrays with reasonable factors. The array factor is a function of the position of the antennas in the array and the weights used. By adjusting these parameters, the performance of the antenna array is optimized and the desired properties are achieved. The spectrum in higher windows is accompanied by greater absorption loss. One can use high frequency up to 10 THz at the expense of beyond hardware realization challenges.

In 6G networks, the 5G paradigm will be refined and expanded. One possibility is the virtualization of (critical) end-to-end connectivity from the device over the cellular network to the packet data network and to the cloud. Under the 6G paradigm, the network seeks to maximize the quality of experience (QoE) through intelligent traffic management, edge computing, and policies set by the user either proactively or per transaction or through traffic orchestration. The latter may, for example, use policies set by the user or by the operator for a group of subscribers who are each treated equally within that group. It is network neutral in the sense that it treats all applications in the network and all users with the same subscription type equally.

Recently, there has been a growing interest in machine learning (ML) and artificial intelligence (AI). ML relies on Big Data that is mined to gain information and knowledge. There are other needs in networking that require intelligence such as self-configuration or complexity management. In addition to Big Data, AI relies on an abundance of computing power. 6G uses the growing computing power for coping with the higher transmission rates, but also to achieve additional flexibility. However, power consumption increases dramatically, which requires further research.

14.4 CONCLUDING REMARKS

The sixth-generation (6G) mobile network is envisaged to be commercially deployed around 2030, which will profoundly change people's lifestyles and accelerate the digitalization of society. Each usage scenario is related to a set of key capabilities. To ensure that the requirements of 6G can be achieved, it is essential to establish a set of key performance indicators (KPIs). This chapter comprehensively assesses the KPIs not only from the service requirements but also from the technical feasibility points of view.

The 6G wireless communications support services in various usage scenarios and applications, including enhanced mobile broadband (eMBB), low-latency communications (uRLLC), and machine mass communication (mMTC) with stringent performance dimensions such as increased system capacity, low latency, high reliability, higher spectral efficiency, as well as enabling the massive IoTs (mIoTs) into a fully connected, intelligent digital world. Furthermore, 6G communication networks are the first generation of AI native networks. This means that AI will not just be an application, but an inherent part of infrastructure, network management, and operations.

In conclusion, we have an exciting future ahead of us. The road to overcoming the challenge is filled with challenges, but there are enough insights to start research toward promising directions that will serve as motivation for researchers in the near future. The network complexity could potentially be overcome in convergence with 6G and AI, paving the way to a sustainable ecosystem. Challenges remain on how to tailor AI on edge nodes and systematically work toward green 6G.

BIBLIOGRAPHY

3GPP TR37.885, *Study on NR vehicle-to-everything*, Mar. 2019.

3GPP TSG Services and System Aspects, Management and Orchestration, *5G end to end key performance indicators (KPI)*. TS 28.554 Release 17 V17.7.0, Jun. 2022.

6G Flagship White Paper, *Key drivers and research challenges for 6G ubiquitous wireless intelligence*. 6G Research Visions 1, Sept. 2019.

A. Benjennour, K. Kitao, Y. Kakishima, C. Na, 3GPP defined 5G requirements and evaluation conditions. *NTT DOCOMO Technical Journal*, vol. 19, no. 3, pp. 13–23, Jan. 2018.

A. Benjennour, K. Kitao, Y. Kakishima, C. Na, History of 5G initiatives. *NTT DOCOMO Technical Journal*, vol. 22, no. 2, pp. 4–12, Oct. 2020.

A. Clemm et al., Toward truly immersive holographic type communication: Challenges and solutions. *IEEE Communications Magazine*, vol. 58, no. 1, pp. 93–99, Jan. 2020.

A. Yastrebova et al., Future networks 2030: Architecture and requirements. In *Proc. 10th International Congress on Ultra Modern Telecommunications and Control Systems (ICUMT)*, 2018.

Beyond 5G Promotion Consortium, *Beyond 5G– Message to 2030*. White Paper, Mar. 2022.

C. Zhang et al., Breaking the blockage for Big Data transmission: Gigabit road communication in autonomous vehicles. *IEEE Communications Magazine*, vol. 56, no. 6, pp. 152–157, Jan. 2018.

C.-X. Wang, M.D. Renzo, S. Stanczak, S. Wang, E.G. Larsson, Artificial intelligence enabled wireless networking for 5G and beyond: Recent advances and future challenges. *IEEE Wireless Communication*, vol. 27, no. 1, pp. 16–23, Feb. 2020.

D. Blinder et al., Signal processing challenges for digital hologram video display systems. *Signal processing Image Communication*, vol. 70, no. 2, pp. 114–130, Feb. 2019.

D. Patel, Y. Sachs, *5G E2E technology to support vertical uRLLC requirements*. White Paper 5G-Smart, EU Horizon, 2020.

D.M. Botín-Sanabria et al., Digital twin technology challenges and applications: A comprehensive review. *MDPI Remote Sensing*, vol. 14, no. 1335, pp. 1–25, 2022.

ETSI TS 128 554 Management and Orchestration, *5G end to end key performance indicators (KPI)*, May 2019.

G.P. Fettweis, The Tactile Internet: Applications and challenges. *IEEE Vehicular technology Magazine*, vol. 9, no. 1, pp. 64–71, 2014.

H. Tataria et al., 6G wireless systems: Vision, requirements, challenges, insights and opportunity. *IEEE Proceedings of the IEEE*, vol. 109, no. 7, pp. 1166–1199, Jul. 2021.

H.-J. Song, T. Nagatsuma, Present and future of terahertz communications. *IEEE Transactions on Terahertz science and technology*, vol. 1, no. 1, pp. 256–263, Sept. 2011.

H.X. Nguyen, R. Trestian, D. To, M. Tatipamula, Digital twin for 5G and beyond. *IEEE Communications Magazine*, vol. 59, no. 2, pp. 10–15, 2021.

IMT-2030 (6G) Promotion Group China, *6G vision and candidate technologies*. White Paper, Jun. 2021.

ITU-R Recommendation M.2083, *IMT vision – Framework and overall objectives of the future development of IMT for 2020 and beyond*, Feb. 2012–Sept. 2015.

ITU-R Report M.2320, *Future technology trends of terrestrial IMT systems*, Nov. 2014.

ITU-R Report M.2410, *Minimum requirements related to technical performance for IMT-2020 radio interface(s)*, Nov. 2017.

ITU-R Working Party 5D (WP 5D), *Progress of recommendation ITU-R IMT vision for 2030 and beyond*. Workshop on IMT for 2030 and Beyond, Jun. 2022.

Evaluation of Representative 6G Use Cases

ITU-R WP5D Contribution 1232, *Considerations on usage scenarios and capabilities of IMT for 2030 and beyond for the development of working document towards preliminary draft new recommendation IMT vision for 2030 and beyond*, Jun. 2022.

ITU-R WP5D Meting 41st, Doc.1232, *IMT vision – Framework and overall objectives of the future development of IMT for 2030 and beyond*, Jun. 2022.

ITU-R WP5D SVG Recommendation, *IMT vision – Framework and overall objectives of the future development of IMT for 2030 and beyond*, Mar. 2021 – Preliminary Draft Jun. 2022–Jun. 2023.

ITU-T, Focus Group on Technologies for Network 2030, *A blueprint of technology, applications and market drivers towards the year 2030 and beyond*. White Paper, May 2019.

ITU-T, Focus Group on Technologies for Network 2030, *Additional representative use cases and key network requirements for Network 2030*. Technical Report, Jun. 2020.

ITU-T, Focus Group on Technologies for Network 2030, *Description of demonstrations for Network 2030 on sixth ITU workshop and demo day*. Technical Report, Jan. 2020.

ITU-T, Focus Group on Technologies for Network 2030, *Gap analysis of Network 2030 new services, capabilities and use cases*. Technical Report, Jun. 2020.

ITU-T, Focus Group on Technologies for Network 2030, *Network 2030 architecture framework*. Technical Specification, Jun. 2020.

ITU-T, Focus Group on Technologies for Network 2030, *Representative use cases and key network requirements*. Technical Report, Feb. 2020.

ITU-T, Focus Group on technologies for Network 2030, *Representative use cases and key network requirements*. Technical Report, Feb. 2020.

ITU-T, Focus Group on Technologies for Network 2030, *Terms and definitions for Network 2030*. Technical Specification, Jun. 2020.

J. Finkelstein, Broadband network evolution. In *Prog. 1st ITU Workshop Network 2030*, pp. 1–23, Oct. 2018.

K. Leppänen, M. Latva-aho, Key drivers and research challenges for 6GFlagship. In *Prog. 5th ITU Workshop Network 2030*, pp. 1–18, Oct. 2019.

K. Liolis et al., Use cases and scenarios of 5G integrated satellite-terrestrial networks for enhanced mobile broadband: The SaT5G approach. *International Journal of Satellite Communications and Networking*, vol. 37, no. 2, pp. 91–112, 2018.

K. Makhijani, Holographic type communication: Delivering the promise of future media by 2030. In *Prog. 5th ITU Workshop Network 2030*, pp. 1–24, Oct. 2019.

M. Giordani et al., Toward 6G networks: Use cases and technologies. *IEEE Communications Magazine*, vol. 58, no. 3, pp. 55–61, Mar. 2020.

M. Latva-aho, Radio-access networking challenges towards 2030. In *Proc. 1st ITU Workshop Network 2030*, pp. 1–22, Oct. 2018.

M. Li, Enabling technologies for future networks. In *Prog. 1st ITU Workshop Network 2030*, pp. 1–13, Oct. 2018.

M. Shahbazi, S.F. Atashzar, R.V. Patel, A systematic review of multilateral teleportation system. *IEEE Transactions on Haptics*, vol. 11, no. 3, pp. 338–356, 2018.

M.A. Usitalo et al., 6G vision, value, use cases and technologies from European 6G Flagship project Hexa-X. *IEEE Access*, vol. 9, pp. 160004–160020, 2021.

M.N.A. Rahim et al., 6G for vehicle-to-everything (V2X) communications: Enabling technologies, challenges, and opportunities. *Proceeding of the IEEE*, vol. 110, no. 6, pp. 712–734, Jun. 2022.

N.H. Mahmood et al., A functional architecture for 6G special purpose industrial IoT networks. *IEEE Transactions on Industrial informatics*, 2022. (Early access).

P. Varga et al. 5G support for industrial IoT applications – Challenges, solutions, and research gaps. *MDPI Sensors*, vol. 20, no. 828, pp. 1–43, 2020.

R. Dilli, Design and feasibility verification of 6G wireless communication systems with state of the art technologies. *International Journal of Wireless Information Networks*, vol. 29, pp. 93–117, 2022.

R. Kantola, Trust networking for beyond 5G and 6G. In *Proc. 2nd 6G Wireless Summit 2020*, pp. 1–7, 2020.

R. Li, Networks 2030: Market drivers and proposals. In *Prog. 1st ITU Workshop Network 2030*, pp. 1–21, Oct. 2018.

R. Saccaro, Digital twins: Bridging physical space and cyberspace. *IEEE Computer*, vol. 52, no. 12, pp. 58–64, Dec. 2019.

R.W.L. Coutinho, A. Boukerche, Design of edge computing for 5G-enabled Tactile Internet-based industrial applications. *IEEE Communications Magazine*, vol. 60, no. 1, pp. 60–66, Jan. 2022.

S.A.A. Hakeem, H.H. Hussein, H.W. Kim, Security requirements and challenges of 6G technologies and applications. *MDPI Sensors*, 22, 1969, pp. 1–43, 2022.

The ITU-T Focus Group Technologies for Network 2030, *(FG NET-2030) was established by ITU-T Study Group 13 at its meeting in Geneva, 16–27 July 2018 and concluded its activity on July 2020.*

V. Petrov et al., Last metric indoor terahertz wireless access: Performance insights and implementation roadmap. *IEEE Communications Magazine*, vol. 58, no. 6, pp. 158–165, Jun. 2018.

Y. Huang et al., 6G mobile network requirements and technical feasibility study. *China Communications*, vol. 19, no. 6, pp. 123–136, Jun. 2022.

Index

Note: **Bold** page numbers refer to tables and *italic* page numbers refer to figures.

3GPP 5GC core network 16
    control and user plane separation 19, 87
    mobility management 18, 142
    network functions 15, 18
    point-to-point reference architecture *14*, 15, *18*
    security 34
    service-based reference architecture *18*
    session management 15–18, 33
    virtualized network functions 19, 87
3GPP 5G NR (new radio) network
    BS (base station) radio transmission and reception 13
    CP protocol stack *20*
    data transport 15
    F2 interface connect lower/upper parts of PHY layer 16
    IAB-node (integrated access and backhaul) wireless rallying 42, 90
    NG interface connect gNodeB and 5GC 13
    procedures (IAM/Beam management/Power control/HARQ) 21, 31, 90, 389
    scheduler operation 278
    signalling transport 15
    UE (user equipment) radio transmission and reception **10**, 81
    UP protocol stack *20*
    Xn interfaces interconnect gNodeB 13
3GPP 5G NR Layer 1 (PHY layer)
    carrier aggregation 89
    downlink transmission 21, 26, 105
    encoding, decoding, modulation, multiantenna processing, signal mapping 21
    hybrid automatic repeat request (HARQ) 21, 27
    LDPC codes 142, 329
    PHY channels and modulation 21
    PHY channels multiplexing and coding 25
    PHY layer measurements 31
    PHY layer procedures for control/data 15, 53
    polar codes 346
    transport channels (BCH, DL-SCH, UL-SCH, SL-BCH) 16, 20
    uplink transmission 21
3GPP 5G NR Layer 2 (data link layer)
    control channels (BCCH, PCCH, CCCH, DCCH) 89, 328

data radio bearers (DRBs) 145
HARQ (hybrid automatic repeat request) 21, 27
logical channels mapping onto transport channels (DTCH) 16, 279
MAC (medium access control) multiplexing/demultiplexing 16, 104
PDCP (packet data convergence protocol) 20, 104
PDUs (protocol data units) 18
QoS flows (QFs) 33, 88, 104
RLC (radio link control) transmission modes (TM, UM, AM) 16, 21
SDAP (service data adaptation protocol) 20, 104
SDUs (service data units) 141
3GPP 5G NR Layer 3 (network layer)
    QoS management 22
    RRC (radio resources control) protocol 16, 22, 104
    security functions (key management) 22
3GPP 5G NR support of services and verticals
    enhanced mobile broadband (eMBB) 6
    industrial IoT (IIoT) 6
    private networks (NPN) 39
    ultra-reliable and low latency communications (URLLC) 6
    vehicle-to-everything (V2X) 6, 37
3GPP 5GS system *13*
    5GC core network **14**, 16
    charging 13, 18
    codecs and media 111
    connectivity 37
    interoperability 13–15
    mobility and roaming 14
    network management 54
    NG interface 13
    NG-RAN nodes **14**
    operations and maintenance 10
    security 34
    UE user equipment 8, 18, 141
    user services 101
3GPP conformance testing 8
    minimum level of performance 41
3GPP CU-DU functional split 15, *16*; *see also* 3GPP 5G NR
    baseband processing 50, 68
    centralized unit (CU) 5, 56

481

Index

3GPP CU-DU functional split (cont)
  C-RAN centralized virtualized architecture 29, 66
  distributed unit (DU) 5, 56
  F1/F2 interface 16
  options 1-8, *16*
  radio unit (RU) 56
3GPP MTC (machine-type communication) 152
  eMTC 152, *431*
  FeMTC 229, 431
  NB-IoT **37**, 152, 365, 431
3GPP SA1 use cases requirements 111
3GPP SA4 codecs and media 111
3GPP Technical Specification Groups (TSG) 8
  core network and terminals (CT) 8
  cycle of activities (Stage 1, Stage 2, Stage 3, ASN.1) 112
  initial study (SI) 8
  radio access networks (RAN) 8
  services and systems aspects (SA) 8
  working groups (WG) 34, 44
  working item (WI) 8
3GPP Technical specifications 4
  37.XXX, 38.100, 38.200, 38.300, 38.400 series 8
  feasibility study (FS) 8, 105, 111, 438
  release (R) 8
  technical report (TR) 8, 96, 110
  technical specification (TS) 7, **12**, 86 (see also ETSI, ITU-R IMT)
3rd Generation Partnership Project (3GPP) 159
  organizational partners (ARIB, ATIS, CCSA, ETSI, TSDSI, TTA, TTC) 8
  PCG (project coordination group) 45
  standardization framework 7, 153

5G-Advanced **12**, 44, 55, 138
  3GPP release 15 **12, 37, 38**
  3GPP release 16 **12, 37, 38**
  3GPP release 17 **12, 38**
  3GPP release 18 **12**, 18, *144*, 259
5G enabling technologies 156, 359
  device-to-device (D2D) 373
  full-duplex 42
  green communication 147
  machine-to-machine (M2M) 372
  millimeter Wave 66, 477
  network function virtualization (NFV) 141
  quality of service (QoS) 33
  vehicle-to-everything (V2X) 6, 37
5G MBS multicast and broadcast services 97
  broadcast (delivery mode 2) 104
  mixed-mode multicast 98
  multicast (delivery mode 1) 104
  multicast/broadcast service function (MBSF) 103
  multicast/broadcast service transport function (MBSTF) 104
  overall high-level architecture 97
  point-to-multipoint/point-to-point (PTM/PTP) switching 105
  stand-alone broadcast 98
5G MS media streaming 93
  application function (AF) 94
  application server (AS) 95
  common media application format (CMAF) 95, 100
  overall high-level architecture 94
  profiles 95
  UE internal function client 95
5G multimedia production and distribution 84, 88
  5G-MAG multimedia action group 107
  DVB-I over 5G 107
  EBU 5GCP 108
5G New radio (NR) *see* 3GPP 5G NR
5G NR protocol stack 8, **105**
  control plane protocol stack 20, 143
  user plane protocol stack **20**, **143**
5G NR-Light a.k.a. RedCap (reduced capability) **12**, 44–48; *see also* UE
5G QoS model (*see* QoS)
  flow forwarding 34
  service parameters 113
  trade-off capability *vs.* complexity **145**
5G security 34
5G services 139, *358*; *see also* KPI
  key metrics (capacity, latency, connection density) 139
5G system architecture 12, **36**, *46*, 87, **103**
  core network (CN) 16
  NSA (non-standalone) mode deployment 7, 12
  radio access network (RAN) 8
  reference architecture 14
  SA (standalone) mode deployment 10, **12**, *17*
  transport network (TN) 5
5G system scheduler
  channel scheduling *92*
  data traffic management 293
  packet scheduler 294
  RB allocation 279
  resource scheduler 299
  scheduling in MAC layer 297
  scheduling in massive MIMO 302
  scheduling in PHY layer 304
  traffic profile 295
  TTI (time transmission interval) 279
  Type A (slot based scheduling) **26**
  Type B (mini-slot based scheduling) **26**
  uplink/downlink scheduling 21
5G technology ecosystems 106
  commercialization **107**
  innovation **107**
  PoC **107**
  standardization **107**
  system trials **107**

Index

vision **107**
5G terrestrial broadcast
    FeMBMS (future evolved multimedia broadcast and multicast services) 97
5G vertical domains **12**, **64**
    automotive industries (V2Xs) 37, 159
    industrial Internet of things (IIoT) 36
    vertical expansion **12**
5G XR immersive communication 120
5GS reference architecture
    interface names
    point-to-point model *14*, *15*, *18*
    reference points 15
    service-based model *18*

6G service classes
    control, localization, and sensing and energy services (3CLS) 58
    human-centered services (HCS) 57, **59**
    immersive, intelligent, and ubiquitous usage scenarios **471**
    key network requirements 474
    massive URLLC (mURLLC) 57, 59
    mobile broadband reliable communication with low-latency (MBRLLC) 57
    multipurpose (MPS) 58, **59**
6G systems
    3D nature **60**
    candidate technologies **472**
    research directions 60, 64, 464
    space–terrestrial integrated networks (STIN) 468
    system capabilities 474
    Terahertz communication **472**

AAS (advanced antenna system) 48; *see also* beamforming
    antenna gain 25
    antenna port 87
AI (artificial intelligence) 51, 68, 374, 408
    decisions 158
    intelligent networks 8, 170, 375, 404
    RAN intelligence 158 (*see also* SON)
    self-sustaining network 68
    solutions 49, 146
AI-enabled massive IoT 169
AI/ML-based solutions
    load balancing 146
    mobility optimization 157
    network energy savings 157
    services 46
    system-level optimization 68
AI/ML industry initiatives **49**, 57
AI/ML support 153, 157, 171
    network intelligence 54, 382
    NR air interface 146
    PHY enhancements 49

RAN architectures 55, *158*
AIoT (artificial intelligence of things) 262–267
API (application programming interface) 5, 85
    distribution services 107
    exposure of capabilities, events, and analytics 18
    northbound interface 222
    southbound interface 222
    Web application 432

base station (BS) 20, 284; *see also* 3GPP CU-DU split
    backhaul transport networks 22
    baseband unit (BBU) 15
    cellular 9, 373
    gNodeB parameters **301**
    remote radio unit (RRU) 15
beamforming 25, 48, **65**, 87, 142; *see also* MIMO
    3D performance metrics 9, **60**

CAPEX (capital expenditures) 56, 153, 222; *see also* OPEX
CDN (content delivery network) 85, 96
cellular systems 139, 151
cloud computing (CC) 14, 254, **381**
    cloud service models 383
    edge computing (EC) 32, 221
    fog computing (FC) **64**, 263
cloud RAN (C-RAN) 150
    centralized cloud 377
    virtualized vRAN 52
control and data plane separation 222; *see also* SDN
control and user plane separation 19, 87
CP (control plane); *see also* UP
    AF (application function) 88
    AMF (access and mobility management function) 18, 88, 142
    NEF (network exposure function) 18, 88
    SMF (session management function) 18, 33, 88
CR (cognitive radio) 170
CRAS (connected robotics and autonomous systems) **59**, **61**
critical communications 278, *371*, 429, 438
CSI (channel state information) 146, 302

D2D (device to device communication) 26, 373
DC (data center) 5, 121, 231, 290; *see also* cloud computing
DC (double connection) 13, **47**
degrees of freedom (DoFs)
    3DoF 109, 146
    3DoF+ 109, 116
    6DoF 109, 116, 146
digital twins (DT) 260
    cyber-physical systems (CPS) 265, 466

digital twins (DT) (*Cont*)
   industry DT 270
   models 265, 270
   network DT 269
   network emulation 269
DNN (deep neural network) 126, 257, 317, 333
duplexing schemes 39
   FDD (frequency-division duplex) 25–28, 147, 259
   TDD (time-division duplex) 25–28, 147, 259

efficiency indicators
   cost efficiency 438, 441
   energy efficiency 40, **67**, 156
   spectrum efficiency 10, 140, 259
eMBB (enhanced mobile broadband) 6, 48, 57, 88, 159, *167*, 328, 469
end-to-end (E2E) communication 118, 154
   latency 141, 149, 164, 168
   reliability 4, 37
   slice management 440
   video coding 124, 127
energy efficiency (EE) 40, 156, 160
   EC (energy consumption) 147, 156
   ES (energy saving) 47, 147, 156, 372
   metrics 158
ETSI (European Telecommunications Standards Institute) 107, 145, 156
   ISG ARF augmented reality framework 122
   ISG ENI experimental network intelligence 54
   ISG NFV network functions virtualization 222

flow forwarding 34
frame structure (FS) 27–29, 142; *see also* 3GPP 5G NR Layer 1
   bandwidth part (BWP) 27
   numerology 27
   OFDM symbols 27
   resource block (RB) 21, *29, 30*
   resource element (RE) 27, 30
   resource grid (RG) 29
   slots 27
   sub-frame 27
frequency range 1 (FR1 NR operations 425 MHz - 7.125 GHz) 25, 86, *87*
frequency range 2 (FR2 NR operations 24.25 - 52.6 GHz) 25, *87*; *see also* mmWave
frequency range 2-2 (FR2-2 24.25 - 71 GHz) 90; *see also* mmWave

generic communication services 6, 139
green networks 147

HetNet 373
holographic communication 465
   DVB-VR CM 122
   holographic radio **61**
   IEEE Digital reality 122
   JPEG Pleno Holography 122
   ITU ILE (immersive live experience) 122
   MPEG Immersive video (MIV) 122

immersive communication
   augmented reality (AR) 407
   camera calibration 409
   detection and tracking 410
   displaying virtual objects 411
   extended reality (XR) 403, 415
   interaction in the virtual world 414
   mixed reality (MR) 407
   pose estimation 409
   processing requirements and resource constraints 417
   simultaneous localization and mapping (SLAM) 118, 410
   virtual reality (VR) 108, 407
immersive media 111–117, 465, 469
   MPEG-I representation of immersive media 116
   NBMP (network-based media processing) 117
   OMAF (omnidirectional media format) 117, 118
   point cloud compression (PCC) 117
IMT broadband mobile systems 7, 62, 464
IMT-2020 6, **41**, 145, 258
IMT-2030 62
IMT-2020 standardization framework
   call for proposal and consensus building 6
   ITU-R vision and definition 6
   minimal prerequisites and evaluation criteria 6
   technical specification 4
IMT-2020 terrestrial radio interface specifications; *see also* 3GPP 5G NR (New Radio)
   global core specifications (GCS) 6
   LTE/E-UTRA (Long Term Evolution / Evolved-Universal Terrestrial Radio Access) 21
   radio and device testing 7
   RIT (radio interface technologies) 7
   Set of RIT (LTE+NR, DECT-2020) 145
IMT-2030
   vision for 2030 and beyond 62, 464
indoor localization *366*
   high accuracy 122
industrial internet of things (IIoT)
   Industry 4.0 116, 121, 157
   use cases (critical IoT, massive IoT, broadband IoT) 159
intelligent connectivity (ICon) 390–395
intelligent transportation system (ITS) 166
IoE (internet of everything) 10, 259

Index

ITU Radiocommunication Sector (ITU-R)
ITU-R IMT (international mobile telecommunications) standard
  global standardization timeline 9
  independent evaluation groups (IEGs) 145
  selected test environments **41**
  service and spectrum requirements 7
  submission and evaluation process **41**, 258
  usage scenarios (eMBB, URLLC, mMTC) 6
ITU-T (telecommunication standardization sector)
  ML5G (machine learning for future networks) **49**
  NET-2030 Focus Group 62
ITU WRC (world radiocommunication conference) 63
  ITU-R Radio Regulations (RR) 7

joint sensing and communication **10**
joint communication and control **61**
joint optimization **61**
joint learning–communication co-design 68
joint distribution 188, 192
joint probability density function (JPDF) 193

KPIs (key performance indicators) 40, 475
  category (eMBB, URLLC, mMTC) 328, **140**
  minimum performance targets 41
  test case requirements 41

LDPC (low-density parity check) codes 27, 142, 281
link
  downlink rate 25, 26
  radio link 16, 66
  uplink rate **25**, 26
load
  balancing 49, 158, 380
  download 89
  offload 120
  prediction 39, 157
  upload 89
low-latency communication 37, 137, 159, 222, 328; *see also* uRLLC
  3GPP general timing model 119
LTE (long-term evolution) 7
  eNodeB base station 17
  EPC core 17
  LTE-Advanced 100, 369

machine learning (ML); *see also* AI/ML
  cross-checked validation 128
  deep learning (DL) 126–129, 257, 291–302, 310–316, *322*, 333, 375
  federated learning (FL) 171
  inference 146
  model **49**, 146, 170, 258

reinforcement learning (RL) 291, 375
supervised learning 285, 374
testing 146, 258
training data set 11, 51, 123, 258
unsupervised learning 289, 375
validation 146, 258
machine type communications (MTC)
  critical MTC (cMTC) 258
  massive MTC (mMTC) 258
  ultra-reliable MTC (uMTC) 6
MDAF (management data analytics function) 157; *see also* NWDAF
MEC
  mobile edge computing 9, 68, 151, 254
  multiaccess edge computing 141, 263
media delivery platform 86
  broadcast 44, 84
  multicast 44, 85
  unicast 84, 94
MIMO (multiple-input multiple-output); *see also* beamforming
  cell-free massive 67
  holographic 61
  large antenna arrays 317
  massive 302
  multiantenna techniques 26
  multiuser 280
  ultra-massive **67, 472**
mmWave 66, 477
  5G radio spectrum FR2 band 12, 144
  tiny cells **60**
mobile broadband systems
  3GPP system 162
  IMT system 7
mobile communication 4, 58, 63, 86
  5G system 11
  6G system 68
mobile networks 149
  capacity 81, 115
  coverage 10, 81
  latency 153, 311
  operator (MNO) 85, 114, 138
multimedia communications 68, 92, 106

network functions (NFs) *18*, 52
  AF (Application Function) 18
  AMF (access and mobility management function) 18
  AUSF (authentication server function) 18
  NEF (network exposure function)
  NRF (NF repository function) 18
  NSSF (network slice selection function) 18
  PCF (policy control function) 18
  SMF (session management function) 18
  UDM (unified data management) 18
  UPF (user plane function) 18
network slicing (NS) 31–33, 141, 224, 299

network slicing (NS) (*cont*)
  cross-slice isolation 32
  network functions (NF) 18
  service-level agreement (SLA) 31, 439
NFV (network function virtualization) 141, 222, 372; *see also* SDN
  centralized/distributed model 228, 233
  NFV infrastructure (NVFI) 224
  NFV orchestration 224
  queuing theory 227
  reference architecture 223
  VM-based virtualized network functions (VNFs) 157, 226, 87
non-public network (NPN) 155
  DNN (data network name) 154
  private networks 39
  public network integrated PNI-NPN 40, 154
  standalone SA-NPN 40, 153
NTN (non-terrestrial networks) 47
  NTN IoT 145
  satellite networks 44
  UAV (unmanned aerial vehicle) **47**, 185, 278
NWDAF (network data analytics function) 5, 18, 54; *see also* MDAF
  3GPP eNA (enablers for network automation) 46
  data analytics function 18, 54
  network automation 44, **49**, 55, 146
  operations, administration, and maintenance (OAM) *55*, 156

OAM (operations, administration, and maintenance) 54, 156
  procedures 54
OAM (orbital angular momentum) **472**
OFDM (orthogonal frequency division multiplexing) 21; *see also* 3GPP 5G NR Layer 1
  baseband signal 87
  cyclic prefix CP-OFDM 26
  DFTS-OFDM 26
  grid 91
  scalable SC-OFDM 26
  subcarrier spacing 26, 148
  symbols 27, 280
OPEX (operational expenditure) 153, 222; *see also* CAPEX
O-RAN (open RAN) **49**
  A1 interface 54
  alliance 52
  E2 interface between near-RT and DU 54
  near-RT (near-real-time) xApp 49
  non-RT (non-real-time) rApp 49
  reference architecture 53
  RIC (RAN radio intelligent controller) 53

performance indicators 40, **140**, **475**

connection density 41, 58
energy efficiency 41, 59
latency 41
mobility 41
peak data rate 278, 328, 406, 475
traffic density **41**, 58, 164
user data rate **41**
privacy and security issues
  AI regulation and accountability 458
  data privacy by design 455
  DPIA (data protection impact assessment) 449
  GDPR (general data protection regulation)
    concepts 445
    privacy requirements 445
  separation of personal data from non-personal data 450
proof of concepts (PoC) 66, 106

QAM (quadrature amplitude modulation) 26
QoE (quality of experience) 4
  metrics reporting 96
  quality of experience index (5GQI) 34
QoPE (quality of physical experience) 57–60
QoS (quality of service) 33
  5QI flow identifier 88
  policies 88, 96
QPSK (quadrature phase-shift keying) 26
quantum communication 66

radio access technology (RAT) 11, 117, 137, 228
radio spectrum
  dynamic spectrum sharing (DSS) 44, 48
  licensed 25, **47**
  range of frequencies 278
  unlicensed 25, 39
RAN (radio access networks) 8, *13*, 20, 151
research programs 63
  5G Public Private Partnership (5G PPP) 150, 159, 227, 261
  focus areas 475
RISs (reconfigurable intelligent surfaces) **47**, 185, **472**
  holographic MIMO **61**
  intelligent reflecting surface (IRS)
  large intelligent surface (LIS)
  optimal deployment and location 60
  transmitters vs. reflectors 60

SAE (Society of Automotive Engineers)
  levels of automation (LoA) 166
SAPs (service access points) 16
SBA (service based architecture); *see also* 3GPP 5GC
  IDL (interface description language) 5
  SBI (service based interface) 5, 54
SDAE (sparse denoising autoencoder) 344
SDN (software-defined networking) 5, 19, 222

Index

network functions virtualization (NFV) 5, 222
openflow (OF) protocol 222
separation CP (control plane) and DP (data plane)
service-based interfaces (SBIs)
NBI (northbound interface) 222
SBI (southbound interface) 222
virtualized network functions (VNFs) 19, 41
SDO (standards development organizations) 6, 11, 266
SE (spectral efficiency) 89, 153, 160, **406**
SL (sidelink) operation **47**, 146, 169; *see also* D2D
PC5 interface 52
positioning and ranging 144
V2X communication and public safety 90
Smart City (SC) 255, 357, *371*, 393
Smart Grid (SG) 360, 367, 425
advanced metering 427
real-time situational awareness 427
smart grid architecture model (SGAM) 437
virtual PMU (phasor measurement unit) 427, 438
wide-area monitoring system (WAMS) 434
wide-area monitoring, protection, and control (WAMPAC) 426
state estimation (SE) 427
SON (self-organizing networks); *see also* O-RAN
automate RAN deployment 160
self-configuration BS functions **48**
self-optimisation BS functions **48**

tactile/haptic-based communication 468
terrestrial mobile networks 7, 39; *see also* IMT-2020
TSC (time-sensitive communications) 42, 162

UE (user equipment) **18**, 91
UP (user plane) (*see* CP)
user plane functions (UPF) 18, 88
data network (DN) 88
data protocol unit (PDU) session 18, 33
URLLC (ultra-reliable low-latency communications) **12**, 57, **64**, 149, 159
usage scenarios
5G use cases 4, 428
6G evolution *469*, *471*

V2X applications; *see also* ITS
ADAS (advanced driving assistance) 166
radar, LIDAR, cameras 165

road/public safety 38, 44
vehicle status management 166
VRU (vulnerable traffic users) 165
V2X communication
channel modelling 185
cooperative direct links 210
cooperative dual-hop communication 195
high-bandwidth, low-latency, highly reliable 165
performance evaluation 183
V2I (vehicles to infrastructure) 38, 169
V2P (vehicle to pedestrians) 38, 184
V2V (vehicles to vehicles) 38, 169, 189, 195, 198, 201, 210
V-Comm systems 184
video codec
AI-based tools 123
DNNVC (deep neural network video coding) 126
DVC (deep video compression) 127
HEVC (high efficiency video coding) 111
HLVC (hierarchical DL-based video compression) 127
RLVC (recurrent learned DL-based video compression) 128
VCM (video coding for machine) 129
VVC (versatile video coding) **111**
video formats
360-degree omnidirectional video 110
3D-video 4, 121, 167
high dynamic range (HDR) 110
ultrahigh-definition video (UHD) 4, 108
volumetric video 84, 112, 117, 122
wide color gamut (WCG) 110
wireless channel
multi-path fading 317
propagation models 60, 63
radio wave propagation loss 25, 87
time-varying 166
wireless communication systems
coverage vs. capacity 45
latency vs. reliability 148
energy efficiency vs. latency 148
WUS (wake-up signal) 148
low-power **46**, 147

zero interruption time 148
zero tolerance 456
zero trust model 156
zero-energy devices 58
zero-touch cognitive networks 375